STATISTICS

BASIC PRINCIPLES AND APPLICATIONS

Second Edition

William J. Adams
Irwin Kabus
Mitchell P. Preiss

Second Edition, Revised
William J. Adams

Mathematics Department
Pace University
With Illustrations By Ramunė B. Adams

Library of Congress Control Number:		2007908900
ISBN:	Hardcover	978-1-4363-0181-7
	Softcover	978-1-4363-0180-0
	eBook	978-1-4691-0731-8

Statistics, Basic Principles and Applications, Revised Second Edition, is available on the web at webpage. pace. edu/wadams.

Print information available on the last page.

Rev. date: 04/06/2018

To order additional copies of this book, contact:
Xlibris
1-888-795-4274
www.Xlibris.com
Orders@Xlibris.com
576178

To

Onutė, Andrius, Ramunė, and Rasa Adams

Judy, Alan, Robert, and David Kabus

Lucille Preiss and the Memory of Charles Preiss

CONTENTS

Preface to the Second Edition, Revised

Still another book on statistics when the weight of available books on the subject must come close to a ton? What's the point?, you would be justified in asking.

The rationale for this addition to the ton is founded on our conviction that the statistics literature on introductory statistics that we are familiar with does not adequately address the ten fundamental guidelines noted below. These ten commandments, as we term them, should, in our judgment, be given high priority in an exposition of basic statistics.

- The focus should be placed on a clear development of basic ideas and principles.
- The exposition of these basic ideas and principles should be streamlined so as to avoid having the undergrowth get in the way of the statistical forest.
- High priority should be given to the assumptions that underlie the application of statistical principles.
- Understanding of abuses, misuses, and misunderstandings which have arisen from the application of statistics is essential for a correct understanding of statistics.
- The coverage should provide students with sufficient preparation for continued study of intermediate and advanced level statistics or disciplines which use statistical methodology.
- The exposition should be readable and understandable by students without sacrifice of mathematical accuracy.
- The organization should clearly distinguish mainstream topics inherent in every basic level statistics course, irrespective of applied interests, from topics of special interest to particular audience segments.
- The computation dimension should not be given equal billing with statistical principles and ideas. Statistics is the master and, important as it is, the computation tool is the servant.

- Exercises to provoke-thought—exercise the little grey cells, as Hercule Poirot would put it—should be a prominent part of the exposition.

- Exercise banks to help the student see statistics as a whole are important.

To realize the second of our commandments we ruthlessly cut back on peripheral topics to satisfy this condition.

Much attention is given to the assumptions that underlie statistical practice to put into perspective the conditions required before statistical techniques can be applied and to combat the disturbing attitude that statistical computation is condition free.

Beginning with the first chapter much attention is given to the misleading or incorrect use of statistics. Much attention is also given to the issue of how statistics may be interpreted.

The exposition does not cater to any one group of applied interests. Rather it would be appropriate to say that it caters to them all, for as is well known, basic principles of statistics underlie the needs of all applied interests.

Illustrative examples and applications have been chosen to avoid unnecessary technical features that detract from the main point, be accessible to the intended student audience without special background, and clearly illustrate the statistical point in question. We desire applications to be interesting but, needless-to-say, this quality is as much in the mind of the reader as in the eye of the writer.

It is our conviction that all applications are important when it comes to gaining insight into the scope of statistical principles in operation. There is nothing more limiting to a sound understanding of statistics than the parochial view that applications of interest are defined by one's discipline of interest.

This book is a suitable text for a variety of audiences, including students majoring in a business discipline, economics, education, the life and health sciences, psychology, and the social sciences. It is also suitable for a non-calculus introduction to probability and statistics for students taking a major in computer science, mathematics, or one of the physical sciences.

Our other concern is with the student who, apart from future course work and professional needs, wants to be a well-educated person. Statistics and its wide spectrum of applications is one of the great triumphs of the human intellect and, like fine art, fine music, and fine literature, brings us together as human beings interested in partaking of the best of the human spirit.

The depth and rate at which topics are covered will, of course, depend on the audience at hand and its level of mathematics preparation. Strictly speaking, there are no mathematical prerequisites for this exposition apart from some knowledge of algebra, but the rate at which one may progress depends on the mathematical maturity and sophistication of the audience. This book is suitable for a one or two semester introductory course in statistics or probability and statistics, with the main focus on statistics, depending on course structure and time parameters.

Fundamental statistical principles and the probability foundation that underlies them are developed in the first four parts of the book, which form the mainstream for any course in statistics. Part 5 presents selected topics from which one may pick and choose. The basic structure of this book is shown by the tree diagram that follows.

In the revised second edition a number of refinements in the exposition have been made to add clarity, sharpen focus, and update background data.

Acknowledgments

We should like to express our appreciation to Ramunė Adams for preparing the illustrations, to Pace University's Council of Deans for its support in the form of sabbaticals and a summer research grant, to our colleague Dr. Michael Kazlow for his insights on the computation dimension of statistics, to the Pace Word Processing Department for its assistance in preparing the manuscript, to our editor Bill Walsh for his support and encouragement, and to Joan van Glabek, whose queries to the authors saved us from a number of embarrassing missteps.

Most of all, we wish to acknowledge our debt to our students, who played a major role in shaping and refining our ideas on the presentation of statistics.

W. J. A.
I. K.
M. P. P.

Availability

To make this book widely available I have put it on the web at webpage.pace.edu/wadams.

W. J. A.

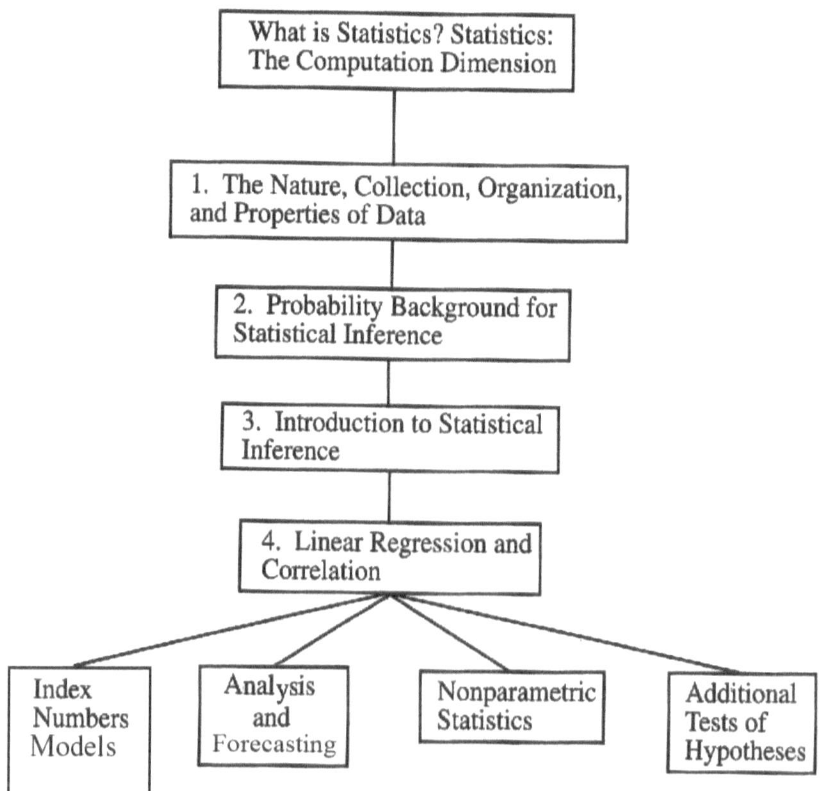
What is Statistics? Statistics:
The Computation Dimension
1. The Nature, Collection, Organization, and Properties of Data
2. Probability Background for Statistical Inference
3. Introduction to Statistical Inference
4. Linear Regression and Correlation
Index Numbers Models
Analysis and Forecasting
Nonparametric Statistics
Additional Tests of Hypotheses

Perspective on the Chapters and Exercises

Chapter 1: Since data are the basic raw materials of statistical analysis—the food for statistical analysis, if you will—it is essential that their quality be given suitable attention. At the very least this vital concern warrants a chapter in its own right.

Chapter 2: The nature of data that must be taken into account (data scales) prior to employing statistical analysis and sampling methods for obtaining data are addressed.

Example 1 introduces the "Heavy" *Basic Statistics* text situation which we return to a number of times throughout the book, each time adding another dimension of interest to it.

Chapters 3, 4, and 5: With attention having been given to the reliability, relevance, and tools for obtaining data in Chapters 1 and 2, the stage is set for developing tools for making data manageable and visible, and obtaining useful numerical descriptions for them. This is undertaken in Chapters 3, 4, and 5. The "Heavy" *Basic Statistics* situation is employed to illustrate the concepts of frequency distribution, stem-and-leaf plot, histogram, frequency polygon, and ogive in Chapter 3, and the concepts of mean, median, and standard deviation of a frequency distribution in Chapter 5.

Needless-to-say, statistics has a "significant" computation dimension, and the issue of computation accuracy unavoidably arises. This issue is given attention in Section 4.2, which should be considered a most important background section.

Chapters 6 and 7: Chapter 6 is a linchpin for the introductory study of probability and statistics. Properties of relative frequency are used as a springboard for developing the concept of finite probability model. Probability models are developed for random processes with much attention being paid to the role of assumptions in formulating such models. The subjective probability interpretation of probability is introduced and a careful distinction is made between probability and its relative frequency and subjective probability interpretations.
Chapter 7 is a short but important chapter which introduces the pivotal concepts of random variable, expected value and probability distribution.

Chapters 8 and 9: The normal curves, or distributions, play such an important role in probability and statistics that they deserve a major chapter in their own right. Chapter 8 on the normal curves sets the stage for consideration of their central role in statistical inference. It is illuminating to look at the historical setting which gave rise to the normal curves and this is treated in Chapter 8 as well. The "Heavy" *Basic*

Statistics situation makes an appearance in connection with the question of whether a population's distribution can be described by a normal curve.

The short but pivotal Chapter 9 introduces the key concept of sampling distribution, which underlies statistical inference, and a "central limit theorem" which establishes a link between sampling distributions and the normal curves.

With these foundation stones in place we are ready to undertake a study of statistical inference, which is initiated in the next two chapters.

Technique vs. Theorems

With calculators and computers so readily available, crunching the numbers has almost become a trivial exercise. A blessing? Certainly. The downside, however, is that many, dare we say most, students and users of statistics come to see statistics as a calculation free-for-all. Throw the data into the computer and let it do its thing, is well on its way to becoming standard operating procedure for life with statistics.

The hypothesis of theorems—strings attached, as we refer to them—guide number crunching and technique and to the extent possible in what is generally viewed as a number crunching course we pay a good deal of attention to strings attached. This point of view is reflected in a number of exercises. We strongly believe that to instruct a student in the use of statistics without emphasizing conditions that make possible its meaningful application would be as irresponsible as instructing a student in the use of firearms without emphasizing safety conditions that make possible their safe use. In both instances the potential exists for inflicting serious injury and damage.

Chapters 10 and 11: Both chapters are concerned with defining the framework of statistical inference, confidence interval analysis in Chapter 10 and hypothesis testing in Chapter 11. While this is the case, it is with these chapters that one begins to see substantial payoff in terms of applications. Although some applied payoff is to be seen in Chapters 6-9, their major role is to set in place the foundation needed for statistical inference defined in Chapters 10 and 11 with their extensions in Chapters 12, 13, 16, and 17.

The "Heavy" *Basic Statistics* text situation reappears in Exercise 9 of Section 10.3 (page 281) in connection with obtaining a confidence interval for the population mean. This setting illustrates the unity of ingredients obtained from Sections 2.3 (a sample drawn at random by using a table of random numbers), 5.2 (the mean obtained from a frequency distribution), and 5.3 (the standard deviation obtained from a frequency distribution). It also serves as a specific setting for addressing the question of whether a population is normally distributed, a condition required for constructing the confidence interval asked for in the aforenoted Exercise 9. This question is first considered in

Example 1 of Section 8.4 and re-examined from the point of view of hypothesis testing in Example 1 of Section 16.4 by means of the Kolmogorov-Smironv test and Example 3 of Section 17.3 by means of the χ^2 goodness of fit test.

Both are "large" chapters, especially Chapter 11, and a question might arise as to which sections should be given priority should time or other considerations not allow all to be covered. For Chapter 10 we recommend Sections 10.1, 10.2, 10.3, 10.4, 10.6 and 10.7; for Chapter 11 we recommend 11.1, 11.2, 11.7, 11.8, 11.10, 11.11, 11.13-11.16.

Chapters 12 and 13: The Daniel company, a new entry in the fine baked goods market, wants to predict sales volume on the basis of advertising expenditure. This setting and problem are introduced in Example 1 of Chapter 12 and serves as a vehicle for defining and illustrating ideas needed for the development of linear regression and correlation analysis in Chapters 12 and 13.

Chapters 14 and 15: A parochial view has it that index numbers and time series are the exclusive domain of business students. A more enlightened view has it that index numbers and time series deal with serious down-to-earth life issues that touch all of us, which makes them of general interest irrespective of major. We strongly subscribe to this more enlightened view.

Chapter 16: What do you do when you desire to test a hypothesis but the conditions required of the test you have in mind are not satisfied or the nature of your data render the test inapplicable? Nonparametric methods, an interesting spectrum of which are taken up in Chapter 16, may be just the answer.

Chapter 17: Chapter 17 provides an introduction to additional hypothesis testing techniques with a wide spectrum of interesting and important applications. In this chapter the χ^2 distributions dominate the scene.

Questions that Challenge "the Little Grey Cells: All questions challenge to some extent "the little gray cells," as Hercule Poirot describes them. Some questions are especially noteworthy in being thought provoking and helpful in leading to insights. In the spirit of the show 20-questions, the following twenty questions are among our favorites in being thought provoking and insightful.

1. Sec. 1.5, page 24, No. 2
2. Sec. 2.4, page 57, No. 4
3. Sec. 2.5, page 57, No. 10
4. Self-Tests for Part 1, Self-Test 2, page 143, No. 1

5. Sec. 6.6, page 176, No. 17
6. Sec. 8.4, page 236, No. 4
7. Self-Tests for Part 2, Self-Test 3, page 253, No. 1
8. Sec. 10.2, page 268, No. 10
9. Sec. 10.2, page 271, No. 15
10. Sec. 10.3, page 281, No. 9
11. Sec. 10.4, page 286, No. 1
12. Self-Tests for Parts 1-4, Self-Test 5, page 470, No. 3
13. Sec. 10.8, page 305, No. 3
14. Sec. 11.2, page 327, No. 20
15. Sec. 11.8, page 360, No. 5
16. Sec. 11.13, page 396, No. 3
17. Self-Tests for Part 3, Self-Test 4, page 405, No. 1
18. Self-Tests for Part 4, Self-Test 2, page 463, No. 2
19. Sec. 16.3, page 579, No. 3
20. Sec. 17.3, page 650, No. 6

To Our Students

Dear Student:

As a student taking a course in statistics you have three major resources available to you: the course instructor, the text, and you yourself. Course instructors, many of us have experienced, range in quality from a hinderance to truly helpful and even inspiring.

Course texts also range in quality from a hinderance to most useful. Our primary objective in writing this book was to write an exposition that would be an ally to you in your encounter with statistics. It is intended to be user useful and friendly.

The third major resource that you bring to the course, you yourself, includes your commitment to the course and willingness to make best use of the other resources. It is the most important of all; if it is weak, then the others will be of little or no use; if it is strong, then weaknesses in the others can be overcome.

This book presupposes your serious commitment to the study of statistics, which is not to say that we view the audience as composed of budding statisticians and mathematicians; quite the opposite. This text should not be viewed as just a collection of homework exercises. The level and style of the exposition reflect more than half a century of our collective teaching experience with students like you, and it was written under the assumption that it will be read. To do this effectively we urge you to read each section at least twice, the first time to get the general idea and the second time to nail down details. A third careful reading should also be undertaken. Basic definitions should be given careful attention and written down separately. Definitions provide the backbone of the exposition and must be mastered for the exposition itself to make sense. After completing a section we urge you to write a short summary, including a listing of technical terms, in your own words. We deliberately refrained from doing this to provide you with the opportunity to do so. This should be thought of as a standard exercise applicable to all sections.

As to homework exercises in general, they are an integral part of the text which is essential for its understanding. Some exercises are simple, straightforward, and not particularly interesting, but intended to help you get basic concepts and technique under control. The exercises progress to more challenging and interesting with the intent of bringing together a number of ideas with a context for their application. The number of exercises that should be done to achieve an understanding of the material will vary from person to person, but there is no doubt that everyone must do a fair number of exercises to achieve that understanding. We believe that it is helpful to work together with fellow students and exchange ideas, as long as everyone in a group effort is contributing. At the same time, examinations are solo efforts, and beyond a certain point one should be prepared to fly solo. As an additional resource, answers to selected exercises are given at the end of the book.

Self-tests have been included to help you bring blocks of material under control. We suggest that you do these self-tests under conditions similar to those under which actual tests are taken. Prepare yourself as you would for an actual test and then do a self-test at one sitting. You will be your own monitor and taskmaster.

As to formulas, there are many formulas in this subject. The challenging part is to know which applies to which situation. It is not a matter of memorizing formulas, but of obtaining a sense of the context from which they come and apply to. This calls for an understanding of the subject rather than rote memorization. Some simple basic formulas should be known cold, so-to-speak, and more complex ones should be known well enough so that you know what to look for on a formula sheet and how to use it in a particular situation. As you work through the book we suggest that you prepare your own formula sheet for your use.

Finally, there is a question of attitude toward statistics. Some would suggest that it is both hard and boring with numerous calculations. One of our colleagues, in another department, jokingly describes statistics as sadistics to his students. This is a joke which is clever, but it is neither funny nor accurate and contributes to a counterproductive poisoning of the well. What is hard and boring must, like beauty, be left to the eyes of the beholder. There are calculations, but the ones required in this text are not onerous and can be carried out with the assistance of a rather primitive hand-calculator. The more challenging, interesting, and essential components of statistics are to be found in its concepts and principles. It is these components which we emphasize along with an indication of their wide spectrum of applications. Many people have been fascinated by these aspects of statistics and we hope that you will be too.

With regards and best wishes for a successful undertaking,

W.J.A.
I.K.
M.P.P.

What Is Statistics?

Until the middle of the nineteenth century the answer would have been that statistics is a collection of numerical data concerning economic, military, and political affairs of state. The term is still sometimes used or understood in this restricted sense to refer to data arising from some enterprise of interest. The enterprise might be governmental, business, economic, industrial, scientific, medical, or of a sporting nature, to take a few examples, so that we often hear mention of medical statistics, baseball statistics, and so on.

The modern, more general use of the term refers to a discipline which may be viewed in terms of two components called Descriptive Statistics and Inferential Statistics. **Descriptive Statistics** is concerned with methods of organizing masses of numerical data, graphically displaying them, and obtaining numerical values which help us better understand and deal with the data that we have. To take an example, suppose President Marx of Huxley College requests data on the mathematics placement exam scores of all entering freshmen. The freshman class consists of 5000 students, so that the president would be confronted by a list of 5000 numerical values should his request be answered literally. The forest of data would be so overwhelming that he would be at a loss to make sense of the individual trees. An organization of the data which shows how they fall in various intervals of interest (70-79, 80-89, etc.), graphs that show these features visually, and numerical measures, such as the mean and median values of the data, would be more meaningful and useful to the president than a forest of 5000 values. The problems of organizing the data into suitable groups and obtaining numerical measures of the data are problems of Descriptive Statistics.

A statistical population is a collection of data viewed as a whole in its own right, whereas a sample is a part of this all-inclusive whole. **Inferential Statistics** is concerned with making and testing judgments about population values of interest on the basis of "suitably selected" samples and obtaining measures of reliability for these judgments. The anthropologist Alice Williams claims that recently discovered humanoid fossil remains in Central America are one million years old, let us say. The problem of conducting a statistical test on a suitably chosen sample of these remains to test this claim is a problem of inferential statistics. We would also require a measure of how reliable is the test result, which is an accompanying problem of statistical inference. And then there are related issues of how "suitably chosen" sample is defined and in fact chosen, and deceptive statistics that arise through carelessness, ignorance, or deliberate intent.

Statistics may be likened to a river such as the powerful and majestic Mississippi. Like the Mississippi, it is fed by many tributary rivers and streams, two major ones being what we have identified as descriptive and inferential statistics. The other tributaries consist of mathematical methods, particularly from probability, and problems and issues which arose and continue to arise from such disciplines as agriculture, biology, business, economics, meteorology, the physical sciences, political science, psychology, and sociology which are statistical in nature. They have all fed the mighty river and benefitted from it.

It is our objective in this text to examine some of the major tributaries which have fed the mighty river statistics, obtain a sense of the river as a whole, and issue warning signs where the river is in danger of being polluted through misuse and contamination.

Statistics: The Computation Dimension

Statistics has a significant computation dimension and the question naturally arises as to whether a computer is essential for life in the world of statistics. The answer to this question depends on which part of the world of statistics one inhabits. The highest priority of students taking a first course in statistics is to master basic principles and concepts, which the computation dimension should support and not interfere with. The computation requirements of this exposition can be satisfactorily carried out with a basic calculator (by today's standards) with the four arithmetic operations and a key which allows one to take square roots. Such a machine leads to a "reasonable amount" of manual computation which gets one's hands mildly dirty, in an arithmetic sense, but is healthy in a learning sense. In first encountering the concept of standard deviation, for example, working through the steps required of its definition helps one to obtain a real sense of this concept which cannot be achieved by only pressing a key on a sophisticated calculator or computer. Such is the case for all basic definitions, principles, and techniques.

Computation Allies: An Overview

At some point, which cannot be pinpointed exactly, the pedagogical benefits of working out standard deviations and the like step by step no longer accrue, and it makes good sense to seek more powerful tools to get the computation job done as quickly and painlessly as possible. The following overview might be useful for those seeking an enhanced computation or computer dimension. In considering enhanced computation options one's primary objectives and the trade-offs inherent in these options should be kept in mind.

Calculators

Calculators range from basic to sophisticated with a rich variety of push-button options, programs, and memory. Basic calculators are satisfactory for small scale needs, relatively inexpensive, and occupy little space. Such a calculator meets the needs of the exposition in this book.

As to computer options, there are three main categories of statistical packages.

Mainframe Packages

The big four basic statistics packages are SPSS (Statistical Package for the Social Sciences), SAS (Statistical Analysis System), BMDP (Biomedical Computer Programs), and MINITAB, which began as a teaching tool and was extended to a research tool. Mainframe packages have the following advantages: One may have common data sets on the mainframe for student use. Security constraints permitting, these packages may be accessed from remote locations. One can have confidence in their numerical accuracy. Numerical results are generally not worked out by a computer by means of the nice computation results cited in statistics books because the round off errors are too large. Computer packages employ different algorithms which must undergo extensive testing to ensure their accuracy. The aforementioned packages are based on well known algorithms whose numerical accuracy is assured.

Two disadvantages of mainframe packages are that a fair amount of start-up time must be invested to become comfortable with their use and that response times are sometimes slow.

IBM Compatible or PC/MS DOS

In part, this category contains mainframe packages that have been rewritten for the IBM PC environment. The core of such mainframe packages is carried over, but certain tests and functions may have to be obtained separately. Graphics on PC packages tend to be much stronger than on their mainframe counterparts.

IBM compatibles includes SPSS, MINITAB (both of which have minimal graphics), SAS, STAT-GRAPHICS, SYSTAT, and EXECUSTAT. The last noted states the assumptions that must be satisfied for a technique to be executed. Many MS-DOS statistics packages have student versions with restricted features and data sets.

Macintosh—Post Apple II

There are versions of SPSS and MINITAB for the Macintosh environment. Their graphics capability is not high. There is also JMP, SYSTAT, and DATA DESK, which have high graphics capability. These packages employ an exploratory data analysis approach to statistics. Macintosh packages have student versions with restricted features and data sets.

Unquestioned Acceptance of Computer Generated Statistics

Unfortunately, the attitude that computer generated statistical results are infallible because they are computer generated has become widespread. Those who do not appreciate the GIGO (garbage in, garbage out) principle and those who desire quick results without investing the needed time and effort to understand the discipline are particularly vulnerable to this kind of computer virus. While the computer has made it possible for many to address problems that heretofore had been intractable, it has also made it possible for more and more people having less and less understanding of statistics to generate more nonsense more quickly than ever before.

Henry Clay's sage observation, "Statistics are no substitute for judgment," may be profitably extended to, computer generated statistics are no substitute for an understanding of statistics. Achieving an understanding of the principles and ideas of statistics is our only defense against errors that we ourselves might fall into and those that others have fallen into.

PART ONE

THE NATURE, COLLECTION, ORGANIZATION, AND PROPERTIES OF DATA

- On Guard!
- The Basic Raw Materials: Data
- Making Mountains of Data Manageable
- Descriptive Measures for Ungrouped Data
- Descriptive Measures for Grouped Data

1

On Guard!

1.1 ■ How "Good" Are the Data?

Data are the fundamental raw material of statistical analysis. For a meaningful and useful statistical analysis of a situation under study the data must be relevant to the objective of the study and reliable. As the examples in this chapter make clear, relevance and reliability are not characteristics that we may simply take for granted. We must always be on alert.

Obtaining well-chosen data for a study is what might be called a pre-statistical concern, pre-statistical in the sense that the methods of descriptive and inferential statistics are not directed at this concern. They presuppose as their starting point that the data are reliable, relevant and go on from there. Obtaining well-chosen data requires insight into the phenomenon under study which enables one to distinguish primary factors from secondary and irrelevant ones. If the mark is missed at this point, the application of no amount of statistical machinery to organize and portray the data, compute its numerical characteristics, and infer conclusions will bail us out. The finest statistical tower of Pisa built on a shaky foundation is doomed to topple. It is unfortunate that the number of people having real insight into what they are doing is small, which is one reason why we have been favored by the proliferation of so much statistical junk.

1.2 ■ Are the Data Reliable?

Case 1. Are Statistically Dangerous Schools Necessarily Dangerous? Are Statistically Safe Schools Necessarily Safe?

In July 1986 the New York City Board of Education issued a list of its most dangerous schools based on incident and crime reports that it had received. One respondent notes that he never felt unsafe or threatened in teaching at the fifth listed "most dangerous" junior high school. [Spector; 28] Another respondent comments: "I believe that the scorecard . . . names not the most dangerous schools but the schools whose administrators have the courage to report what is really happening." [Richman; 25]

In June 1994 New York City Schools Chancellor Ramon Cortines rejected data on violence in the city's schools, saying that he suspected school administrators were underreporting acts of violence to make their schools appear less turbulent. [Dillon; 8].

In September 1995 Edward Costikyan, the chairman of the Mayor's Commission on School Safety, observed: 'There seems to be a total absence of any reliable numbers on anything. How can you manage anything without knowing what you're dealing with.' [Toy; 30]. Are things better now? Well . . .

In September 2007 New York City comptroller William C. Thompson Jr. stated that an audit showed that the city had not ensured that all principals accurately report violence in their schools, making it difficult for the public to assess their safety. [Gootman; 13]. Are things better now? Well . . .

Case 2. Are American Students Really That Bad in Math and Science?

Every few years another study appears which shows again that American students are worse in math and science than their counterparts in even the poorest countries. But are they being compared with their counterparts? Some say no. Without denying that there is much to be improved in American education, a growing number of critics have argued that the test results are flawed because American students in total are consistently being compared with the elite students of other countries. [Kolata; 9]

Case 3. These Data May Give You Nightmares

Halcion, manufactured by the Upjohn Company and introduced in the United States in 1983 is one of the world's best known sleeping pills. Its main advantage over competing products, Upjohn has claimed, is in encouraging nighttime sleep without daytime drowsiness.

How safe is Halcion? It received Food and Drug Administration approval and its manufacturer claims that it is just as safe as other drugs of its kind. Dissenters argue that Halcion is more likely to cause symptoms such as amnesia, paranoia, and depression and that Upjohn engaged in data manipulation to conceal its side effects. This view emerged from a law suit filed by Ilo Grundberg, who killed her mother the day before her mother's 83rd birthday and placed a birthday card in her hand. Mrs. Grundberg claimed that Halcion had made her psychotic, and charges against her were eventually dismissed. Upjohn settled the lawsuit with Mrs. Grundberg before it was to go to trial in August 1991, but in preparation for the suit it had to make available a good deal of data about Halcion to the plaintiff's attorneys.

Dr. Ian Oswald, who was head of the department of psychiatry at the University of Edinburgh and spent 30 years doing research on sleep, was obtained as an expert witness. Dr. Oswald spent two years going over Upjohn's data and concluded that

Upjohn had known about the extent of the drug's adverse effects for 20 years and concealed these data. He concluded that "the whole thing had been one long fraud." [Kolata; 18]. Dr. Graham Dukes, former medical director of the Dutch drug regulatory agency, who examined some of Upjohn's data, believed that the data on Halcion had been organized in such a way as to minimize the drug's adverse effects and that this could not have occurred accidentally.

In reaction to the criticisms voiced, Britain, the Netherlands and Belgium were led to remove the sleeping pill from the market. A report issued in April 1994 by F.D.A. investigators stated that the Upjohn Company had engaged in ongoing misconduct with Halcion. The F.D.A. will investigate, it was announced. We have not heard anything further.

Case 4. Is This Food Survey Healthy?

A report issued in the fall of 1991 by a scientific panel and the General Accounting Office criticized the latest National Food Consumption Survey, carried out in 1987-88, as so flawed that its data are probably useless. The major problem is the survey's low response rate of 34 percent, making it questionable whether the data are representative of the population. Follow up studies are required of those who do not respond, but no follow up studies were conducted.

The flawed data are used for making major Government policy decisions involving school breakfast and lunch programs, food stamp allotments, setting pesticide levels in foods, calculating nutrient consumption levels, and determining the public's exposure to pesticides and toxic metals. [Burros; 5]

The significance of low response rates when a survey is taken is discussed in Section 2.5 Polls, Surveys, and Questionnaires.

Case 5. How "Solid" Are Those Figures?

A number of figures have been bandied about in the campaign to reduce youth smoking. Here are some of them.

1. President Clinton warned that 1 million people would die prematurely if Congress did not pass tobacco legislation in 1998.

2. Senator John McCain urged lawmakers to stop 3000 kids a day from starting this life-threatening addiction.

3. After a $368.5 billion settlement proposal between state officials and tobacco producers was agreed to in 1997, the American Cancer Society stated that a 60 percent decrease in youth smoking could reduce premature deaths from diseases caused by tobacco by 1 million in coming years.

4. Deputy Treasury Secretary Lawrence Summers cited studies saying that every 10 percent increase in the price of a pack of cigarettes would produce up to a 7 percent reduction in the number of children who smoke.

5. Richard Kluger, author of *Ashes to Ashes*, a history of the battle between smoking and health in the United States, notes: 'I think this whole business of trying to prevent kids from smoking being the impetus behind legislation is great politics. Nonsense in terms of anything you can put numbers next to.' [Meier; 22]

Case 6. Top of the Line Deception

In 1992 the General Accounting Office audited seven "Star Wars" tests conducted between 1990 and 1992. It found that four of the test results described to Congress as successes were false whereas the three tests that were described as complete or partial failures were correct. [Weiner; 31]

Case 7. Spin Versus Counterspin

Speaking on television on Tuesday night of 3 August 1993, President Clinton described the budget legislation then before Congress as "the largest deficit reduction in history." Almost immediately after the President spoke, Senator Robert Dole, Republican leader in the Senate, described the legislation as "the largest tax increase in world history."

Who is right? Neither; when the dollar amounts are adjusted for inflation so that dollar comparisons are meaningful, 1993's budget bill is neither the biggest reduction measure nor the biggest tax increase in recent years. In 1993 dollars, the bill would lower the annual deficit by a projected total of $496 billion over five years; $241 billion of this would come from tax increases. The bill signed by George Bush in 1990 contained $532 billion in deficit reduction in terms of 1993 dollars. The bill signed by Ronald Reagan in 1982 raised taxes by $286 billion over five years in terms of 1993 dollars.

For discussion of how to take inflation into account in comparing dollar amounts in different time periods, see Section 14.7 Determining "Real" Dollar Amounts.

Case 8. Can We Trust TV Ratings?

The life span of a television program is determined by the public's reaction to it, which is measured by TV ratings. These ratings, produced by the Nielsen Company, estimate the audience in terms of the percentage of those sets in use which are turned to each channel, called a share, or in terms of the percentage of the total possible audience, sets on or off, called a rating. Shares and ratings are further broken down according to the sex and age of viewers so that advertisers can better focus their advertising campaigns. These numbers determine the buying and selling of billions of dollars of television air time. They mean life or death to television programs. The half-hour comedy *Good & Evil*, which had promising ingredients in terms of writing, acting and production talent, had a short life after its premiere in the fall of 1991 because of low initial ratings. In March 1992 NBC announced that they were dropping two successful shows, *Matlock* and *In the Heat of the Night*, because the demographic numbers favored older viewers while the network wished to build around a more youthful audience.

Since 1986 the data which underlie the ratings have been collected by a device called a people-meter. The remote control part of a people-meter rests on top of the television set. When the set is turned on, the meter prompts viewers to enter their identification number. Information is provided on what channels are being beamed into the household and who is watching them. Nielsen puts its people-meter into 4000 households selected at random—that is, without bias—from the approximately 93 million homes in America with television.

The people-meter data gathering system produced lower ratings for the networks than had been expected and a serious question arose as to whether this was because of the increased or decreased accuracy of this system over the method it replaced. The networks commissioned a study of the Nielsen methodology and two years later this Committee on Nationwide Television Audience Measurement (CONTAM) issued a nine-volume report that was highly critical of the Nielsen system. The report found evidence of button fatigue—that over time people did not push the buttons that would insure data accuracy as they did in the beginning. CONTAM was highly critical of Nielsen's sampling procedures for obtaining the 4,000 households that make up their sample; random sampling was envisioned in the methodology, but the actual sampling deviated significantly from this requirement. From this came ratings which were highly suspect. David Poltrack, senior vice president of research at CBS, observes that: "The whole business is crazy. I don't think there's an advertising agency in the United States that could get up in front of its clients and justify the way business is done right now. It's being bought on narrow based demographics, demographic targets which are not representative of product consumption in the United States." [24]

Random sampling was called for in theory, but not delivered in practice. This yielded highly suspect ratings.

The problem of achieving random sampling in practice is discussed in Sections 2.2 Survey of Sampling Methods and 2.3 More on Random Sampling.

Nielsen has overcome the statistical sampling problem, but it is still plagued by the problem of getting "honest" data from viewers in the sample selected. Its people-meter system for eliciting viewing data has been described as too mechanical and as not being user friendly. The problem of obtaining accurate viewer data remains. Matters came to a head in March 1997 with the results of what is termed the February sweeps, an intense ratings period that determines television's winners and losers in terms of how $46 billion in advertising money will be allocated.

> The problem now is to get honest viewer data from the sample of viewers chosen.

According to the Nielsen ratings, the average number of American households watching prime-time television fell by over one million in February 1997 compared to February 1996. This was the fourth decline in the last five years. The networks do not find Nielsen's numbers credible. As Don Ohlmeyer of NBC put it: 'I don't trust their numbers at all. They're trying to measure 21st-century technology with an abacus.' [Carter; 6]

Nielsen's response is that the networks are engaging in the time-dishonored practice of blame the messenger. It is safe to assume, however, that they are working to improve their data collection system.

1.3 ■ Slippery Statistics: An International Dimension

The Slippery Statistics Society (SSS) has an international clientele. Here are a few examples.

Brazil

In a frank conversation between television interviews that was inadvertently broadcasted across his country, Brazil's finance minister Rubens Ricupero expressed the sentiments of many kindred spirits when he confessed of economic indicators: "I have no scruples, what is good we take advantage of. What is bad, we hide" [Brooke; 3]. Minister Ricupero was immediately dismissed, but was this because of his performance or indiscretion?

Britain

In the past *The Economist* has been critical of the U.K. Central Statistical Office as having 'figures often tasting of fudge.' [Duncan and Gross; 9, p. 66], [Economist; 10, p. 88], [Economist; 11, p. 65].

China

Chinese government statistics have run a gamut of slipperiness. After the Communist Party assumed control in 1949, government statistics were systematically distorted to serve the wishes of the new political establishment. During the period of the Cultural Revolution of the late 1960s and early '70s data-gathering was abandoned as unscientific.

Since the passing of the Cultural Revolution, data-gathering and the publication of state statistics has resumed and other pressures have developed. In May 1994 Zhang Sai, director of the State Statistical Bureau, 'warned that distorted statistics are increasing tensions between Beijing and localities.' [Tefft; 29]

Foreign investors in China are wary of Chinese statistics and many have taken to generating their own.

Japan

By late July 1998 American financial experts reached the conclusion that the magnitude of Japan's banking crisis was far worse than had been publicly acknowledged. The bad debts were estimated as being on the order of $1 trillion, nearly twice the official estimate. The true amount, financial experts emphasized, is hard to pin down because Japanese banks have been using accounting tricks to conceal debts that are not being paid. [Sanger; 26]

Soviet Union and Russia

From the beginnings of the Soviet State, Soviet statistics have acquired a reputation of being unreliable. (See [Clark; 7], [Shaffer; 27].) Writing in 1990, V.N. Kirichenko, Chairman of the USSR State Committee on Statistics, expressed a hope to 'ensure the accuracy of the data . . . restore the trust in such data on the part of the Soviet and international public. The country can no longer afford to seek the right way with the help of trick mirrors.' [Duncan and Gross; 9, p. 66] and [Kirichenko; 16, pp. 50-57].

Since the breakup of the Soviet Union, Russia has continued to have problems with government statistics, but for different reason. Rather than exaggerating output the statistical pendulum has swung to the extreme of underestimating it. In June 1998 Russia's top statisticians were arrested on charges of manipulating data to underestimate the production of Russian businesses to help them minimize their tax obligations. [Gordon; 14]

United States

In June 1998 the thrust of the Republican majority in the House of Representatives was to cut taxes beyond what was called for in the earlier balanced-budget agreement. But then there are the spending cuts needed to achieve balance. The Congressional Budget Office did not produce the numbers needed for this to work out, which prompted the Republican leadership to address a letter to the Appropriations subcommittee warning that if the C.B.O. did not begin to produce better numbers, 'we must review [its] structure and funding.' [*The New York Times;* 23, A22]

1.4 ■ Are the Data, Their Characteristics, and the Framework Generating Them Well-Chosen?

Reliable data are not always relevant, that is, well-chosen in connection with the intent of a study undertaken. If interest centers on the heights of those in a certain community and we are presented with their weights, very carefully obtained, then the data are reliable, but hardly well-chosen in connection with the focus of the study. Subsequent mathematical refinements or conclusions obtained from poorly chosen data might have the aura of precision, but are no better than its basic stating point.

Data may be described by various numerical characteristics (mean and median, for example) and the problem of determining which characteristic is "most suitable" for the situation at hand is both challenging and serious. This issue is explored in Case 10 and considered in Example 6 of Section 4.3 (page 112).

Another dimension takes us a further step back. A decision-making framework is to be set up for some situation, let us say. The framework requires data. The data obtained may be consistent with the decision-making framework, but if that framework is poorly formulated, the data it leads to cannot be viewed as "well-chosen." This dimension of suitability of data is explored in Case 13.

Case 9. Which Data "Best" Reflect Airline Reliability?

The long time standard measure of an airline's reliability is its percentage of on-time arrivals, where a flight is deemed on-time if it arrives within 15 minutes of its scheduled arrival time. Such data are widely trumpeted by airlines in their advertising campaigns.

But is this statistic the "best" measure of an airline's reliability? According to Julius Maldutis, an airline analyst with Salomon Brothers, the answer is no. Maldutis argues that a much better measure of reliability is the percentage of flight-miles completed. Look at the cancellation rate, which is indicative of a more troublesome situation to travelers than that indicated by the artificial on-time statistic, says Maldutis. [Bryant; 4]

Case 10. Temperature vs. Wind Chill Factor

Richard Browne was informed by Metro Weather that the temperature was 60°. He put on his jacket and stepped outside, intending to take a walk, but within two minutes he was back inside, shivering. Nobody said anything about that wicked wind, he thought, and, turning on the Weather Channel, learned that the wind chill factor was 28°.

The reliability of these numbers is not at issue. The wind chill number is clearly more relevant to the question of what kind of outerwear Richard should use.

Case 11. Can Andy Afford a $200,000 Porsche?

For many of us $200,000 is a considerable sum, and if Andy's financial state were anything like ours, we would probably be inclined to say no. There's a big "if" here, that points to the question which is at the heart of the matter: What is Andy's financial state? Since we don't know, we can only proceed by making some assumptions.

Scenario 1.

Andy's after tax assets from salary and investments are approximately $50,000 per year. With a $200,000 obligation against $50,000 in assets per year, it's difficult to see how Andy would be able to avoid defaulting down the road.

Scenario 2.

Andy's after tax assets from salary and investments are approximately $2 million per year. With a $200,000 obligation against $2 million in assets per year, we would probably be inclined to tell Andy to go for it.

The lesson to be learned from Andy's situation is that when it comes to a debt to be carried, focusing on the amount of the debt in absolute terms ($200,000 or whatever) is not the appropriate number trail. The appropriate number trail consists of the ratio of debt to ability to pay expressed by a measure of assets. What about countries? One may ask.

Case 12. The National Debt: Public Nuisance or Menace?

By the beginning of 1993 the gross national debt of the United States, it was generally agreed, was in the neighborhood of $4.2 trillion, a staggering figure which boggles the mind. If you had to transfer this amount of money in $100 bills from one location to another, you would have to deal with a stack of bills 2670 miles long.

The figure sounds ominous, but here is where disagreement begins. One point of view argues that the figure itself and the rate at which it has been increasing portend catastrophic consequences in the offing. When the debt grows faster than the country's ability to carry it, a breakdown with social, political, and economic upheaval is inevitable, and we are coming dangerously close to this state, this view has it.

But is the total size of the debt the figure we should be giving our first priority? Another view argues that in terms of the state of the economy, we are looking at the wrong figure and that, while debt reduction is desirable, it should not be given top priority and carried out in a "mindless" way since this will severely damage the economy. Its proponents focus on the ratio of publicly held debt to Gross Domestic Product (GDP).

In principle, how different is this situation from the one Andy finds himself in?

Case 13. How Good Is This Decision-Making Framework?

The president of Ecap University charged his Dean of Administrative Affairs, Michael Russell, with the task of setting up a criterion for running or cancelling course sections that would take into account student needs, be sensitive to maintaining academic quality, address the cost dimension, and be simple to use.

Dean Russell started with the assumption that each section, with perhaps a few exceptions, should pay its own way. He set up a course section run-cancel criterion, RC-1, based on the difference between the tuition revenue generated for the section based on student enrollment and the salary cost of the instructor for the section. RC-1 says run the section if revenue minus cost equals or exceeds $5000.

$$R - C \geq 5000$$

Otherwise, it is to be canceled, unless a compelling student need for the course could be established.

How well does RC-1 satisfy the requirements for a section run/cancel criterion stated by the president of Ecap University?

Implementation of RC-1 requires that the tuition revenue R and instructor cost C be determined for each section. Consider one such section, Stat 200 A1, let us say. If $R-C=6000$ we run Stat 200 A1; if $R-C=4000$, we cancel Stat 200 A1, unless a "compelling" reason to offer it can be presented.

RC-1 gets the job done, but there is a serious question about whether it's the "best" system for handling course run/cancel decision making in a way that satisfies the president's mandate.

1. **Academic Quality.** Department heads are not free to assign the "best" person to teach a section because the "best" person might be too costly in terms of the $R-C\geq 5000$ condition. A senior faculty member assigned to a course will often carry a much higher cost than a junior or adjunct colleague. If circumstances lead to changes in teaching assignments, courses which would run under one assignment framework might well have to be cancelled under another. With RC-1 the running or cancellation of courses depends more on how the game of academic musical chairs is played out than on the academic needs for running them, which is an unsatisfactory condition. It is in the overall interest of Ecap U. and, in particular, its students, that department heads be able to assign the most suitable faculty to courses and make changes when circumstances dictate. RC-1 is not compatible with this condition.

2. **Tunnel View.** RC-1 does not take into account the total revenue—total cost picture since reorganization of faculty teaching assignments by itself neither changes the total tuition revenue nor the total cost of faculty salaries.

3. **Data Collection.** The seeming simplicity of RC-1 in terms of data needed for its implementation is deceptive. A good deal of data has to be obtained and evaluated for each section in terms of a particular faculty assignment to the section before a run/cancel decision can be make. This is time consuming, but must be carried out within a tight time frame after class registration has taken place, but prior to the beginning of classes.

The methods of descriptive statistics can help us to organize and, along with computer technology, attractively present the data needed for the implementation of RC-1. The more skillful we are in accomplishing this, with judicious use of margins, boldface print, color, etc., the more impressive will the presentation

---

Descriptive statistics can help us present data, but it does not transform poorly-chosen data into well-chosen data, or unreliable data into reliable data.

---

make RC-1 appear to be as a decision making criterion. But this is illusionary. Presentation of data is important, but we must be careful not to equate it with higher-order substance. There is no magic for converting poorly-chosen data into well-chosen data, or unreliable data into reliable data. To distinguish between poorly-chosen and well-chosen data we must employ the most difficult to obtain and exacting tool of all—good judgment—to look behind the scene at the mechanism generating the data (RC-1 in this case).

We return to Ecap University's section run/cancel decision making problem in Exercises 12 and 13, cited after Section 1.6, to examine whether Dean Russell has succeeded in improving on RC-1.

1.5 ■ There's More to It Than What the Statistics May Suggest

Statistics, as numerical values, do not lie, but what the best of them give us cannot exceed the rules and limitations that we impose to define the framework from which they arise. If that framework is overly simplistic, then so too will be the statistics which it yields.

And then there is the problem of statistics overload. Different statistics seem to be indicating very different pictures and the problem is to determine which ones are key to the situation under study.

Case 14. There's Less to Baseball Statistics Than Meets the Bat

In this book *The Last Yankee: The Turbulent Life of Billy Martin,* David Falkner [12] concludes that Martin was the best manager of his era, possibly of many eras. Falkner's judgment was strongly influenced by baseball statistics compiled by the Elias Sports Bureau and a formula which claims to show which managers' teams won more games than they were reasonably expected to win.

In his review of Falkner's book George F. Will [32; p. 17] disputes Falkner's conclusion which, he argues, the rest of the book refutes. "In fact," Will notes:

> The Last Yankee might usefully be made required reading for graduate students in the social sciences and all others who need to be immunized

against the seduction of numbers .... There are limits—and Mr. Falkner's reporting shows that Elias passed them regarding Martin—to the ability to capture messy reality in tidy formulas.

Case 15. Is the Recession Over if the Statistics Say So?

In early 1992 President Bush was, as modern parlance would put it, an unhappy camper. Government statistics showed a mild recession and strong economic fundamentals. Yet business and consumer confidence in the economy had been shaken to an extent that seemed way out of proportion to the statistical signs, and many were blaming George Bush for having missed the wake up call.

"The problem," observes Charles McMillion [21], "is that many of those statistics are wildly misleading."

One statistic concerns **unemployment.** During the 1982 recession, the worst since World War II, unemployment reached 10.8 percent. During the 1991-92 recession it reached 7.8 percent. McMillion notes that the 1991-92 unemployment number looks good by comparison because it mixes two factors, jobs and the size of the labor force, and neglects the fact that the labor force has contracted sharply. "A better gauge," he argues, "is the number of actual jobs." Three hundred thousand more jobs were lost in the 1991-92 recession than the 1982 recession, June 1981-January 1983. A larger portion of the jobs lost this time involved higher-wage white collar workers, with ramifications throughout the economy. People working or seeking employment has declined by 1.2 million people in the first 19 months of this recession as opposed to 125,000 in the first 19 months of the 1982 recession. These features are not revealed by the unemployment rate.

Manufacturing output is another statistical indicator of economic health. According to this statistic, using constant output values, manufacturing has remained near 22 percent of America's gross national product since World War II. But, McMillion notes: "Even Commerce Department officials who assigned these values admitted—in the Survey of Current Business last year—that 'only a substantial research effort over many years holds any promise of overcoming . . . formidable statistical problems' with these figures."

The rapid pace of technological change makes it virtually impossible to measure 'constant' output over time.

U.S. Competitiveness: A comparison of the gross domestic product per worker of the United States against that of other major industrial competitors shows the United States to be well ahead of such rivals as Germany and Japan. "The tally depends," McMillion observes, "on the value assigned to the dollar . . . . Most comparisons use theoretical—so called 'domestic purchasing power parity' values that vastly overvalue the dollar."

Clearly, the assumptions underlying such statistical economic indicators as unemployment, manufacturing output and U.S. competitiveness must be watched. What must also be watched are the limitations of such indicators, and what they omit which is relevant.

1.6 ■ What Do the Statistics Say?

How statistics with the limitations we have imposed are interpreted will, at best, depend on the judgment and capacity of the interpreters and, at worst, on the cleverness of the spin doctors entrusted with putting on them the best possible spin. It is often the case that "experts" see very different things in the same statistics and that different statistics concerning the same issue seem to be contradictory.

Case 16. Discrimination or Difference?

Concerned about charges of subtle patterns of bias at its executive levels, The United Federation of Worlds set up a Commission to investigate. During its hearings Lork from Mork pointed out that while Morkians are 30% of the Federation's work force at its lower levels, they make up only 1% of its executive staff. "Good faith recruitment efforts have been made," observed Lork, "but the stastistics show subtle patterns of discrimination against Morkians." Tallia from Talos I disagreed. "The statistics show discrepancies," countered Tallia. "The subtie patterns of discrimination are your interpretation of the discrepancies

Who is right? We would have to agree with Tallia that the statistics show discrepancies. How the discrepancies are interpreted is another matter. Lork is offering one interpretation of the discrepancies that the statistics reveal.

What other explanations might account for the discrepancies?

Case 17. How Successful Was the Patriot?

During the Gulf War television viewers were moved by scenes of Patriot missiles streaking across the sky to intercept Iraqui Scud missiles that had been launched against Israel and Saudi Arabia. The Patriot's success seemed to epitomize the success of a high tech, low causality military campaign.

Apart from very successful military public relations, how successful was the patriot as a military tool? Different statistics have been given and, as is almost always the case, it is important to look further than the statistics if a realistic picture is to emerge.

The Army originally stated that Patriots "intercepted" 45 of 47 incoming Scud missiles, and President Bush revised that to 41 of 42. What does this mean? Brigadier General Robert Drolet of the Army's Missile Command testified that "a Patriot and a Scud passed in the sky." There are other statistics of interest. Before Patriots were employed in Israel, 13 Scuds fell near Tel Aviv. There were no deaths, but 115 people were wounded and 2,698 apartments were damaged. After Patriots were employed in this region, 11 Scud attacks left 1 dead, 168 injured and 7,778 apartments damaged. (see [Jagger; 15] and [Marshall; 20]) This is explained by the fact that successful hits led to more deadly debris being sprayed over a larger area than otherwise would have been the case and that the Patriots tended to strike the bodies of Scuds, leaving their warheads armed and able to cause significant damage on landing. But then it should also be kept in mind that the Patriot was not designed as an antimissile weapon, but to defend against fast-flying aircraft.

Case 18. The Reagan Economic Boom: Blessing or Disaster?

Martin Anderson, former advisor to President Reagan and senior fellow at the Hoover Institution, employs statistics to support his view that the Reagan economic boom was the greatest ever [2]:

Anderson's View

The two key measures that mark a depression or expansion are jobs and production. Let's look at the records that were set.

Creation of Jobs

From November 1982, when President Ronald Reagan's new economic program was beginning to take effect, to November 1989, 18.7 million new jobs were created. It was a world record: . . . The new jobs covered the entire spectrum of work, and more than half of them paid more than $20,000 a year. As total employment grew to 119.5 million, the rate of unemployment fell to slightly over 5 percent, the lowest level in 15 years.

Creation of Wealth

The amount of wealth produced during this seven year period was stupendous—some $30 trillion worth of goods and services. Again, it was a world record . . . . According to a recent study, net asset values—including stocks, bonds and real estate went up by more than $5 trillion between 1982 and 1989, an increase of roughly 50 percent . . . .

Income Tax Rates, Interest Rates and Inflation

Under President Reagan, top personal income tax rates were lowered dramatically from 70 percent to 28 percent. This policy change was the prime force behind the record breaking economic expansion . . . .

The Stock Market

Perhaps the key indicator of an economy's booms and busts is the stock market, the bottom line economic report card . . . . starting in late 1982, just as Reaganomics began to work, the stock market took off like a giant skyrocket. Since then, the Standard & Poor's index has soared, reaching a record high of 360, almost triple what it was in 1982.

There were other consequences of the expansion. Annual Federal spending on public housing and welfare, and on Social Security, Medicare and health all increased by billions of dollars. The poverty rate has fallen steadily since 1983.

When you add up the record of the Reagan years, and the first year of President Bush . . . the conclusion is clear, inescapable and stunning. We have just witnessed America's Great Expansion.

Leontief's View

In a reply, Nobel Prize winning economist Wassily Leontief [19] concedes some of Anderson's statistics but goes on to look at a number of cost thorns in his statistical rose garden.

True, the long recovery from the deep depression that brought President Reagan to power carried this country to the high point of the usual cyclical wave characterized by a low rate of unemployment and a high gross national product. It is most likely that wholesale tax cuts inaugurated by Mr. Reagan have made the level of the G.N.P., as measured by the Government statisticians, several billion dollars higher than it would otherwise have been. But at what a cost.!

Drastic cuts in public spending (except for military purposes) left the physical infrastructure of this country in ruin. City streets and transportation facilities, water-supply and sewage systems, particularly in large metropolitan areas, are collapsing, the once glorious interstate highways are crumbling, and cramped airports are incapable of handling the rapidly increasing traffic. Despite the valiant effort of the underfinanced, underpowered Environmental Protection Agency, our lakes, rivers and forests are succumbing to deadly acid rain.

What is even worse, the intellectual, cultural and social infrastructure of the country has suffered even more during this greater-than-ever boom than its physical counterpart. Primary and secondary schooling have been so weakened that a whole generation of boys and girls can hardly read, write or count, while the soaring price of higher education makes it impossible for many young people to take advantage of it.

No wonder the competitiveness of the United States is rapidly declining; many of our high technology industries are losing one battle after another in the struggle for their share of the foreign and even their own domestic market. At the same time, the rich are getting richer, and the poor are getting homeless.

Let us hope that contrary to Mr. Anderson's expectations the "Reagan boom" will not continue in its present form for four or eight more years. If it does, the United States will find itself entering the 21st century as the richest country (in total value of stocks and bonds traded on the stock exchanges), but culturally and socially less advanced than other developed countries.

EXERCISES

1. A crucial question facing a patient in need of a heart operation is, how good is my surgeon? One quantitative approach that has been advanced for rating the quality of heart surgeons involves keeping statistical score cards for them which state their surgical success rates. [Altman; 1] Discuss the issues of reliability and how well-chosen such data are in ascertaining the quality of heart surgeons.

2. Ellen Ames and Ann O'Neil, professors of economics at Huxley College, saw a grade of 50 on Professor Ames's last economics exam in different terms. "A grade of 50 is an F," noted Professor Ames. "But it's the highest grade in the class," replied Professor O'Neil, "and as such an F does not make sense." Discuss the significance of 50 as the highest grade on an exam, and as an F.

3. "80 percent of the crimes in this city were committed by Plutonians," remarked Oscar to his wife Janet. "They're a bad lot." Are there other interpretations of this figure? Are there other figures that might be relevant? Discuss.

4. Which provides a better measure of a company's earnings, Generally Accepted Accounting Principles (GAAP) or Standard & Poors Core Earning's Measure (CEM)? N. Byrnes, M. Derhovanesian, "Earnings: A Closer Look," *Business Week*, May 27, 2002; 34-37; P. Krugman, "America's Poor Standards," *The New York Times*, May 17, 2002; A25.

   Why is the proposed Standard & Poors Core Earnings Measure a big deal? P. Coy, "Why Better Numbers Really Matter *Business Week*, May 27, 2002; 36-37; "Editorial: A Good Idea About Earnings," *Business Week*, May 27, 2002; 114.

5. Which is the better measure of a basketball player's scoring efficiency, shooting percentage or points per shot? How so? T. Keegan, "New Stat Creates Great Divide: Points Per Shot, A Terrific Gauge," *The New York Post*, Feb. 9, 1997; 92.

6. The trade deficit is way up. Should we push the panic button? M. Weidenbaum, "Trade Deficit: Obsessing Over a Misleading Indicator," *The Christian Science Monitor*, June 18, 1998; June 18, 1998; 11. R. Eisner, *The Great Deficit Scares* (New York: The Century Foundation Press, 1997), Ch. 3.

7. Concerning the rise in inflation, is the government looking at the wrong prices? D. Ranson, "Inflation May Be Worse Than We Think", *The Wall Street Journal*, Feb. 27, 2008; A17.

8. Should we accept statistics on how many lives air bags have saved in automobile crashes at face value? A. Q. Nomani, J. Taylor, "Shaky Statistics Are Driving the Air Bag Debate," *The Wall Street Journal*, Jan. 27, 1997; B1.

9. Should we accept the Census Bureau's counts on ethnicity at face value? L. Belsie, "Census Nods to New Views of Ethnicity," *The Christian Science Monitor*, July 28, 1999; 1.

10. Should we accept official counts of the number of poor at face value? J.T. Allen, "The Politics of Poverty," *U.S. News & World Report*, Nov. 1, 1999; 40.

11. Should we accept FBI crime statistics at face value? J. DiLulio, A. M. Piehl, "What the Crime Statistics Don't Tell You," *The Wall Street Journal*, Jan. 8, 1997.

12. Dean Michael Russell of Ecap University was asked to develop other criteria for running or cancelling course sections which takes into account student needs, is sensitive to maintaining academic quality, addresses the cost dimension, and is simple to use. The dean came up with Russell criterion 2, RC-2, which operates as follows: For a given department, English, for example, determine the average salary cost of the faculty in the English department. Run English section A, let us say, if the tuition revenue based on section A's enrollment equals or exceeds the average salary cost of the English department for section A by $5000. Otherwise, cancel section A unless compelling student or academic needs can be established. The same system would apply to courses in other departments.

(a) What data would be needed for the implementation of RC-2?

(b) What are the merits and disadvantages of RC-2?

13. Another criterion developed by Dean Russell, RC-3, operates as follows: For a given department, determine the average class size for all sections being run by the department for the session in question, excluding such one-on-one activities as independent study and thesis supervision. If this average class size is at least 25, then run all sections with an enrollment of at least 10. If the enrollment of a section is less than 10, then go to RC-2 to make a determination for the section. If the average class size is less than 25, then go to RC-2 for all sections.

Address the same questions (a) and (b) for RC-3 as those posed for RC-2 in 12.

(c) Is RC-3 an improvement over RC-2? Explain.

14. You have been hired as a consultant to help Dean Russell develop a criterion for running and cancelling course sections which reflects the conditions stated by the president of Ecap University (see Case 13 or Exercise 12).

(a) Provide a statement describing your run-cancel criterion.

(b) What data would have to be obtained to implement your criterion?

(c) Discuss the merits and disadvantages of your criterion.

(d) How could your run-cancel criterion be improved so as to overcome the disadvantages cited in (c)?

15. Locate two articles which take issue with data presented from some source from the point of view of underlying assumptions, reliability or relevance. Write a commentary for each article explaining their authors' points of view. Possible sources for such articles include newspapers and magazines which address serious domestic and international issues, such as *The New York Times, The Christian Science Monitor, The Wall Street Journal, Newsweek, Business Week, Time,* and *U.S. News & World Report,* as well as professional journals concerned with your area of professional interest.

REFERENCES

1. L. Altman, "Surgical Scorecards: Can Doctors Be Rated Just Like Ball players?" *The New York Times,* Jan. 4, 1992.
2. M. Anderson, "The Reagan Boom—Greatest Ever," *The New York Times,* Jan. 17, 1990.
3. J. Brooke, "In Brazil, Slip of the Tongue Makes Campaign Slip," *The New York Times,* Sept. 5, 1994.
4. A. Bryant, "A Different Gauge for Rating Airlines," *The New York Times,* March 7, 1995.
5. M. Burros, "Major U.S. Survey on Food Use and Pesticides Is Drawing Fire," *The New York Times,* Sept. 11, 1991.
6. B. Carter, "Watching the Watchers," *The New York Times,* March 10, 1997.
7. C. Clark, *A Critique of Russian Statistics* (London: MacMillan and Co., 1939).
8. S. Dillon, "Report Finds More Violence in the Schools," *The New York Times,* July 7, 1994.
9. J. Duncan, A. Gross, *Statistics for the 21st Century* (Chicago: Irwin, 1995).
10. *The Economist,* "The Good Statistics Guide," Sept. 7, 1991.
11. *The Economist,* "The Good Statistics Guide," Sept. 11, 1993.
12. D. Falkner, *The Last Yankee: The Turbulent Life of Billy Martin* (New York: Simon & Schuster, 1991).
13. E. Gootman, "Undercount of Violence in Schools: Defective Reporting is Found at 10 sites", *The New York Times,* Sept. 20, 2007.
14. M. Gordon, "Moscow Statisticians Accused of Aiding Tax Evasion," *The New York Times,* June 10, 1998.
15. J. Jagger, "Why Patriot Didn't Work as Advertised," *The New York Times,* June 9, 1991.
16. V. Kirichenko, "Return Credibility to Statistics," *Business Economics,* Oct. 1990.
17. G. Kolata, "Which Students Are Worst at Science," *The New York Times,* Dec. 24, 1991.
18. G. Kolata, "Maker of Sleeping Pill Hid Data on Side Effects, Researchers Say," *The New York Times,* Jan. 20, 1992.
19. W. Leontief, "We Can't Take More of this 'Reagan Boom,' *The New York Times,* Feb. 4, 1990.
20. E. Marshall, "Patriot's Scuds Busting Record is Challenged," *Science,* May 3, 1991.
21. C. McMillion, "Facing the Economy's Grim Reality," *The New York Times,* Feb. 23, 1992.
22. B. Meier, "Politics of Youth Smoking Fueled by Unproven Data," *The New York Times,* May 20, 1998.

23. *The New York Times*, "Rigging the Numbers," June 15, 1998.
24. NOVA, "Can You Believe TV Ratings? WGBH, Boston, 1992.
25. E. Richman, Letter, *The New York Times*, Aug. 12, 1986.
26. D. Sanger, "Bad Debt Held by Japan's Banks Now Estimated Near $1 Trillion," *The New York Times*, July 30, 1998.
27. H. Shaffer (ed.), *The Soviet Economy: Western and Soviet Views* (New York: Appleton-Century-Crofts, 1963).
28. R. Spector, Letter, *The New York Times*, Aug. 12, 1986.
29. S. Tefft, "China Is Under Pressure to Clean Up Its Statistics," *The Christian Science Monitor*, June 9, 1994.
30. V. Toy, "Draft Audit Says Board of Education Underrates Crime in Schools," *The New York Times*, Sept. 2, 1995.
31. T. Weiner, "General Details Altered 'Star Wars' Test," *The New York Times*, Aug. 27, 1993.
32. G. Will, "Paranoid in Pinstripes," Review of [12], *The New York Times Book Review*, April 5, 1992.

1.7 ■ The Thrust to Quantify

Our age may be termed the age of quantitative methods because the respectability of a presentation or argument is more and more being seen as enhanced, if not unequivocally defined, by numbers, statistics, and mathematical arguments. The thrust of our age is to quantify.

But then we may ask, which is preferable: numbers which give us a sense of substance and precision about the situation under study, but are considerably off reality's mark, or no numbers at all. If you agree with humorist Artemis Ward that, "It ain't so much the things we don't know that get us in trouble, it's the things we know that ain't so," then the latter state is clearly the lesser of two evils. But this lesser of two evils is not a viable option in our quantitatively oriented age, and the only alternative is to develop a sense of what we should beware of when tantalizing statistics are floated our way.

1.8 ■ Ask Yourself

Pushed by the flow, more and more people, knowing less and less about the correct use of statistics, are generating more nonsense more quickly than ever before. The answer to this phenomenon does not lie in discarding the statistical baby with the bathwater, but in becoming knowledgeable about the basic principles underlying statistical applications, irrespective of our applied area of interest, and sensitive to statistical abuses and misuses.

Application of the following statistical ten commandments to an article or presentation which relies on statistics or has statistical implications might put us in a better position to detect misleading statistics and statistical fraud.

Test Questions

1.	Are the numbers/statistics well-chosen, relevant, to the issue in question?

While it is often difficult to answer this question in general, we should nevertheless keep it in mind. If the answer is no, questions about their reliability become moot . Richard Browne (Case 10) and Andy and his Porsche (Case 11) are simple examples to keep in mind.

2.	Is the source given for the data/statistics?

If the answer is no, then questions 3, 4, 5 and 10 are not viable. Be on guard.

3.	Is the source unbiased and reliable?

What experience have I had with the source in the past? Has it earned membership in the Slippery Statistics Society, a reputation for reliability or is its reliability status unknown?

4. How were the data/statistics obtained?

Does the method for obtaining the data/statistics have weaknesses that may compromise their integrity?

5. How current are the data/statistics?

6. Is there less, or more, to the issue than the data/statistics are being interpreted as meaning?

There often is. Baseball statistics (Case 14), according to George Will, presents us with an example of less than and statistics and the recession (Case 15), according to Charles McMillion, presents us with a case of more than.

7. Are there other data/statistics that support the conclusion reached?

Jim felt ill and sought medical advice. "Your temperature reading is 98.6°; you're ok", he was told. If his blood pressure reading, pulse rate, and data that a blood test would reveal fall within a normal range, this would support the conclusion reached.

8. Are there other data/statistics that contradict the conclusion reached? How is the contradiction to be resolved?

If some of the afore data do not fall within a normal range, this would contradict the conclusion that Jim is ok.

9. Do the data/statistics admit contradictory interpretations? If so, how are these contradictions to be resolved?

In 1986 a team led by Dr. Charles Bluestone submitted a paper to the *New England Journal of Medicine* which concluded that the antibiotic amoxacillin was effective in treating middle-ear infection. Dr. Erdem Cantekin, a member of the team, dissented from this conclusion and submitted a separate report to the *Journal* which he claimed was a re-analysis and re-interpretation of the paper that Bluestone had submitted a month earlier.

The *Journal's* editor saw the issue as determining which group had the right to publish and, on the advice of officials at the University of Pittsburgh and the hospital where many patients in the study were treated, decided to publish the Bluestone paper only. The papers were not sent out for review. Shortly after submitting his paper for publication in 1986, Dr. Cantekin was removed as director of Pittsburgh University's center for studies of middle-ear infections. Dr. Cantekin persisted, and saw his report published in the The Journal of the American Medical Association in December 1991.

Open, honest discussion obviously would have been a preferable way of resolving conflicting interpretations.

10.	What assumptions underlie the data/statistics and the conclusion(s) reached? Are these assumptions realistic?

If the assumptions made are unrealistic, the conclusions reached from them must be suspect in terms of being realistic.

Charles McMillion's argument (Case 15) is that the assumption that government statistics which led to the conclusion of a mild recession with strong economic fundamentals was unrealistic.

Dean Michael Russell (Case 13) was charged with the task of setting up a criterion for running or canceling course sections that would take into account student needs, be sensitive to maintaining academic quality, address the cost dimension, and be simple to use. His basic assumption that each section should pay its own way yielded RC-1, which did not satisfy any of the afore requirements.

1.9 ■ Suggestions for Further Reading

The following books are especially useful for helping us become more aware and sensitive to number pollution and slippery statistics.

1. W.J. Adams with R.B. Adams, *Get a Grip on Your Math* (Dubuque: Kendall/Hunt, 1996)

    Part One, "Get a Grip on Your Numbers" (Chapters 1-9), discusses sources of slippery numbers with the objective of helping the general reader become more aware and sensitive to their origins and nature. More generally, *Get a Grip* examines such questions as: What can mathematics do for us? What are its limitations? In what sense is mathematics precise?

2. W. J. Adams with R. B. Adams, *Get a Firmer Grip on Your Math,* (Dubuque: Kendall/Hunt, 1996)

This sequel to *Get a Grip* is intended for readers who wish to dig deeper. It provides food-for-thought questions to help us obtain a more concrete understanding of slippery numbers, and a more in-depth discussion of ideas taken up in *Get a Grip*.

3. J. Duncan, A. Gross, *Statistics for the 21st Century*, (Chicago: Irwin, 1995).

Explores "strengths, weaknesses, opportunities, and threats posed by statistical data generated by government bureaus, associations, international agencies, and corporations."

It is a useful companion to every text on statistics. Its attitude and spirit are invaluable for all who study, apply, and make use of statistical methodology.

4. D. Huff with I. Geis, *How to Lie with Statistics*, (New York: W. W. Norton & Co., 1954).

"This book is a sort of primer in ways to use statistics to deceive. It may seem altogether too much like a manual for swindlers. Perhaps I can justify it in the manner of the retired burglar whose published reminiscences amounted to a graduate course in how to pick a lot and muffle a footfall. The crooks already know these tricks; honest men must learn them in self-defense."

A highly readable, insightful, engaging book on slippery statistics which must be at least in its twelfth printing. Although it's more than half a century since *How to Lie with Statistics* appeared, its illustrations are timeless.

5. A. Jaffe, H. Spirer, *Misused Statistics: Straight Talk for Twisted Numbers*, (New York: Marcel Dekker, 1987).

"*Misused Statistics* describes sophisticated statistical issues so that nonstatisticians can understand them . . ."

In this it succeeds. It's in the class of books pioneered by Huff's classic, but if we had both books before us and had to choose which to read first, we would go by author in alphabetical order. But then we would continue with this insightful work.

2

The Basic Raw Materials: Data

2.1 ■ Data Scales

At the end of the academic year each student at Ecap University is asked by the student government to fill out a questionnaire. The following are examples of some of the questions asked.

Personal Data: Please indicate the following:

1. Sex: (1) Male, (2) Female

2. Marital status: (1) Married, (2) Single, (3) Separated, (4) Divorced

3. Age: (1) Under 18, (2) 18-19, (3) 20-24, (4) Over 24

4. Work status: (1) Employed full time, (2) Employed part-time, (3) Not employed

5. Highest SAT score (total): (1) Under 900, (2) 900-1000, (3) 1001-1100, (4) Over 1100

University Related Data

6. Class designation: (1) Freshman, (2) Sophomore, (3) Junior, (4) Senior

7. Area of major: (1) Business, (2) Humanities, (3) Science, (4) Computer Science, (5) Mathematics, (6) Education, (7) Engineering, (8) Undecided

8. Distance of residence from campus: (1) Under 1 mile, (2) 1-3 miles, (3) Over 3 miles

9. Preferred class room temperature (Fahrenheit): (1) 66-68, (2) 69-70, (3) 71-72 (4) 73-74

10. Satisfaction with teaching quality this past academic year (largest number indicates highest degree of satisfaction): 5 4 3 2 1

The data arising from questions 1, 2, 4, and 7 are said to have a **nominal scale** because the responses express categories for which no ordering is implied. In question 7, for example, eight categories are cited, but there is no ordering implied as stated which would put any one category ahead of any other. Numbers are assigned to the categories to help record the responses, but this is their only role. Nominal scaling is the weakest form of measurement.

1. The **nominal** measurement level classifies data into non-overlapping categories on which no ranking or order is imposed. Each datum falls into one of the categories.

Suppose we obtain the information from 400 students in response to question 2 shown in Table 1.

Table 1

Category	No. of Responses
Married	100
Single	240
Separated	40
Divorced	20

If of interest, we could calculate proportions for data on a nominal scale to determine the percentage of respondents in each category. Twenty-five percent of the responding students are married, 60% are single, 10% are separated, and 5% are divorced.

Suppose that the first five respondents answered (1), (1), (2), (4), (2) to question 2. While we could take the average of the numbers 1, 1, 2, 4, and 2, to obtain 2, it would be meaningless to do so in terms of the background situation. To say that the average respondent is single makes no sense.

> We can compute the proportion of the total who are in each category arising from a nominal scale. It's meaningless to compute the average of the numbers used to designate the categories themselves.

The data arising from questions 6 and 10 are said to have an **ordinal scale** because there is an implied order which allows us to speak about one category being better or preferable to another. In question 10 a rating of 5 is better than or superior to a rating of 4 or any of the other ratings. Ordinal scaling is stronger than nominal scaling, but it is still weak. We cannot meaningfully talk about differences of ratings in an ordinal scale because there are many ways to choose such numbers and differences depend on which numbers are chosen. In question 10 the ratings could just as well have been 20, 10, 5, 2, −7 or A, B, C, D, F. In the social and behavioral sciences such factors as power, prestige, and emotional stability might be measured on an ordinal scale.

> 2. The **ordinal** measurement level classifies data into non-overlapping exhaustive categories, as does the nominal scale, but categories that can be ranked or ordered. The ordinal rankings may or may not involve the use of numbers.

Suppose we obtain the information from five students who had Professor X for statistics shown in Table 2.

Table 2

Rating	No. of Responses
20	0
10	2
5	1
2	1
−7	1

20 corresponds to Outstanding, 10 to Good, 5 to Satisfactory, 2 to Poor, and −7 to Terrible. If we average the results cited in Table 2, we obtain 4, which does not correspond to any of the ratings available. How is 4 to be interpreted? We know that a class consensus value of 4, as defined by averaging the ratings, puts Professor X between Satisfactory and Poor, which is all that can be said. Because differences

between ratings have no meaning, we cannot meaningfully talk about how much below satisfactory and above Poor the consensus value 4 places Professor X.

> **Ordinal Scale:** Values are assigned, at our discretion, to establish a ranking of categories. Differences between rank values have no meaning and because of this an arithmetic average of rank values which differs from the rank values has no clear interpretive meaning.

The data arising from questions 5 and 9 are said to be measured on an **interval scale.** Interval data are more than rankings; they are numerical values for which differences are meaningful. There is a meaningful difference of one degree between each unit. On the Fahrenheit temperature scale the difference between the freezing point (32°F) and boiling point (212°F) of water is divided into 180 intervals representing equal amounts of heat, so that it takes the same amount of fuel to raise the temperature of a fixed quantity of water 1F degree on the scale. Differences are meaningful because of the consistency in the behavior of nature; it's not a matter open to our discretion.

One property lacking with an interval scale is that there is no natural, meaningful zero. The 0°F mark, for example, is an artificially chosen value. It does not signify no heat, heat being a property of molecular motion. 0°F does not signify absence of molecular motion.

As a result, ratios have no meaning in an interval scale. It is meaningless to say, for example, that an object with a temperature of 60°F is twice as hot as one with a temperature of 30°F.

> 3. The **interval** measurement level employs numbers for which differences are meaningful. One can talk about how much more or less of the characteristic the quantity has. There is no meaningful zero, however.

The data arising from questions 3 and 8 are said to be measured on a **ratio scale.** Ratio scale data have the properties of interval scale data and in addition there is a natural, zero point so that it is meaningful to consider ratios of measurements. A distance of 4 miles, for example, is twice the distance of 2 miles; an age of 30 years is three times the age of 10 years.

> 4. The **ratio** measurement level employs numbers, has the properties of the interval measurement level, and there is a meaningful zero so that ratios can be formed.

In order of strength, we have the nominal, ordinal, interval, and ratio scales. For further discussion of data scales, see [1].

There is a very important reason for the fuss, namely:

> The kind of statistical inference technique that we can meaningfully employ on the data depends on the data scale.

REFERENCE

1. S. Stevens, "Mathematics, measurement, and psychophysics," Ch. 1 of *Handbook of Experimental Psychology*, S. Stevens (ed.), (New York: Wiley and Sons, 1951).

EXERCISES

1. (a) What is the difference between nominal and ordinal data?

   (b) What is the difference between interval and ratio data?

2. What makes SAT score data interval data?

3. Classify the following examples of data as nominal, ordinal, interval, or ratio. Explain the reason for your classification in each case. (a) Eye color, (b) Religious affiliation, (c) Course grade (A, B, C, D, F), (d) IQ score, (e) Height, (f) Weight, (g) Ranking of twelve football teams in a citywide competition, (h) Ratings in a beauty contest, (i) Temperature in the Celsius scale.

4. Bob received 1200 on an SAT exam, while Joe scored 600. Bob then turned to his friend and proclaimed: "I've twice the aptitude for college as you." Is this a meaningful statement? Explain.

5. Joe found that he had $20 while Bob had only $10. He then proclaimed: "I've twice as much money as you." Is this a meaningful statement? Explain.

6. The Ecap University Student Association holds an Uglyman-on-Campus contest each year in which ratings are determined and prizes given. This year we have:

Robert J.	12	Joseph R.	6
Howard K.	8	Anthony B.	4

Which of the following statements are meaningful? Explain.

(a) The winner is Robert J.

(b) Robert J. is twice as ugly as Joseph R. and Howard K. is twice as ugly as Anthony B.

(c) The average of the uglyman scores is 7.5.

7. In a *New York Post* online poll on who was the most evil person of the millennium, Adolf Hitler came in first with 8.67% of the vote and Hillary Clinton came in sixth with 3.99% of the vote. A. Soltis, "Post Readers: Hitler Was Most Evil," The New York Post, Nov. 17, 1999; 2.

Does this mean, according to the poll, that Adolf Hitler is 2.17 times as evil as Hillary Clinton? Explain.

Does it mean that Adolf Hitler is 6 times as evil as Hillary Clinton? Explain.

8. The administration of Ecap University used the following system to determine merit raises for the faculty.

Step 1. Each department chairperson rates each department colleague on Teaching, Scholarship, and Service. The following five performance categories are to be used. 5 = Outstanding, 4 = Very Good, 3 = Good, 2 = Needs Improvement, 1= Unacceptable. An overall rating is obtained by taking the numerical average of the three ratings.

Step 2. Each chairperson's assessment is reviewed by the appropriate dean.

Step 3. Merit increases (percentage of salary) are determined on the basis of the overall rating as follows. 2.9 or below: 0; 3.0-3.4: 2.8; 3.5-3.9: 3; 4.0-4.4: 3.5; 4.5-4.9: 4, 5: 5.5.

(a) On what scale is the performance ratings of faculty? Explain.

(b) Is it meaningful to average the performance ratings to obtain an overall rating? Explain.

(c) Is it appropriate to determine salary increments as described in Step 3? Explain.

Sources of Data

Data which come from records maintained by agencies, departments of government, businesses, and organizations of the most diverse nature are called **internal data.** Many concerns will take one to an organization's internal data.

A number of public and private organizations publish data for the use of interested parties. Examples of such organizations include the Departments of Commerce and Labor, the Federal Reserve Board, the United Nations, Center for Responsive Politics, Common Cause, Federal Election Commission, state and city agencies, journals, trade magazines, and newspapers. It is wise to consider all data in the light of the ten questions noted in Section 1.8. And then there are situations where the data required are not available and we must generate our own. This is the focus of the remaining sections of this chapter.

As we noted in the brief overview *What is Statistics*? a statistical **population** is a collection of data viewed as a whole in its own right, whereas a **sample** is a part of the all-inclusive whole. Whether a set of data is viewed as a population or sample depends on circumstances. The final exam grades of Professor Oscar Ford's statistics class, for example, are a population from the point of view that it is a totality in its own right and the grades of other statistics classes are not relevant. This set, from the point of view of the chairperson of the mathematics department, is a sample of the population of grades of all statistics classes in the department.

A population of interest emerges, the population of life-times of 100 watt light bulbs made by the Maxwell Company, for example. The company would like to obtain an estimate of their average life-time, but it is clearly in a quandary. It can determine an individual bulb's life-time, in the course of which the bulb is destroyed, but if it should proceed in this way for its entire output it would end up with nothing to sell. The simple idea that emerges is that of taking a "representative" sample of light bulbs, determining their average life-time, and using this value as an estimate for the population average life-time. Sampling may be desirable because of the difficulty or expense involved in extracting data. Apart from these considerations, the feeling often emerges that a "well-chosen" sample is sufficient to do the job and that it would be a waste of time, energy, and money to examine the entire population.

Carrying out the seemingly simple idea of taking a "representative" sample and using its characteristics to infer population characteristics is not so simple a matter, which is the subject of most of this book. Our starting point is the subject of sampling itself, a discussion of which is initiated in the following three sections.

Then there is the subject of what people believe, what they think, and what their characteristics are. If you want to know, ask them, which leads us to consideration of polls, questionnaires, and surveys in Section 2.5.

2.2 ■ Survey of Sampling Methods

In class, on the job, and in statistics books, among others, we hear or read comments of the following sort: Take a sample of size 30 from the population; survey public opinion on this or that issue. Easily said, but far from easily done. In this section we briefly look at some methods of sampling and problems that arise in carrying them out.

Sampling methods fall into two general classes, **probability sampling**, which includes random sampling (or simple random sampling, as it is sometimes called), systematic selection, stratified sampling, and cluster sampling, and **nonprobability sampling**, which includes quota sampling and judgment sampling.

In the following we assume that the underlying population is finite, that is, that the number of elements or units in the population can be described by a positive integer; the integer might be small, such as 2, or fairly large, such as 2 trillion.

Random Sampling

A sample of a certain size, 10 units, let us say, is to be chosen from the underlying population. When we say that the sample is to be chosen at **random** we have in mind the idea that there is to be no bias, deliberate or inadvertent, which favors certain samples of size 10 being chosen over others. The sampling procedure is to be an equal opportunity procedure. Whether the randomly chosen sample is to be of size 10 or 2 or 1500 is immaterial; the sample is to be chosen in an unbiased manner; no favoritism.

Andy Arunas, freshman reporter for *The Huxley College Press*, wants to interview a sample of Huxley's faculty concerning their views on students, administration, and the issues of cheating, grading, tenure, teaching, and research. Andy is considering choosing a random sample of 10 of Huxley's 300 faculty. To do this he is thinking of proceeding in the following way.

(1) Obtain a list of the 300 faculty and number them 1 through 300;

(2) Obtain 300 ping-pong balls from the local sport equipment store and number them 1 through 300;

(3) Put the balls in a bag, shake the bag, turn it upside down several times, and then (without peeking) draw 10 balls from the bag, one at a time. The faculty corresponding to the 10 numbers drawn would be the faculty Andy would seek to interview if he decides to choose his interviewees by random sampling.

Does this device ensure that random sampling will be carried out? It is easy to declare that random sampling is what we desire or assume to be the case, but achieving it in practice is another matter which is the focus of Section 2.3. The probability structure that the concept of random sampling leads to is discussed in Section 6.5.

> The assumption of random sampling underlies a good deal of statistical theory. It is the main focus of the methods developed in this book.

Systematic Selection with a Random Start

The **idea of systematic sampling** is to choose every twentieth name on a list, or every hundredth number in a telephone directory, or every fifth house in a residential neighborhood. To add an element of randomness to this procedure one may choose the starting point at random.

If Andy were considering systematic selection with a random start as a mechanism for choosing a sample of 10 of the 300 faculty, he would begin by determining the **sampling interval** k by taking the ratio of the population size to the sample size and rounding to the nearest integer. Andy's sampling interval is $k = 300/10 = 30$. The next step is to choose an integer at random from 1, 2, . . . , 30. Suppose 11 were obtained; then Andy would seek to interview the faculty numbered 11, $11+30=41$, $11+2(30)=71$, . . . , $11+9(30)=281$ should he decide on this approach.

An appealing feature of this sampling procedure is that it is easy to carry out and spreads the sample through the population. It may yield unrepresentative results, however. If every thirtieth faculty number on the list starting with the eleventh comes from Huxley's Business School, for example, the sample obtained would consist entirely of business faculty.

If the underlying population is listed at random, then systematic selection with a random start is equivalent to random sampling.

Stratified Random Sampling

Sometimes a population can, with respect to a defining characteristic, be "meaningfully" divided into relatively homogeneous subpopulations or *strata* such that each member of the population belongs to one and only one *stratum*. Figure 1 shows a population Q which has been partitioned into five *strata* denoted by $S_1, S_2, \ldots, S_5$.

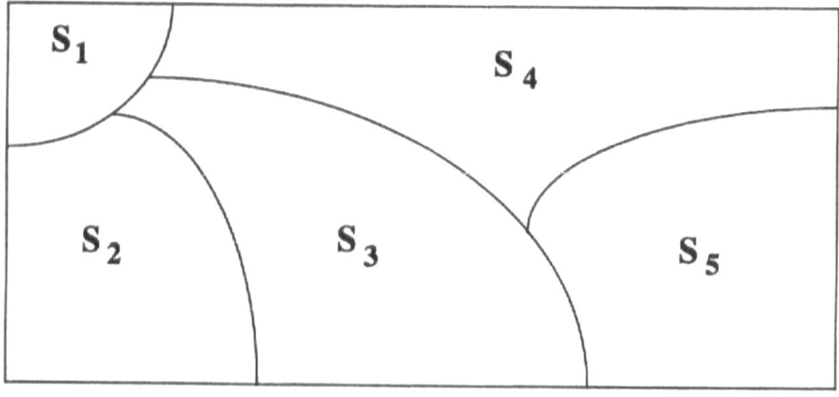

Figure 1

Stratified random sampling consists of "meaningfully" partitioning the population into *strata* and then taking a random sample from each *stratum*. To employ stratified random sampling a listing must be available for each stratum. The overall sample size *n* may or may not be allocated to the *strata* in proportion to the *strata* sizes. If *n* is allocated in this way, the sampling is called **proportional sampling.**

Andy Arunas, for example, might consider partitioning the Huxley College Faculty according to school: Arts and Sciences, Business, Education, Mathematics and Statistics, and Equine Studies. With an envisioned total sample size of 10, he might follow up by choosing 2 faculty at random from each school, but this is not his only option. If all the schools have the same number of faculty, then the subsample allocation of 2 per school constitutes proportional sampling.

Stratified random sampling is effective in reducing sampling error when the population is diverse, appropriate *strata* can be defined which are relatively homogeneous, and it is desirable to have each *stratum* represented in the overall sample. An interesting readily accessible application of stratified sampling to a problem in accounting involving the Chesapeake and Ohio Freight Study is discussed by John Neter in [1].

Cluster Random Sampling

In **cluster random sampling** the population is partitioned into relatively small subunits called **clusters** (rather than *strata*). A random selection of clusters is chosen and a subsample (possibly 100 percent) of each chosen cluster is taken.

Huxley College's Faculty are organized into departments (accounting, art, economics, etc.) which may be thought of as clusters. To select the faculty to be interviewed by cluster sampling using the departments as clusters, Andy would proceed by choosing a sample of departments at random and then interviewing a sample of the faculty from each of the departments from the sample chosen.

Estimates based on random sampling are generally more reliable than those based on cluster random sampling, when both can be carried out, but cluster sampling is generally more cost effective to implement and is sometimes a viable option when random sampling is not feasible. Suppose, for example, that an organization wants to study family income levels in the Boston area and has decided to interview 1500 families. Random sampling may not be feasible because a complete list of the families (the target population) may not be available. Moreover, the cost of interviewing families scattered over a large area may be prohibitively high. An alternative approach would be to divide the region into city blocks, the clusters in this case, and interview all or a sample of families in a number of randomly selected blocks.

In some applied situations it may be necessary to use a combination of the afore-discussed sampling procedures. If an education organization, for example, wanted to study education problems faced by elementary school teachers in the United States, its statisticians might proceed by stratifying by states, partitioning the states into counties, and using cluster sampling to obtain a sample of counties. The last stage might involve using random sampling to emerge with schools whose personnel or sample of personnel would be interviewed.

Nonprobability Sampling

Nonprobability sampling includes any sampling method which does not include probability tools in its design. Two main methods of nonprobability sampling are judgment sampling and quota sampling.

Judgment Sampling

As its name suggests, a **judgment sample** is one selected on the basis of someone's judgment. If Andy decides to interview those faculty who he had as instructors, then this would clearly be judgment sampling.

Judgment sampling is useful in carrying out a small scale survey and in pilot studies that precede major surveys. A judgment sample might involve the choice of a few "typical" cities to test market a new product. It might involve choosing a small sample on which to try out a questionnaire to detect unforeseen difficulties prior to launching a major survey.

As is clear from its nature, if the judgment exercised in choosing the sample is good, the results will be insightful; if poor judgment is exercised, the results might have disastrous consequences.

Quota Sampling

In a **quota sample** the interviewer is instructed to interview an assigned number, or quota, of individuals in groups defined by specified characteristics. Andy, for example, might be told by his editor to interview 3 faculty from the history department, 2 from the English department, 2 from the economics department, and 3 from the marketing department. Which faculty should be interviewed is left to Andy's discretion, as long as the above quota is satisfied. Quota sampling is a form of judgment sampling. It is vulnerable to the interviewer's biases in the selection of interviewees. The interviewer may focus on those who are readily available and seek to avoid those with an "unsavory" look, factors which are not controlled by a quota.

Random sampling, on the other hand, is merciless in that the interviewer has no choice; if an individual has been pinpointed through a random sample, it is the interviewer's obligation to locate that party, whether the trail is difficult or not, whether the party is unsavory or not.

Quota sampling has sometimes been used in marketing surveys and public opinion polling. It is cheaper per sample unit than random sampling and, when carefully done, can yield good results.

Inherent in probability sampling is the means for objectively estimating the precision of the sample results, which is its major advantage over nonprobability sampling.

REFERENCE

1. J. Neter, "How Accountants Save Money by Sampling," *Statistics: A Guide to the Unknown*, ed. J.M. Tanur et al (San Francisco: Holden-Day, Inc., 1972), 203-211.

EXERCISES

1. (a) What is the difference between probability and nonprobability sampling methods?

   (b) Are probability methods always preferable to nonprobability methods? Explain.

2. What is the difference between stratified and cluster sampling?

3. What is the difference between cluster and quota sampling?

4. Describe two situations in which stratified sampling would be advantageous.

5. Andy Arunas has been asked to interview a sample of 10 members of Huxley College's administration about the challenges and problems of administration. Huxley's administration consists of five major divisions: (1) Administrative Affairs, which numbers 15 and includes the president, provost, 8 who are concerned with internal affairs of the college (finance, purchasing, and human resources, for example) and 5 who are concerned with external affairs of the college (fund raising, for example); (2) Academic Affairs, which numbers 45 and includes 5 deans of the earlier mentioned schools, 35 academic department heads, and 5 directors of special programs (honors, international studies, etc.); (3) Athletics, which numbers 3; (4) Student Services, which numbers 10 (bursar, registrar, financial aid director, etc.); and (5) Academic Support Services, which numbers 7 (directors of library, computer services, etc.)

   (a) Describe in reasonable detail how Andy might implement the probability and nonprobability sampling methods discussed in this section to carry out his task.

(b) Discuss the advantages and disadvantages of these methods.

6. The faculty, administration, and other workers at Huxley College may be viewed as consisting of four non-overlapping groups: Faculty (300), Administration (80), Supervisory staff (20), Nonsupervisory staff (150). A sample of those working at Huxley is to be drawn and their average salary used as an estimate of the average salary of all who work at the college. The question is, how should the sample be drawn? Discuss the merits and drawbacks of random sampling, stratified sampling, and quota sampling for this problem.

7. Andy Arunas has been asked to conduct a survey of Huxley's students (estimated at 6000) to determine the facilities they would prefer (exercise room, swimming pool, food service, game room, etc.) in the new student union building that is being planned. The question here too is, how should the sample be drawn? Discuss the merits and drawbacks of random sampling, systematic sampling with a random start, stratified sampling, cluster sampling, and quota sampling for this problem.

2.3 ■ More on Random Sampling

> As was noted in the preceding sections, when we say that a sample is to be chosen at **random** we have in mind the idea that there is no bias, deliberate or inadvertent, which favors certain samples of the same size being chosen over others—a level playing field, so-to-speak.

It is easy to declare, "take a random sample," or "we assume that a sample is chosen at random," but achieving random sampling in practice is not a simple matter. It is perhaps best viewed as an ideal which we strive to realize in practice.

Achieving Random Sampling in Practice

To raise money for Huxley College's scholarship fund the student Social Science Society sold tickets at the College's homecoming affair. The prize was a 32-inch television set. The tickets were placed in a bowl and mixed. About midway through the festivities Huxley's President, blind folded, reached into the bowl and drew the winning ticket. What everyone expected and assumed was that the drawing was fair in the sense that there was no bias in the drawing which favored some tickets being drawn over others—that is, that it was drawn at random.

On the face of it the procedure seems reasonable enough for the task at hand, but how close does it come to satisfying the requirements of a random drawing? The problem is with the physical stirring of the tickets to achieve a "thorough mix." Obtaining a "thorough mix" becomes more and more difficult to achieve as the number of tickets increases, and it is not clear whether early, middle, or late ticket buyers might be favored and to what extent. Still, for the task at hand the degree of randomness achieved might be random enough, except possibly for those individuals who are willing to go to war over a TV set or what they consider the "principle of the matter."

As the Nielsen Company learned, much to its distress, implementing random sampling in practice is not as straightforward as it might seem in theory (Case 8, Section 1.2). And then there is the problem of implementing conscription in a "fair" way.

Going to War

When it comes to going to war based on the outcome of a random drawing, the stakes are raised considerably. The European phase of the Second World War and Japan's aggressiveness in Asia greatly alarmed the United States, and on September 6, 1940 Congress passed a conscription law, America's first peacetime draft. To implement the draft each eligible man in a Selective Service District was assigned a number which was put into a capsule. The capsules were put into a bowl, stirred, and then drawn one at a time. The order of the drawing determined the order of the draftees. The properties of the resulting sequence prompted questions about the randomness of the drawing, which gets back to the thoroughness of the mixing (see, for example, [3] and [4]).

In 1969 the administration of the draft in the United States to determine the order in which men born in 1950 would be drafted was changed to a lottery system. Three hundred sixty six capsules were prepared (for a leap year), each containing a birthdate. Each month's capsules were put into a separate box. The boxes were emptied into a drum, first those for January, followed by those for February, and so on for the subsequent months. The drum was rotated a few times, the capsules were poured into a bowl, and on December 1, 1969 the drawing was made. Those with birthdays on the capsules drawn first would be drafted first, and so on. If your birthday fell among those drawn last, there was a good chance that you would not be drafted at all. The results of the drawing are given in Table 3, from which we see that the earlier months, January through June got the larger share of the last-to-be drafted numbers and the later months, July through December, got the larger share of the first-to-be drafted numbers.

Table 3

Month	1-122 (First-Drafted)	123-244 (Middle)	245-366 (Last Drafted)
January	9	12	10
February	7	12	10
March	5	10	16
April	8	8	14
May	9	7	15
June	11	7	12
July	12	7	12
August	13	7	11
September	10	15	5
October	9	15	7
November	12	12	6
December	17	10	4

The results suggest the possibility that the earlier months' capsules were concentrated at the bottom of the bowl, while those of the later months were concentrated at the top and were more accessible for picking. Formal hypothesis tests of randomness did not support the hypothesis that a random drawing had been carried out (see, for example, [2] and [3]).

This period was one of great turbulence in American history and slogans such as "Draft Beer, not Students" and "Hell No, We Won't Go," were a prominent part of the scene. Perceptions of an unfair draft lottery on top of what was considered by many to be an indefensible draft for an indefensible war added fuel to a raging fire. In response to criticism the Selective Service modified its number selection mechanism for the draft lottery conducted in 1970. What these examples serve to make clear, however, is the difficulty of achieving a random selection in practice.

Another way of obtaining a random sample which avoids the difficulties of obtaining a sufficiently thorough mix of tickets or capsules is by use of a table of random numbers.

Random Number Generation

The idea behind random number generation is that the digits from 0 to 9 are selected by a process such that each digit selected is independent of any other digit selected and all digits have the same likelihood or chance of being selected. The process, usually based on a computer program, is set in motion and thousands of digits are generated and recorded in the order in which they are generated, Table 4 is a short table of random digits; a larger table is given as Table G (page 659).

Table 4

Rows	Columns							
	1-2	3-4	5-6	7-8	9-10	11-12	13-14	15-16
1	24	81	11	56	37	19	01	87
2	18	09	27	32	67	53	19	03
3	57	10	58	83	94	12	53	52
4	04	32	54	08	81	19	01	09
5	77	07	71	02	09	53	13	20
6	64	48	59	26	71	49	78	24
7	72	78	84	39	12	98	61	68
8	20	09	61	93	41	13	78	38
9	53	14	53	63	18	17	30	20
10	37	32	17	10	07	86	94	65
11	13	69	68	76	33	24	19	48
12	98	15	39	13	23	22	49	05
13	63	08	52	46	36	21	30	22
14	50	32	05	50	70	07	42	19
15	10	29	20	53	49	93	16	41

Example 1 The "Heavy" Basic Statistics Text

Statistics texts have not only become more expensive in recent years, but bigger and heavier as well. This fact has weighed heavily on student minds, wallets, and wrists. A recent edition of *Basic Statistics*, which weighed in at 5 pounds with dimensions 8.5 by 11.5 inches, was the focus of a wrist strength test at Ecap University. Table 5 gives the length of time, in minutes, that the population of 156 students at Ecap University who used the text were able to hold it in a reading position in one sitting before their wrists gave out. Choose a sample of 5 values at random from this population.

To employ a table of random numbers we must number each value in Table 5. One way of doing this is by numbering the entries row by row, we obtain:

row 1: 001-013	row 5: 053-065	row 9: 105-117
row 2: 014-026	row 6: 066-078	row 10: 118-130
row 3: 027-039	row 7: 079-091	row 11: 131-143
row 4: 040-052	row 8: 092-104	row 12: 144-156

Table 5

	Column												
Row	1	2	3	4	5	6	7	8	9	10	11	12	13
1	1.7	4.3	6.0	4.7	1.7	7.2	3.1	3.9	3.3	7.0	2.5	4.9	5.2
2	2.3	4.5	6.9	3.4	3.8	6.6	3.6	3.5	5.5	5.4	5.9	6.1	4.4
3	7.9	4.7	5.1	6.0	5.9	8.4	5.1	6.3	5.2	4.4	5.2	3.0	4.4
4	4.3	3.3	5.0	3.8	7.0	2.9	3.0	4.1	4.5	5.2	3.3	4.2	2.5
5	5.1	2.1	4.6	2.9	4.1	5.9	4.7	4.8	4.1	5.5	5.0	2.6	4.4
6	6.1	5.0	6.3	4.6	8.1	4.6	6.1	6.6	3.2	5.9	1.8	3.2	1.5
7	5.7	7.3	4.3	3.2	3.4	4.1	5.0	1.7	4.3	4.2	3.5	4.1	4.5
8	4.8	3.7	3.0	4.9	7.2	4.5	4.9	6.5	3.1	5.2	4.8	3.9	4.9
9	7.5	3.8	4.5	6.8	4.1	0.3	5.9	2.4	3.6	3.4	1.9	3.4	1.6
10	3.5	3.4	3.6	2.6	5.9	5.1	4.9	6.2	5.0	3.9	6.1	5.6	4.6
11	7.5	6.7	4.4	5.8	4.5	5.1	5.3	4.4	5.8	4.7	2.5	3.5	7.6
12	4.3	4.7	5.1	5.8	3.1	2.3	5.7	8.2	0.8	2.5	7.2	2.6	6.7

Our next task is to choose a starting point in the table of random numbers. One way is to close your eyes and place a finger anywhere in the table. Let us suppose that this approach pinpointed row 5, columns 3-4 of Table 4. The entry is 07. Since our numbering of the data given in Table 5 required three digits, we take the adjacent value in the next column, column 5. This gives us 077. We next move our finger (or card) down columns 3-4-5 and pull out the first three digit numbers from 000 to 156, the scope of Table 5. If 077 should reoccur, we pass it by since we have obtained the value at this position. We obtain 096, 145, 153, and 085. The translation from these listings to their corresponding entries in Table 5 is shown in Table 6. The random sample determined consists of 3.2, 7.2, 4.7, 2.5, and 5.0.

Table 6

Random nu. listing	Table 5 row and column		Table 5 data value
077	Row 6,	col. 12	3.2
096	Row 8,	col. 5	7.2
145	Row 12,	col. 2	4.7
153	Row 12,	col. 10	2.5
085	Row 7,	col. 7	5.0

> We return to the "Heavy" *Basic Statistics* text situation a number of times, each time adding to it another dimension of interest. The random sample obtained here is used in Example 1 of Section 16.4 to conduct a test of the hypothesis that the population of book holding times has a particular distribution of interest.

The study of many complex phenomena requires the generation of large streams of random numbers. It came as quite a shock when three scientists showed that five of the most often used computer programs for generating random numbers induced errors in the study of the behavior of atoms in a magnetic crystal because the numbers produced were not random, despite the fact that they passed several statistical tests for randomness [1]. The deviations from randomness were subtle and, although the pseudorandom numbers produced were satisfactory for many purposes, they were not satisfactory for the problem at hand. Is it possible that no machine based system can produce truly random numbers? John von Neumann, regarded as the father of the modern computer, thought that the answer is yes. In an observation made in 1951 von Neumann expressed the view that anyone who believed a computer could produce truly random numbers was living is a state of sin. It may be that the best we can hope to do is produce pseudorandom numbers which are satisfactory for the purpose at hand, and that the truly random number is a mathematical ideal which cannot be attained. The question that arises in an applied situation then is, how random is random enough?

REFERENCES

1. M. Browne, "Coin-Tossing Computers Found to Show Subtle Bias," *The New York Times*, Jan. 12, 1993, p. C1.
2. C. Hawkins, J. Weber, *Statistical Analysis: Applications to Business and Economics* (Harper and Row, 1980), 297-303.
3. J. Rosenblatt, J. Filliben, "Randomization and the Draft Lottery," *Science*, 171 (1971), 306-308.
4. S. Stouffer, W. Bartky, statement, *Chicago Tribune*, Nov. 2, 1940, p. 4.

EXERCISES

1. To carry out a marketing survey on consumer preferences for kitchen appliances Elias Marketing Research Associates placed two interviewers on the busiest street in town to interview passersby. Does the sample of opinions obtained qualify as a random sample? Explain.

2. Huxley College's Student Government Council (SGC) consists of 28 members who are listed in Table 7.

Table 7

Sharon J.	Duncan M.	Alex R.	Cathy B.
Bill W.	Horace C.	Dina S.	Russell F.
Jenny K.	Conrad B.	Linda J.	Paula G.
Mary S.	Julio S.	Peter D.	Meryl G.
Harry A.	Annette V.	Aramis D.	Sue L.
John M.	Danielle T.	Ozlem C.	Allison J.
Pat M.	Jim L.	Tania R.	Yolanda S.

(a) Select 4 members of the SGC at random to represent the SGC on the Huxley College Senate.

(b) Select 3 members of the SGC at random to serve on Huxley's Student Union Building Planning Committee. Members selected may serve on both committees.

3. The Top Value Company is to receive a shipment of 500 30-inch televisions numbered TV001-TV500. Their quality control procedure requires that a sample of 5 sets be randomly chosen from the shipment and tested before the full shipment is accepted. All 5 sets must be judged satisfactory if the full shipment is to be accepted. Choose a sample of 5 sets at random from the shipment.

4. Ecap University's Council of Deans has authority to grant 8 sabbaticals for next year. Two hundred applications, listed S001-S200, have been received. The Council is at a loss as to how to make a judgment based on the merit of the applications and has decided that a random selection would offend the least number of people. Choose 8 sabbatical applications at random from the 200 submitted.

5. Elias Marketing Research Associates wants to select a random sample of 25 households from a list numbered H001 through H700. Those selected will be sent a questionnaire about beverage preferences. Choose a random sample of 25 for Elias Marketing Research Associates.

6. The local chapter of the Ordinal Number Society (ONS) consists of 24 members who are listed in Table 8. A delegation of 5 members is to be sent to the national conference of the ONS to be held in San Antonio next year. Choose 5 members at random from the local chapter to be sent to the national conference.

Table 8

R. Kaplan	J. O'Brien	S. Jamis	P. Koo
J. De Rosa	R. Lam	L. Morales	A. Schlinder
R. Coscun	N. Eller	F. Chan	V. Martov
F. Chen	V. Sari	R. Man	S. Kirnan
M. Catania	A. Walsh	D. Mitchell	V. Santora
R. Alonzo	R. Levy	A. Morgan	J. Valentine

2.4 ■ Problems of Sampling

Two samples chosen from a population differ in some respects from each other and from the population they are taken from. The extent to which a sample characteristic differs from its population counterpart defines an error. In the following we discuss some major potential sources of error.

Sampling Error

Sampling error refers to the random or chance error inherent in estimating a population characteristic by its sample counterpart. The good news about sampling errors is that if a probability sample is the type taken and it is "properly" taken, the sampling error can be measured and controlled by employing the tools provided by the theory of probability. We shall have more to say about this in our discussion of confidence intervals and hypothesis testing in Chapters 10 and 11.

There are a number of other potential sources of error, broadly classified as nonsampling errors, two of which we turn to here.

Deviation in Practice from the Requirements of Theory

As we appreciate from our daily experience, deviations from a stated norm may have consequences ranging from negligible to catastrophic. If a "little more or less" of a delicate herb than specified is added in the preparation of a gourmet meal, for example, the effect would in all likelihood be inconsequential. If we are off the norm by more than a "little," a gourmet delicacy could easily be turned into a gourmet disaster. Medication that is life saving when taken as specified might become life threatening when the prescribed dose is exceeded.

Deviation from mathematical or statistical requirements may, in a similar vein, have negligible or profound consequences, depending on the situation. The requirement of random sampling, for example, is frequently encountered in statistical analysis. As was noted in the preceding section, it is a requirement which is often difficult, if not impossible, to satisfy "perfectly." When it is not satisfied "perfectly," a sampling error is induced. Whether that error is negligible or serious depends, as we have seen, on circumstances.

> The profound lesson to be gleaned from these considerations is that when conditions are stipulated, whether they be in connection with sampling or of a more general nature, it is our obligation to insure, as best we can, that they are met if the conclusions derived from theory are to be in close agreement with the results obtained from practice.

Sampling from the Target Population

The target population is the population about which we desire to make inferences. In seeking information about a target population it is important that the population being sampled be the target population. Any difference between the two is a potential source of bias. If the *Society for the Prevention of Statistical Nonsense* (SPSN) seeks to sample its current membership, but samples from the available list which is three years old, then the target population is clearly not the same as the one being sampled. The population being sampled does not include members who have joined the organization in the last three years and does include those former members who have died or who, for other reasons, have not renewed their membership. The error introduced by the difference in the target and sampled populations clearly depends on the extent of the difference.

EXERCISES

1. *The Huxley College Press*, which wants to obtain a sense of faculty opinion about the College's budget allocations for the forthcoming year, sent a reporter to the School of Business to interview a randomly chosen sample of faculty in the accounting, finance, marketing, management, and business law departments about this matter. Will doing this give the *Press* what it wants? Explain.

2. To obtain a sense of the business community's opinion of what ails American business, *Business Tomorrow* interviewed a randomly chosen sample of CEO's of Fortune 500 companies. Will doing this give *Business Tomorrow* what it desires? Explain.

3. The Harold Institute for Urban Studies is seeking to obtain a sense of the opinion of Baxter City's residents about the problems of the poor. To obtain this they interviewed a random sample of Baxter City's residents whose names appeared in a list of those who voted in last year's municipal elections. Will doing this give the Harold Institute what it seeks? Explain.

4. Tickets were sold at Ecap University's graduation celebration to help raise funds for the University's new library. The tickets sold were placed in a bowl as soon as they were sold. At the end of the graduation festivities the University's Library Director, Harriet Warren, reached into the bowl and choose a ticket at random. The ticket holder was awarded a newly published edition of Charles Dickens's collected works. Kevin Reynolds, who was among the first to purchase a ticket, protested that the drawing procedure was biased and demanded that ticket purchasers be given a refund or that the drawing be held again. "Your claim is not justified Mr. Reynolds," replied the Dean of Student Affairs. "Ms. Warren was blindfolded and the ticket was chosen at random." Who is right? Explain.

2.5 ■ Polls, Surveys, and Questionnaires

If you want to know what people think, ask them. This simple proposition offers difficulties at both ends. In many societies and situations the leaders, managers, bosses, what have you, do not want to know what "the people" think; they want the people to do what they are told, preferably without thinking or without thinking too much. Once this formidable hurdle is surmounted the problem of reliably determining what people think emerges and with it modern polling and survey analysis. The term survey refers to a method of collecting information from a "suitably chosen" sample of individuals to learn something about the population from which the sample was chosen. Today the practice of conducting surveys, or polls as they are also referred to, is widespread with an enormous range.

Political preference polls are perhaps the best known, but the majority of surveys that are conducted are concerned with specific social, commercial, or administrative concerns that do not occupy the public's spotlight with the same intensity. Thus, for example, government agencies conduct surveys to obtain information about employment, who uses food stamps, and who uses the national parks. TV networks employ surveys to obtain information about the popularity of their programming; businesses employ surveys to determine consumer preferences and reactions to their products; and interest groups conduct surveys to obtain a sense of the public's reaction to such matters as the state of education, the economy, crime, race relations, health costs, environmental issues, gun control, foreign policy issues, and the war against drugs.

Polls have become so numerous and so often referred to as reflecting or defining opinion that we may often feel sympathetic with Russell Baker's call (possibly with tongue in cheek) for a poll-ban treaty [1]. It might be interesting, in fact, to have a poll on that. Nevertheless, surveys and polls have come to play such a major role in reflecting and defining opinion and reaction to it that it is especially important to obtain a sense of their power, limitations, and misuses.

America's democratic political institutions provided a fertile soil for the emergence of the practice of asking people what they think in some sort of organized fashion. If people are expressing their preferences on election day, the day they would be asked their opinions beforehand could not be far behind. The first published presidential poll was published in *The Harrisburg Pennsylvanian* on July 24, 1824. Andrew Jackson won with 335 votes to 169 for John Quincy Adams. (Jackson subsequently carried the popular vote by a wide margin, but not the electoral vote. The election went to the House of Representatives which elected Adams President.) The practice of polling was thus launched. It was to undergo many extensions and refinements.

Pre-election Polls as a Laboratory

A major advantage of political election polling is that it provides a real-world laboratory from which one can study and refine polling techniques by comparing the predictions of theory with the experience handed down by reality. By the early twentieth century three methods for conducting polls had emerged.

1. **Newspaper Polls:** One method, the least reliable, was to print the ballot in a newspaper or magazine and invite the reader to send it in. Some publications encouraged people to buy more than one copy so that they could vote more than once. The *Sheboygan Press'* poll, for example, gave Herbert Hoover 25 percentage points more support in 1928 in Sheboygan, Wisconsin, than he actually received. More serious is the problem arising from those who do not respond, leading to a nonresponse error. Those who respond, the self-chosen, care enough about the issue to make sure that their voices are heard. What about the others? It is a highly questionable assumption that nonrespondents would split the vote 50-50, for example, or in any preconceived proportion. The best one can say is that the responses received are representative of those who cared to respond.

Although discredited rather early, this method has continued to be used in various versions in modern times. Telephone polls in which the listener to a television or radio program is invited to call a 900 number to express an opinion about a situation are of this type. Talk shows that invite listeners to call in and speak with the host are of this type. Campaigns to flood the telephone exchange or mail room of your governor, senator, or the President with expressions of support or opposition for a person or issue are employing this approach. Ross Perot's organization *United We Stand America* employed this method in the form of a questionnaire published by *TV Guide* in its March 20-26, 1993 issue, which invited reader response.

2. **Mail Polls:** Another method for conducting polls was the mail ballot. Names were obtained from such lists as telephone directories, voter lists, automobile registrations, and lists supplied by commercial addressograph companies. A sample was often chosen by systematic sampling in which every tenth, twentieth, or whatever name was chosen.

3. **Personal Interviews:** The personal interview approach had interviewers go out to interview voters. The idea was to approach a prospective voter and courteously request that he mark a ballot and, anonymously, place it in a box held by the interviewer. Some newspapers became rather sophisticated in their approach and achieved a high degree of precision by instructing interviewers to choose prospective voters so as to take into account such factors as occupation, party, nationality, and religion. For Ohio, for example, *The Cincinnati Enquirer* conducted six presidential polls between 1908 and 1928 and obtained an average difference of less than two percentage points between the poll results and presidential results for the state.

The Polling Goliath and Three Davids

The polling Goliath that emerged in the period between the two world wars was *The Literary Digest*, a weekly publication similar to today's *Harper's Magazine*. In early 1920 the Digest mailed over eleven million ballots to obtain a sense of the public's view on possible presidential candidates. In 1924 it conducted its first presidential poll, sending out some sixteen and a half million ballots to people in all 48 states. For the 1928 and 1932 elections even more ballots were sent out. The mailings were massive, which by itself convinced many about the reliability of the poll. The hoopla accompanying the polls was deafening, and their overall success in predicting the presidential winners with a small margin of error was stunning. Some doubts were, however, raised. The mailing list for the ballots came largely from telephone directories and automobile registrations and some critics suggested that there is a class bias in the list which favors Republican candidates. This bias was pointed to as explaining why the predicted Republican vote was considerably higher than it turned out to be in a number of states. In 1932 the *Digest* poll predicted Roosevelt would win the popular vote with a margin of 59.85% and carry 41 states with 474 electoral votes. Roosevelt received 59.14% of the popular vote and carried 42 states with 472 electoral votes. It's hard to argue with success.

The 1936 presidential election was on the horizon with Franklin D. Roosevelt, seeking reelection, facing the Republican nominee Alfred E. Landon and minor party candidates. Riding high, *The Literary Digest* geared up. It mailed some ten million ballots to prospective voters and eagerly awaited the returns.

Against this background it seemed foolhardy in the extreme for George H. Gallup to take on Goliath. Gallup had founded the *American Institute of Public Opinion* in 1935 and initiated a weekly column called *America Speaks*. To attract newspaper subscriptions he offered a money back guarantee that his prediction of the presidential winner would be more accurate than that of *The Literary Digest*.

Gallup appreciated that a survey sample should be representative of the voting population at large and he employed quota sampling to obtain an appropriate sample mix. He also appreciated the problem of nonresponse bias and sought to minimize it. Gallup predicted a Roosevelt victory with about 54% of the popular vote. In response to its poll the *Digest* received 2,376,523 responses, with 1,293,669 for Landon, 972,897 for Roosevelt (42.9% of the two-party vote) and the remainder for third party candidates. The *Digest* had Landon carrying 32 states with 370 electoral votes and Roosevelt carrying 16 states with 161 electoral votes.

The actual vote gave Roosevelt a landslide victory with 62.5% of the two-party popular vote. He carried 46 instead of 16 states and received 523 instead of 161 electoral votes. Already in financial trouble, *The Literary Digest* suffered a devastating blow to its credibility and folded the following year.

Independently of Gallup, two other researchers, Elmo Roper and Archibald Crossley, using similar methods, had also predicted a Roosevelt landslide. The three Davids, particularly Gallup, and their methods stood tall over the fallen Goliath humbled by reality.

Lessons

The 1936 Roosevelt landslide carried with it a number of lessons for the practice of polling in general and election polling in particular.

Sample Size: The accuracy of a poll is not determined by enormous sample sizes, impressive as they might appear to be. The major point is that the sample must be "properly" chosen. Pre-election poll samples these days tend to be between one and two thousand in size.

Target Population versus the Sampled Population: As we have discussed, it is most important that the population actually being sampled be the one about which we seek to draw inferences. The failure of *The Literary Digest's* 1936 poll was in part due to the sampled population being significantly different from the target population of prospective voters. It was generally appreciated that samples drawn largely from telephone directories and automobile registrations favored wealthy Americans who tended to vote Republican, but the extent to which this might be the case was not generally appreciated. After all, the *Digest's* spectacularly successful 1932 poll drew from the same sort of lists.

What went wrong? In 1932 voters of all economic *strata* tended to vote against President Hoover and the Republicans, holding them responsible for the Great

Depression and its aftermath. Economic class differences were obscured, and *The Digest* got lucky in its poll. In 1936 the upper economic *strata*, disturbed by the direction of President Roosevelt's *New Deal*, were much more willing to return to the Republicans, as reflected by the *Digest's* 1936 poll. Spectacular success without an understanding of its origins bred spectacular arrogance followed by spectacular failure.

Nonresponse Error: Of 10 million odd ballots sent out by *The Digest* 2,376,523 returns were received for a 76% nonresponse rate. What were the nonrespondents thinking? As we have observed, it is hazardous to predict one way or another. Those who did respond wanted to make sure that their opinion was counted. Those who did not respond did not care about the poll. In some locations, such as Allentown, Pennsylvania, the sampling list was not drawn from telephone and automobile registration lists, but from voter registration lists with no inherent Republican bias. The data show a clear response bias in that the proportion who favored Landon in the poll was much higher than the proportion who favored him in the election. Those voters who came from the lower economic *strata* tended not to respond to the *Digest's* poll when invited to do so and they strongly favored Roosevelt.

Mail Polls: A mail poll carries with it a seed with the potential to render it useless; the seed is nonresponse bias. Gallup was much closer to the mark than the *Digest*, but his predictions were wrong on six states and he had predicted a Roosevelt victory with about 54% of the popular vote, seven percentage points off the mark. One lesson for Gallup was that mail balloting was too unreliable for pre-election polling.

Other lessons were to come, but the laboratory of reality required more time to make clear that additional fine tuning was required. Gallup, Crossley, and Roper were pioneers in the quota method of sampling which they carried out effectively enough to obtain far more accurate predictions than the *Digest* achieved by its mail poll, but still left them significantly short of the mark. It might not matter when the outcome is strongly one-sided, but in a close election it could make a big difference. The Dewey versus Truman election of 1948 provided a decisive test case.

Thomas E. Dewey versus Harry S. Truman

Vice President Harry S. Truman became President on the death of President Franklin D. Roosevelt on April 12, 1945. It was a difficult post-war period and when Truman was nominated by the Democratic Party as its presidential candidate in 1948 he faced an uphill fight against the Republican challenger Thomas E. Dewey, Governor of New York. The polls showed Dewey comfortably ahead and he adopted a strategy of caution and platitudes ("Our future lies before us."), seeking to avoid making commitments and enemies. The political establishment, with the exception of Truman, felt that Dewey had it in the bag. Truman vigorously counterattacked, delivering 300 odd "give'em hell" speeches on a whistle-stop tour of the country.

Truman went to bed early election day evening and woke up to find himself President for another term. The pollsters and political pundits had missed the boat. The actual popular vote and the predictions of Crossley, Gallup, and Roper are shown in Table 9.

It was a close election, no doubt. In electoral vote terms, Dewey would have won had he carried California, Illinois, and Ohio, each of which he lost by less than 1%. The size of Gallup's and Crossley's error in predicting Dewey's victory in 1948 (5.4 and 4.7 percentage points, respectively) was smaller than the error in predicting Roosevelt's victory in 1936 (7 percentage points for each). But in terms of the bottom line wrong is wrong, and the exaggerated faith that had come to be placed in "modern scientific polling," as it had come to be called, was shattered. What happened?

Table 9

	Dewey (%)	Truman (%)	Other Candidates (%)
Actual vote	45.1	49.5	5.4
Crossley	49.9	44.8	5.3
Gallup	49.5	44.5	5.5
Roper	52.2	37.1	10.7

Further Lessons

Stability of Voter Opinion

By early September the polls showed Dewey ahead of Truman by at least ten points. The pollsters believed that public opinion was pretty well set by this time and would change little in the time remaining before the election. At this point they closed down their polling operations. Reality, of course, proved them wrong.

The Undecided Vote

How is the undecided vote to be handled? Gallup split the undecided vote in the same proportion as those who had expressed a preference for Dewey and for Truman, which strongly favored Dewey. Subsequent analysis showed that 14 percent of the voters made up their minds in the last two weeks of the campaign and that 74 percent of these went for Truman. It is clearly hazardous to decide for the undecided.

Quota versus Probability Sampling

Gallup and other pollsters were criticized in 1944 for continuing to use the quota method in their polling. Their failure in 1948 brought matters to a head. Stratified random sampling was the technique favored by many critics, but it was more complicated and much more expensive to implement. Gallup had serious doubts that the increased accuracy achieved would be worth the expense, but after the 1948 election he and other pollsters switched to stratified random sampling for pre-election polls.

Aftermath of the 1948 Election

Faith in polling as a reliable instrument for measuring public opinion was shaken by the scope of its failure in the 1948 election. But, as we see from the scope of its use in current affairs, that faith has been restored with some reservations. A better awareness of difficulties on the part of professional pollsters who are concerned with accuracy and reliability has led to important refinements in the art and science of survey design, interviewing, and poll taking. We should keep in mind, however, that some folks, well-meaning that they might be, are not well informed about the difficulties of polling and that others, not so well-meaning, are more concerned about manipulating opinion than measuring it.

The Affect of the Instrument

The problem of obtaining an accurate poll reading is analogous to that of obtaining an accurate blood pressure reading. The presence of the instrument itself affects that which is being measured. In determining a person's blood pressure the white-coat syndrome refers to the tendency of blood pressure to rise when the person is being examined by a medical professional. There are many white coats in survey or poll taking and we look at several of them here from the point of view that to be aware of problems is the first step towards dealing with them.

The Poll Itself

Being polled itself may put people on guard if they feel that they might loose benefits or be penalized in some way if they give the "wrong" answers. In the 1930's for example, many people on welfare were afraid they would be thrown off the welfare rolls if they gave an undesirable response. The election of Violeta Chamorro as President of Nicaragua in early 1990 was contrary to poll predictions which had Daniel Ortega with a substantial lead. One reason that the polls were so inaccurate was that the intimidation factor was not accurately taken into account. Nicaragua had been under authoritarian rule and in a state of civil war for a number of years, and many voters were not about to freely express their political preferences to pollsters. In such situations it is especially important to go all out to win the trust of respondents.

Interviewer Induced Bias

Conducting a "successful" interview, whether in-person or by telephone, is not a simple matter. People respond to interviewers as well as questions and the interviewer must strike a balance in being personable, respectful and considerate of the person being interviewed, and professional. The type of person who sets your teeth on edge by the way he says Good Morning is not likely to be a successful interviewer.

The Questions

Wording: The wording of a question is, with the best of intentions, a delicate matter. In an experiment conducted in 1940, for example, of those asked if the U.S. should forbid public speeches against democracy, 46% replied "no." Of those asked if the U.S.

should allow public speeches against democracy, 25% said "yes." Support for free speech was much greater when the term "forbid" was used rather than "allow." "Do you think President Nixon should be impeached and compelled to leave the Presidency, or not?" In a Gallup poll held in July 1974 24% of the respondents said "yes." When the question was posed as, do you "think there is enough evidence of possible wrong doing in the case of President Nixon to bring him to trial before the Senate, or not?," 51% said "yes." This was in response to a Gallup poll held at the same time. A case of fickle respondents? No; the second wording makes clear that Nixon would be brought to trial, which was not as clear to respondents from the first wording.

Loaded Questions: And then there are loaded questions worded to elicit a desired response for political purposes. Ross Perot's mail poll on National Referendum—Government Reform which appeared in the March 20-26, 1993 issue of TV Guide contains questions of this type. Question 13, for example, reads: "Should laws be passed to eliminate all possibilities of special interests giving huge sums of money to candidates?" The term "special interests" as it is used carries with it the ominous suggestion of special interests taking over the country. When stated in this form in a Time/CNN poll, it received an approval rating of 80 percent. The Time/CNN poll also put the question as follows: "Should laws be passed to prohibit interest groups from contributing to campaigns, or do groups have a right to contribute to the candidates they support?" Forty percent of the Time/CNN respondents stated they would prohibit interest groups from contributing, while 55 percent stated that they had a right to contribute.

Overloaded Questions: A study released by the Education Department in September of 1993 concluded that half of the adults in the United States cannot read or handle arithmetic. This is certainly an alarming figure, but is it accurate? That, to a large extent, depends on the success of the test makers in developing questions free of cultural bias, verbal ambiguity, and distracting irrelevancies. The following arithmetic question, for example, was posed: "The price of one ticket and bus for 'Sleuth' costs how much than the price of one ticket and bus for 'On the Town'? A charter bus will leave from the bus stop (near the Conference Center) at 4 p.m., giving you plenty of time for dinner in New York. Return trip will start from West 45th Street directly following the plays. Both theaters are on West 45th Street. Allow 1 1/2 hours for the return trip. Time: 4 p.m., Saturday, November 20. Price: On the Town, Ticket and bus: $11.00; Sleuth: Ticket and bus: $8.50. Limit: Two tickets per person."

The question itself raises the question of to what extent it is intended to test arithmetic and to what extent it is intended to test one's ability to successfully negotiate a verbal maze, particularly when many of those participating in the test were foreign born adults whose first language is not English. With tests of this sort, the underlying question of how well the questions achieve their intended objective must be given most careful consideration before meaningful conclusions can be drawn about the population being studied.

Direct Questions: Asking a direct question may provoke a misleading response. Do you intend to vote in the forthcoming election? Many people would answer yes rather than run the risk of being thought an irresponsible citizen.

Questions with Technical Terms: Questions involving terms which might be understood one way in everyday usage, but have a specific technical meaning, must be handled carefully. Robbery, involving confrontation between victim and offender, for example, is technically different from burglary, which does not involve personal confrontation. Such a distinction must be made clear in any question involving such terms.

Ordering of Questions: Not only may the wording of questions affect the response, but so may the order in which they are asked. A January 1984 prepresidential *New York Times*/CBS poll found that voters preferred incumbent President Ronald Reagan to Democratic challenger Walter Mondale by 16 percentage points; the question was posed at the beginning of the interview, which favored the better known Reagan. Gallup and *Washington Post*/ABC News polls taken around the same time posed the question near the end of the interview after questions about Reagan's policies had been asked; this helped Mondale because he had less of a record to defend. These polls had Reagan and Mondale about even.

Personal Questions: Questions on very personal matters such as sexual behavior, orientation, or preference where the responses, to one degree or another, carry social disapproval often yield inaccurate data.

Response Options

The two-category response option of the form Agree or Disagree, for example, is much more restrictive than the four-category response option Agree, Disagree, Not sure, Not enough information. The number of response options available may profoundly influence the response given, including the possibility of nonresponse.

It is interesting to compare voter reactions in two July 1992 opinion polls on President George Bush and Democratic Party candidate Bill Clinton. The responses for the two polls are shown in Tables 10 and 11.

Table 10

	Bush	Clinton
Favorable	40%	63%
Unfavorable	53%	25%
Don't know	7%	12%

Table 11

	Bush	Clinton
Favorable	27%	36%
Unfavorable	49%	24%
Undecided	22%	31%
Haven't heard enough	2%	9%
No answer	1%	1%

KISS vs. COMPLEX

There is merit to the KISS (Keep it simple, stupid) philosophy in that keeping it simple avoids potential bias and misunderstanding which may arise from complex questions. It is one matter to ask interviewee Arnold Mount who he favors in the forthcoming election and another to present Mount with a list of a hundred questions on domestic and foreign policy. Of course, the less one asks, the less one will learn, and there is an important balance between simple and simplistic that must be kept in mind.

The Unavoidable Gap

"There's many a slip between the cup and the lip," the old saying goes, and this is especially the case with the gap between the last poll taken and the election itself. A poll may be likened to a picture taken by a camera at a point in time. Looking at a sequence of polls is analogous to looking at a sequence of still—photos, a videotape, if you will. While this might be strongly indicative of what will happen, it is not the same as what does happen.

Designing and Carrying Out a Survey

The following describes considerations common to designing and carrying out all surveys.

1. What are the objectives of the study? What kind of information is needed? The rest will be a follow-up on these starting points.

2. Given the objectives, a methodology for achieving them must be developed. A questionnaire must be developed, the target audience must be identified, a method for reaching the target audience must be determined.

3. The sample size to be used must be set. There is no simple rule for determining sample size. The margin of error within which the result of the study must fall when they are projected onto the larger population is a factor. This, in turn, will depend on how the results are to be used. The budget allocated for the study should also be expected to have an impact on the sample size taken.

Potential Pitfalls

All of us, to one extent or another, are subjected to time and budget pressures. The temptation to cut corners, which may undo the whole undertaking, is often present.

1. **Pre-testing:** It is only by pre-testing the questionnaire that unforeseen sources of misunderstanding and bias can be pinpointed and dealt with.

2. **Nonrespondents:** As we have discussed, a "significant" nonresponse rate makes projection of the results of the study highly questionable. Plans must be made to contact nonrespondents and try to convince them to participate in the study. This may require returning to households where the party sought was unavailable,

returning telephone calls when necessary, convincing those who are reluctant to participate because of confidentiality considerations that their confidentiality will be preserved. Needless to say, a plan for doing this must have been worked out in the first place.

3. **Special Training:** If interviews are to be conducted, interviewers must be selected and given "appropriate" training concerning the survey's concepts and procedures. The training may also include interviewing techniques.

4. **Mechanics:** There are tedious tasks of checking to maintain quality control throughout the survey process which if not tended to can have disastrous consequences.

2.6 ■ Qualitative vs. Statistical Studies

Sexuality By the Numbers, or Not?

In 1987 Shere Hite published *Women and Love: A Cultural Revolution*, her third book on human sexuality. In her first two books, *The Hite Report* (1976) and *The Hite Report on Male Sexuality* (1981), Hite restricted herself to telling what she had learned from women and men who replied to the extensive questionnaires concerning their sexual problems and attitudes she had circulated. For her first book she circulated approximately 100,000 questionnaires, from which she received 3019 responses for a response rate of about 3%. Four different versions of the questionnaire were sent to women's organizations that were asked to circulate them. A similar methodology was employed for her second book. Approximately 119,000 questionnaires were distributed with a response rate just under 6% being obtained.

Shades of the *Literary Digest's* debacle come to mind, but Hite was not claiming that her sample was representative of women and men in general. Hers was a qualitative rather than statistical study. In statistical studies, the same questions must be asked of all prospective respondents with the same response options being available to them all. Uniformity of the underlying conditions and the choosing of a "representative" sample so that the results obtained could be projected onto the population at large are essential for a statistical study. Qualitative studies, on the other had, focus on the special qualities of each individual potential respondent. Capturing the diversity inherent in individuals takes priority over ensuring uniform underlying conditions. It is not a matter of one kind of study being superior to the other, but rather of which methodology is appropriate to the study being undertaken. Although her first two books raised much controversy, Hite was on safe methodological ground.

In her third book, Hite attempted to cross the bridge from the qualitative results she had obtained to statistical generalizations about sexual attitudes of women in America. The bridge collapsed. Her methodology was almost universally criticized. ABC News in conjunction with *The Washington Post* conducted a telephone poll for October 15-19, 1987 to see if they could duplicate her results. They could not; their results were sharply at variance with those projected by Hite.[8] To take two examples, she found 84% of women as not being satisfied emotionally with their relationships; ABC/WP found 7% of married women and single women in a relationship as not being emotionally satisfied. Hite found 78% of women feeling they are only occasionally treated as equals most of the time. The ABC/WP figure was 9%. There were differences in the way questions were posed, but they were not startling. As to who is closer to the mark, ABC/WP clearly takes the Trustworthy Prize because of its sound statistical methodology.

REFERENCES

1. R. Baker, "Observer: Uncle Pete: Perot," *The New York Times*, July 11, 1992.
2. G. Gallup, "Opinion Polling in a Democracy," *Statistics: A Guide to the Unknown*, ed. by J.M. Tanur, et al. (Holden Day, 1972), 146-152.
3. G. Gallup, S.F. Rae, *The Pulse of Democracy* (New York: Greenwood Press, 1968)
4. D. Moore, *The Super Pollsters: How They Measure and Manipulate Public Opinion in America* (New York: Four Walls Eight Windows, 1992).
5. L. Rogers, *The Pollsters; Public Opinion, Politics, and Democratic Leadership* (Alfred A. Knopf, 1949).
6. M. Traugott, P. Lavrakas, *The Voter's Guide to Election Polls* (New Jersey: Chatham House Publishers, Inc., 1996).
7. M. Wheeler, *Lies, Damn Lies, and Statistics: The Manipulation of Public Opinion in America* (New York: Liverright, 1976).
8. "Hite/ABC Poll Comparison Analysis," news release by ABC for 6:30 pm, EST, Mon., Oct 26, 1987.

2.7 ■ How Trustworthy Are These Poll/Survey Results?

Keep the following in mind:

1. Context and Basic Information

Focusing on poll results without an appropriate context is misleading. News accounts of poll results should give information about the date the poll was taken, sample size, survey design, percentage of respondents among those contacted, response options, random sampling error and what it means. Out-of-context poll results are best viewed with questioning skepticism.

2. Questions.

The complete wording of questions should be provided. We may then ask ourselves: Are the questions clearly posed? In addition to the questions posed, are there questions that should have been posed to avoid overall bias in the questioning? Are there leading questions whose coloring would favor a certain kind of response? Are there personal questions that a respondent might be reluctant to answer truthfully? At first thought it might not seem like a big deal to ask a person his preference in an upcoming election, but where an atmosphere of repression or intimidation exists it is indeed a most sensitive question.

3. The Response Rate.

What was the response rate? A major lesson of *The Literary Digest* 1936 presidential is that a low response rate may render the results obtained untrustworthy as a basis for predicting the attitudes of the target population. This issue is more complex than it might seem to be at first sight. For further discussion, see G. Langer, "About Response Rates: Some Unresolved Questions," *Public Perspective*, May/June 2003; 16-18.

4. Self-Selected Respondents.

A popular, but seriously flawed, polling technique is the mail-survey, magazine, online, or telephone poll in which the public is invited to respond to a written questionnaire, or call one number to register approval of a candidate or position and another number for disapproval. Polls of this sort lend themselves to gross manipulation by individuals and pressure groups and give us no handle on the opinions of non-respondents. Self-selected respondents represent only themselves.

Any claim or suggestion that results obtained from polls of this sort express the views of a wider population is totally without merit.

5. Online Polls.

The first stage of online polling does not differ from mail-in or telephone-response polls. Pseudo-poll would be a more appropriate label for such techniques rather than poll, which has come to suggest a rigorous, scientific framework not possessed by the first stage of online polls and their ilk.

The advantage of online pseudo-polls over their kind is that the internet dimension permits the development of this pseudo-poll into a "legitimate" poll, sometimes termed an interactive online poll. This development is in progress.

When presented with the results of an online poll there is no way of knowing whether they are from a pseudo-poll or interactive online poll unless the methodology is identified. This is often not done.

Beware.

6. Who Commissioned The Poll/Survey and Who's Doing It?

As has been observed, many honest polls/surveys are commissioned by interested parties, but there are fair-minded interested parties and not-so-fair-minded interested parties who are more concerned with manipulating public opinion than obtaining an objective assessment of it. It is useful to know who wants to know and who's doing the survey to help provide some perspective for the results obtained.

7. Qualitative vs. Statistical Studies.

Each type, in its own right, may yield valuable information and insights. Equating the results of the two types of study results in nonsense and muddled thinking.

EXERCISES

1. State and discuss in appropriate detail the problems that arise from newspaper polls and their modern equivalents.

2. Does increasing the sample size by itself insure greater reliability of the poll to be taken? Explain.

3. **Bell City and MERT:** The City Council of Bell City wants to obtain a sense of the public's view of the effectiveness of the City's Medical Emergency Response Team (MERT). A random sample of 200 residents was chosen from the City's home owners listing and sent a questionnaire. One hundred twenty five responses were received; 95 gave MERT a favorable rating and 30 gave MERT an unfavorable rating. The City Council concluded that MERT has a 76% favorable rating by the public and, with much satisfaction, announced this result to the news media. Do you agree or disagree with their conclusion. Explain.

4. Mayor Keith Joos of Masters-on-the-Mississippi is planning to run for re-election. On a recent talk show he invited the public to call a 900 number and express a favorable or unfavorable rating of his administration. Five hundred calls were received; 350 callers gave his honor an unfavorable rating and 150 gave him a favorable rating. With a 70% unfavorable rating, he strongly considered not running for re-election. Is his pessimism over the results of this call-in warranted? Explain.

5. What factors contributed to the failure of the Crossley, Gallup, and Roper polls to correctly predict the outcome of the 1948 presidential election? Explain.

6. "I don't understand what went wrong," lamented an editor of *The Literary Digest*. "We were so on-target with our 1932 poll and so off-target with our 1936 poll, for which we used the same methods." What would you tell him?

7. Many hotels have a practice of leaving post card questionnaires in guests's rooms inviting comments and ratings. Discuss the pros and cons of this method for obtaining guests's reactions.

8. "Rate Your Professor" is an online means for students to state and obtain views on professors. Although intended for students anyone who chooses to may participate. Discuss the pros and cons of this means for obtaining information about professors.

9. In the article "Powell in The Middle," by Michael Hirsh and Roy Gutman, (*Newsweek*, Oct. 1, 2001, 26-29), the following assertion is made in boldface without further information being given:

   > "71% favor striking terrorists bases even if civilians die, but 59% say we should take time to plan a response that will work."

   (a) Would you accept the figures as accurate without additional information being given? Answer Yes or No.

   (b) If you answered Yes, explain the basis for your view.

   (c) If you answered No, what additional information would you want provided before you would feel comfortable reaching a decision about the cited figures? Explain.

10. "The votes are in. Adolf Hitler is the most evil person of the millennium and Thomas Edison is the most influential, according to a poll of *The Post's* online users." (A. Soltis, "Post readers: Hitler Was Most Evil," *New York Post*, Nov. 17, 1999; 2.) NYPost.com conducted seven separate polls. "Voters were asked to select up to 10 people in each category-either by picking names from dozens of candidates on *The Post's* web site or by writing in others."

    (a) Could the results of this online poll be considered representative of the population in general? Explain.

    (b) If your answer to (a) is no, whose views do the respondents represent? Explain.

11. To obtain a "sense" of what an MBA is "really" worth *Business Week* surveyed the class of 1992 from the "Top 30" MBA programs. BW reached about 4800 alumni out of a total graduate pool of approximately 5700; one thousand four hundred ninety six (1496) responded. For discussion of BW's conclusions and a description of how the survey was carried out see J. Merritt, K. Hazelwood, "What's an MBA Really Worth?," "How We Conducted the Survey," *Business Week*, Sept. 22, 2003; 90-98.

    Do you have concerns about the reliability of the picture painted by the responses received? Explain.

12. A National Household Survey on Drug Abuse stated that 9% of those 12 to 17 "reported" using an illicit drug in the survey taken in 1999 compared with 11.4% in the one taken in 1997. For the 18 to 25 group 18.8% "reported" illicit drug use in the 1999 survey compared with 14.7% in the one in 1997. It was noted that the 1999 study did not cover active-duty military personnel, people in prison or drug-treatment centers or homeless people not in shelters when the survey was conducted. D. Stout, "Use of Illegal Drugs Is Down Among Young, Survey Finds," *The New York Times*, Sept. 1, 2000; A18.

    Would you accept the figures presented as accurate without additional information being given? Answer Yes or No; if you answered Yes, explain the basis for your view; if you answered No, what additional information would you want provided before you would feel comfortable reaching a decision about the cited figures? Explain.

13. Does use or non-use of the big A words, Affirmative Action, in a question make a "significant" difference in how respondents reply? L. Harris, "Affirmative

Action and the Voter," *The New York Times*, July 31, 1995; A13. J. Leo, "Hold the 'Wrong' Story," *U.S. News & World Report*, Aug. 10, 1998; 12.

14. The Center for Critical Thinking on Domestic and World Affairs is preparing a questionnaire to obtain a sample of public opinion on domestic and world affairs. The following is a sample of some of the questions being posed.

    1. Would you vote for a presidential candidate who was willing to take more out of your pocket by raising taxes?

    2. Do you want the nation's defense capability reduced by budget cuts in an age of rampant terrorism?

    3. Do you believe that very high priority should be given to reducing the crushing budget deficit that has been imposed on our country?

    4. Do you believe that we should continue to squander money on foreign aid while there are so many urgent domestic needs that require attention?

    Two response options were to be allowed: Yes, No.

    The Center is engaged in pre-testing its questionnaire. If you were approached, what opinion would you give them about the four questions noted and the response options?

15. As we recall from Exercise 7 of Section 2.2, (page 47), Andy Arunas, freshman reporter for *The Huxley College Press*, was asked by his editor to conduct a survey of Huxley's students (estimated at 6000) to determine the facilities they would prefer (exercise room, swimming pool, food service, game room, etc.) in the new student union building that is being planned. Andy needs help with developing a questionnaire.

    (a) What questions would you suggest that he include in the questionnaire? Explain.

    (b) What response options would you suggest that he allow? Explain.

    (c) What advice or suggestions would you give Andy in connection with conducting the survey?

16. Based on the results of *Christian Science Monitor*/TIPP poll conducted Oct. 7-13, 2002, it was concluded that "seventy-five percent of Americans say it's important that the U.S. take military action against Iraq by April."[1] The question posed and the results were the following:

How important do you think it is for the U.S. to take military action within the next six months in order to remove Saddam Hussein from power in Iraq?

Very important:	46%
Somewhat important:	29%
Not very important:	13%
Not at all important:	9%
Not sure/Refused:	3%

(a) Do you Agree or Disagree with the view that this being the only question posed concerning Iraq and the nature of its wording rig the poll to favor going to war with Iraq? Explain.

(b) Are there questions and response options that you would recommend be added to the survey to obtain a more accurate assessment of the public's views? If so, state them.

(c) Would you accept the afore figures as accurate without additional information being given? Answer Yes or No and explain the basis for your view.

(d) If you answered No, what additional information would you want provided before you would feel comfortable reaching a decision about the cited figures? Explain.

17. In failing to present non-military options to the public, are the pollsters reflecting public opinion or shaping it? F. Solop, K. Hageb, "War or War? 9/11 Surveys Restricted the Options," *Public Perspective*, July/August, 2002; 36-37. What did Thomas Friedman find when he talked to a number of people about Iraq as opposed to what the polls suggested? T. Friedman, "Iraq Upside Down," *The New York Times*, Sept. 18, 2002; A31.

---

1 B. Knickerbocker, "Americans Back Iraq War - Warily," The Christian Science Monitor, Oct. 17, 2002; 1)

3

Making Mountains of Data Manageable

3.1 ■ Frequency Distributions

Descriptive statistics deals with techniques that enable us to get a handle on large quantities of data. Without such techniques we would be left to "staring" at the mountain of data, an approach which most of us do not find insightful; the data trees obscure the nature of the statistical forest.

One of the most often used devices for summarizing a large amount of data, sometimes called raw data, is the frequency distribution. This is a two column table which groups the data into categories called classes in the first column, and states the frequency with which the data fall into the classes in the second column. To appreciate the advantages of a frequency distribution, suppose for a moment that you were interested in learning about the caliber of students in your school and decided that inspection of their GPA's (Grade Point Averages) would satisfy your curiosity. Let us assume that these data were available in the Registrar's Office and that when you obtained them you found yourself staring at 5,000 numbers. What could you really learn from staring at such a mountain of numbers? Virtually, nothing! You might spot some 4.0's on the list and you know that would be the highest GPA. Unless you spotted a zero, it might be difficult to be sure which was the lowest GPA. As far as getting some idea about what range most of the GPA's fell into or how many students had high GPA's or low GPA's, it would be a hopeless task to attempt by visual inspection of the 5,000 numbers.

Suppose that a frequency distribution were available as shown in Table 1. From this frequency distribution we can obtain the following information.

Table 1

GPA's	Frequency
1.00-1.49	80
1.50-1.99	200
2.00-2.49	1300
2.50-2.99	2100
3.00-3.49	900
3.50-3.99	400
4.00-4.49	20
	5000

1. No student has a lower GPA than 1.00.

2. There are 20 students with a 4.00 GPA. We can determine this because 4.00 is the highest possible GPA.

3. It seems reasonable to conclude that the average GPA is a value between 2.50 and 2.99.

4. Most students have GPA's between 2.00 and 2.99.

5. Not many students have GPA's lower than 2.00 or above 3.50.

6. It is rare that a GPA falls below 1.50 and more rare to find a 4.00.

The Mechanics of a Frequency Distribution

1. **Number of Classes:** It should neither be too small, which gives up too much information, nor too large, which would defeat the purpose of delivering a concise message. Generally, frequency distributions have neither less than 5 nor more than 15 classes.

2. The classes should be **collectively exhaustive.** That is, they should allow a place for every data value. Had the last class in Table 1 been omitted the table would not have allowed a place for a GPA of 4.00.

3. The classes must be **mutually exclusive.** That is, they cannot overlap. For example, had the second class been 1.25-1.99, then a GPA of 1.37 would belong to both classes 1 and 2 and Table 1 would not define a frequency distribution.

4. The numbers defining the classes must have enough decimal places to allow for every data value. If the GPA's had been carried to one decimal place, then class 1 would have been 1.0-1.4, class 2 would have been 1.5-1.9, etc. However, had the GPA's been carried to three decimal places, class 1 would have been 1.000-1.499, class 2 would have been 1.500 to 1.999, etc.

Basic Terms

1. **Classes:** The categories that the data are grouped into. The frequency distribution defined by Table 1 has 7 classes.

2. **Frequency:** The number of data values that fall into a class.

3. **Class Limits (CL):** Each class has two limits; a **lower class limit (LCL)** and an **upper class limit (UCL).** These are the smallest and largest numbers that can belong to a class in the frequency distribution. Thus, for example, for the GPA distribution defined by Table 1, 1.00 is the lower class limit of class 1 and 1.49 is its upper class limit; 1.50 is the lower class limit of class 2 and 1.99 is its upper class limit.

4. **Class Boundaries (CB):** Each class has two boundaries, a **lower boundary (LB)** and an **upper boundary (UB).** With the exception of the lower boundary of the first class and the upper boundary of the last class, the boundaries of a class are defined by the midpoints of the upper class limit of one class and lower class limit of the next class. Thus, for example, the upper class boundary of the first class in the GPA distribution is 1.495 which is also the lower boundary of the second class. The second class has upper class boundary 1.995. The gap between the upper class limit of one class and the lower class limit of the next one is 0.01 in this distribution so that the class boundaries move up and down by a factor of $0.01/2 = .005$.

   To obtain the lower boundary of the first class we move the same distance below the lower class limit of the first class as the upper boundary is above the upper class limit of the first class. Likewise, the upper boundary of the last class is the same distance above the upper class limit as the lower boundary is below the lower class limit. This gives us $1.00-0.005=0.995$ as the lower class boundary of the first class of the GPA distribution and $4.99+0.005=4.995$ as the upper class boundary of its last class.

5. **Class Marks (CM):** These are the **midpoints of the classes.** Their values are found half way between consecutive class limits or consecutive boundaries. For example, for the GPA distribution the class mark of class 1 is $(1.00+1.49)/2=1.245$ or equivalently, $(0.995+1.495)/2=1.245$

Figure 1 gives us a geometric view of the class marks, limits, and boundaries of the classes of the GPA frequency distribution.

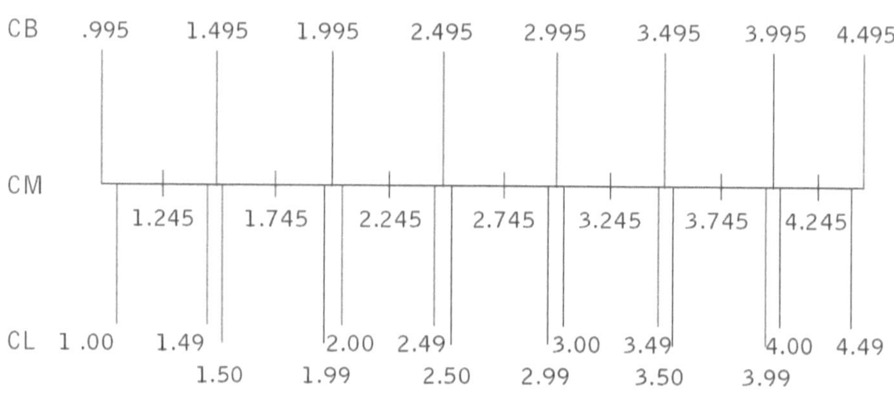

Figure 1

6. **Class Width (CW):** The **width of a class** is the distance between the upper and lower boundaries of the class, For example, the width of class 1 of the GPA distribution $1.495-0.995=0.05$. All the classes in this frequency distribution have a width of 0.5.

If all classes have the same width, this value is called the **width of the frequency distribution.** The width of the GPA frequency distribution defined by Table 1 is 0.5. If the width of a frequency distribution is defined, the class marks of the classes may be obtained by adding this value to the first class mark and successive class marks. By adding 0.5 successively to 1.245, the class mark of the first class of the GPA distribution, we generate the other class marks, 1.745, 2. 245, etc.

The basic data for the GPA frequency distribution is summarized in Table 2. There is an advantage in terms of simplicity to frequency distributions whose classes have the same width. There are also computation advantages to such distributions that we will make use of in Chapter 5.

Table 2

Class	Frequency	LCL	UCL	LB	UB	CM	CW
1.00-1.49	80	1.00	1.49	0.995	1.495	1.245	.50
1.50-1.99	200	1.50	1.99	1.495	1.995	1.745	.50
2.00-2.49	1300	2.00	2.49	1.995	2.495	2.245	.50
2.50-2.99	2100	2.50	2.99	2.495	2.995	2.745	.50
3.00-3.49	900	3.00	3.49	2.995	3.495	3.245	.50
3.50-3.99	400	3.50	3.99	3.495	3.995	3.745	.50
4.00-4.49	20	4.00	4.49	3.992	4.495	4.245	.50

Percentage Distribution

We are often interested in the percentage distribution of the data. Table 3 shows the frequency distribution of GPA's expressed in percentages. These percentages are obtained in the usual manner. For example, since 80 data values fall into class 1, $(80/5000)\times 100 = 1.6\%$ of the values fall into class 1, etc.

Table 3

GPA's	Frequency (%)
1.00-1.49	1.6
1.50-1.99	4.0
2.00-2.49	26.0
2.50-2.99	42.0
3.00-3.49	18.0
3.50-3.99	8.0
4.00-4.49	0.4
	100.0

Cumulative Frequency Distributions

We are often interested in answers to questions concerning how many (or what percentage) of the data lies above, below, or between certain values. For example, with respect to the GPA's, suppose you were interested in how many (or what percentage of) students have passing GPA's in your school (i.e., a GPA ≥ 2.00) or how many (or what percentage of) students have GPA's less than 3.00. The structure that readily provides answers to such questions is called a **cumulative frequency distribution.** Consider, for example, the following description of cumulative classes: less than 1.49; less than 1.99; less than 2.49; etc.

This is not a suitable description of cumulative classes. Inspection of Table 1 shows that there is no way to determine how many students have GPA's less than 1.49 since Table 1 only indicates that 80 students have GPA's between 1.00 and 1.49, inclusive, but it does not indicate how many, if any, equal 1.49. Without this information it is impossible to tell how many data are less than 1.49. However, we immediately see that 80 data are less than 1.50. Generally, cumulative frequency distributions are less than distributions. Table 4 shows the less than cumulative frequency distribution for the GPA data.

Table 4

GPA'S	Cumulative Frequency	Cumulative Frequency (%)
< 1.00	0	0
< 1.50	80	1.6
< 2.00	280	5.6
< 2.50	1580	31.6
< 3.00	3680	73.6
< 3.50	4580	91.6
< 4.00	4980	99.6
< 4.50	5000	100.0

Note that the first class is defined by "less than 1.00". It is important to include this class since it gives more information than if the first class had been "less than 1.50". For then we would have known that there were 80 students with GPA's less than 1.50, but we would not have known how far down these GPA's dropped. Knowing that there are no students with GPA's below 1.00 tells us that the lowest GPA is not less than 1.00.

Example 1 The "Heavy" Basic Statistics Text

As noted in Example 1 of Chapter 2 (page 42) statistics texts have not only become more expensive in recent years, but bigger and heavier as well. A recent edition of *Basic Statistics* was the focus of a wrist strength test at Ecap University. Table 5 gives the length of time, in minutes, that the population of 156 students at Ecap University who used the text were able to hold it in a reading position in one sitting before their wrists gave out.

Table 5

1.7	4.3	6.0	4.7	1.7	7.2	3.1	3.9	3.3	7.0	2.5	4.9	5.2
2.3	4.5	6.9	3.4	3.8	6.6	3.6	3.5	5.5	5.4	5.9	6.1	4.4
7.9	4.7	5.1	6.0	5.9	8.4	5.1	6.3	5.2	4.4	5.2	3.0	4.4
4.3	3.3	5.0	3.8	7.0	2.9	3.0	4.1	4.5	5.2	3.3	4.2	2.5
5.1	2.1	4.6	2.9	4.1	5.9	4.7	4.8	4.1	5.5	5.0	2.6	4.4
6.1	5.0	6.3	4.6	8.1	4.6	6.1	6.6	3.2	5.9	1.8	3.2	1.5
5.7	7.3	4.3	3.2	3.4	4.1	5.0	1.7	4.3	4.2	3.5	4.1	4.5
4.8	3.7	3.0	4.9	7.2	4.5	4.9	6.5	3.1	5.2	4.8	3.9	4.9
7.5	3.8	4.5	6.8	4.1	0.3	5.9	2.4	3.6	3.4	1.9	3.4	1.6
3.5	3.4	3.6	2.6	5.9	5.1	4.9	6.2	5.0	3.9	6.1	5.6	4.6
7.5	6.7	4.4	5.8	4.5	5.1	5.3	4.4	5.8	4.7	2.5	3.5	7.6
4.3	4.7	5.1	5.8	3.1	2.3	5.7	8.2	0.8	2.5	7.2	2.6	6.7

(a) Construct a frequency distribution expressing frequencies in magnitudes and percentages and interpret your answer.

(b) Construct a cumulative less than distribution expressing frequencies in magnitudes and percentages.

(c) Construct a summary table of basic values similar to Table 2 for the GPA frequency distribution.

(a) We first determine the smallest and largest data values. Inspection of the data shows that the smallest value is 0.3 and the largest value is 8.4. Thus, the spread between them is $8.4-0.3=8.1$. Calculating this spread gives us an indication of how many classes might be appropriate. Since 8.1 happens to be the product of 9 and 0.9, this suggests the possibility of forming 9 classes; 0.0-0.9, 1.0-1.9, etc., as shown in Table 6.

Table 6

Time Interval	Tally	Frequency	Percentage (%)
0.0-0.9	II	2	1.3
1.0-1.9	~~IIII~~ II	7	4.5
2.0-2.9	~~IIII~~ ~~IIII~~ III	13	8.3
3.0-3.9	~~IIII~~ ~~IIII~~ ~~IIII~~ ~~IIII~~ ~~IIII~~ ~~IIII~~ I	31	19.9
4.0-4.9	~~IIII~~ ~~IIII~~ ~~IIII~~ ~~IIII~~ ~~IIII~~ ~~IIII~~ ~~IIII~~ ~~IIII~~ II	42	26.9
5.0-5.9	~~IIII~~ ~~IIII~~ ~~IIII~~ ~~IIII~~ ~~IIII~~ ~~IIII~~ II	32	20.5
6.0-6.9	~~IIII~~ ~~IIII~~ ~~IIII~~ I	16	103
7.0-7.9	~~IIII~~ ~~IIII~~	10	6.4
8.0-8.9	III	3	1.9
		156	100.0

One of the desired properties of a frequency distribution is that it be easy to read. There are many frequency distributions that can be constructed for a given set of data. In this situation we could have classes such as 0.3-1.2, 1.3-2.2, 2.3-3.2, and so on up to 8.3-9.2 or we could have 0.2-1.1, 1.2-2.1, 2.2-3.1, and so on up to 8.2-9.1. However, the classes of the frequency distribution defined by Table 6 are simpler and easier to read. In addition, they allow for simpler tallying to determine the frequencies of the classes. Thus, by going through each row of Table 5, number by number, we need only look at the ones digit to identify which class it belongs to. For example, 1.7 belongs to the class 1.0-1.9; 4.3 belongs to class 4.0-4.9, etc. Each number quickly yields a tally mark as shown in Table 6, which also gives percentages.

In summary, the *Basic Statistics* text holding time frequency distribution is shown in Table 7 and the percentage distribution is shown in Table 8.

Table 7

Time Interval	Frequency
0.0-0.9	2
1.0-1.9	7
2.0-2.9	13
3.0-3.9	31
4.0-4.9	42
5.0-5.9	32
6.0-6.9	16
7.0-7.9	10
8.0-8.9	3
	156

Table 8

Time Interval	Percentage (%)
0.0-0.9	1.3
1.0-1.9	4.5
2.0-2.9	8.3
3.0-3.9	19.9
4.0-4.9	26.9
5.0-5.9	20.5
6.0-6.9	10.3
7.0-7.9	6.4
8.0-8.9	1.9
	100.0

We see that most of the holding times (about two thirds) lie between 3 and 5.9 minutes with an average time that appears to be between 4.0 and 4.9 minutes (about 4.5). A small number of students can hold the book from 1.0 to 2.9 minutes and a small number can last as long as 6.0-7.9 minutes. Almost all the students can hold the book for at least 1.0 minute and hardly any can hold it for 8 or more minutes.

(b) Table 9 shows the cumulative less than distribution in magnitudes and percentages.

Table 9

Time Interval	Cumulative Frequency	Cumulative Percentage (%)
< 0.0	0	0
< 1.0	2	1.3
< 2.0	9	5.8
< 3.0	22	14.1
< 4.0	53	34.0
< 5.0	95	60.9
< 6.0	127	81.4
< 7.0	143	91.7
< 8.0	153	98.1
< 9.0	156	100.0

(c) Table 10 summarizes basic values. Let us note that this frequency distribution has width 1 and that by successively adding 1 to the first class mark, 0.45, we generate the other class marks.

Table 10

Class	Frequency	LCL	UCL	LB	UB	CM	Width
0.0-0.9	2	0.0	0.9	-0.05	0.95	0.45	1.0
1.0-1.9	7	1.0	1.9	0.95	1.95	1.45	1.0
2.0-2.9	13	2.0	2.9	1.95	2.95	2.45	1.0
3.0-3.9	31	3.0	3.9	2.95	3.95	3.45	1.0
4.0-4.9	42	4.0	4.9	3.95	4.95	4.45	1.0
5.0-5.9	32	5.0	5.9	4.95	5.95	5.45	1.0
6.0-6.9	16	6.0	6.9	5.95	6.95	6.45	1.0
7.0-7.9	10	7.0	7.9	6.95	7.95	7.45	1.0
8.0-8.9	3	8.0	8.9	7.95	8.95	8.45	1.0

Stem-and-Leaf Plots

It would be helpful in many cases to have a systematic way of exhibiting data. The stem-and-leaf plot provides us with such a technique. Basically, a stem-and-leaf plot organizes the data by exhibiting the numbers according to some number of leading digits (the stem) and writing next to the stem all the remaining digits that correspond to it.

For example, inspection of the *Basic Statistics* holding time data given in Table 5 shows us that no number has a ten's value. Thus, let us pick the stem to be the units digit and the leafs to be the first decimal place. Table 11 shows the corresponding stem-and-leaf plot obtained by going through the data row by row and recording the leaf for each first digit stem.

Table 11

Stem	Leaf
0	3, 8
1	7, 7, 8, 5, 7, 9, 6
2	5, 3, 9, 5, 1, 9, 6, 4, 6, 5, 3, 5, 6
3	1, 9, 3, 4, 8, 6, 5, 0, 3, 8, 0, 3, 2, 2, 2, 4, 5, 7, 0, 1, 9, 8, 6, 4, 4, 5, 4, 6, 9, 5, 1
4	3, 7, 9, 5, 4, 7, 4, 4, 3, 1, 5, 2, 6, 1, 7, 8, 1, 4, 6, 6, 3, 1, 3, 2, 1, 5, 8, 9, 5, 9, 8, 9, 5, 1, 9, 6, 4, 5, 4, 7, 3, 7
5	2, 5, 4, 9, 1, 9, 1, 2, 2, 0, 2, 1, 9, 5, 0, 0, 9, 7, 0, 2, 9, 9, 1, 0, 6, 8, 1, 3, 8, 1, 8, 7
6	0, 9, 6, 1, 0, 3, 1, 3, 1, 6, 5, 8, 2, 1, 7, 7
7	2, 0, 9, 0, 3, 2, 5, 5, 6, 2
8	4, 1, 2

There are a number of advantages to a stem-and-leaf plot. It provides us with a systematic way of sorting the data so that it can be set up into a frequency distribution; it yields a representation which shows us what values occur most and least frequently; and it displays the data in such a way that if it were necessary to order the data from low to high, it would be relatively easy to do so.

For a given set of data, the choice of the stem and the leaf is at our discretion. However, one should pick the stem so that the break down is neither too brief nor too voluminous. For example, suppose we had a set of 1,000 data ranging from a low of 26 to a high of 1140. If we chose the stem based on increments of 10 we would have too many classifications to deal with; if we chose the stem based on increments of 1000, we would only have two classes to deal with and far too many leaves. Choosing the stem based on increments of 100, however, would yield the 12 classes 000, 100, . . . , 1100 which achieve the balance of providing a good representation for the data and being manageable.

Open Ended Classes

On occasion we encounter data that are concentrated within a certain range, but some values fall far outside of this range. In such cases it is awkward and confusing to construct a frequency distribution with classes of equal width for we would have many empty classes with frequencies of zero. Suppose, for example, that in constructing a frequency distribution of the IQ's of 1,000 students attending a high school which had classes for the mentally retarded it turned out that 980 students had IQ's ranging from 84 to 128; 10 had IQ's ranging from 30 to 38, and 10 had IQ's ranging from 170 to 178. Putting the 980 students into classes of 80-89, 90-99, 100-109, 110-119, and 120-129 would no doubt work fine.

However, if we wanted to complete the frequency distribution with classes of equal widths, we would have to start with 30-39 and go up to 170-179. This would yield a frequency distribution of 15 classes, 8 having frequency 0. Unless other considerations discussed in Chapter 5 enter the scene, it would be simpler to define the first class by "less than 80" and the last class by "greater than or equal to 130". This would reduce the number of classes to seven with none having a frequency of zero. It is, however, generally preferable to choose close ended classes.

Categorical Distributions

The frequency distributions shown so far are quantitative distributions because the classes are described by numerical intervals. Often, however, the classes are described by non-numerical classifications such as colors, nationalities, makes of cars, etc. In such cases the distributions are called **categorical** (or **qualitative**). Table 12 shows a frequency distribution summarizing the choices of 200 college students with respect to their favorite college sport.

Table 12

Sport	Frequency	Percentage (%)
Baseball	20	10
Football	100	50
Basketball	70	35
Tennis	10	5
	200	100

EXERCISES

For the situations described in Exercises 1-5, construct or determine the following:

(a) A frequency distribution. Classes for a distribution are specified and the following questions should be answered for it, but also specify your own distribution and answer for it the questions that follow.

(b) A summary table giving class limits, class marks, class boundaries, and class widths. If defined, state the width of the distribution.

(c) The percentage distribution.

(d) The cumulative less than distribution and cumulative less than percentage distribution.

(e) Stem-and-leaf plot.

(f) What do the frequency distributions and derived distributions tell us about the data in these situations?

1. **MERT:** The City Council of Bell City wishes to obtain a sense of how quickly the city's Medical Emergency Response Team (MERT) was responding to calls. A response time is the time elapsed between the time a call is received and the arrival time at the destination, in minutes, rounded off. A random sample of 36 response times chosen from the response time records of the past year is given in Table 13. Use classes 4-6, 7-9, 10-12, 13-15, 16-18, 19-21.

Table 13

12	13	17	9	10	15
6	8	20	10	8	5
10	5	8	8	11	19
12	7	16	11	8	10
9	10	21	18	5	7
6	9	10	4	6	4

(g) Could we use the following classes for a frequency distribution? Explain. 4-6, 6-8, 8-10, 10-12, 12-14, 14-16, 16-18, 18-21. If so, what are its advantages and disadvantages compared to the one specified?

(h) Could we use the following classes for a frequency distribution? Explain. 5-8, 9-12, 13-16, 17-21. If so, what are its advantages and disadvantages compared to the one specified?

(i) Could we use the following classes for a frequency distribution? Explain. 4-12, 13-21. If so, what are its advantages and disadvantages compared to the one specified?

(j) Could we use the following classes for a frequency distribution? Explain 3-5, 6-9, 10-15, 16-18, 19-22. If so, what are its advantages and disadvantages compared to the one specified?

2. The Water Quality Control Division of Cameron City's Health Department is responsible for monitoring the chlorine levels of the city's reservoir system. A randomly drawn sample of 34 readings of chlorine levels from the main reservoir on a specified day are shown in Table 14. The readings describe parts per million chlorine per ten gallons of water. Use classes 4.8-5.0, 5.1-5.3, 5.4-5.6, 5.7-5.9, 6.0-6.2.

(g) Could we use the following classes for a frequency distribution? Explain. Less than 5.0, 5.1-5.3, 5.4-5.6, 5.7-5.9, greater than 6.0. If so, what are are its advantages and disadvantages compared to the one specified?

(h) Could we use the following classes for a frequency distribution? Explain. 4.5-4.9, 5.0-5.4, 5.5-5.9, 6.0-6.4. If so, what are its advantages and disadvantages compared to the one specified?

Table 14

4.8	5.3	5.2	4.8	5.0	6.1
5.1	5.4	4.9	6.2	5.8	5.2
5.0	4.9	6.1	6.2	5.7	5.6
6.1	6.0	5.7	4.8	4.9	5.2
5.5	5.2	4.9	5.7	5.6	5.3
3.0	5.1	5.3	5.1		

3. Mary's Gifts For All Occasions does a large mail-order business. Table 15 gives a sample of the size of orders (in dollars) received last summer.

Table 15

18.92	14.67	22.67	35.41	19.14	42.91
13.17	24.22	47.49	59.19	62.13	21.46
19.16	31.14	10.50	13.29	25.14	27.18
32.47	38.47	46.42	48.19	51.42	63.46
15.16	22.18	29.37	32.42	61.14	35.60
69.90	52.14	18.19	14.16	52.14	55.19
52.16	44.60	60.37	16.18	22.52	39.15
61.75	52.14	33.20	24.16	35.18	42.16
33.18	21.14	62.14	67.18	41.14	52.18

Use classes 10.00-19.99, 20.00-29.99, 30.00-39.99, 40.00-49.99, 50.00-59.99, 60.00-69.99.

(g) Could we use the following classes for a frequency distribution? Explain. 10-19, 20-29, 30-39, 40-49, 50-59, 60-69. If so, what are its advantages and disadvantages compared to the one specified?

(h) Could we use the following classes for a frequency distribution? Explain. Less than 20.00, 20.00-35.99, 36.00-50.00, greater than 50. If so, what are its advantages and disadvantages over the one specified?

4. The final examination grades of the 165 students who took statistics at the Vilnius Technical Institute in a recent semester are shown in Table 16. Use classes 30-39, 40-49, 50-59, 60-69, 70-79, 80-89, 90-99.

(g) Could we use the following classes for a frequency distribution? Explain. Less than 60, 60-80, greater than 80. If so, what are its advantages and disadvantages over the one specified?

Table 16

76	79	60	58	62	51	34	47	73	33	66
49	59	81	92	52	56	76	76	44	81	70
89	71	88	70	66	41	85	72	77	65	84
54	76	34	53	72	60	72	72	76	69	83
50	90	62	64	92	901	53	53	60	84	71
90	40	58	90	41	59	70	67	56	88	71
75	61	86	77	90	81	30	77	50	79	67
51	65	78	60	54	89	69	84	56	53	57
43	51	52	72	74	85	61	57	68	40	64
56	42	79	78	80	82	87	73	60	82	68
60	72	63	52	90	67	77	60	64	93	91
45	49	79	85	55	61	51	81	51	53	80
35	60	78	78	50	96	41	82	61	91	49
52	66	60	55	64	84	82	47	83	81	61
56	67	77	54	48	51	65	72	89	71	85

5. The before tax income (rounded to the nearest thousand dollars) of a random sample of 40 households in Halifax County are shown in Table 17. Use classes 0-49, 50-99, 100-149, 150-199, 200-249, 250-299, 300-349, 350-399, 400-449, 450-499, 500-549.

Table 17

120	122	102	90	50	45	530	500
410	90	86	120	150	205	290	90
250	157	45	310	300	180	192	210
275	538	475	510	86	70	95	45
175	92	105	75	60	65	510	520

(g) Could we use the following classes for a frequency distribution? Explain. 10-39, 40-89, 90-139, 140-189, 190-239, 240-289, 290-339, 340-539. If so, what are its advantages and disadvantages over the one specified?

6. Refer to the frequency distribution for Mary's Gifts For All Occasions (Exercise 3) based on the given classes. Is it possible to determine from this distribution the number of orders received with values subject to the following conditions? If your answer is yes, state the number of orders; if no, explain why not.

   (a) less than \$50.00; (c) more than \$50.00;
   (b) \$50.00 or less; (d) \$40.00 or more.

7. Refer to the statistics examination grade frequency distribution (Exercise 4) based on the given classes. Is it possible to determine from this distribution the number of grades subject to the following conditions? If your answer is yes, state the number of grades; if no, explain why not.

   (a) less than C (defined by less than 70);

   (b) more than C; (c) at most 80;

   (d) passing (defined by 60 or more); (e) equal to 90.

3.2 ■ Seeing Is Believing

Seeing is believing is based on the old adage, "a picture is worth a thousand words."

Histograms

The **histogram of a frequency distribution** is a sequence of rectangles representing the classes of the distribution which are joined at the sides. The base of each rectangle is labeled to indicate the data class and the rectangle constructed on the base is inscribed with a number (magnitude or percentage) to show how many (or what percentage) of the data fall into the class associated with the rectangle. The base of each histogram rectangle is labeled with either its class limits, class boundaries, or class mark. Figure 2 shows the histogram for the *Basic Statistics* text holding time distribution (Table 7; page 87) with the three possible options for labeling the base positioned accordingly.

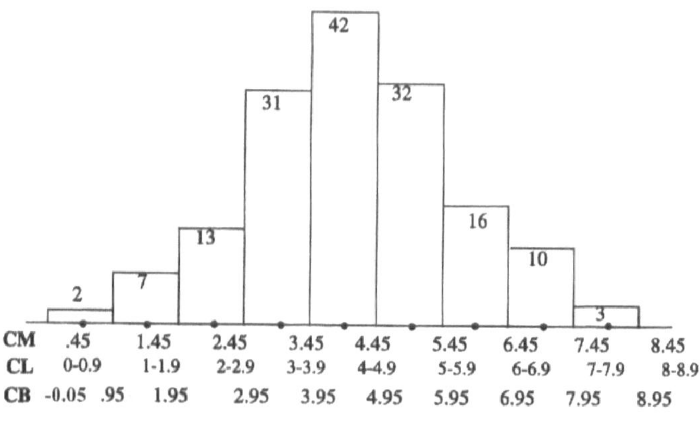

Figure 2

Frequency Polygons

The **frequency polygon of a frequency distribution** is obtained by plotting the frequency of each class against the class mark of the class (midpoint of its rectangle base) and joining these points with line segments. To tie down the frequency polygon to the horizontal axis we introduce two points on the horizontal axis at the beginning and end which are given by the width of the distribution. Figure 3 shows the frequency polygon for the *Basic Statistics* text holding time distribution given in Table 7.

The frequency polygon tells a picture story about how the class frequencies change in going from one class to another.

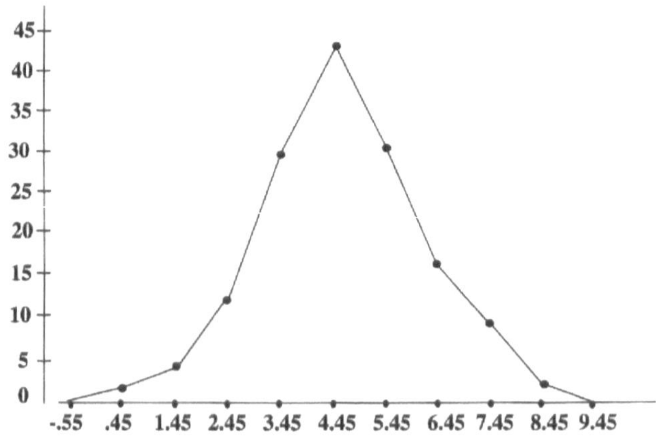

Figure 3

Ogive

An **ogive** is the graph of the cumulative less than frequency distribution. The lower class limits are written along the *x*-axis. The points that are plotted consist of a horizontal coordinate (lower class limit) and a vertical coordinate which represents the sum of the frequencies less than that specific *x* coordinate. These points are joined by line segments between the plotted points. Figure 4 shows the ogive for the *Basic Statistics* text cumulative less than distribution given in Table 9.

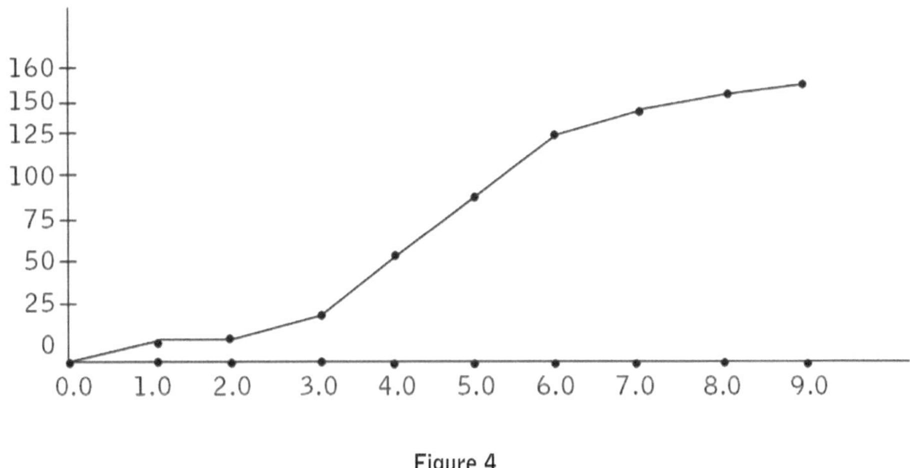

Figure 4

It should be noted that we also could have taken the *y* coordinates to represent the percentages shown in Table 9 and the *y*-axis scaled in percentages from 0 to 100. Inspection of Figure 4 shows that there is a steep rise in the amount of data between 3 and 6, indicating that most of the data are in this interval, The nearly horizontal line segments between 0 and 3 and between 6 and 9 indicate that a relatively small percentage of the data are in these intervals.

Pie Charts and Percentage Bar Charts

A commonly employed method for displaying categorical data is the pie chart, Figure 5 shows the pie chart for the college sport distribution given in Table 12 (page 69).

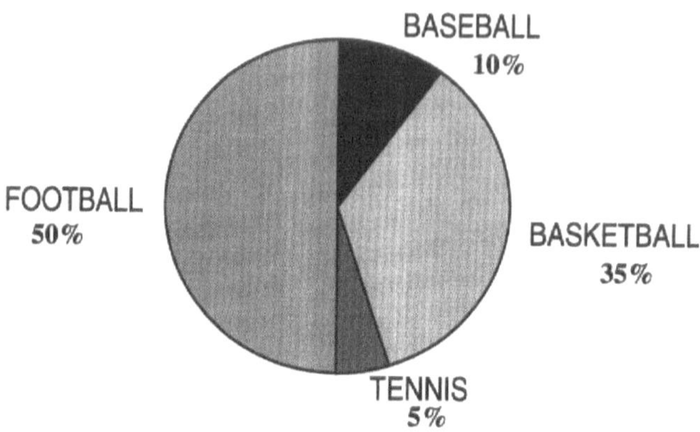

Figure 5

Another commonly employed method of presenting categorical data is the percentage bar chart. Figure 6 shows the percentage bar chart for the data in Table 12.

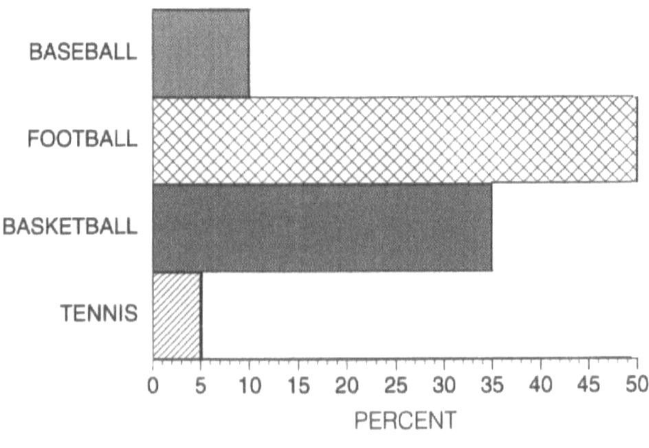

Figure 6

Exercises

1. **MERT:** Refer to the frequency distribution for Bell City's Medical Emergency Response Team (Sec. 3.1, Exercise 1; page 91) based on the given classes. Construct the: (a) histogram, (b) frequency polygon, (c) ogive for the cumulative less than distribution.

2. Refer to the frequency distribution for Mary's Gifts For All Occasions (Sec. 3.1, Exercise 3) based on the given classes. Construct the: (a) histogram, (b) frequency polygon, (c) ogive for the cumulative less than distribution.

3. Refer to the frequency distribution for Ecap University's mathematics department (Sec. 3.1, Exercise 4) based on the given classes. Construct the: (a) histogram, (b) frequency polygon, (c) ogive for the cumulative less than distribution.

4. Refer to the frequency distribution for the statistics final examination (Sec. 3.1, Exercise 4) based on the given classes. Construct the: (a) histogram, (b) frequency polygon, (c) ogive for the cumulative less than distribution.

3.3 ■ Seeing Is Misleading

Seeing may be misleading when the "picture" does not do proper justice to the structure it is intended to portray. Variation in the width of a histogram rectangle or the scale of an ogive, not properly accounted for, can leave an unsuspecting reader with a distorted message. Consider, for example, the *Basic Statistics* holding time frequency distribution shown by Table 7 (page 87). Suppose it were decided to consolidate the intervals 5.0-5.9 and 6.0-6.9 into one interval 5.0-6.9, yielding the frequency distribution shown by Table 18.

Table 18

Time Interval	Frequency	Percentage (%)
0.0-0.9	2	1.3
1.0-1.9	7	4.5
2.0-2.9	13	8.3
3.0-3.9	31	19.9
4.0-4.9	42	26.9
5.0-6.9	48	30.8
7.0-7.9	10	6.4
8.0-8.9	3	1.9
	156	100.0

If the histogram for this distribution were constructed as shown in Figure 7, a quick inspection might lead one to believe that the rectangle defined by 5.0-6.9 dominates the scene. The fact is, however, that the class 5.0-6.9 has double the width of any of the other classes. For a proper perspective this should be compensated for by cutting the height of the rectangle in half. The proper histogram representation is shown in Figure 8.

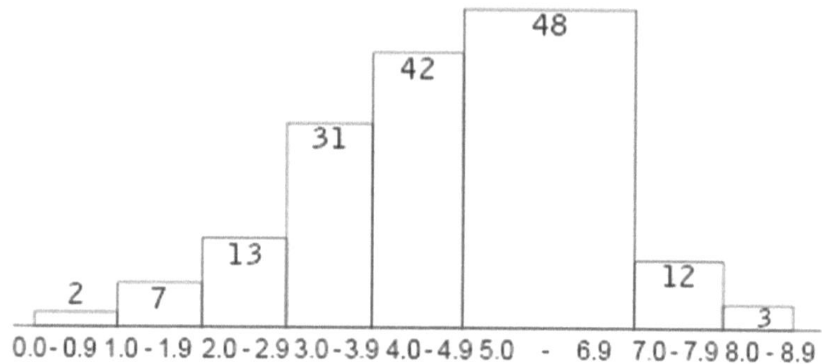

Figure 7

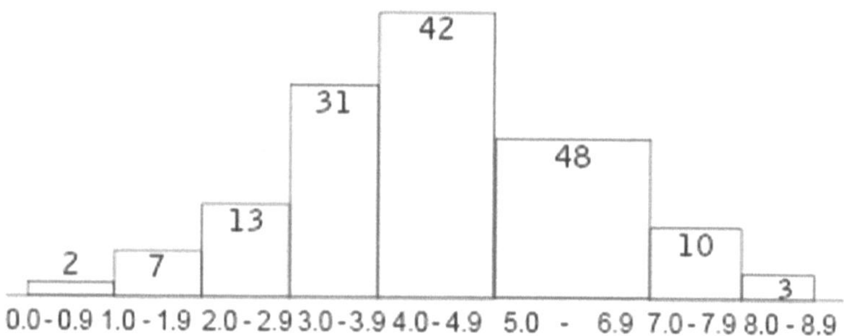

Figure 8

Consider the cumulative distribution for the *Basic Statistics* text holding time data of Example 1 with the combining of the two classes. Table 19 shows the revised distribution. If the ogive were constructed as in Figure 9, one might get the incorrect impression that the jump in frequency from 5.0-6.9 was the same as the jump from 4.0-4.9 because the line segments joining the points have nearly the same slope. This, of course, is not the case.

Table 19

Time Interval	Cumulative Frequency	Cumulative Percentage (%)
< 0.0	0	0
< 1.0	2	1.3
< 2.0	9	5.8
< 3.0	22	14.1
< 4.0	53	34.0
< 5.0	95	60.9
< 7.0	143	91.7
< 8.0	153	98.1
< 9.0	156	100.0

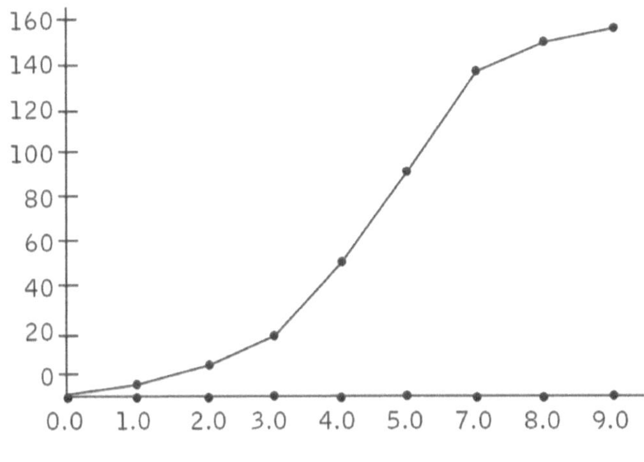

Figure 9

For a proper perspective, one must take into account the feature that the interval 5.0-6.9 has twice the width of the interval 4.0-4.9; it is easy to overlook this point when glancing at the ogive. The correct ogive appears in Figure 10 where the double spread has been accounted for. This representation was constructed by leaving a space along the holding time axis where 6.0 would have gone if Table 14 had included the class < 6.0. Figure 9 leads one to believe that there was a sharp increase in frequency between holding times of 5.0 and 7.0 because of the slope of the line segment joining the cumulative frequencies corresponding to holding times of < 5.0 and < 7.0, respectively. However, Figure 10 correctly reflects the rate of increase of frequency between holding times of 5.0 and 7.0 by leaving a space for < 6.0 to account for the fact that it was not included in Table 19 (page 101).

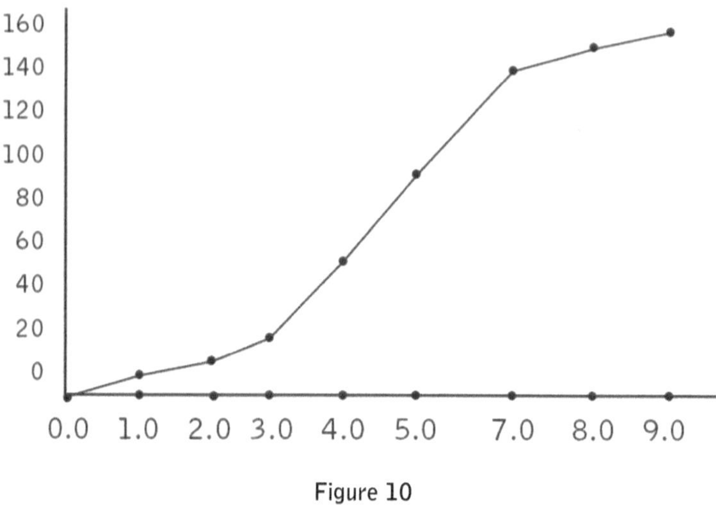

Figure 10

EXERCISES

Closely examine the histogram given in Exercise 1 for the stated situation. If there are features to this histogram that are misleading, describe their nature and make appropriate adjustments.

1. Figure 11 gives the histogram for the GPA frequency distribution specified in Table 1 at the beginning of this chapter (page 80).

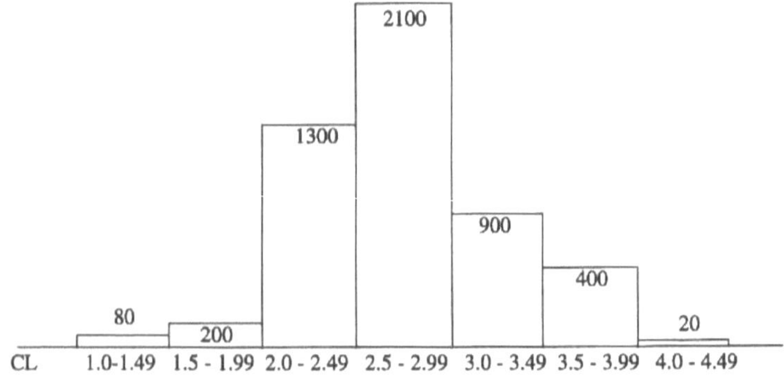

Figure 11

4

Descriptive Measures for Ungrouped Data

4.1 ■ Notation

The first problem that arises concerns notation. Devising a suitable notation and using it consistently may not seem like something worth a lot of bother, but it lies at the root of sound communication.

Population Parameters and Sample Statistics

As was noted, a **statistical population** is a collection of data viewed as a whole in its own right, whereas a **sample** is a part of this all inclusive whole. **Inferential Statistics** is concerned with making and testing judgments about population values of interest, called **parameters**, on the basis of "suitably obtained" sample values, called **statistics.** To formulate a statement involving a sample statistic and its population counterpart without the risk of ambiguity it is desirable to have a notation which allows us to distinguish the two. We shall follow the practice of using Latin letters for sample statistics and Greek letters for their population parameter counterparts. Thus, for example, the population mean (sum of the population values divided by the number of such values) is denoted by the Greek letter μ (pronounced moo), whereas the mean of a sample is indicated by $\bar{x}$ (pronounced x bar). From the notation one can easily see which mean, population or sample, is at issue in a statement involving the term.

It is also standard practice to use a capital letter, such as N, to denote the number of data in a population and a small letter, such as n, to denote the number of data in a sample.

Summation Notation

Sums of quantities make an appearance in many areas of mathematics, statistics in particular, and it is useful to have a notation for sums. The Greek letter Σ (sigma) is used to denote sums.

To illustrate, consider the sum:

$$1^2 + 2^2 + 3^2 + 4^2 + 5^2$$

First, we look for a pattern in the way the terms of the sum change. Here the exponent is the same value, 2, for all terms being added and the base is 1 for the first term and increases by 1 as we go from term to term. This prompts us to introduce a symbol to denote the quantity which is changing from term to term. Letters such as *i*, *j*, and *k* are often used for this purpose. Letting *i* denote the base value, we see that a typical term in the sum can be denoted by i^2. In terms of notation, the preceding sum may be indicated by

$$\sum_{i=1}^{5} i^2 \text{ or } \sum i^2 .$$

The first form is more complete in that it specifically states that the sum is to begin with $i = 1$ and end with $i = 5$, with *i* taking all integer values between 1 and 5, inclusive. This yields 1^2, 2^2, 3^2, 4^2, and 5^2 which the Σ symbol tells us must be added. The second form does not explicitly indicate the initial and terminal values for *i* and may be used when the context makes clear what these values are.

The sum

$$w_1 x_1 + w_2 x_2 + \cdots + w_n x_n$$

is expressed by

$$\sum_{i=1}^{n} w_i x_i \text{ or } \sum w_i x_i$$

if it is understood that the sum is be taken from 1 to n.

The notation

$$\sum_{i=1}^{n} x_i \text{ or } \sum x_i$$

expresses the sum

$$x_1 + x_2 + \cdots + x_n .$$

4.2 ■ Accuracy

With a world of computation facing us, it might be useful to give thought to computation accuracy.

Many people believe that the accuracy of a numerical result is increased by expressing the value in terms of a large number of decimal places. That this is not necessarily the case is illustrated by the following experiment. Take a tub of water and place seven empty pails on the floor. Spill the water from the tub into the pails, giving each an equal share, as nearly as possible. It would be appropriate to say that each pail contains roughly 1/7 of the tub of water. While this may not seem very accurate, it's the best that can be said. If, however, we divide 1 by 7 to obtain 0.1428 . . . , and proclaim that each pail contains about 14.28% of the water of the tub, one might be deluded into believing that this is a more accurate description of the amount of water in each pail if one equates accuracy with number of decimal places. From this point of view the accuracy of the description would be further enhanced by carrying the division of 1 by 7 to still more places, 0.142857 . . . , for example. It's nonsense, of course; we are no better off than when we said that each pail contains about 1/7 of the tub of water.* The important point is:

Real accuracy depends on the accuracy of the data.

> In calculating with approximate numbers we should keep in mind that the accuracy of a result cannot exceed the accuracy of the least accurate number in the data.

The sum of the approximate numbers 2.34 (known to two places) and 1.131 (known to three places), for example, should be recorded as 3.47 (to two places) rather than 3.471. The number 3.471 would give a misleading sense of accuracy.

The holding-time numbers arising from the "Heavy" Basic Statistics text situation (Example 1, Section 2.3, page 51) are approximate values correct to one place. The result of a series of computations with such data cannot be expected to be accurate

* M. Richardson, Fundamentals of Mathematics, Revised Edition. (New York: The Macmillan Co., 1958), 402.

beyond one place. In carrying-out the computations it is advisable to carry one more place and round off at the end.

This counsel, as with all advice, should not be followed blindly, but take circumstances into consideration and be exercised with discretion. If your pet cat, for example, weighed in at 5 pounds and then at 6 pounds the following month (approximate values), it makes good sense to record its average weight as 5.5 pounds. To round off this average to 6 pounds would give us a highly distorted view of the underlying situation. On the other hand, if your pet pony weighed in at 500 pounds and later at 501 pounds, for an average weight of 500.5 pounds, considering the magnitude of these values, it would not distort the sense of the situation to round off the average to 501 pounds.

If the numbers are considered to be exact, then all places may be retained when computation is done. If 2.34 and 1.131 are considered exact, then recording their sum as 3.471 does not yield a false sense of accuracy.

If 2, for example, is considered to be exact, then the accuracy of an approximation of $\sqrt{2}$ is increased by expressing its value in terms of more decimal places. (Note: the symbols ⊔ and ≈ mean approximately in a computation setting.)

$$\sqrt{2} \approx 1.41$$

More accurate: $\sqrt{2} \approx 1.414$

Still more accurate: $\sqrt{2} \approx 1.414214$

How accurate an approximation of $\sqrt{2}$ should be taken in a computation situation depends on the setting of the problem.

If, on the other hand, 2 is considered approximate, the accuracy of an approximation of its square root is not enhanced by taking more and more decimal places.

4.3 ■ Measures of Location

To gain insight into the nature of collections of data we seek to define numerical values, called measures of location, which describe how the data are concentrated and might also serve to "characterize" the data. The term **ungrouped data** means that we are dealing with the raw data themselves. They have not been grouped into classes of a frequency distribution.

The Arithmetic Mean

In our everyday language the terms average and mean are used in an ambiguous way. To meet the needs of a technical discipline we must take care to be precise in our use of such terms so that there is no misunderstanding.

The arithmetic mean of a set of numerical data is their sum divided by the number of values in the sum. If a **population** consists of N data $x_1, x_2, \ldots, x_N$, the **population mean** is defined by

$$\mu = \frac{x_1 + x_2 + \cdots + x_N}{N},$$

which in summation notation is expressed by:

$$\frac{\sum_{i=1}^{N} x_i}{N} \text{ or } \frac{\sum x}{N}$$

If a **sample** consists of n data, $x_1, x_2, \ldots, x_n$, the **sample mean** is defined by

$$\bar{x} = \frac{x_1 + x_2 + \cdots + x_n}{n},$$

which in summation notation is expressed by:

$$\frac{\sum_{i=1}^{n} x_i}{n} \text{ or } \frac{\sum x}{n}$$

The term mean used without qualification is understood to refer to arithmetic mean.

Example 1 Mean of Population vs. Mean of Sample

The nine students in an introductory psychology class at Ecap University had the following midterm exam scores:

85, 78, 74, 77, 75, 82, 80, 85, 84

(a) Suppose we regard this class as a whole in its own right, without regard to other introductory psychology classes; calculate the mean grade.

(b) Suppose we regard this class as part of the inclusive whole consisting of all sections of introductory psychology at Ecap University; calculate the mean grade.

In this situation we obtain the same numerical answer for both (a) and (b). The difference is notational and conceptual. With respect to notation, the answer to part (a) is:

$$\mu = \frac{85+78+74+77+75+82+80+85+84}{9} = 80$$

The symbol μ is used because the class is considered a population. For (b) we perform the same calculation, except that μ is replaced by $\overline{x}$.

$$\overline{x} = \frac{85+78+74+77+75+82+80+85+84}{9} = 80$$

Conceptually, in (a) the class is a population because we are looking at the class as an isolated unit without regard to a larger inclusive whole. In (b) the class is considered a sample because it is viewed as part of the larger whole of all introductory psychology sections.

Example 2 The "Heavy" Basic Statistics Text

This takes us back to Example 1, Section 2.3 (page 51). $\Sigma x = 714.0$ and $N = 156$. Thus:

$$\mu = \frac{714.0}{156} = 4.6$$

Dividing 714.0 by 156 yields 4.5769. Would 4.5769 be a more accurate description of μ than 4.6? Why is μ being used as the symbol for the mean as opposed to $\overline{x}$?

Example 3 A Rating Situation

The nine students in the introductory psychology course noted in Example 1 were asked to express their satisfaction with their instructor's teaching quality in terms of the ratings 20 (Outstanding), 10 (Good), 5 (Satisfactory), 2 (Poor), -7 (Terrible). The results obtained are shown in Table 1.

Table 1

Bob S.	2	Ken M.	2	Nancy R.	5
June K.	20	Amy N.	10	Irene S.	2
Adam G.	5	Bill B.	-7	Vera N.	10

Determine the instructor's mean rating and interpret the result.

Let us first observe that the ratings are ordinal data (see Section 2.1) and that the mean is not meaningful for ordinal data; not meaningful in what sense? one might ask.

There is no law which prohibits us from determining the mean of numbers on hand; the question is, what sense can we make of the value? Adding up the ratings and dividing by 9 yields 5.4, which does not correspond to any of the ratings. Can 5.4 be interpreted as a consensus value of the class, as defined by the mean, which places their instructor above Satisfactory but below Good? So far, okay; it's when we attempt to go further and draw a conclusion about how much better than Satisfactory and how short of Good the class consensus rates their instructor that we enter the realm of the meaningless. This is because differences are not defined when ordinal data are employed.

The Weighted Arithmetic Mean

Conrad McBain invested his money for one year by investing \$1,000 at 10% and \$100 at 20%. What is his mean rate of return for the year on the \$1,100 that he invested? The answer is not 15%, the average of 10% and 20%. The only way 15% could be the answer is if equal amounts of money were invested at each rate. On the \$1,000 he earned \$100 and on the \$100 he earned \$20. Since he earned \$120 for the year on his total investment of \$1,100, his rate of return is $120/1100 = 0.1091$ or 10.91%.

We can think of the situation from the point of view of investing 1,100 one dollar bills into 1,100 investments. The first 1,000 investments paid a 10% return and the last 100 investments paid a 20% return. Since the same amount of money ($1) was put into each investment, to calculate the average return on investment we would add the 1,000 ten percent returns to the 100 twenty percent returns and divide by the total number of investments. This yields:

$$\begin{aligned}\text{Average Rate of Return} &= \frac{1000(0.10)+100(0.20)}{1000+100} \\ &= 0.1091\end{aligned}$$

In effect, we have calculated a **weighted average rate of return.** That is, we weighted each return by the amount of money associated with that return, added the results, and divided by the total amount of money invested.

More generally, if we have data $x_1, x_2, x_3, \ldots, x_n$ and a corresponding set of weights, $w_1, w_2, w_3, \ldots, w_n$, the **weighted average** of the data is defined by:

$$\bar{x}_w = \frac{w_1x_1 + w_2x_2 + \ldots + w_nx_n}{w_1 + w_2 + \ldots + w_n} = \frac{\sum wx}{\sum w}$$

If the data are viewed as a population, n should be replaced by N and $\bar{x}_w$ by μ_w.

A commonly made mistake is to confuse the x's and the w's. To avoid such an error we must take care to read the problem at hand carefully and keep in mind that the x's stand for the data whose weighted average is being calculated.

Other Means

There are other concepts of mean value—the geometric and harmonic means—which do not play a role in statistical inference. The geometric mean is briefly discussed in Exercise 12.

The Median

For a set of numerical data arranged in increasing order, the **median**, denoted by M, is that number such that half the data are less than or equal to it and half the data are greater than or equal to it. It is the middle value in this sense.

If n is the number of data values that have been arranged in increasing order, then $(n+1)/2$ gives us the position of the median value, not to be confused with the value of the median.

The symbol M does not by itself indicate whether the data comprise a sample or a population. This would have to be determined from the context of the problem.

Example 4 Median: Odd number of Data.

Determine the median of the nine psychology midterm grades given in Example 1: 85, 78, 74, 77, 75, 82, 80, 85, 84.

Arranging the 9 grades in increasing order gives us:

74, 75, 77, 78, 80, 82, 84, 85, 85.

Since $n = 9$, the position of the median value is given by

$$\frac{9+1}{2} = 5.$$

Thus the median is the fifth data value, 80. Half the grades are less than or equal to 80 and half are greater than or equal to 80. In this case the mean and median values agree.

Example 5 Median: Even Number of Data

Suppose there had been eight students in the psychology class with midterm grades (in increasing order) 74,75,77,78,80,82,84,85. What would the median value be in this case?

The position indicator $(n+1)/2$ gives us 4.5, which puts us between the fourth and fifth terms, 78 and 80 (see Figure 1). In such a case the median is defined as the mean of the two adjacent terms singled out by the position locator $(n+1)/2$. In this case we obtain $M = 79$.

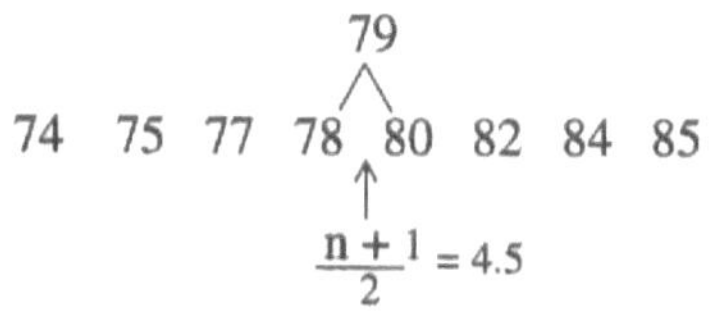

79

74 75 77 78 80 82 84 85

↑

$$\frac{n+1}{2} = 4.5$$

Figure 1

Example 6 Appropriate Number: Mean vs. Median

The seven full time faculty members in Ecap University's social science department have salaries (in thousands of dollars) 30, 30, 30, 30, 30, 135, 135. When one of the department members complained to the dean about the low salaries in the department the dean pointed out that there is no justification for complaint because the mean salary level is $60,000. What would be an appropriate clarification for the dean's information?

This situation takes us back to Section 1.3 (page 10) on pinpointing the characteristic of data that is most appropriate for the situation under consideration.

Mean values are strongly influenced by extreme values. When such are present, as is the case here, the mean does not serve as a good indicator of the nature of the data. In such cases the median, which is not influenced by extreme values, is a better indicator of the nature of the data. The faculty's median salary of $30,000 is a more reliable indicator of the social science department's salary state of affairs.

When the median for a large, but not too large, set of data is to be found the stem-and-leaf device is a useful tool for listing the data in increasing order.

Example 7 The "Heavy" Basic Statistic Text

This takes us back to Example 1, Section 2.3 (page 51). With N = 156 data, the construction of a stem-and-leaf plot of the holding-time data is developed in Table 11 of Chapter 3 (page 69) and reproduced in Table 2.

Table 2

Stem	Leaf
0	3, 8
1	7, 7, 8, 5, 7, 9, 6
2	5, 3, 9, 5, 1, 9, 6, 4, 6, 5, 3, 5, 6
3	1, 9, 3, 4, 8, 6, 5, 0, 3, 8, 0, 3, 2, 2, 2, 4, 5, 7, 0, 1, 9, 8, 6, 4, 4, 5, 4, 6, 9, 5, 1
4	3, 7, 9, 5, 4, 7, 4, 4, 3, 1, 5, 2, 6, 1, 7, 8, 1, 4, 6, 6, 3, 1, 3, 2, 1, 5, 8, 9, 5, 9, 8, 9, 5, 1, 9, 6, 4, 5, 4, 7, 3, 7
5	2, 5, 4, 9, 1, 9, 1, 2, 2, 0, 2, 1, 9, 5, 0, 0, 9, 7, 0, 2, 9, 9, 1, 0, 6, 8, 1, 3, 8, 1, 8, 7

6	0, 9, 6, 1, 0, 3, 1, 3, 1, 6, 5, 8, 2, 1, 7, 7
7	2, 0, 9, 0, 3, 2, 5, 5, 6, 2
8	4, 1, 2

The position indicator $(N+1)/2$ gives us 78.5, which puts us between the 78th and 79th terms. From Table 2 we see that the 78th and 79th terms terms are the 25th and 26th terms in the row with stem 4, which are, when arranged in increasing order, 4.5 and 4.6.

Therefore, $M = 4.55$. 78 of the data are less than 4.55 and 78 are greater than 4.55.

Example 8 A Rating Situation

We return to Example 3 which had nine students in an introductory psychology course who were asked to express their satisfaction with their instructor's teaching quality in terms of the ratings 20 (Outstanding), 10 (Good), 5 (Satisfactory), 2 (Poor), -7 (Terrible). The results obtained are reproduced in Table 3.

Table 3

Bob S.	2	Ken M.	2	Nancy R.	5
June K.	20	Amy N.	10	Irene S.	2
Adam G.	5	Bill B.	-7	Vera N.	10

Determine the instructor's median rating, interpret the result, and compare these with the mean rating and its interpretation obtained in Example 3.

Arranging the 9 ratings in increasing order yields:

-7, 2, 2, 2, 5, 5, 10, 10, 20

The median is the fifth value, 5; half of the students judge the instructor's teaching as satisfactory or below satisfactory; half judge his teaching as satisfactory or better than satisfactory.

There is no question about the interpretation of the median of ordinal data being meaningful. This cannot be said about the mean of ordinal data.

It is worthy of note that while the median is independent of the rating system used, this is not the case with the mean.

The Mode

Another measure of central location is the **mode**, defined by the data value that occurs most often. While it is true that the data value that occurs most frequently often lies near the center of the data, many times this is not the case. The salary data of Example 6 has mode 30,000; the grade data of Example 5 has no mode since all values occur with the same frequency. If the grade data of Example 4 had an additional grade of 74, we would say that the data is bimodal with means 74 and 85.

EXERCISES

1. Robert Franks, an experimental psychologist, is studying a species of friendly furry creature by timing them as they run through a maze. A randomly chosen sample of 10 friendly furry creatures traversed the maze in the following times, in seconds: 52, 48, 59, 60, 55, 51, 51, 50, 49, 58. Determine the (a) mean, (b) median, and (c) mode of these data.

2. To determine whether a new fertilizer is more effective than the one in current use for growing corn Carl Cairns selected 5 test plots on his farm at random which he treated with the fertilizer in current use and another 5 plots at random which he treated with the new fertilizer. The following results (in bushels per acre) were obtained.

   Current fertilizer: 70.5, 80.2, 90.3, 60.2, 85.4
   New fertilizer: 73.2, 82.1, 93.1, 63.6, 87.1

   (a) Determine the (i) mean, (ii) median, and (iii) mode for the two data sets.

   (b) Should the data sets be viewed as samples or populations? Explain?

3. A fierce competition has developed between the manufacturers of Longlife and Neverdie batteries. An independent testing agency chose 6 items at random from a list of battery operated items and operated them by means of Neverdie batteries. Another 6 randomly chosen items were operated by Longlife batteries. The following lifetimes data (in hours) were obtained.

Neverdie:	53.0, 48.2, 38.6, 40.1, 54.2, 42.6
Longlife:	52.2, 47.4, 37.8, 39.4, 53.5, 41.8

(a) Determine the (i) mean, (ii) median, and (iii) mode for the two sets of data.

(b) Should the data sets be viewed as samples or populations? Explain.

4. Professor Ann Martin's students received the following midterm exam grades in financial accounting: 85, 70, 72, 85, 75, 82, 93, 91, 62, 93, 68, 87, 65, 85, 93.

   (a) Determine the (i) mean, (ii) median, and (iii) mode.

   (b) Should the data be viewed as a sample or population? Explain.

5. **MERT:** Refer to Bell City's Medical Emergency Response Team's response time situation described in Exercise 1 of Section 3.1 (page 91).

   (a) Determine the (i) mean, (ii) median, and (iii) mode.

   (b) Should the data be viewed as a sample or population? Explain.

6. Refer to Cameron City's water quality control situation described in Exercise 2 of Section 3.1 (page 92).

   (a) Determine the (i) mean, (ii) median, and (iii) mode.

   (b) Should the data be viewed as a sample or population? Explain.

7. One of the participants in a panel discussion boldly asserted that half the families in New Jersey make less than the median family income. Flabbergasted, the other panelists fell silent. Had you been on the panel, how would you have replied?

8. Kristin Haas invested $2,000 at 8% per year and $500 at 12% per year. What is her average rate of return for the year?

9. Mark Clark bought 100 shares of Xcel at $50 per share, 150 shares of Zeron at $25 per share, and 70 shares of Tobin Motors at $75 per share. What is the average cost per share?

10. The exam grades of the nine students in an electrical engineering course were 49, 50, 52, 52, 55, 86, 95, 100, 100. Asked how well the class did by one of the students, the instructor replied: "not bad at all; the mean is 71."

    (a) Is the mean the best indicator of how well the class did? Explain.

    (b) If your answer to (a) is no, then what do you believe would be a more suitable indicator? Explain.

11. A panel of five faculty of the English department of Ecap University is to read and rank student essays submitted for the Ecap Prize.The ranking values available are 50 (Outstanding), 40 (Excellent), 35 (Very Good), 25 (Good), 10 (Satisfactory).Two highly thought of contenders for the prize are Leslie R. (whose essay received rankings 40, 40, 40, 50, 50) and Morgan J. (whose essay received rankings 25, 35, 50, 50, 50). Let us assume that the awarding of the prize comes down to these two students.

    (a) Would you award the prize to Leslie R. because the mean of his rankings, 44, exceeds the mean of Morgan J.'s rankings 42? Explain.

    (b) If your answer to (a) is no, then on what basis would you award the prize? Explain.

12. Given a set of data, $x_1, x_2, \ldots x_n$, its geometric mean is defined as

$$G = \sqrt[n]{x_1 x_2 \cdots x_n}$$

Thus, G is that number whose nth power is equal to the product of the n numbers. Consequently, G should lie close to the center of the data. For example, consider the numbers 3, 8, 11, 18, 25. Then, $G = \sqrt[5]{3 \cdot 8 \cdot 11 \cdot 18 \cdot 25} = 11.27$, which is fairly centrally located.

The geometric mean is used in the world of finance to calculate the constant periodic growth rate equivalent to growth rates that are changing periodically. Suppose, for example, that the annual percentage changes of an investment over the past five years were 15%, 12%, -4%, 7.7%, and 19.8% and we would like to know the constant annual percentage change equivalent. At the end of the first year the original investment increased by a factor of 1.15, then that amount increased by a factor of 1.12 by the end of the second year. Thus, for the five years the original investment increased by a factor of $(1.15)\ (1.12)\ (.96)\ (1.077)\ (1.198) = 1.61$ and, $G = \sqrt[5]{1.61} = 1.1$.This tells us that the five annual percentage changes were equivalent to a constant annual increase of 10%.

(a) Find the geometric mean of the values 2, 4, 8, and 18.

(b) The percentage rate of population growth in Halifax County was 1.5%, 3%, 5%, 9%, and 12.5%. Find the geometric mean of these growth rates.

4.4 ■ Measures of Variation

As important as the mean is in describing a set of data, it does not by itself give us a complete picture. Many sets of data with different appearances, thereby telling different stories, can have the same mean. Consider, for example, the following quiz grades obtained by Mary and John in an accounting course.

Mary: 74, 75, 77, 78, 75, 74, 75, 72
John: 47, 92, 92, 97, 45, 55, 74, 98

Both students averaged 75 on the quizzes, but it is clear that we are looking at very different stories. Mary is consistent in her performance on the quizzes. If we had to guess what her next grade would be, 75 would be a "reasonable" prediction. On the other hand, a guess of 75 for John's next grade would be more of a wild guess than a reasonable prediction because of his erratic performance on the quizzes. To help us distinguish between situations of this sort we need a measure of the "spread" of the data about their mean.

The Range

The **range** of a set of data is the difference between the largest and smallest data values. For Mary's quizzes the range is $78-72=6$, whereas for John's quizzes the range is $98-45=53$.

The range is easy to compute, but it gives us very limited information about a set of data. It gives us no information on how the intermediate values are distributed, and very different data sets may have the same range. For example, consider the following distributions of math SAT scores.

(a) 200 200 200 200 200 800 800 800 800 800

(b) 200 250 300 400 500 500 600 700 750 800

(c) 200 500 500 500 500 500 500 500 500 800

All three data sets have a mean of 500 and range of 600. However, they reflect very different student groups. Situation (a) reflects a group where half the students are very poorly prepared and the other half are among the top in the country. Situation (b) reflects a very diverse group in its mathematics preparation, whereas situation (c) reflects a group having better than "average" mathematics preparation, with one student having poor mathematics preparation (or being sick when he took the exam) and one being among the top in the country.

Average Deviation from the Mean

In seeking a more useful measure of spread of the data about their mean, it would seem natural to consider the average deviation of the data from the mean. For three values, x_1, x_2, x_3, let us say, this leads to:

$$\text{Average Deviation} = \frac{(x_1 - \mu) + (x_2 - \mu) + (x_3 - \mu)}{3}$$
$$= \frac{\sum x}{3} - \frac{3\mu}{3} = \mu - \mu = 0$$

It matters not whether we have 3 or 3 million data. The same analysis shows that the average deviation from the mean is zero. This result is due to the cancellation effect that data on opposite sides of the mean have on each other.

The Mean Deviation

To get around the cancellation effect we observed in the average deviation calculation, we might try working with the absolute value of the deviation from the mean, which leads to the concept of **mean deviation.**

$$\text{Mean Deviation} = \frac{\sum |x - \mu|}{N}$$

The mean deviation is a natural development, but it turns out not to blend well, mathematically speaking, with other ideas. It has limited use, so we must return to the drawing board.

Variance and Standard Deviation

Another way of getting around the cancellation effect is to square each of the differences $(x_1 - \mu)$, $(x_2 - \mu)$, and so on, and take the mean of these squared deviations.

This leads to the concept of **population variance**, denoted by the Greek letter σ^2 (small sigma squared). The square root of the population variance is called the **population standard deviation**, denoted by .

$$\sigma^2 = \frac{(x_1-\mu)^2+(x_2-\mu)^2+\cdots+(x_N-\mu)^2}{N}$$
$$= \frac{\sum(x-\mu)^2}{N}$$
$$\sigma = \sqrt{\frac{\sum(x-\mu)^2}{N}}$$

Here $x_1, x_2, \ldots, x_N$, are the data, viewed as a **population**, and μ is their mean.

As to defining sample variance, one option is to proceed in the same way, except for the notation. It is an option favored by some statisticians and writers. Another option is to proceed in the same way to the point of taking the sum of the squared deviations from the mean— $(x_1-\bar{x})^2+(x_2-\bar{x})^2+\cdots+(x_n-\bar{x})^2$ —but then divide this sum by $n-1$ rather than n. The major point in favor of this option is that the sample variance defined in this way gives us, in general, a better estimate of the population variance. In many situations we will be using the sample variance for this purpose. (Further discussion of estimators and their properties is given in Section 10.9.)

We elect the second option in defining sample variance, denoted by s^2, and sample standard deviation, denoted by s.

$$s^2 = \frac{(x_1-\bar{x})^2+(x_2-\bar{x})^2+\cdots+(x_n-\bar{x})^2}{n-1}$$
$$= \frac{\sum(x-\bar{x})^2}{n-1}$$
$$s = \sqrt{\frac{\sum(x-\bar{x})^2}{n-1}}$$

Here $x_1, x_2, \ldots, x_n$ are the data, viewed as a **sample**, and $\bar{x}$ is their mean.

Example 1 Mary's Variance

Find the variance and standard deviation of Mary's eight quiz grades, 74, 75, 77, 78, 75, 74, 75, and 72, assuming that they were a sample of her grades.

Table 1 shows the required calculations. Note the arrangement in Table 1. The

Table 1

x	$(x-\overline{x})$	$(x-\overline{x})^2$
74	-1	1
75	0	0
77	2	4
78	3	9
75	0	0
74	-1	1
75	0	0
72	-3	9
600	0	24

$$\overline{x} = \frac{600}{8} = 75$$

$$s^2 = \frac{24}{7} = 3.43 \text{ and}$$

$$s = \sqrt{3.43} \approx 1.9$$

data are first listed in column form, yielding column 1. Next we subtract the mean $\overline{x}$ from each data value, yielding column 2. As a double check against error, note that the sum of the entries in column 2 should be zero. We then square each entry in column 2 to obtain column 3. The variance is the sum of the entries in column 3 divided by $n-1$ $(8-1=7)$. With this sort of column arrangement it is hard to go wrong in computing a variance or standard deviation.

Derived Theorem for Variance

The definition of variance involves the data and the mean, which is defined in terms of the data. Thus, we can express the variance in terms of the data alone. While we can do this, do we want to? What kind of mathematical monster would we obtain? It turns out that by employing basic algebra coupled with patience we obtain a rather useful result which is not at all a mathematical monster. We have:

$$\sigma^2 = \frac{N(x_1^2 + x_2^2 + \cdots + x_N^2) - (x_1 + x_2 + \cdots + x_N)^2}{N^2}$$

$$= \frac{N\sum x^2 - [\sum x]^2}{N^2}$$

$$s^2 = \frac{n(x_1^2 + x_2^2 + \cdots + x_n^2) - (x_1 + x_2 + \cdots + x_n)^2}{n(n-1)}$$

$$= \frac{n\sum x^2 - [\sum x]^2}{n(n-1)}$$

Example 2 John's Variance

Find the variance and standard deviation of John's eight quiz grades, 47, 92, 92, 97, 45, 55, 74, and 98 assuming that they were a sample of his grades.

Table 2 shows the required calculations. Note the arrangement in Table 2. The data are first listed in column form, yielding column 1. We next square each data value, yielding column 2. The sum of column 1 gives us $\sum x$ and the sum of column 2 gives us $\sum x^2$.

As expected, John's standard deviation is larger than Mary's, 22.9 versus 1.9, indicating in numerical terms the degree to which John was more inconsistent in his performance on the accounting quizzes than Mary.

Table 2

x	x^2
47	2209
92	8464
92	8464
97	9409
45	2025
55	3025
74	5476
98	9604
600	48676

$$s^2 = \frac{8(48676) - (600)^2}{8(8-1)} = 525.1, \quad s = \sqrt{525.1} \approx 22.9$$

Let us note that had $\bar{x}$ not been a "nice" number like 75, as is the case in Example 1, but rather a number like 75.384, the calculations in Example 1 would have been much more cumbersome to carry out due to the $x-\bar{x}$ calculations. In employing the derived theorem for variance it matters not whether the mean is a "nice" or "messy" number. The same amount of work is involved and we avoid the problem of rounding off values.

Let us also note that variance and standard deviation only make sense for interval and ratio data.

Example 3 The "Heavy" Basic Statistics Text

Refer to Example 1, Section 2.3 (page 42); we have $\Sigma x = 714.0$, $N = 156$. Moreover, $\Sigma x^2 = 3650$.

Thus:

$$\sigma^2 = \frac{156(3650) - (714)^2}{(156)^2}$$

$$= 2.45$$

$$\sigma = \sqrt{2.45} \approx 1.6$$

EXERCISES

1. Compute Mary's variance by employing the derived theorem for variance.

2. Compute John's variance by employing the definition of variance.

3. Suppose that the eight quizzes that Mary and John took in the accounting course were the only quizzes that were given.

   (a) What effect, if any, would this have on the variance of their quiz grades? Explain.

   (b) If your answer to (a) is that there is an effect on the variance, then what are the correct variance values?

4. Find the range of (a) Mary's grades, (b) John's grades.

5. Find the (a) range, (b) variance, and (c) standard deviation of the friendly furry creature data given in Exercise 1 of Section 4.2 (page 91).

6. Find the standard deviation of (a) the current fertilizer data and (b) the new fertilizer data given in Exercise 2 of Section 4.2 (page 92).

7. Find the standard deviation of the (a) Neverdie and (b) Longlife battery data given in Exercise 3 of Section 4.2 (page 93).

8. **MERT:** Find the standard deviation of Bell City's Medical Emergency Response Team's response time data given in Exercise 5 of Section 4.2 (page 94).

9. Find the standard deviation of Cameron City's Water quality control data given in Exercise 6 of Section 4.2 (page 95).

10. Data $y_1 = x_1 + c$, $y_2 = x_2 + c, \ldots, y_n = x_n + c$ are obtained by adding the same value c to $x_1, x_2, \ldots, x_n$. Show that the y's and x's have the same variance.

5

Descriptive Measures for Grouped Data

5.1 ■ Preface

When we speak of **grouped data** we mean raw data not viewed directly, but viewed through the lens of a frequency distribution. When, by choice or circumstances, our access to the data is through a frequency distribution, we have in effect given up the data. If we wish to obtain numerical characteristics of the data, we will have to do so through the frequency distribution available to us, which is the subject of this chapter.

The following situation is useful as a bridge from the world of raw data to that of grouped data. Suppose that we have n data, with mean $\overline{x}$, that have been organized into a frequency distribution with, let us say, 2 classes. f_1 of the data belong to class 1 and have mean $\overline{x}_1$ and the remaining f_2 of the data $(n = f_1 + f_2)$ belong to class 2 with mean $\overline{x}_2$. This being the case we can express the overall mean of the data $\overline{x}$ in terms of the class means $\overline{x}_1$ and $\overline{x}_2$ in the following manner.

By definition of $\overline{x}$:

$$\overline{x} = \frac{\Sigma_1 + \Sigma_2}{n} \tag{1}$$

where Σ_1 is the sum of the f_1 data in class 1 and Σ is the sum of the f_2 data in class 2.

$$\overline{x}_1 = \frac{\Sigma_1}{f_1}, \qquad \overline{x}_2 = \frac{\Sigma_2}{f_2}$$

or equivalently,

$$\bar{x}_1 f_1 = \Sigma_1, \qquad \bar{x}_2 f_2 = \Sigma_2$$

Replacing Σ_1 and Σ_2 in (1) by $\bar{x}_1 f_1$ and $\bar{x}_2 f_2$, respectively, give us:

$$\bar{x} = \frac{\bar{x}_1 f_1 + \bar{x}_2 f_2}{n}$$

It does not matter, except in the notation and amount of writing, whether we have 2 or 3, or more generally, *k* classes, with $f_1, f_2, \ldots, f_k$ data, respectively. The analogous result giving the mean $\bar{x}$ of the *n* data in terms of the means $\bar{x}_1, \bar{x}_2, \ldots, \bar{x}_k$ of the data in the *k* classes holds:

$$\bar{x} = \frac{\bar{x}_1 f_1 + \bar{x}_2 f_2 + \cdots + \bar{x}_k f_k}{n} \qquad (2)$$

If we do not know the means $\bar{x}_1, \bar{x}_2, \ldots, \bar{x}_k$ of the data in the *k* classes, then it makes sense to seek good approximations for them, which is precisely what we will do in the next section.

5.2 ■ Measures of Location

The Arithmetic Mean: Frequency Distribution

Consider a set of *n* data which have been organized into a frequency distribution with *k* classes. The *k* classes contain $f_1, f_2, \ldots, f_k$ data, respectively, and have class marks $x_1, x_2, \ldots, x_k$. This framework is shown by the first three columns of Table 1.

Let us assume that the data in each class are, to a "close" approximation, **symmetrically distributed** about the class mark of the class (that is, for the data values above the class mark there are a comparable number, approximately the same distance, below the class mark). A special case of this of interest occurs when the data in a class are, to a close approximation, **uniformly** (or evenly) distributed throughout the class. Then the class mark of each class is a good approximation of the mean of the data in the class:

$$x_1 f_1 + x_2 f_2 + \cdots + x_k f_k \approx \bar{x}_1 f_1 + \bar{x}_2 f_2 + \cdots + \bar{x}_k f_k$$

$$\bar{x} = \frac{x_1 f_1 + x_2 f_2 + \cdots + x_k f_k}{n}$$
$$= \frac{\sum x \cdot f}{n},$$

is defined as the **mean** $\bar{x}$ **of the frequency distribution.**

It is a close approximation of the mean of the n data expressed, from (2), by:

$$\frac{\bar{x}_1 f_1 + \bar{x}_2 f_2 + \cdots + \bar{x}_k f_k}{n}$$

assuming that the data in each class are to a close approximation, symmetrically distributed about the class mark of the class or the data are, to a close approximation, uniformly distributed throughout the class. Otherwise we can't say for sure.

If the data constitute a population, then the notation $\bar{x}$ is replaced by μ. The organization shown in Table 1 makes the computation of $\bar{x}$ (or μ)

Table 1

Class	Class Mark x_i	Frequency f_i	Product of entries in columns 2 and 3 $x_i f_i$
class 1	x_1	f_1	$x_1 \cdot f_1$
class 2	x_2	f_2	$x_2 \cdot f_2$
	$\vdots$	$\vdots$	$\vdots$
class k	x_k	f_k	$x_k \cdot f_k$
		$n = \sum f$	$\sum x \cdot f$

straightforward. Multiply the corresponding entries in columns 2 and 3 to obtain column 4. Add the entries in column 4, divide by the total number of data n (sum of column 3), and we have the mean of the frequency distribution.

Potential Confusion Alert: In the context of grouped data (frequency distribution) x_1 ..., x_k (note subscript k) are the **class marks** of the k classes of the frequency distribution. In the context of ungrouped data x_1, ..., x_n or x_N (note subscript n or N, depending on the context of sample on population) are **data values.**

Example I The "Heavy" Basic Statistics Text

Find the mean of the "Heavy" *Basic Statistics* text distribution considered in Example 1 of Chapter 3 (page 85).

This distribution is reproduced in the first two columns of Table 2. Our first task is to obtain the class marks of the classes. The class mark of the first class is 0.45. By adding 1, the width of the distribution, to this and successive values we generate the class marks shown in column 3. Taking the product of corresponding values in the frequency and class mark columns gives us the key column 4, the sum of whose values is 714.2.

Table 2

Time Interval	Frequency f_i	Class Mark x_i	$x_i f_i$
0.0-0.9	2	0.45	0.90
1.0-1.9	7	1.45	10.15
2.0-2.9	13	2.45	31.85
3.0-3.9	31	3.45	106.95
4.0-4.9	42	4.45	186.90
5.0-5.9	32	5.45	174.40
6.0-6.9	16	6.45	103.20
7.0-7.9	10	7.45	74.50
8.0-8.9	3	8.45	25.35
	$N = 156$		714.20

$$\mu = \frac{714.2}{156}$$
$$= 4.6$$

Thus, on average, students were able to hold the book in a reading position in one sitting before their wrists gave out for 4.6 minutes.

Let us observe that the mean of this distribution equals the mean of the raw data obtained in Example 2, Section 4.3 (page 108).

EXERCISES

1. **MERT:** Refer to Bell City's Medical Emergency Response Team's frequency distribution introduced in Exercise 1 of Section 3.1 (page 91).

(a) Determine the mean of this distribution. Interpret the result obtained.

(b) If the mean of this distribution is to serve as a good approximation to the mean of the data obtained from the data, what condition must be satisfied?

(c) How does the mean of the distribution compare with the mean obtained from the data?

(d) Determine the mean of the frequency distribution based on the classes 4-8, 9-12, 13-16, 17-21.

2. Determine the mean of the grade point average (GPA) frequency distribution defined in Table 1 of Section 3.1 (page 80) and interpret the result obtained.

3. Refer to Cameron City's chlorine level frequency distribution introduced in Exercise 2 of Section 3.1 (page 92).

(a) Determine the mean and interpret the result obtained.

(b) Could we use the frequency distribution of part (g) of that exercise to obtain an estimate for the mean of the data? Explain.

4. Refer to Mary's Gifts order size frequency distribution introduced in Exercise 3 of Section 3.1 (page 93).

(a) Determine the mean of this distribution.

(b) What conditions would insure that the mean of the distribution is a good estimate of the mean of the data? Explain.

5. Refer to the final exam grade distribution introduced in Exercise 4 of Section 3.1 (page 93).

(a) Determine the mean of this distribution.

(b) Interpret the value obtained in answer to (a).

The Median

In defining the concept of median for grouped data in terms of a frequency distribution it is useful and insightful to use a specific setting as a vehicle. We thus return once more to the "Heavy" *Basic Statistics* holding time frequency distribution, which we reproduce in Table 4. The histogram of this distribution is reproduced in Figure 2 (see Section 3.2, Figure 2, page 96).

Table 4

Time Interval (class limits)	Class Boundaries	Frequency f
0.0-0.9	-0.05, 0.95	2
1.0-1.9	0.95, 1.95	7
2.0-2.9	1.95, 2.95	13
3.0-3.9	2.95, 3.95	31
4.0-4.9	3.95, 4.95	42
5.0-5.9	4.95, 5.95	32
6.0-6.9	5.95, 6.95	16
7.0-7.9	6.95, 7.95	10
8.0-8.9	7.95, 8.95	3
		156

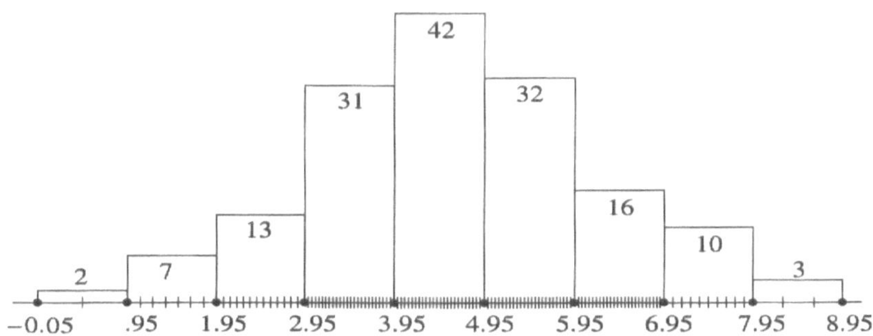

Figure 2

> The **median M of a frequency distribution** is the number which breaks the histogram of the distribution into two equal pieces in terms of area.

Median M of the "Heavy" Basic Statistics Text Distribution

Our starting point for determining M for the *Basic Statistics* text distribution is to treat the data of each class as evenly laid out throughout the class with the largest value of the class coinciding with the upper class boundary of the class. For the first class, for example, we view the two data as dividing the base into two segments of equal length; for the second class we view the seven data as dividing the base into seven segments of equal length. From this point of view the 2nd data value is 0.95, the 9th is 1.95, the 22nd is 2.95, the 53rd is 3.95, the 95th is 4.95, and so on; the 156th is 8.95. The median, indicated by the $156/2 = 78\text{th}$ value, is between 3.95, the 53rd value, and 4.95, the 95th value (see Figure 3).

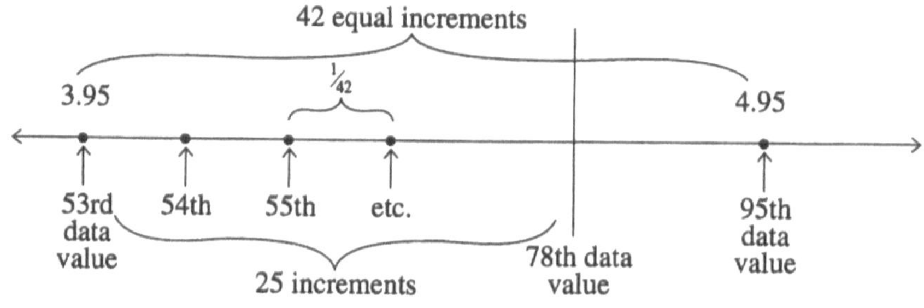

Figure 3

To reach the 78th value we must add $78 - 53 = 25$ of the 42 increments to 3.95. Each increment has length 1/42. We need 25 of them, which means we must add 25/42 to 3.95 to obtain the 78th or median value of the distribution.

$$\begin{aligned} M &= 3.95 + 25\left(\frac{1}{42}\right) \\ &= 4.55 \end{aligned}$$

Let us observe that the median of this distribution equals the median of the raw data obtained in Example 7, Section 4.3 (page 112).

To obtain a procedure which can be used to obtain the median value in terms of any frequency distribution we restate the preceding analysis in the form of a sequence of steps.

Steps to Obtain the Median of a Frequency Distribution

1. Determine as target value *n*/2, where *n* is the total number of data. Our objective is to find the number that corresponds to this target value. It is defined as the median of the frequency distribution. For the *Basic Statistics* text distribution $n = 156$.

2. *n*/2 determines the class within which the median falls, called the **class of the median.** The median *M* equals the lower class boundary *L* of the class of the median plus an increment. The class of the median for the *Basic Statistics* text distribution has lower class boundary 3.95 and frequency 42.

3. The number of unit increments *j* that we must move into the class of the median is given by the target value *n*/2 minus the total frequency value expressed by *L*. For the *Basic Statistics* text distribution $j = 78 - 53 = 25$.

4. The size of each unit increment is equal to the width of the class of the median *c* divided by the frequency f_M of the class of the median (i.e., the number of pieces the class of the median is to be broken into), c / f_M. For the *Basic Statistics* text distribution $c = 1$, $f_M = 42$ and $c / f_M = 1/42$.

5. The increment to be added to *L* is the product of the number of unit increments required, *j*, with the length of each, c / f_M, yielding $j \cdot (c / f_M)$. For the *Basic Statistics* text distribution we have 25(1/42).

In summary, the median *M* is defined by:

$$M = L + j \cdot \frac{c}{f_M}$$

Example 2 Return to the GPA Frequency Distribution

Determine the median of the grade point average (GPA) frequency distribution introduced in Table 1 of Section 3.1 and reproduced in Table 5 with class boundaries. Interpret the value obtained.

Table 5

	GPA Class	Class Boundaries	Frequency	
	1.00-1.49	0.995, 1.495	80	sum = 1580 (80 + 200 + 1300)
	1.50-1.99	1.495, 1.995	200	
	2.00-2.49	1.995, 2.495	1300	
Class of M	2.50-2.99	2.495, 2.995	2100	
	3.00-3.49	2.995, 3.495	900	
	3.50-3.99	3.495, 3.995	400	
	4.00-4.49	3.995, 4.495	20	
			5000	

1. $n = 5000$; $\frac{n}{2} = 2500$, which is our target value.

2. We can come as close to 2500 as $80 + 200 + 1300 = 1580$. Thus, the class of M is the next class, 2.50-2.99. The class of M has lower class boundary $L = 2.495$, which corresponds to the 1580th value.

3. Our target value is the 2500th value, so that $j = 2500 - 1580 = 920$ is the number of unit increments needed.

4. The width c of the class of M is $2.995 - 2.495 = 0.5$. The frequency of the class of M is $f_M = 2100$. Thus, the length of each unit increment is $\frac{c}{f_M} = \frac{0.5}{2100}$.

5. The increment to be added to $L = 2.495$ is given by:

$$j \cdot \frac{c}{f_M} = 920\left(\frac{0.5}{2100}\right)$$

6. Thus:

$$\begin{aligned} M &= 2.495 + 920\left(\frac{0.5}{2100}\right) \\ &= 2.495 + 0.219 \\ &= 2.71 \end{aligned}$$

The median M is 2.71 in terms of the given frequency distribution. In defining *M* we have worked under the assumption that the data in each class are uniformly distributed in the class.

If the GPA data realistically reflect this assumption, then 2.71 will closely approximate the median of the data and approximately 2500 of them are less than 2.71 and the remaining 2500 exceed 2.71.

Observations

Suppose that the *Basic Statistics* text situation gave rise to an odd number of data, say, 157 values. Then the target value $\frac{n}{2}$ is 78.5, which may strike some as strange. We are not seeking the 78.5th value, but rather the half-way point between the 78th and 79th values. To obtain the median *M* we follow the steps as presented.

Quartiles, Deciles, and Percentiles

Other measures of location of interest are quartiles, deciles and percentiles. There are three quartiles $Q_1, Q_2 = M$, and Q_3, values such that 25% of the data are less than Q_1, 50% of the data are less than $Q_2 = M$, and 75% of the data are less than Q_3. Nine deciles $D_1, D_2, \ldots, D_9$ break the data into tenths, with 10% of the data less than D_1, 20% of the data less than D_2, and so on. Ninety-nine percentiles $P_1, P_2, \ldots, P_{99}$ break the data into hundredths, with 1% of the data less than P_1, 2% of the data less than $P_2, \ldots, 99\%$ of the data less than P_{99}. To say that your exam grade, GPA, salary or what have you is in the 99th percentile, for example, means that your value is in the top 1% of the values; 99% are below.

Quartiles, deciles, and percentiles of a frequency distribution are defined and determined in a similar way as that for the median. The difference is in the way the target value is defined. For Q_1 the target value is 0.25*n*, for D_4 it is 0.4*n*, and for P_{95} it is 0.95*n*.

Example 3 Second Return to the GPA Frequency Distribution

Find Q_1, D_4, and ${}_{95}$ for the GPA distribution, reproduced in Table 6 with class boundaries.

Table 6

	GPA Class	Class Boundaries	Frequency
	1.00-1.49	0.995, 1.495	80
	1.50-1.99	1.495, 1.995	200
Class of Q_1	2.00-2.49	1.995, 2.495	1300
Class of D_4	2.50-2.99	2.495, 2.995	2100
	3.00-3.49	2.995, 3.495	900
Class of P_{95}	3.50-3.99	3.495, 3.995	400
	4.00-4.49	3.995, 4.495	20
			5000

For Q_1, $0.25n = 1250$. The class of Q_1 is 2.00-2.49, with $L = 1.995$. $j = 1250 - 280 = 970$, $c = 0.5$, $f_{Q_1} = 1300$. Thus

$$Q_1 = L + j \cdot \frac{c}{f_{Q_1}} = 1.995 + 970\left(\frac{0.5}{1300}\right) = 2.37 .$$

For D_4, $0.4n = 2000$. The class of D_4 is 2.50-2.99, with $L = 2.495$. $j = 2000 - 1580 = 420$, $c = 0.5$, $f_{D_4} = 2100$. Thus:

$$D_4 = L + j \cdot \frac{c}{f_{D_4}} = 2.495 + 420\left(\frac{0.5}{2100}\right) = 2.60 .$$

For P_{95}, $0.95n = 4750$. The class of P_{95} is 3.50-3.99, with $L = 3.495$. $j = 4750 - 4580 = 170$, $c = 0.5$, $f_{P_{95}} = 400$. Thus:

$$P_{95} = L + j \cdot \frac{c}{f_{P_{95}}} = 3.495 + 170\left(\frac{0.5}{400}\right) = 3.71 .$$

If the GPA's are, to a close approximation, uniformly distributed in each class, then approximately 25% of them are less than 2.37, 40% are less than 2.60, and 95% are less than 3.71.

The Modal Class

It is sometimes useful to make note of the class with the largest number of data, called the modal class of the frequency distribution. The modal class of the GPA distribution is 2.50-2.99, with 2100 data.

EXERCISES

6. **MERT:** Refer again to Bell City's Medical Emergency Response Team's frequency distribution introduced in Exercise 1 of Section 3.1 (page 91).

   (a) Determine (i) Q_1, (ii) D_7, (iii) M.

   (b) How are the values obtained in answer to (a) to be interpreted?

   (c) What assumptions underlie the calculations of (a) and interpretations of (b)?

7. Refer again to Mary's Gifts order size frequency distribution introduced in Exercise 3 of Section 3.1 (page 93).

   (a) Determine (i) M, (ii) Q_3, (iii) P_{65}.

   (b) How are the values obtained in answer to (a) to be interpreted?

   (c) What assumptions underlie the calculations of (a) and interpretations of (b)?

8. Refer again to the final exam grade frequency distribution introduced in Exercise 4 of Section 3.1 (page 93).

   (a) Determine (i) Q_1, (ii) M, (iii) D_8.

   (b) How are the values obtained in answer to (a) to be interpreted?

   (c) What assumptions underlie the calculations of (a) and interpretation in answer to (b)?

(d) A student was informed that his grade was in the 95th percentile. What does this mean?

(e) How does the median compare with the mean (see Exercise 5)? What does this tell us about the grade distribution?

9. Do the definitions of median, quartiles, deciles, percentiles presuppose that all classes in the frequency distribution have the same class width? Explain.

5.3 ■ Measures of Variation

Variance and Standard Deviation

As we have seen in working with ungrouped data, the mean of a set of data tells a partial story. It is somewhat like one key to a safe deposit box; without the second key, we cannot gain entry to the box. For ungrouped data the most successful second key is to be found in the concepts of variance and standard deviation. This is also the case for grouped data. To define comparable concepts of variance and standard deviation for grouped data we return to the setting of a set of n data which have been organized into a frequency distribution with k classes.

The k classes contain $f_1, f_2, \ldots, f_k$ data, respectively, and class marks $x_1, x_2, \ldots, x_k$. This framework, originally summarized in Table 1, is reproduced here in Table 7.

Table 7

Class	Frequency f	Class Mark x	$x \cdot f$	$(x-\bar{x})^2 f$
class 1	f_1	x_1	$x_1 \cdot f_1$	$(x_1-\bar{x})^2 f_1$
class 2	f_2	x_2	$x_2 \cdot f_2$	$(x_2-\bar{x})^2 f_2$
	$\vdots$	$\vdots$	$\vdots$	$\vdots$
class k	f_k	x_k	$x_k \cdot f_k$	$(x_k-\bar{x})^2 f_k$
	$n=\Sigma f$		$\Sigma x \cdot f$	$\Sigma(x-\bar{x})^2 \cdot f$

$$\bar{x} = \frac{\Sigma x \cdot f}{n}$$

To define sample variance s^2 we proceed in a way analogous to that used to define s^2 for ungrouped data, with appropriate modifications. We do not have the data, so that for each class we take its class mark as a representative value, subtract the mean $\overline{x}$, and square the result, yielding $(x_1-\overline{x})^2$, $(x_2-\overline{x})^2,\ldots,(x_k-\overline{x})^2$ for the *k* classes. It is reasonable to weight these values by the frequencies $f_1, f_2,\ldots, f_k$ respectively, since it does not make sense for a class with few data to carry the same weight as one with many data.

In summary, we emerge with the entries in column 5 of Table 7.

The **sample variance** s^2 **of the frequency distribution** is defined as the sum of the entries of this column divided by $n-1$.

$$s^2 = \frac{(x_1-\overline{x})^2 f_1 + (x_2-\overline{x})^2 f_2 + \cdots + (x_k-\overline{x})^2 f_k}{n-1} \qquad (3)$$

$$= \frac{\Sigma(x-\overline{x})^2 f}{n-1}$$

The **sample standard deviation** *s* **of the frequency distribution** is, as before, defined as the square root of the variance.

If the data are viewed as a population, replace s^2 and *s* by σ^2 and , $\overline{x}$ by μ, *n* by *N* in denoting the number of data, and $n-1$ in the denominator of the expression defining variance by *N*.

Derived Theorem for Variance and Standard Deviation of a Frequency Distribution

While (3) is our starting point in that it defines variance, we are led to seek a result of the sort obtained for ungrouped data which would simplify the computation of variance and standard deviation.

By employing algebra mixed with a healthy dose of patience we are able to derive the following form for **sample variance:**

$$s^2 = \frac{n(x_1^2 f_1 + \cdots + x_k^2 f_k) - (x_1 f_1 + \cdots + x_k f_k)^2}{n(n-1)}$$

$$= \frac{n(\Sigma(x^2 f) - (\Sigma xf)^2}{n(n-1)}$$

The standard deviation is obtained by taking the square root of the variance.

If the data are viewed as a **population**, s^2 and *s* are replaced by σ^2 and σ, *n* by *N* in denoting the number of data, and $n(n-1)$ in the denominator of the derived form for sample variance by N^2.

To establish a simple computation routine for variance and standard deviation we introduce an additional column into Table 7 by multiplying each $x \cdot f$ entry in column 4 by the class mark *x*, yielding column 5 in Table 8: $x_1^2 f_1, x_2^2 f_2, \ldots, x_k^2 f_k$.

Table 8

Class	Frequency f	Class Mark x	$x \cdot f$	$x^2 f$
class 1	f_1	x_1	$x_1 \cdot f_1$	$x_1^2 f_1$
class 2	f_2	x_2	$x_2 \cdot f_2$	$x_2^2 f_2$
	$\vdots$	$\vdots$	$\vdots$	$\vdots$
class *k*	f_k	x_k	$x_k \cdot f_k$	$x_k^2 f_k$
			$\Sigma x \cdot f$	$\Sigma x^2 f$

The derived formula for sample variance then takes the following form:

$$s^2 = \frac{n(\text{sum of column 6}) - (\text{sum of column 4})^2}{n(n-1)}$$

Example 1 The "Heavy" Basic Statistics Text

Find the variance and standard deviation of the "Heavy" *Basic Statistics* text distribution considered in Example 1 of Chapter 3 (page 85).

This distribution is stated by the first two columns of Table 9. Also reproduced in columns 3 and 4 from Table 2 of Example 1 (page 127) of the previous section are the class marks and corresponding products *xf*. These were needed to determine the mean of the distribution and are needed here as well.

Table 9

Time Interval	Frequency f	Class Mark x	$x \cdot f$	$x^2 f$
0.0-0.9	2	0.45	0.90	0.41
1.0-1.9	7	1.45	10.15	14.72
2.0-2.9	13	2.45	31.85	78.03
3.0-3.9	31	3.45	106.95	368.98
4.0-4.9	42	4.45	186.90	831.71
5.0-5.9	32	5.45	174.40	950.48
6.0-6.9	16	6.45	103.20	665.64
7.0-7.9	10	7.45	74.50	555.03
8.0-8.9	3	8.45	25.35	214.21
	$N = 156$		714.20	3679.21

Our next step is to generate $x^2 f$ values by multiplying corresponding entries in columns 3 and 4. This yields column 5, the sum of whose terms is 3679.21. Thus:

$$\begin{aligned}\sigma^2 &= \frac{N(\Sigma(x^2 f) - (\Sigma xf)^2}{N^2} \\ &= \frac{156(3679.21) - (714.20)^2}{(156)^2} \\ &= 2.62 \\ \sigma &= \sqrt{2.62} \\ &\approx 1.6\end{aligned}$$

The more closely satisfied is the underlying assumption that the data in each class are symmetrically distributed about the class mark of the class, the more confident we can be that $\sigma^2 = 2.62$ and $\sigma = 1.6$ approximate the variance and standard deviation obtained from the raw data.

Let us observe that $\sigma^2 = 2.62$ and $\sigma = 1.6$ for the frequency distribution are in good agreement with $\sigma^2 = 2.45$ and $\sigma = 1.6$ obtained for the raw data in Example 3, Section 4.4 (page 122).

EXERCISES

For the frequency distributions noted in Exercises 1-5 determine the (a) variance, (b) standard deviation.

1. **MERT:** Refer to Bell City's Medical Emergency Response Team's frequency distribution introduced in Exercise 1 of Section 3.1 (page 91).

   (c) What assumption about the data underlies your answer to (a) and (b)?

   (d) If the standard deviation of this distribution is to serve as a good approximation to the standard deviation of the data obtained from the population, what condition must be satisfied?

   (e) How does the standard deviation of the distribution compare with the standard deviation obtained from the data?

   (f) Determine the variance and standard deviation based on classes 4-8, 9-12, 13-16, 17-21.

2. Refer to the grade point average (GPA) frequency distribution defined in Table 1 of Section 3.1 (page 80).

   (c) What assumption about the data underlies your answer to (a) and (b)? Explain.

3. Refer to Cameron City's chlorine level frequency distribution introduced in Exercise 2 of Section 3.1 (page 92).

4. Refer to Mary's Gifts order size frequency distribution introduced in Exercise 3 of Section 3.1 (page 93).

   (c) What conditions would insure that the standard deviation of the distribution is a good estimate of the standard deviation of the data?

5. Refer to the final exam grade distribution introduced in Exercise 6 of Section 3.1 (page 95).

(c) What assumption about the data underlies your answer to (a) and (b)?

SELF-TESTS FOR PART ONE

Allow 90 or so minutes for each self-test. Go over the first before going on to the second.

Self-Test 1

1. The number of personal computers sold by 25 salespersons of the San Francisco branch of Norden Computer Products last month is given by Table 1.

Table 1

7	36	25	47	3
10	32	24	38	15
21	42	22	27	20
28	18	29	4	39
30	42	45	14	32

NOTE:
$\sum x_{i} = 650$
$\sum x_{i}^{2} = 20,770$

(a) Determine the mean.

(b) Determine the variance and standard deviation.

(c) What point of view underlies your calculation of the variance and standard deviation? Explain.

(d) Construct a stem-and-leaf plot of the data.

(e) Determine the median.

(f) What does the median value tell us?

(g) Find the mode

(h) Find the range.

2. Refer to the data given in Table 1.

(a) Construct the frequency distribution based on the classes 0-9, 10-19, 20-29, 30-39, 40-49.

(b) Find the class limits, class marks, and the class width of the classes of the aforenoted frequency distribution.

(c) Find the mean in terms of this distribution.

(d) What assumption underlies your answer to (c)?

(e) Find the variance and standard deviation in terms of this distribution.

(f) What assumption underlies your answer to (e)?

(g) Determine the median.

(h) Determine P_{60}.

(i) What assumption underlies your answers to (g) and (h)?

(j) What does the value obtained in answer (h) tell us?

(k) Construct the histogram for the aforenoted distribution.

(l) Construct the cumulative less than and cumulative less than percentage distributions.

(m) Construct the ogive for the cumulative less than percentage distribution.

(n) In terms of the ogive, what percentage of sales were for less than 25 PC's?

3. Classify the following examples of data as nominal, ordinal, interval, or ratio. Explain the reason for your classification in each case.

(a) Political affiliation

(b) Grade in Psychology 101

(c) IQ score

(d) Distance a student must travel from his residence to reach Ecap University

4. Continuing their discussion of what statistics show (Case 16, sec. 1.6, ch. 1), Talia brought up the statistics concerning the Universe Games,. "Consider the Universe Games held every hundred yeas. For the last two thousand years 75% of the participants chosen by the trials have been Morkians. Does this mean that the trials were biased in favor of the Morkians?" "Certainly not," answered Lork, "they earned the right to be there in the trials that were held."

(a) Do the statistics mean that the trials were biased in favor of Morkians? Explain.

(b) Are there other "explanations" that might account for the statistics? Explain.

Self-Test 2

1. Abel Fisher, a financial analyst, was hired to analyze the operations of the accounting department of Arley College and make recommendations on how to improve its financial efficiency. The income of the department is the tuition income of the students being serviced minus costs, primarily salary costs. Mr. Fisher collected data on the class size of each instructor and each instructor's rank and salary. He found that a number of the full professors at the top of the salary scale were teaching classes with a small number of students.

   To improve the income of the department he recommended that teachers at the top of the salary scale be assigned classes which can be expected to have a large number of students.

(a) Is the data collected appropriate to the problem under study? Explain.

(b) Do you agree that implementation of Abel Fisher's recommendation will improve the accounting department's efficiency from an income-cost point of view? Explain.

2. It was recently reported in the local paper that a statistician had drowned in Lake Walton. Lake Walton has an average depth of one foot. How could a person drown in such a shallow lake?

3. Leo Jansen invested $20 at 10% interest and $30 at 30% interest. What is the average rate of return?

4. Classify the following examples of data as nominal, ordinal, interval, or ratio. Explain the reason for your classification in each case.

   (a) Social security number,

   (b) Annual income,

   (c) Zip code,

   (d) Class ranking at graduation,

   (e) Graduate Management Admissions Test (GMAT),

   (f) Student ranking of course instructors (Excellent, Superior, Average, etc.),

   (g) Length of a steel rod,

   (h) Classroom temperature.

5. Andy Arunas of *The Huxley College Press* has been asked to conduct a survey of Huxley's students (estimated at 6000) on education matters concerned with the quality of the education they are receiving at Huxley.

   (a) One problem centers on the question of how the sample should be drawn. Discuss the merits and drawbacks of random sampling, systematic sampling with a random start, stratified sampling, cluster sampling, and quota sampling for this problem.

(b) A second problem concerns the questionnaire. Advise Andy on the questions that should be posed in the questionnaire.

6. One thousand senior high school students in Ralph City took a college level math course, with 80% of them passing the statewide standard exam. The following year 3000 senior high school students in Ralph City took the course, with 50% of them passing the statewide standard exam. Do these figures gives us cause for pessimism or optimism concerning Ralph City's education system?

7. "Get ready to get dirty. Cosmo surveyed thousands of guys in an online poll to find out exactly what they want from you in bed," stated the opening statement of "Cosmo's Sexiest Survey Ever," Cara Birnbaum, *Cosmopolitan*, March 2001, 192-199.

   (a) Could the results of this online poll be considered representative of the attitudes of men in general? Explain.

   (b) If your answer to (a) is no, whose views do the respondent represent? Explain.

Self-Test 3

1. The following data represent the number of homes sold by 16 real estate agents at the Harmon Company in a yearly period.

15	36	25	9
10	32	24	38
21	30	22	28
28	18	28	4

NOTE:
$\sum x = 368$
$\sum x^2 = 9888$

   (a) Find the mean.

   (b) Find the standard deviation.

   (c) What assumption underlies your calculation of the standard deviation? Explain.

(d) Construct a stem-and-leaf plot.

(e) Determine the median.

2. Construct a frequency distribution for the preceding data with at least four classes. The following questions pertain to *this frequency distribution.*

(a) For the first class (*only*) state the (i) class limits, (ii) class boundaries, (iii) class mark, (iv) class width.

(b) Determine the mean.

(c) Under what conditions is the mean of the distribution a good approximation of the mean of the data? Explain.

(d) Find the standard deviation.

(e) On what assumption is your standard deviation calculation based? Explain.

(f) Determine the median.

3. Does increasing the sample size by itself ensure greater reliability of a pre-presidential poll to be taken? Explain.

4. "I don't know what you're complaining about," said the baseball owners' representative to the players' counterpart. "The players make, on average, $1.2 million." Is the average salary the best indicator of player salaries? Explain.

5. A panel of five members of the Slippery Statistics Society is to rank the achievements of nominees recommended for the Slippery Statistics Achievement Award. The ranking values available are 100 (outstanding), 80 (superb), 70 (very slippery), 50 (slippery). Determination of the Award's recipient has come down to two candidates, Joseph S. (who received rankings 50, 70, 70, 80, 100) and Vera S. (who received rankings 50, 70, 80, 80, 80).

(a) Should Joseph S. receive the SSA Award because the mean of his rankings, 74, exceeds that of Vera S., 72? Explain.

(b) If your answer to (a) is no, then on what basis would you determine the recipient of the SSA Award? Explain.

6. The Baldwin Insurance Company hired marketing analyst Arnold W. Williamson to develop a strategy to make its car insurance policies more attractive. Mr. Williamson suggested that to reward and encourage safe driving the company offer a 5% discount to policy holders who had been with the company for five years and had not been in an accident. An additional 1% discount would be given for each additional year that had been accident free up to a maximum of 15%.

   Do you agree that the number of years of safe driving is the number to focus on as a measure of safe driving?

7. The City Council of Southchester wants to obtain a sense of the public's view of the city's school system. A random sample of 1000 of Southchester's residents was sent a questionnaire. Of the 100 responses that were received, 32 gave the school system a favorable rating and 68 gave it an unfavorable rating.

   Dismayed by this overwhelming show of lack of confidence in Southchester's school system, the City Council voted to dismiss the head of the board of education. Does the reliability of the 68% unfavorable rating support this action being taken? Explain.

8. "I don't understand what went wrong," lamented an editor of *The Literary Digest*. "We were so on-target with our 1932 poll and so off-target with our 1936 poll, for which we used the same methods." What explanation would you give him?

PART TWO

PROBABILITY BACKGROUND FOR STATISTICAL INFERENCE

- Uncertainty and Probability
- Random Variables
- The Remarkable Normal Curves
- Sampling Distributions

6

Uncertainty and Probability

6.1 ■ Preface to Probability

Probability theory is a mathematical discipline with two major dimensions. One dimension is concerned with its direct application to such diverse disciplines as physics, engineering, biology, psychology, economics, accounting, and management. The other dimension is in the fundamental role that probability plays as a foundation for statistical inference. The probability that we develop in Part Two is intended for this dimension.

By a **random process** we mean a process which gives rise to an outcome, but which outcome cannot be predicted with any certainty in advance. The processes of tossing a coin, tossing a die, and choosing a sample from a shipment of items are simple examples of random processes. Although the outcome of any repetition or occurrence of a random process cannot be predicted with any certainty in advance, if we focus on an event of interest over a long series of repetitions of the random process we find stability in the relative frequency with which the event occurs. This stability can serve as a basis for understanding and predicting the behavior of the random process.

To develop this point of view we introduce the following definition.

> Let A denote an event connected with a random process. The **relative frequency of A**, denoted by $R(A)$, is the ratio
>
> $$R(A) = \frac{\text{number of times } A \text{ occurs}}{\text{number of times the process is repeated}}.$$

Thus, if a coin is tossed 5 times and head shows on 2 of the 5 tosses, the relative frequency of heads for this series of 5 tosses is 2/5 or 0.40. We say that head showed 40 percent of the time in the 5 tosses.

Over the short run, that is, for a small number of repetitions of a random process, the behavior of $R(A)$ for a single toss is greatly influenced by a unit increase in its denominator while the numerator remains the same or increases by one. As a result $R(A)$ is unstable and fluctuates considerably. In Figure 4 the relative frequency of the event head shows is plotted (vertical axis) against the number of tosses of a certain nickel for the first 15 tosses (horizontal axis). The relative frequency of this event is rather erratic and varies from 0 to 0.533. In Figure 5 the relative frequency of head shows is shown for the last 15 of a series of 200 tosses of the nickel. The relative frequency of this event is rather stable at this point, varying from 0.536 to 0.546, since the denominator of the relative frequency ratio is large and the change brought by an additional toss of the nickel is small.

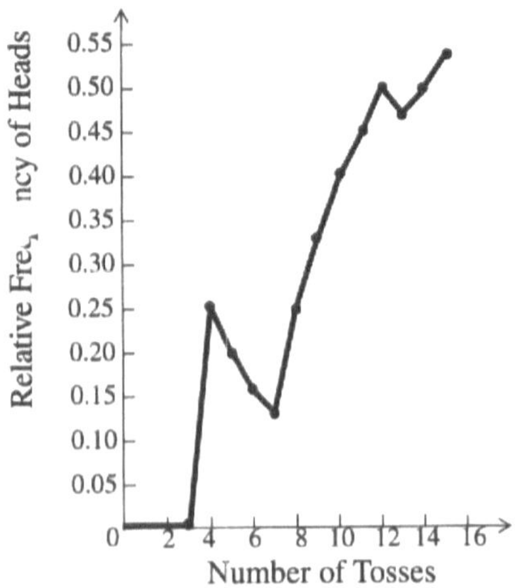

Figure 4

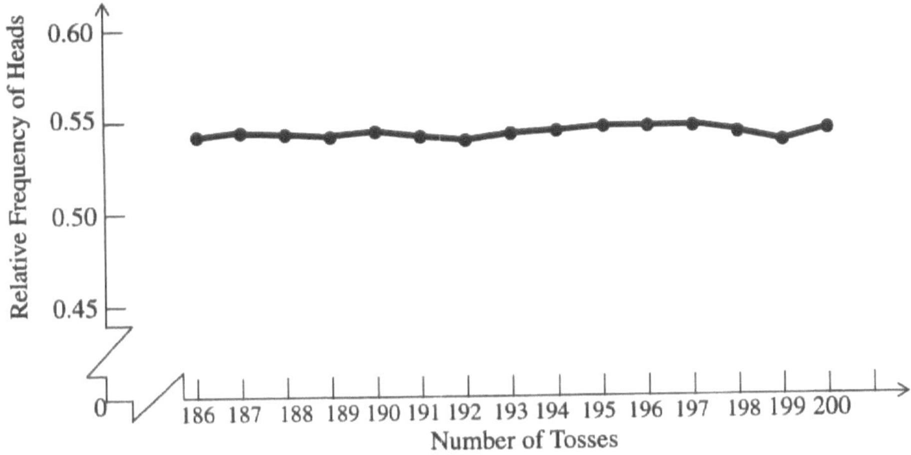

Figure 5

The desire to develop a mathematical structure that would help us understand and predict the long term relative frequency behavior of events connected with random processes played a leading role in the development of probability as this mathematical instrument.

Sample Space for a Random Process

Since our study focuses on the behavior of events, it is desirable to express the nature of the many events that arise in the study of a random process in terms of a set of comparatively few events which is adopted as a foundation. This leads us to the concept of sample space for a random process. To develop this concept consider the process of tossing a die. Some of the events connected with this process are 1 shows, 2 shows, an even number shows, a number between 2 and 5 shows, a number less than 4 shows. To describe events of interest concerning the number which shows when a die is tossed, consider the set of events

$$S = \{1 \text{ shows, } 2 \text{ shows, } 3 \text{ shows, } 4 \text{ shows, } 5 \text{ shows, } 6 \text{ shows}\},$$

which we further abbreviate by writing

$$S = \{1, 2, 3, 4, 5, 6\}.$$

This set of events has the property that whenever a die is tossed a unique event in *S* is determined; *S* is complete in the sense that some event in *S* occurs when a die is tossed, and unambigious in the sense that only one event in *S* occurs when a die is tossed. Other events involving the character of the number showing when the die is tossed can be described in terms of subsets of *S*. For example, an even number shows is described by {2, 4, 6}; a number between 2 and 5 shows is described by {3, 4}; a number greater than 4 but less than 3 shows is described by the humble, but important, empty set ∅.

$S = \{1, 2, 3, 4, 5, 6\}$ is a **sample space**, for the die tossing process because of its fundamental property that whenever a die is tossed, one and only one of the events in *S* occurs. *S* is further said to be a **finite sample space** because the number of events in it, 6, is a positive integer.

More generally, a collection of events

$$S = \{s_1, s_2, \ldots, s_n\},$$

is said to be a **finite sample space** for a random process if whenever the process is repeated, one and only one of the events in *S* occurs. The *n* events $s_1, s_2, \ldots, s_n$ which make up *S* are called **sample points**. The subsets of *S* describe events which can be analyzed in terms of *S*.

In general there are many possible sample spaces which can be given for a random process since all we need for a sample space is a collection of events with the fundamental defining property cited. However, not all sample spaces that may be given for a process are useful in helping us to further understand the process. $S_1 = \{O, E\}$, where O is the event an odd number shows and *E* is the event an even number shows, is also a sample space for the die tossing process since exactly one of these events occur when a die is tossed; S_1 is not a very useful sample space because there is very little which can be described in terms of it.

Example 1 Two Samples Spaces

Specify two sample spaces for the tossing of a pair of dice.

To easily distinguish the dice, we shall assume that one of them is red and the other is green. One **sample space**, which we shall call S_1, is the set of events listed in Table 1. The first number in each ordered pair specifies the number that shows on the red die and the second number specifies the number that shows on the green die. (4,2), for example, is the event that red shows 4 and green shows 2; (2,4) is the event that red shows 2 and green shows 4. S_1 is a sample space because it has the property that whenever the dice are tossed exactly one of the events in S_1 occurs.

Table 1

$$S_1 = \begin{bmatrix} (1,1) & (1,2) & (1,3) & (1,4) & (1,5) & (1,6) \\ (2,1) & (2,2) & (2,3) & (2,4) & (2,5) & (2,6) \\ (3,1) & (3,2) & (3,3) & (3,4) & (3,5) & (3,6) \\ (4,1) & (4,2) & (4,3) & (4,4) & (4,5) & (4,6) \\ (5,1) & (5,2) & (5,3) & (5,4) & (5,5) & (5,6) \\ (6,1) & (6,2) & (6,3) & (6,4) & (6,5) & (6,6) \end{bmatrix}$$

Another **sample space** S_2 is

$$S_2 = \{2,3,4,5,6,7,8,9,10,11,12\}$$

where 2 denotes the event that the sum of the numbers showing is 2, 3 denotes the event that the sum of the numbers showing is 3, and so on. S_2 is also a sample space because it too has the property that whenever the dice are tossed exactly one of the events in S_2 occurs.

Properties of the Relative Frequency Function $R(A)$

In developing a mathematical structure, called probability model, that would help us understand and predict the behavior of random processes it would make sense to look at properties of $R(A)$ to obtain a sense of direction. Let us recall that for an event A connected with a random process,

$$R(A) = \frac{\text{number of times } A \text{ occurs}}{\text{number of times the process is repeated}}.$$

Properties

1. $R(A) \geq 0$. Both numerator and denominator of $R(A)$ are non-negative.

2. $R(A) \leq 1$. The numerator of $R(A)$ cannot exceed its denominator.

3. $R(A) = 0$ if A is an event whose occurrence is not possible (which happens when A is defined by incompatible conditions). Such an event A is identified with $\varnothing$.

4. $R(S) = 1$, where $S = \{s_1, s_2, \ldots, s_n\}$ is a sample space for the random process. Whenever the process is repeated one of the sample points in S occurs, and thus S occurs. Therefore, the numerator and denominator of $R(S)$ are equal.

   In connection with die tossing consider the event an even number shows. Let us observe that since an even number shows as often as 2, 4, and 6 show, the relative frequency of occurrence of this event is the sum of the relative frequencies of the events 2 shows, 4 shows and 6 shows. If, for example, 2, 4, and 6 are observed to show with relative frequencies 19/100, 27/100, and 14/100, respectively, then the relative frequency of an even number's occurrence is $19/100 + 27/100 + 14/100 = 60/100 = 60$ percent. Such is the case in general, and we state this observation as follows:

5. **$R(A)$ = sum of the relative frequencies of the sample points that describe A.**

6. $R(S) = R(s_1) + R(s_2) + \cdots + R(s_n)$, where $S = \{s_1, s_2, \ldots, s_n\}$ is a sample space for the process. This property is obtained by applying property 5 to S.

7. $R(s_1) + R(s_2) + \cdots + R(s_n) = 1$; the sum of the relative frequencies of all sample points is 1. This follows from properties 4 and 6.

EXERCISES

1. Consider the process of tossing a pair of dice. Let *A* denote the event that the sum of the numbers showing is less than 5, *B* denote the event that the sum of the numbers showing equals 5, and *C* denote the event that the sum of the numbers showing is greater than 5. Is $S=\{A,B,C\}$ a sample space for this process? Explain.

2. Set up two sample spaces for the process of tossing a coin twice in succession.

3. Set up three sample spaces for the process of dealing a card from a standard deck of 52 cards.

6.2 ▯ Finite Probability Model

Since the concept of probability model is intended to be applicable to the study of the long term relative frequency behavior of events connected with random processes, the manner in which we define this concept is guided by properties of relative frequency.

A **finite probability model** for a random process consists of two components:

i) A sample space $S=\{s_1,s_2,\ldots,s_n\}$.
ii) A function *P*, called a **probability function**, which assigns to each subset *A* of *S* a value, denoted by *P*(*A*) and called the **probability of** *A*, subject to the conditions listed below. We state these conditions on the sample points and then extend them to subsets of *S*.

1. $P(s_1)\ \geq 0,\ P(s_2)\ \geq 0,\ldots,\ P(s_n)\ \geq 0$

2. $P(s_1)\ \leq 1,\ P(s_2)\ \leq 1,\ldots,\ P(s_n)\ \leq 1$

1. and 2. require that probability values be between 0 and 1, inclusive.

3. $P(s_1)+P(s_2)+\ldots+P(s_n)=1$

That is, however probability values are assigned, the sum of the probabilities assigned to all sample points must be 1.

4. If A is a subset of S, $P(A)$ = sum of the probabilities of the sample points describing A; if $A=\varnothing$, $P(A)=0$. (For example, if $A=\{s_1,s_5\}$, $P(A)=P(s_1)+P(s_5)$.)

The structural requirements of a probability model for a random process are somewhat analogous to a community's building code requirements for building a house. A building code does not tell us how to build a house. It tells us that however we build our house, for it to be legitimate in terms of the building code it must satisfy such and such conditions which are spelled out in the code. The concept of probability model does not tell us how to define a sample space and probability function for a random process; it tells us that in building a specific probability model for a random process, which can be done in many ways, we must satisfy the conditions stated in the definition of probability model (the building code in this case) in order for the model to be mathematically legitimate.

To take an example, Janet James was presented with the following structure which was claimed to be a probability model for the process of tossing a die.

$$S=\{O,E\},\quad P(O)=1/3,\quad P(E)=2/3,$$

where O is the event that an odd number shows and E is the event that an even number shows on a throw of the die. Is this structure a probability model?

S is a sample space for the die tossing process since it satisfies the requirement that exactly one of its events occur when the die is tossed. The assignment by P of the numerical values to O and E satisfies the requirements of a probability function since 1/3 and 2/3 are non-negative, less than 1, and sum to 1. This establishes the mathematical legitimacy of S, P as a probability model for the die tossing process; whether or not this model realistically describes the behavior of any die in Janet's possession is another question entirely. We turn to the issue of formulating a specific probability model for a specific random process in the next two sections.

At first sight the concept of probability model seems no different from the properties of relative frequency that were listed, except for the notation used. This is not the case. To be sure, the conditions required of a probability function closely mirror properties of relative frequency, but relative frequency is defined in a very specific way while probabilities can be assigned to events in a wide variety of ways as long as the conditions cited are satisfied. Probability assignments reflect properties of relative frequency, but go beyond them in much the same way that a son may reflect properties of his father, but goes beyond them.

The Special Case of Equally Likely Outcomes

Many situations, particularly those that arise in statistical inference, lead us to take as our probability function the one that assigns the same value to all sample points in the sample space.

> **Theorem 1.** Let $S = \{s_1, s_2, \ldots, s_n\}$ denote a sample space with *n* sample points. Let us suppose that for one reason or another we are led to the probability function *P* which assigns the same value *x* to all of the sample points in *S*; then this value is $x = 1/n$; that is, 1 over the total number of sample points.

Proof: From the definition of probability model, we have:

$$P(s_1) + P(s_2) + \ldots + P(s_n) = 1$$

Since each of $P(s_1)$, $P(s_2), \ldots, P(s_n)$ is assigned the same value, call it *x*, we obtain:

$$\underbrace{x + x + \ldots + x}_{n \text{ terms}} = 1$$

A probability model in which all sample points are assigned the same probability value is called **an equally likely outcome model.**

> **Theorem 2.** If $S = \{s_1, s_2, \ldots, s_n\}$, $P(s_1) = \ldots = P(s_n) = 1/n$, and *A* is an event that is described by *k* sample points, then:
>
> $$P(A) = \frac{k}{n} = \frac{\text{Nu. of sample points describing A}}{\text{Total nu. of sample points}}$$

Proof: For the sake of simplifying our discussion, let us suppose that *A* is described by the first *k* sample points $s_1, s_2, \ldots, s_k$. Then we have:

$$\begin{aligned} P(A) &= P(s_1) + P(s_2) + \cdots + P(s_k) \\ &= \underbrace{\frac{1}{n} + \frac{1}{n} + \cdots + \frac{1}{n}}_{k \text{ terms}} = \frac{k}{n} \end{aligned}$$

In the special case of equally likely outcomes probability questions reduce to counting questions. To determine the probability of *A* in this framework, count the number of sample points that describe *A*, count the number of sample points, and take their ratio. If *S* contains 10,000 sample points and *A* is described by 1000 of them, then $P(A) = 1000/10,000 = 1/10$ in this model.

6.3 ■ Which is the "Right" Probability Model?

The Adventures of Hasty Harry

What should I get my brother for his birthday, pondered Bob. It's his thirtieth, the big 30, and this calls for something very special. His brother Harry, who was fond of games of chance and had an extensive collection of "unusual" dice. Bob decided to add to it by obtaining for him what would undoubtedly be the crown jewel of his collection, a gold die embedded with diamond chips to show off the spots on its faces. Bob had the die custom made and on the appointed day a very pleased Harry received a very special die.

Harry could hardly wait to show off his new treasure to his friends and, in addition, win some vacation money in a bit of friendly gaming activity that was certain to follow. In preparation for this he went to Martin's Models to obtain a probability model to describe the behavior of the new crown jewel of his collection.

"Mr. Martin, I want you to build me a probability model for the tossing of this die. I'm particularly interested in the probability that an even number shows."

"What can you tell me about your die, Harry? What do you know about its behavior from your experience with it?"

"I just got it as a present and I have no experience with it. I want to be prepared with a probability model before obtaining that experience. A die is a die, nothing special, apart from its being made of gold with diamond chips. What else is there to know?"

"All right Harry, I'll proceed on the **assumption** that it's an ordinary die of uniform composition, a fair die, as we say. This being the case the equally likely outcome model R4 (red) that I have in stock is the most realistic description of the behavior of your die.

R4 (red): $S = \{1, 2, 3, 4, 5, 6\}$

$$P(1) = P(2) = \cdots = P(6) = \frac{1}{6}$$

It follows from R4 that the probability that an even number shows is:

$$\begin{aligned} P(2) + P(4) + P(6) &= \frac{1}{6} + \frac{1}{6} + \frac{1}{6} \\ &= \frac{3}{6} \\ &= 0.50 \end{aligned}$$

"What does this mean in terms of some friendly gaming activity, Mr. Martin?"

"If you toss this die a large number of times, an even number should show roughly half the time. We cannot say when an even number will show, that's a matter of chance, but it should show roughly 50% of the time."

Reality Strikes

Harry proudly showed his die to his friends and all, with the exception of Harry, had a great time playing games of chance in which "friendly" bets were placed on which face would show when the die was tossed. Harry expected an even number to show roughly 50% of the time as predicted by Martin's model, and bet accordingly. It came to pass, however, that after 500 tosses of his die an even number had showed 66% of the time, which is sharply at variance with what he had expected. He had hoped to make a modest profit from this "friendly" gaming activity and now he found himself an unfriendly three hundred bucks in the red. Confused and feeling that he had been cheated, he stormed back to Martin's Models for some answers.

"I don't understand what went wrong Martin. If your mathematics is so precise, how could it happen that an even number showed 66% of the time instead of around the 50% you told me to expect? I'm three hundred bucks down because of this. You sold me a defective model and I want my money back."

"Mathematics, the probability model I gave you in this case, Harry, delivered what it was capable of, namely, a valid conclusion with respect to the assumptions made. Please remember that it was you who provided me with the starting point of the analysis. I quote: 'a die is a die, nothing special.' As it turned out there was something very special about this die which **made my assumption**, based on your information, **unrealistic.** As a result we obtained a valid conclusion from the model whose prediction about how often an even number can be expected to show was at variance with the nature of your die.

When there is a sharp conflict between a math model and the reality it is intended to describe, reality always wins. Didn't your brother say anything to you about the nature of the die, or were you too dazzled by the gold and diamond chips to pay attention?"

"I'm not sure now. I'll have to ask him. Maybe I was too hasty."

A New Model for Hasty Harry's Die

"Mr. Martin, I spoke to my brother and I listened this time. He said he told me that he had the die weighted internally so that the even numbered faces were twice as likely to show as the odd numbered ones. It's not at all an ordinary die in terms of its internal make up."

"Harry, I spoke to Bob after you left and he told me about the die's structure. He spent a lot of money to have the die made in this way and, ironically, it ended up costing you money. I developed another probability model for your die based on the information Bob gave me. This model should be a realistic fit to your die.

I call it B7 (blue). It is defined by,

$$\text{B7 (blue)}\quad S=\{1,2,3,4,5,6\}$$
$$P(1)=P(3)=P(5)=\frac{1}{9}$$
$$P(2)=P(4)=P(6)=\frac{2}{9}$$

It follows from B7 that the probability that an even number shows is:

$$P(2)+P(4)+P(6)=\frac{2}{9}+\frac{2}{9}+\frac{2}{9}$$
$$=\frac{2}{3}=0.67$$

The relative frequency interpretation of this valid conclusion is that if your die were tossed a large number of times, an even number should show about 67% of the time, which I understand, is in agreement with the "friendly" game experience that you had."

"How much do I owe you?"

"The same as for the previous model. There's no additional charge for your hasty action. You've paid that price already."

Hasty Harry's Other Die

"I have another die in my collection Mr. Martin, one that is weighted in such a way that the even numbered faces are three times as likely to show as the odd numbered ones. Do you have a suitable model for this die?" "As a matter of fact I do. It's Y3 (yellow) in my catalog listing and it's defined by:

$$\text{Y3 (yellow)} \quad S = \{1, 2, 3, 4, 5, 6\}$$

$$P(1) = P(3) = P(5) = \frac{1}{12}$$

$$P(2) = P(4) = P(6) = \frac{3}{12}$$

EXERCISES

1. Consider the yellow model Y3 for the process of tossing Harry's other die. $S = \{1, 2, 3, 4, 5, 6\}$, $P(1) = P(3) = P(5) = \frac{1}{12}$, $P(2) = P(4) = P(6) = \frac{3}{12}$

   Let E denote the event that an even number shows.

   (a) Find $P(E)$.

(b) State the relative frequency interpretation of the result obtained in (a).

(c) In tossing the die in question 1000 times, an even number was observed to show 665 times. Does this show that the conclusion obtained in (a) is not valid? Explain.

(d) Is the conclusion reached in (a), interpreted in relative frequency terms, true? Explain.

(e) Is the yellow model Y3 realistic for the die in question? Explain.

2. Consider Martin's Models model G4 (green) for the process of tossing a die. $S=\{1,2,3,4,5,6\}$, $P(1)=P(3)=P(5)=\frac{1}{10}$, $P(4)=\frac{3}{10}$, $P(2)=P(6)=\frac{2}{10}$.

Let A denote the event that an odd number shows.

(a) Find $P(A)$.

(b) State the relative frequency interpretation of the result obtained in (a).

(c) In tossing the die in question 1000 times an odd number was observed to show 302 times. Does this evidence establish that the conclusion obtained in (a) is valid? Explain.

(d) Is the conclusion obtained in (a), interpreted in relative frequency terms, true? Explain.

(e) Is the green model G4 realistic for Hasty Harry's other die? Explain.

6.4 ▯ Probability Models for Random Processes

Consider the process of dealing a card from a standard deck of 52 cards and let us address the problem of setting up a probability model for this process and determining the probability that a picture card is dealt.

For convenience in referring to the cards, let us set up a translation system so that we can refer to the cards as 1, 2, . . . , 52; in this translation system 1 might denote the ace of spades, 2 the king of spaces, etc. We take as our sample space,

$$S = \{C_1, C_2, \ldots, C_{52}\},$$

where C_1 is the event that card 1 is dealt, C_2 is the event that card 2 is dealt, etc.

Our next task is to define a probability function P on S. This can be done in many ways, and the function P that emerges **depends on the assumption** we make about how the card will be dealt from the deck.

Suppose we assume what is usually assumed in such situations, but not always made explicit, which is that the card is dealt from a well shuffled deck in an unbiased way, at random, as we say. The probability function P which best reflects this assumption assigns the same value, 1/52, to each sample point in S. We thus emerge with the following probability model:

$$\text{Model 1: } S = \{C_1, C_2, \ldots, C_{52}\}$$
$$P(C_1) = \cdots = P(C_{52}) = \frac{1}{52}$$

> From Model 1 it follows that the probability that a picture card is dealt is 0.23.

As is well-known, some card dealers are less than honest. Suppose we assume, based on past experience, that the dealer intends to "arrange things" so that the card we are dealt is neither a picture card nor an ace, but that any of the other cards may be dealt without bias. For notational convenience suppose that the picture cards and aces are in the cards we numbered 1,2, . . . , 16, and that the cards 17, 18, . . . , 52 correspond to the remaining cards. This leads to Model 2.

$$\text{Model 2: } S = \{C_1, \ldots, C_{16}, C_{17}, \ldots, C_{52}\},$$
$$P(C_1) = \cdots = P(C_{16}) = 0, P(C_{17}) = \cdots = P(C_{52}) = \frac{1}{36}$$

> From Model 2 it follows that the probability that a picture card is dealt is 0.

Which model is realistic (if either one) will be determined by reality. Observe a card being dealt a large number of times, and see how often a picture card is dealt. If it shows up in the neighborhood of 23% of the time, this evidence would favor Model 1; if it hardly shows up at all, this evidence would favor Model 2; if it shows up a percentage of the time which differs sharply from 23% and 0%, (say 15% or

45%), this evidence would cast doubt on the realism of both Models 1 and 2 and the assumptions on which they are based.

EXERCISES

1. Consider the statement: the event A, connected with a random process, has probability 0.96. Does the following correctly state the relative frequency interpretation of this statement? Explain. "If the process is repeated over and over, in the long run the probability of A will be 0.96."

2. Two marbles are chosen, one after the other (the first marble is not replaced before the other is chosen), from a bag containing one red, one blue, one green and one yellow marble.

   (a) Define two probability models for this process and state the assumptions on which these models are based.

   (b) Find the probability that red and green marbles are chosen in each of your models.

6.5 □ Return to Equally Likely Outcome Models

As noted in Section 6.2, in the special case of an equally likely outcome probability model, wherein all sample points are assigned the same probability value, probability questions reduce to counting questions. If k is the number of sample points describing an event A and n is the total number of sample points, then the probability of A is the ratio $\frac{k}{n}$ (see Theorem 2, Section 6.2). The counting process may range from trivial to extraordinarily complex, and to facilitate the task of counting we look at a basic principle.

A Tool for Counting

> **Multiplication Principle.** Let us suppose that a task is to be performed and that it can be viewed as a sequence of two procedures where the first can be performed in h ways and, after it has been performed, the second can be performed in k ways; then the two procedures can be performed in the stated order in $h \cdot k$ ways.

Since any one of the h ways in which the first procedure can be performed can be coupled with any of the k ways in which the second procedure can be performed, there are h groups of k possibilities, which gives us $h \cdot k$ outcomes.

More generally, the multiplication principle extends to a sequence of any number of procedures which are to be performed in order.

> The challenge to applying the multiplication principle is in seeing a situation from the point of view of a sequence of procedures to be performed in order. Sometimes it is obvious that this is the case, but often this view is more deeply hidden.

Example 1 Book Arrangement

Two books are to be chosen from three, denoted by A, B and C, and arranged next to each other on a bookshelf. (a) How many arrangements are possible? (b) How many arrangements are possible if second place must be filled by book B?

(a) Two positions are to be filled, which we think of as first place and second place. For first place we can choose any of the three available books and, after this has been done, for second place we can choose any of the two remaining books. This yields $3 \cdot 2 = 6$ possible arrangements, namely, AB, AC, BA, BC, CA and CB.

$$\underset{\text{1st place}}{\underline{\quad 3 \quad}} \cdot \underset{\text{2nd place}}{\underline{\quad 2 \quad}} = 6$$

(b) If a certain task is to be handled is a special way, it is best to turn to it first. Second place can be filled in one way since it must be filled with B. Turning to first place, we have two options, fill it with A or C. This yields $2 \cdot 1 = 2$ arrangements, AB and CB.

$$\underset{\text{1st place}}{\underline{\quad 2 \quad}} \cdot \underset{\text{2nd place}}{\underline{\quad 1 \quad}} = 2$$

In both cases the result is obvious, but the approach underlying the analysis is instructive.

Example 2 License Plate Construction

A license plate is to consist of two capital letters followed by three digits. In how many ways can we construct such a license plate?

We can look at this problem from the point of view of five spaces to be filled in order. In first place we may put any of the 26 capital letters, in second place we may put any of the 26 capital letters, in third place we may put any of the 10 digits, as is the case for fourth and fifth places. Thus, there are $26 \cdot 26 \cdot 10 \cdot 10 \cdot 10 = 676,000$ ways in which we can construct such a license plate.

$$\frac{26}{\text{1st place}} \cdot \frac{26}{\text{2nd place}} \cdot \frac{10}{\text{3rd place}} \cdot \frac{10}{\text{4th place}} \cdot \frac{10}{\text{5th place}} = 676,000$$

EXERCISES

1. In how many ways can 2 of 5 books be chosen and arranged next to each other on a shelf?

2. An encyclopedia of science consists of 7 volumes. (a) In how many ways can these volumes be arranged next to each other on a shelf? (b) How many of these arrangements are out of order? (c) In how many of these arrangements will volume 1 occupy first place and volume 2 occupy second place?

3. A traveler is planning to go from New York to Chicago by plane and make the return trip by bus. There are 5 airlines that have flights at the desired time and 3 bus lines that provide Chicago to New York service. In how many ways can the trip be made?

4. Motors are to pass through two inspection stations. At the first station 2 ratings are possible; at the second station 4 ratings are possible. In how many ways can a motor be marked?

5. In how many ways can 5 people line up at a ticket office?

6. An examination consists of 8 true-false questions. How many possible different answer sheets can be turned in?

7. In how many ways can n people line up at a ticket office?

8. A telephone dial has 10 holes. How many different signals, each consisting of seven impulses in succession, can be formed (a) if no impulse is to be repeated in any given signal; (b) if repetitions are permitted?

Permutations

> A **permutation** of a set of objects is an arrangement of these objects in some order in a line. If there are n distinct objects, then any arrangement of r of them in some order in a line is called a **permutation of the n distinct objects taken r at a time.** The number of permutations of n distinct objects taken r at a time is denoted by $P(n,r)$ or ${}_nP_r$.

Example 3 Arranging Books

Returning to Example 1, which involved choosing 2 books from 3, denoted by A, B, and C, and arranging them next to each other on a shelf, there are $P(3,2)=3\cdot 2=6$ permutations of the 3 books taken 2 at a time. The permutations are AB, AC, BA, BC, CA and CB.

> As a computation tool for $P(n, r)$ we have
>
> $$P(n,r)=n(n-1)\cdots(n-r+1),$$

which is the product of the first r integers in decending order starting with n.

To establish this result, we note that there are r places to be filled and n objects from which to choose. First place can be filled with any of these n objects. After it has been filled with one of these n objects, there remain $n-1$ objects available for filling the second place. Once it has been filled, 2 objects will have been used and $n-2$ choices remain for the third place; once first, second, and third places have been filled, 3 objects will have been used, and $n-3$ choices remain for the fourth place. More generally, when we come to the rth place, $r-1$ objects will have been used to fill the previous $r-1$ places, and $n-(r-1)$ or $n-r+1$ choices remain for rth place. From the multiplication principle, the number of ways of filling the r places is $n(n-1)\cdots(n-r+1)$.

For example, $P(7,3)$ is the product of the first 3 integers in decending order beginning with 7, $P(7,3)=7\cdot 6\cdot 5=210$. Also, $P(10,4)=10\cdot 9\cdot 8\cdot 7=5040$, and $P(48,2)=48\cdot 47=2256$.

Since products of consecutive integers arise frequently in counting problems, it is useful to have notation to denote such products. The symbol $n!$ (read "n factorial") is used to stand for the product of all integers from 1 to n inclusive. Thus

$$1! = 1 \qquad 4! = 4 \cdot 3 \cdot 2 \cdot 1 = 24$$
$$2! = 2 \cdot 1 = 2 \qquad 5! = 5 \cdot 4 \cdot 3 \cdot 2 \cdot 1 = 120$$
$$3! = 3 \cdot 2 \cdot 1 = 6 \quad 6! = 6 \cdot 5 \cdot 4 \cdot 3 \cdot 2 \cdot 1 = 720$$
$$n! = n(n-1)(n-2) \cdots 1$$

It is convenient to define 0! by

$$0! = 1$$

This definition, although perhaps strange at first sight, is useful in that certain counting formulas can be more easily stated without a need for considering separate special cases.

When $r = n$, we have the following special case: The number of permutations of n distinct objects (taken n at a time) in a line is:

$$P(n,n) = n! = n(n-1) \cdots 1$$

The computation formulas derived for $P(n, r)$ and $P(n, n)$ follow from the multiplication principle. Many, but not all, counting problems exhibit a structure which permit us to apply these results directly. In other cases which exhibit this structure it is preferable to go back to basic principles to work the problems.

Combinations

Many situations arise in which r distinct objects are to be selected from n without regard to order. A subset or selection of r objects chosen from n distinct objects, without regard to the order in which they were chosen or appear, is called a **combination of n objects taken r at a time.** The number of combinations of n distinct objects taken r at a time is denoted by $C(n, r)$, ${}_nC_r$, or $\binom{n}{r}$.

Example 4 Choosing Books

Returning to Examples 1 and 3 which involved choosing 2 books from 3, denoted by A, B, and C, we saw that there are 6 permutations of the 3 books taken 2 at a time, namely AB, AC, BA, BC, CA and CB.

There are 3 combinations of these 3 books taken 2 at a time, namely, {*A*, *B*}, {*A*, *C*} and {*B*, *C*}. The combination {*A*, *B*} is the set consisting of *A* and *B*, no order implied. This combination gives rise to two permutations, *AB*, which means books *A* and *B* in the order *A* followed by *B*, and *BA*, which means books *A* and *B* in the order *B* followed by *A*.

The number of combinations of *r* objects chosen from *n* distinct objects is expressed by the following formula:

$$C(n,r) = \frac{P(n,r)}{r!} = \frac{n(n-1)\cdots(n-r+1}{r!}$$

To establish this result, consider the related problem of determining $P(n, r)$, and think of the process of forming a permutation of *r* objects selected from *n* as being carried out in two stages. The first stage consists of selecting *r* of *n* objects without regard to order, which can be done in $C(n, r)$ ways.

The second stage consists of arranging the *r* objects chosen in some order, which can be done in $r!$ ways. From the multiplication principle, the number of ways of selecting *r* of *n* objects with regard to order, $P(n, r)$, equals the number of ways of selecting *r* of *n* objects without regard to order, $C(n, r)$, times the number of ways of ordering the *r* objects chosen, $r!$. That is:

$$P(n,r) = C(n,r)\cdot r!,$$

Solving for $C(n, r)$ by dividing both sides by $r!$ yields the desired result.

Thus, for example, the number of combinations of 10 objects taken 3 at a time, is

$$C(10,3) = \frac{P(10,3)}{3!} = \frac{10\cdot 9\cdot 8}{3\cdot 2\cdot 1} = 120\cdot$$

There are 120 ways of choosing 3 of 10 distinct objects **without regard to order.**

In how many ways can no objects be chosen from *n* objects? The answer, of course, is 1; just don't choose any, that's the one way. This leads us to define $P(n, 0)$ and $C(n,0)$ as follows:

$$P(n,0) = C(n,0) = 1$$

A frequently asked question is, when is order important and when is it not important? When is a situation a permutation situation and when is it a combination situation? We have formulas for computing the number of permutations and combinations inherent in a situation, but there is no formula for deciding which is to be used. This is a **matter of judgment** which requires a careful reading and analysis of the situation. It is where we are most on our own.

Example 5 Lines Determined By Points

How many lines are determined by 12 points, no three of which lie on the same line?

The problem reduces to determining the number of ways of choosing 2 of 12 points without regard to order, which is $C(12,2)=66$.

A line is determined by two points, **irrespective of order**. (The line determined by points *P* and *Q* is the same as the one determined by *Q* and *P*.) The condition that no three points lie on the same line is essential to avoiding line duplication. If *P*, *Q* and *R* lie on the same line, *P* and *Q*, *P* and *R*, and *Q* and *R* would determine the same line and 66 would overcount the number of lines determined by the 12 points.

Example 6 Electing a Slate of Officers

The student Math Society at Ecap University has 25 members. An election is to be held to elect a president, secretary and treasurer from its membership. In how many ways can an election slate be formed if no person may hold more than one office?

The problem reduces to determining the number of ways of choosing 3 distinct club members (no repetitions since no one may hold more than one office) **with regard to order**, which is $P(25,3)=25\cdot 24\cdot 23=13{,}800$.

The order feature is determined by the offices to be filled, which must be distinguished.

Example 7 Sampling

Two percent of a lot of 100 items are known to be defective. In how many ways can (a) a sample of 3 items be drawn from the lot; (b) a sample of 3 items, all of which are good, be drawn from the lot; (c) a sample of 3, one of which is defective, be drawn from the lot?

(a) Assuming that our interest in the sample is its composition from the point of view of good versus defective items, and not in any order in which the items may appear, the problem reduces to determining the number of ways of choosing 3 of 100 items **without regard to order**. This is given by $C(100,3)=161{,}700$.

(b) If all items drawn are to be good, they must be drawn from the 98 good ones. There are $C(98,3) = 152,096$ ways to do this.

(c) To form this sample consider two procedures. The first is to choose 1 of the 2 defectives, which can be done in $C(2,1) = 2$ ways. The second is to choose the 2 other items needed to make up the sample of 3 from the 98 good ones, which can be done in $C(98,2) = 4,753$ ways. By the multiplication principle, the number of samples of 3 items containing 1 defective that can be drawn from the lot is $C(2,1) \cdot C(98,2) = 9,506$.

EXERCISES

9. Evaluate $P(7,3)$, $P(12,4)$, $P(6,1)$, $C(16,3)$, $C(18,2)$ and $C(52,4)$. What do these values tell us?

10. A club consisting of 60 members meets to elect 4 officers: president, vice president, secretary and treasurer, from its membership. In how many ways can this slate be formed if no person may occupy more than one position?

11. A guest house with 12 single rooms receives 6 single reservations. In how many ways can these reservations be filled?

12. The committee on sabbaticals at Ecap University has authority to grant 5 sabbaticals for any given year. In how many ways can the sabbaticals be granted if 15 requests are received?

13. How many choices of 3 suits and 4 ties for a trip can be made from a wardrobe of 5 suits and 6 ties?

14. There are 4 vacancies on the state Court of Appeals. In how many ways can these vacancies be filled if 20 names have been placed in nomination?

15. The Alumni Association of Ecap University has organized a one mile race to be run by 2 faculty, W. J. Adams and H. Lurier, and 3 alumni of Ecap University, J. Ross, M. Tilson and *E.* Kapp. (a) How many possible finishes are there? (b) In how many finishes does Adams finish first? (c) In how many finishes do alumni finish in the first three places?

16. From a lot of 50 color television sets, a sample of 3 is selected for inspection. There are 4 defective sets in the lot. (a) How many samples of 3 of 50 sets are

there? How many of these samples contain (b) no defective sets; (c) 1 defective set; (d) 2 defective sets; (e) 3 defective sets?

Two Probability Problems

Example 8 Probability that Adams Finishes First

The setting of this problem is provided by Exercise 15. The Alumni Association of Ecap University has organized a one mile race to be run by 2 faculty, W.J. Adams and H. Lurier, and 3 alumni, J. Ross, M. Tilson and *E.* Kapp.

What is the probability that Adams finishes first?

One approach to this problem is to note that there are $5! = 120$ possible finishes, that Adams is first in $4! = 24$ of them, and conclude that the probability that Adams finishes first is 4! divided by 5!, or 1/5.

This approach, which is based solely on counting, leaves much to be desired. It is based on an underlying assumption of equally likely outcomes, but which outcomes are assumed to be equally likely is not made clear and is left to the imagination of the reader. The absence of an explicitly stated probability model and assumption on which the probability function of the model is based obscures the necessity for a critical examination of the realism of the assumption made and suggests the mistaken view that there is only one probabilistic conclusion possible which is an unassailable truth.

This kind of approach, which is far too commonly seen in applications of probability, should be accompanied by a skull and cross-bones to warn the reader that his perspective and understanding are in danger of being poisoned.

To analyze the question posed, we will have to back up and provide a probability model for the process along with a justification for the model's probability function, which is open to scrutiny.

As to notation, let (ALRTK), to take an example, denote the outcome indicated by the order, Adams (1st), Lurier (2nd), Ross (3rd), Tilson (4th), Kapp (5th). We take as our sample space S the outcomes expressed by all permutations of A, L, R, T and K. There are $5! = 120$ sample points in S.

If all five runners are in comparable physical condition, age, and running experience, then this would make realistic an assignment of equal probabilities of 1/120 to the 120 sample points in S, from which it would follow as a valid conclusion that the probability Adams finishes first is 1/5.

Some observers have argued, however, that the five runners are not in comparable physical condition, that Adams tires quickly when the temperature is over 65°F, that the weather forecast is for an 80°F day when the race is to be run, and that it is therefore unrealistic to assign the same probability value to all finishes.

Example 9 Sampling

The setting for this problem is provided by Example 7. Two percent of a lot of 100 items are known to be defective. A sample of 3 items is chosen at random from the lot. What is the probability that the sample drawn contains no defectives?

For notational convenience, let us think of the 100 items as tagged $I_1, I_2, \ldots, I_{100}$. Our interest in a chosen sample is in the nature of its items (defective versus good), not in the order in which they are drawn or arranged. This leads us to take as our sample space S the events expressed by all combinations of 100 items taken 3 at a time; that is,

$$S = \{(I_1, I_2, I_3), \ldots, (I_{98}, I_{99}, I_{100})\},$$

where (I_1, I_2, I_3), for example, is the event that the sample drawn consists of items I_1, I_2, and I_3. The number of sample points in S is $C(100, 3) = 161,700$.

The sample is envisioned as being drawn at random, which means in a unbiased manner which does not in any way favor certain samples being drawn over others. The probability function *P* which best reflects a random drawing in such a situation assigns the same probability value, $\frac{1}{C(100,3)}$ or $\frac{1}{1/161/700}$, to each sample point in *S*.

$$P(I_1, I_2, I_3) = \cdots = P(I_{98}, I_{99}, I_{100}) = \frac{1}{C(100,3)} = \frac{1}{161,700}$$

In terms of this model, the probability that the sample drawn contains no defectives is:

$$P(\text{no defectives}) = \frac{C(98,3)}{C(100,3)} = \frac{152,096}{161,700} = 0.941$$

Suppose that our criterion for accepting the entire lot of 100 items from a distributer were that a sample of 3 drawn at random from the lot contain no defectives; then the relative frequency interpretation of our conclusion would be that in performing this sampling procedure on lots of 100 items, over the long run we would be accepting such lots with 2% defectives around 94.1% of the time.

EXERCISES

17. In connection with the relative frequency interpretation of the conclusion obtained in Example 11, suppose that samples of 3 items free of defectives are obtained in 75% of the samples drawn in 500 repetitions of the sampling procedure. How are we to account for the discrepancy between the obtained 75% and the predicted 94.1%?

18. With respect to the probability model described in Example 9, determine the probability that the sample drawn (a) contains 1 defective; (b) contains 2 defectives; (c) contains 3 defectives.

Random Selections in Statistical Analysis

Consider a population of size *N*, $Q = \{x_1, x_2, \ldots, x_N\}$, and let us suppose that an unordered sample of size *n* is to be chosen at random from this population. As we have noted, particularly in Section 2.3, when we say that a sample is to be chosen at random we have in mind the idea that there is to be no bias, deliberate or inadvertent, which favors certain samples of size *n* being chosen over others. The sampling procedure is to be an equal opportunity procedure; no favoritism.

The probability assignment P which best reflects random sampling is the one which assigns the same value, $\frac{1}{C(N,n)}$, to each of the $C(N, n)$ unordered samples of size n that can be chosen from Q.

While easily envisioned, random selections are not easily achieved in practice, especially when the population being sampled from is large. Yet, it is essential to closely approximate in reality the random selections envisioned in theory if the results deduced from theory are to be applicable to reality. This is particularly urgent in statistics, a subject whose theoretical framework is to a large extent based on the assumption of random sampling.

See Sections 2.2 and 2.3 for further discussion of random sampling.

6.6 Independent Events

The term independent events suggests events with the property that the occurrence or nonoccurrence of one has no influence on the occurrence or nonoccurrence of the other. Such an intuitive notion of independent events is not precise enough to serve as a mathematical definition of this concept. We use it as a guide to develop a precise condition which captures its spirit.

To get a start on this problem, let us return to the process of tossing a pair of well balanced dice (one red and the other green), and consider two events that, intuitively speaking, seem independent: the red die shows 1 and the green die shows 2. In the sample space shown in Table 2, the first number in each ordered pair specifies the number that shows on the red die, and the second number specifies the number that shows on the green die. The probability function P assigns the same value, 1/36, to each of the 36 sample points. Let A denote the event red shows 1 (described by the sample points in row 1 of Table 2) and B denote the event green shows 2 (described by the sample points in column 2 of Table 2); then $A \cap B = \{1,2\}$. The probabilities of A, B and $A \cap B$ are

Table 2

(1, 1)	(1, 2)	(1, 3)	(1, 4)	(1, 5)	(1, 6)	Red shows 1
(2, 1)	(2, 2)	(2, 3)	(2, 4)	(2, 5)	(2, 6)	
(3, 1)	(3, 2)	(3, 3)	(3, 4)	(3, 5)	(3, 6)	
(4, 1)	(4, 2)	(4, 3)	(4, 4)	(4, 5)	(4, 6)	
(5, 1)	(5, 2)	(5, 3)	(5, 4)	(5, 5)	(5, 6)	
(6, 1)	(6, 2)	(6, 3)	(6, 4)	(6, 5)	(6, 6)	
	Green shows 2					

$$P(A \cap B) = \frac{1}{36}, \quad P(A) = \frac{1}{6}, \quad P(B) = \frac{1}{6}.$$

It is immediately apparent that these probabilities are related in the following way:

$$P(A \cap B) = P(A) \cdot P(B)$$

This relationship is more than coincidental and serves as a basis for a precise definition of independence. Further evidence that this relationship can serve to express in a precise way the idea of independence of events is provided by the relative frequency point of view. Let us envision the dice being tossed a large number of times, say N. Let N_R denote the number of times the red die shows 1, N_G the number of times the green die shows 2, and N_{RG} the number of times the red die shows 1 and the green die shows 2 at the same time. If the occurrence or nonoccurrence of red shows 1 in no way influences the occurrence or nonoccurrence of green shows 2, then we would expect that over the long run the relative frequency of the time that red shows 1 among all tosses of the dice, that is N_R/N, would be approximately the same as the relative frequency of the time that red shows 1 among those tossings of the dice in which green showed 2 at the same time, that is $\frac{N_{RG}}{N_G}$. Expressed in mathematical terms, we have:

$$\frac{N_R}{N} \approx \frac{N_{RG}}{N_G}$$

(Please recall: the symbol $\approx$ means "approximates.") Multiplying both sides of this result by $\frac{N_G}{N}$, the percentage of the time green shows 2 among all tosses of the dice, yields:

$$\frac{N_R}{N} \cdot \frac{N_G}{N} \approx \frac{N_{RG}}{N_G} \cdot \frac{N_G}{N}$$

$$\frac{N_R}{N} \cdot \frac{N_G}{N} \approx \frac{N_{RG}}{N}$$

Over the long run, that is, when N is large, N_R / N approximates $P(A)$, the probability that red shows 1; $\frac{N_G}{N}$ approximates $P(B)$, the probability that green shows 2; and $\frac{N_{RG}}{N}$ approximates $P(A \cap B)$, the probability that red shows 1 and green shows 2. Thus the statement

$$\frac{N_R}{N} \cdot \frac{N_G}{N} \approx \frac{N_{RG}}{N}$$

can be viewed as a translation into relative frequency terms of the probabilistic relationship

$$P(A) \cdot P(B) = P(A \cap B)$$

These observations lead us to the following definition of independence of two events as a precise capturing of the idea that the occurrence or nonoccurrence of one event has no influence on the occurrence or nonoccurrence of the other.

> Two events A and B in a probability model with sample space S and probability function P are said to be **independent** if
>
> $$P(A \cap B) = P(A) \cdot P(B).$$

Example 1

In connection with the process of tossing a pair of dice, assumed to be well balanced, are the events A, an even sum shows, and B, the sum showing is less than 5, independent?

For a view of the sample points that describe A, B and $A \cap B$ (an even sum less than 5 shows), refer to Table 3. A is described by the even numbered diagnois, B={(1, 1), (1, 2), (1, 3), (2, 1), (2, 2), (3, 1)}, $A \cap B$ = {(1, 1), (1, 3), (2, 2), (3, 1)}.

The probabilities of these events are:

Table 3

		2		4		6	
B							
	$A \cap B$	(1, 1)	(1, 2)	(1, 3)	(1, 4)	(1, 5)	(1, 6)
		(2, 1)	(2, 2)	(2, 3)	(2, 4)	(2, 5)	(2, 6)	8
		(3, 1)	(3, 2)	(3, 3)	(3, 4)	(3, 5)	(3, 6)
		(4, 1)	(4, 2)	(4, 3)	(4, 4)	(4, 5)	(4, 6)	10
		(5, 1)	(5, 2)	(5, 3)	(5, 4)	(5, 5)	(5, 6)
		(6, 1)	(6, 2)	(6, 3)	(6, 4)	(6, 5)	(6, 6)	12

$$P(A)=\frac{1}{2},\quad P(B)=\frac{1}{6},\quad P(A\cap B)=\frac{1}{9}.$$

Since $P(A\cap B)=1/9$ is not equal to $P(A)\cdot P(B)=1/12$, events A and B are not independent.

Our interpretation of this is that the occurrence or nonoccurrence of one of these events does influence the occurrence or nonoccurrence of the other.

Three events A, B, and C in a probability model with sample space S and probability function P are said to be **independent** if

$$P(A\cap B\cap C)=P(A)\cdot P(B)\cdot P(C)$$

and all pairs of events selected from A, B, and C are independent. That is:

$$P(A\cap B)=P(A)\cdot P(B),\quad P(A\cap C)=P(A)\cdot P(C)$$
$$P(B\cap C)=P(B)\cdot P(C)$$

> More generally, n events $A_1, A_2, \ldots, A_n$ are said to be **independent** if
>
> $$P(A_1\cap\ldots\cap A_n)=P(A_1)\ldots P(A_n),$$
>
> and this product condition holds for all sub-collections of events chosen from $A_1,\ldots,A_n$ as well.

Our intuitive interpretation of this more general condition is much the same as previously described; namely, that the occurrence or nonoccurrence of any of $A_1, A_2, \ldots, A_n$ does not influence the occurrence or nonoccurrence of the others.

A Property of Independence

Our intuitive interpretation of the independence of $A_1, A_2, \ldots, A_n$ includes the idea that the nonoccurrence of any of these events (which is described by their complements) has no influence on the occurrence or nonoccurrence of the others, but this property is not directly reflected in the definition of independence. While this is the case, the condition involving the nonoccurrence of any of $A_1, A_2, \ldots, A_n$ not influencing the occurrence or nonoccurrence of the others is present in the sense of the following theorem.

> If $A_1, A_2, \ldots, A_n$ is a collection of independent events, then the collection of events obtained by replacing any number of the *A*'s by their complements is also independent.

Thus, for example, if *A* and *B* are independent, with $P(A) = 1/3$ and $P(B) = 1/4$, then A^c and B^c are independent and

$$P(A^c \cap B^c) = P(A^c) \cdot P(B^c) \quad = \frac{2}{3} \cdot \frac{3}{4} = \frac{1}{2}$$

EXERCISES

1. In connection with tossing a well balanced die, are the events an even number shows and a number greater than 3 shows independent?

2. In connection with tossing a pair of well balanced dice (one red and one green), (a) are the events red shows less than 3 and green shows less than 4 independent? (b) Are the events red shows less than 3 and the sum showing is 7 independent?

3. Events *A*, *B* and *C* are independent, where $P(A) = 1/3$, $P(B) = 1/4$ and $P(C) = 1/5$. Find (a) $P(A \cap B^c)$; (b) $P(A \cap B \cap C)$, (c) $P(A \cap B^c \cap C^c)$; (d) $P(A^c \cap B \cap C^c)$).

6.7 ■ Bernoulli Trial Probability Models

It is sometimes natural and productive to view a random process as a sequence of repetitions of some basic process. For example, the tossing of a coin five times in succession may be viewed as a sequence of five repetitions of the basic process of tossing a coin once, firing at a target six times may be viewed as a sequence of six repetitions of the basic process of firing at the target, administering an antibiotic to ten people may be viewed as a ten fold repetition of the basic process of administering the antibiotic. A repetition of a basic process is called a **trial**.

> The problem we address here is that of formulating a probability model for a sequence of trials under the following conditions.
>
> 1. On each trial attention is focused on the occurrence or nonoccurrence of a certain event E. The occurrence of E on a given trial is called **success**; the occurrence of E^c is called **failure**.
>
> 2. The probability of success is the same for each trial; call this probability value p. The probability of failure for each trial is $1-p$, which we shall denote by q.
>
> 3. The outcome of a trial (success or failure) does not influence the outcome of any other trial.

A sequence of repetitions of an experiment carried out under these conditions is sometimes called a sequence of **Bernoulli trials** in honor of Jacob Bernoulli (1654-1705), who initiated a systematic study of such processes in the last part of the seventeenth century.

Example 1 Coin Tossing

Consider the process of tossing a coin, assumed to be well balanced, five times in succession. Take for E the event that head shows on a trial, with assigned probability $p=1/2$. The third Bernoulli trial condition is satisfied since the outcome of any toss of the coin does not influence the outcome of any other toss. We might be influenced about how the coin will fall on a subsequent toss based on its behavior on previous tosses, but the coin has no memory and cannot be influenced by such. The probability assignment of 1/2 for head showing on any toss is realistic for a well balanced coin.

Sequences of coin tossings are realisticly described by Bernoulli trial conditions.

The constancy of p and the independence condition are the ones most likely to give trouble in practice and must be looked at most carefully.

Developing the Bernoulli Trial Probability Model

The problem we now address is that of constructing a probability model for a sequence of *n* Bernoulli trials. If a sequence of *n* repetitions of the underlying experiment results in event *E* occurring on trial 1, not occurring on trial 2, occurring on trial 3, not occurring on trial 4, and so on, then we have a run of successes and failures that can be represented by the sequence

$$\underbrace{SFSF\ldots SF}_{n \text{ terms}}$$

The outcome of any sequence of *n* repetitions of the experiment can be described by a sequence of *n* *S*'s and *F*'s, and every sequence of *n* *S*'s and *F*'s describes a way in which a sequence of *n* repetitions of the experiment could occur. Accordingly, we take as our sample space *T* the events described by all sequences of *n* *S*'s and *F*'s.

To define *p* on the sample points of *T*, let S_1 denote the event success on trial 1, S_2 the event success on trial 2, . . . , S_n the event success on trial *n*, F_1 the event failure on trial 1, F_2 the event failure on trial 2, . . . , F_n the event failure on trial *n*. To reflect the condition that the probabilities of success and failure are the same for each trial, *p* and $q = 1 - p$, respectively, we impose on *p* the condition:

$$P(S_1) = P(S_2) = \cdots = P(S_n) = p \tag{1}$$

$$P(F_1) = P(F_2) = \cdots = P(F_n) = q \tag{2}$$

To reflect the condition that success or failure on any one trial does not influence the outcome on any other trial, we require that any *n* events chosen from $S_1, S_2, \ldots, S_n$, $F_1, F_2, \ldots, F_n$ with different subscripts (that is, when S_k is chosen, F_k is to be left out, and vice versa) be independent. These conditions uniquely define *P* on the sample points of *T*. Consider, for example, the sequence

$$\underbrace{SS\ldots S}_{k \text{ terms}} \; \underbrace{F\ldots F}_{n-k \text{ terms}}$$

whose first *k* terms are *S*'s and whose last $n-k$ terms are *F*'s. Then:

$$\underbrace{SS\ldots S}_{k \text{ terms}} \; \underbrace{F\ldots F}_{n-k \text{ terms}} = S_1 \cap S_2 \cap \ldots \cap S_k \cap F_{k+1} \cap \ldots \cap F_n$$

This condition, the independence of the n events $S_1, S_2, \ldots, S_n$, $F_1, F_2, \ldots, F_n$, and conditions (1) and (2) yield:

$$\begin{aligned} P(SS\ldots SF\ldots F) &= P(S_1 \cap S_2 \cap \ldots \cap S_k \cap F_{k+1} \cap \ldots \cap F_n) \\ &= P(S_1) \cdot P(S_2) \cdots P(S_k) \cdot P(F_{k+1}) \cdots P(F_n) \\ &= \underbrace{p \cdot p \cdots p}_{\substack{k \\ \text{terms}}} \cdot \underbrace{q \ldots q}_{\substack{n-k \\ \text{terms}}} \\ &= p^k q^{n-k} \end{aligned}$$

Each S in the sequence yields a p in the product $p^k q^{n-k}$ and each F in the sequence yields a q in this product. More generally, if Z is any sequence with k S's and $n-k$ F's, we have:

$$P(Z) = p^k q^{n-k}$$

How many sequences have k S's and $n-k$ F's? To obtain such a sequence, we choose k places in the sequence and fill them with S's; the remaining places will be filled with F's. The number of ways of doing this equals the number of ways of choosing k of n places without regard to order, which is $C(n, k)$.

Let X denote the number of successes in a sequence of n trials. Since $C(n, k)$ sequences have k successes and $n-k$ failures, the probability of k successes in n trials is:

$$P(X = k) = C(n,k) p^k q^{n-k}, \text{ for } k = 0,1,2,\ldots,n$$

Thus:

$$\begin{aligned} P(X = 0) &= C(n,0) p^0 q^n = q^n \\ P(X = 1) &= C(n,1) p^1 q^{n-1} \\ P(X = 2) &= C(n,2) p^2 q^{n-2} \\ &\vdots \qquad\qquad \vdots \\ P(X = n) &= C(n,n) p^n q^0 = p^n \end{aligned}$$

In summary, the Bernoulli trial conditions 1, 2 and 3 lead to a **class of Bernoulli trial models.**

To specify a model in this class we must specify the number of trials n, the event E which is the subject of our focus, and the probability p of the occurrence of E on a trial. Since the occurrence of E on a trial is called success, p expresses the probability of success on each trial.

Of particular interest is the event that k successes occur in n Bernoulli trials. If X denotes the number of successes in n Bernoulli trials, then the probability of k successes is written as $P(X=k)$, which is given by:

$$P(X=k)=C(n,k)p^k q^{n-k}\text{, for } k=0,1,2,\ldots,n\text{, where } q=1-p.$$

In many ways this is the easier part of our analysis. The harder part is to apply Bernoulli trial models to real world problems. What makes this harder, at least in certain problems, is that we must make a judgment call as to whether a Bernoulli trial model is a realistic fit for the random process being studied; judgment calls of this type are often difficult and may be rather controversial.

Example 2 Firing at a Target

Records show that a marksman hit the bullseye two-thirds of the time. To win a prize he must hit the bullseye at least 4 times in 5 firings. What is the probability that he will a prize?

Our first task is to set up a probability model for the process, which is to fire at the target five times. Since we have a situation which can clearly be viewed as a sequence of 5 repetitions of the basic process of firing at the target, the Bernoulli trial model defined by $n=5$, $E=$ the event that the marksman hits the bullseye, $p=2/3$ with $q=1/3$, emerges as a strong possibility, depending on the assumptions involved.

The assignment $p=2/3$ for the probability of success on each firing is based on long term statistical evidence of how well the marksman performed in the past. This does assume that the conditions under which the firings take place are uniform and that the marksman is warmed up and performing at his usual level.

As to the independence condition, we are assuming that the outcome of any firing has no affect, or at worst a negligible affect, on any other firing.

If we agree that these conditions are realistic, the big IF, then we emerge with the Bernoulli trial model defined by $n=5$, $E=$ the event that the marksman hits the bullseye, $p=2/3$ with $q=1/3$, as our model for the firings.

Let X denote the number of successes in 5 trials. The probability that the marksman wins a prize is $P(X \geq 4) = P(X = 4) + P(X = 5)$, which is given by:

$$\begin{aligned} &C(5,4)(2/3)^4(1/3) + C(5,5)(2/3)^5 \\ &= 0.329 + 0.132 \\ &= 0.46 \end{aligned}$$

The relative frequency interpretation of this result is that if the sequence of five firings took place a large number of times under the assumed conditions, then the marksman would win a prize approximately 46% of the time.

Example 3 Larry's Apparel Shop

Records kept by Larry's Apparel Shop indicate that 20% of the shoppers who come into the store to browse actually make a purchase. What is the probability that among 100 shoppers who examine Larry's merchandise at least 30 will make a purchase?

We can identify each of the 100 shoppers with a trial, so that a sequence of 100 trials corresponding to the 100 shoppers emerges. We have $n = 100$. Take for E the event that a shopper makes a purchase. From the statistical data available on the percentage of shoppers who make a purchase, it seems reasonable to take $p = 0.20$.

> The probability model we are developing treats the shoppers as identical units who act independently in making purchase decisions. It's one thing to make such an assumption about coins, but people are not coins. How realistic is this assumption? Is a Bernoulli trial model suitable for this situation?
>
> We should keep in mind that shoppers are sometimes acquainted and are influenced by the purchase behavior of friends or relatives. If we are willing to assume that such interaction can be regarded as negligible, then we are led to the Bernoulli trial model defined by $n = 100$, E = the event that a shopper makes a purchase, $p = 0.20$ with $q = 0.80$.

Let X denote the number of successes (purchases made) in 100 trials. We require $P(X \geq 30) = P(X = 30) + \cdots + P(X = 100)$, which is given by:

$$C(100,30)(0.20)^{30}(0.80)^{70}+\cdots+C(100,100)(0.20)^{100}$$

It can be shown that this value is approximately 0.0087. (See Example 1, Sec. 8.2.)

This conclusion is valid with respect to our model, no-more and no-less. If our feeling is that this model does not reflect reality sufficiently closely, then the wise course would be to try to refine the model or develop another which better reflects reality.

Example 4 Flu

Studies have led medical authorities to conclude that 1 out of 1000 people in Pittsfield has the flu. Find the probability that of 100,000 people selected 120 or more will have the flu.

We can identify each of the 100,000 people with a trial, so that a sequence of 100,000 trials emerges. We have $n = 100,000$. Let E denote the event that a person has the flu. Based on the afore studies, take $p = 0.001$, with $q = 0.999$. The Bernoulli trial model assumption that emerges is that whether a person has the flu or not does not have an influence on whether any other person has the flu or not.

If we accept this assumption, let X denote the number of successes (number of people who have the flu). Then we have:

$$\begin{aligned} P(X \geq 120) &= P(X = 120) + \cdots + P(X = 100,000) \\ &= C(100,000;120)(0.001)^{120}(0.999)^{99,880} + \cdots + C(100,000;100,000)(0.001)^{100,000} \end{aligned}$$

which can be shown to be 0.0256 (see Sec. 8.2).

If many groups of 100,000 people are chosen, then in about 2.6% of the cases 120 or more people will have the flu.

It may all seem very precise and convincing, but appearances are often deceptive, as is the case here. This hinges on the **assumption** that whether a person has the flu or not does not have an influence on whether any other person has the flu or not. Flu is highly contagious, which makes this assumption untenable. Thus, the house of mathematical cards founded on this assumption collapses.

Approximation of Bernoulli Trial Probabilities

As Examples 3 and 4 make clear, Bernoulli trial models may give rise to substantial computation problems, even in an era which boasts of powerful computation machinery. Abraham de Moivre (1667-1754) recognized this in the period immediately following the publication of Bernoulli's *Ars Conjectandi* (The Art of Conjecture) in 1713 and set himself the problem of developing a tool to approximate Bernoulli trial probabilities. de Moivre's efforts met with success and his work on this problem first appeared in a privately printed pamphlet dated November 12, 1733. de Moivre's work marks the first appearance of what is today called a normal curve approximation for Bernoulli trial probabilities. We turn to this subject in Section 8.2.

EXERCISES

Set up Bernoulli trial models for the following situations, state the underlying assumptions and comment on their realism, and determine or state (in combinatorial form) the probabilities asked for. In each case express a judgment on the suitability of a Bernoulli trial model for the process in question. Is a more realistic model available for any of the situations described? If so, state the model, comment on its realism, and determine the probabilities required.

1. A coin is tossed 5 times in succession. Find the probability that (a) head shows in 2 of the 5 tosses; (b) head shows in at least 3 of the 5 tosses.

2. Records show that a slot machine, known as stingy to its patrons, has hit the jackpot 1% of the time over its life of ten years. If you play the machine 20 times, what is the probability that you will win at least once?

3. A die is tossed 6 times in succession. What is the probability that an even number shows at least 4 times?

4. Consider an examination that consists of 10 multiple-choice questions. Each question allows four choices, with only one answer being correct. If a student guesses the answer for each question, what is the probability that he will pass the exam (6 or more correct answers)?

5. According to company records, 1% of the output of light bulbs in a certain mass production process are defective. What is the probability that of 1,000 light bulbs produced, no more than 15 are defective?

6. The fruit fly *Drosophila* has been widely used for genetic studies and has furnished much of the evidence for modern genetic theory. It has been shown that mutations can be induced in *Drosophila* by subjecting the flies to X-rays, ultraviolet radiation, or any high-energy radiation, as well as chemical compounds such as those in the mustard gas family. Evidence suggests that 0.01% of large batches of *Drosophila* subjected to high-energy radiation develop a mutation. What is the probability that of 20,000 *Drosophila* subjected to high-energy radiation, at least one will develop a mutation?

7. A recently developed antibiotic has been found to cause an allergic reaction in one out of 200 people injected. Find the probability that fewer than 100 reactions occur among 10,000 people injected with the antibiotic.

8. Studies have led medical authorities to conclude that 1 out of 500 teenagers in Southchester has the measles.

   (a) Find the probability that out of 10,000 teenagers selected at most 25 will have the measles.

   (b) Interpret the result you obtained.

   (c) Are the numbers obtained in answer to (a) and (b) credible? Explain

6.8 ■ Interpretations of Probability

Historically, the study of the stability exhibited by long run relative frequencies of events connected with games of chance played a fundamental role in the birth and early development of the theory of probability. But having been born, probability theory began a separate mathematical life of its own. This mathematical life begins with the concept of probability model, and consists of the theorems and definitions that are built up on the basis of this concept.

Although the relative frequency point of view suggested the formulation of many concepts of the mathematical theory of probability, this interpretation, and any other interpretation for that matter, are not part of the internal structure of this mathematical theory. The relationship between a mathematical theory and an envisioned interpretation of the theory that suggested many of the theory's concepts is somewhat similar to the relationship between parents and child. The parents give birth to the child and influence the child's development. Although we may hear comments about physical and temperament similarities between parents and child, we recognize the child as a separate and distinct individual. In a similar sense, a mathematical theory is a structure separate from an envisioned application or interpretation that may have played the role of parents in giving rise to the theory.

Since a given interpretation of a mathematical theory is a structure separate from the theory itself, it might be possible and useful to interpret the theory's concepts in other ways. Such is the case with probability theory. Another interpretation that has received much attention in recent years is one in which the probability value assigned to an event is interpreted as a quantitative measure of an individual's degree of belief in the occurrence of the event. An ardent New York Mets fan might say that the Mets have a 90 percent chance of taking the pennant next year; translation: the probability that the Mets take the pennant next year is 0.90. A less ardent Mets admirer might say that the Mets have a 10 percent chance of taking the pennant next year. A business man might say that there's an 80 percent chance that the sales volume of the firm will top $5 million this year. Individuals often have beliefs or opinions about possible outcomes connected with situations in which the outcome is not certain. Such an individual sometimes finds it useful to assign a value between 0 and 1, inclusive, to a possible outcome as a quantitative measure of his feelings about the likelihood of occurrence of the outcome. A strong opinion about the occurrence of an event is reflected by the assignment of a value close to 1 to the event; a strong opinion about the nonoccurrence of an event is reflected by the assignment of a value close to 0 to the event.

Probability values that are assigned to an event from this point of view, or that are interpreted in this way, are called **personal** or **subjective probabilities** and are said to express a person's degree of belief in the occurrence of the event. The point of view of subjective probability admits as meaningful probabilistic assertions about events connected with situations that occur once and cannot be repeated; the relative frequency interpretation, on the other hand, is only meaningful for random processes that can be repeated a large number of times. The subjective point of view admits as meaningful such statements as the probability that the Smith Company will spend more than $2 million on advertising this year is 0.8, and the probability that Jack will not study for his math exam is 0.9. Such statements are meaningless from the relative frequency point of view. An important feature, as well as difficulty, of subjective probability is that the assignment of subjective probabilities to outcomes depends very much on the person doing the assigning. Two individuals might assign markedly different subjective probabilities to the same event.

Example 1 Interpretation of the Probability of Bill's Recovery

Bill Bradley, who is suffering from a rare disease, was told by his friend that, since recent medical statistics show that 75 percent of those who have had the disease recovered, the probability that he will recover is 0.75. How is this probability assignment to Bill's recovery to be interpreted?

Although a relative frequency background is involved (recent medical statistics), the focus is on a once-and-only situation, Bill's state of health. This by itself is sufficient to exclude a relative frequency interpretation; to repeat the process a large number of times would entail giving Bill the disease a large number of times and observing how often he recovers, a procedure that does have its difficulties. Bill's friend is using background relative frequency data as a basis for expressing in quantitative terms his degree of belief about his friend's recovery.

EXERCISES

1. It is asserted that the probability of an event *E* is 0.80. State the relative frequency and subjective probability interpretations of this assertion, and describe the main features of these interpretations.

2. After watching a pair of dice being tossed 100 times, an observer commented that the probability of an even sum showing on the 101st toss is 0.85. Is this probability assignment one that is to be interpreted in relative frequency or subjective probability terms? Explain.

3. The following comment appeared in an article on natural gas supplies (*The New York Times*, Feb. 22, 1977, *p.* 14): "How much gas is left to be discovered? . . . The last Geological Survey estimate, made two years ago, was this: Given available technology and current economics, there is a 95 percent probability that 322,000 billion cubic feet can be located and produced; there is a 5 percent probability that 655,000 billion cubic feet can be located and produced." How should these probabilistic statements be interpreted? Explain.

4. In a letter to the editor of *The New York Times* (Feb. 28, 1971) on the background of the atomic bomb, Hans Bethe wrote, "By February 1945 it appeared to me and to other fully informed scientists that there was a better than 90 per cent probability that the atomic bomb would in fact explode . . . ." How should this probabilistic statement be interpreted? Explain.

6.9 ■ Probabilities and Odds

The weather channel reported that there is an 80% chance of rain today. In the more technical language of probability this would read, the probability of rain today is 0.80. In this situation, as well as in others, the terms odds in favor of the event and odds against the event are used.

In this situation it would be said that the odds in favor of it raining today are 0.8 to 0.2 or, equivalently, 8 to 2 or, equivalently, 4 to 1. Translation into monetary terms: it would be considered fair for one to bet $4 against $1 that it rains today.

More generally, if the probability of an event E is p, where p is unequal to 0 or 1, then the **odds in favor of E** are p to 1-p; the **odds against E** are 1-p to p.

We also have the following relationship: if it is considered equitable to bet x dollars against y dollars that a given event E will occur, then E is being assigned the probability $\frac{x}{(x+y)}$. Thus, for example, if odds or 7 to 3 are given that our favorite team will win the pennant this year, this is equivalent to saying that the probability the team will win the pennant this year is $\frac{7}{7+3} = 0.70$.

EXERCISES

1. If if is asserted that the probability of rain today is 0.90, what are the odds in favor of this event?

2. Janet Rodriguez has come to the conclusion that the probability that she will get an A in statistics this semester is 0.95. What are the odds in favor of this event?

3. Find the odds that correspond to the following probability values: (a) 0.82, (b) 0.98, (c) 0.99, (d) 0.10

4. If the odds are 7 to 2 that the home team will win the pennant this year, what is the envisioned probability that this will be the case?

5. An investor feels that the odds are 13 to 5 against the price of his stock going up this week. What is the envisioned probability that this will be the case?

6. Archie Goodwin, confidential assistant to the grand master of detection, Nero Wolfe, likes to state opinions in terms of odds. In *Plot it Yourself*, by Rex Stout, Archie observes that it was 20 to 1, or maybe 30 to 1, that Kenneth Rennert did not write the stories being considered (p. 38). What probabilities correspond to these odds?

7

Random Variables

7.1 ■ Random Variables, Expected Values, and Probability Distributions

Often in connection with a random process events of special concern dominate our interest. To illustrate, we consider a chapter in the adventures of Dapper Dan and Gullible Gus.

Dapper Dan and his friend Gullible Gus have had a friendly on-going competition for years now. The competition involves the process of tossing a die. One day Dan took a die from his pocket and offered Gus the following proposition: "If on a toss of this die 1 shows, you pay me \$4; if any other face shows, I pay you \$1. Since you have the advantage as the arrangement now stands," Dan asserted, "you should pay me 40 cents before each toss of my die in order to make the game fair."

The problem of interest to Gus, which we take up here, concerns whether or not this is the fair price for playing the game.

To begin we must make an assumption about the nature of Dan's die. Since Dan has established a reputation for being honest in such matters as die tossing and no information to the contrary has been forthcoming, we shall assume that Dan's die is well-balanced. This leads us to the probability model with sample space $S = \{1, 2, 3, 4, 5, 6\}$ and probability function P which assigns to each sample point the same value, 1/6, as our model for tossing Dan's die. To describe the amount won by Gus on a toss of Dan's die we define a function Z on sample space S as follows:

$$Z(1) = -4 \qquad Z(4) = 1$$
$$Z(2) = 1 \qquad Z(5) = 1$$
$$Z(3) = 1 \qquad Z(6) = 1$$

Note that to say Gus wins -$4 is to say that Gus pays Dan $4.

Function Z, which assigns to each sample point in S a numerical value, is an **example of a random variable.** A random variable helps us focus on events of interest, in this case the events being Gus loses $4 and Gus wins $1.

> More generally, a **random variable** is a function which assigns a numerical value to each sample point in a sample space S of a probability model. These numerical values are said to be **assumed** by the random variable.

Gus's random variable assumes the values -4 and 1.

A random variable is said to be **discrete** if it assumes isolated values; if the number of possible values can be counted or is known not to exceed some positive integer (100 or 10,000 or 100^{1000} or whatever), then the random variable is called **finite.** Gus's random variable is discrete with a finite number of values, two (-4 and 1). The random variable which expresses the daily number of stock purchases during the year at your favorite stock exchange is a discrete random variable which is finite, as is the random variable which describes the number of atomic particles emitted per minute from a substance undergoing radioactive decay.

A random variable is said to be **continuous** if it can assume all values between isolated values or all real numbers, or all real numbers less than, less than or equal to, greater than, or greater than or equal to a specified value. Random variables that can assume, for example, all values between 1 and 5, or all values greater than or equal to 0 are continuous random variables. The random variable which expresses the emission time of an atomic particle from a substance undergoing radioactive decay is a continuous random variable, as is the random variable which describes the heights of people who live in your city or community. If X stands for the height of a person, measured in inches, let us say, then X may take the value 60 or 65 or any value in between. In measuring the height of an individual we usually round off to the nearest integer or half-integer value, but this is another matter. All values between 60 and 65 inches are possible, and a person who grows from 60 to 65 inches goes through all intermediate values.

Counting situations usually lead to discrete random variables, whereas measurement situations usually lead to continuous random variables. Our primary focus is given over to discrete random variables which assume a finite number of values but, as we shall see, continuous structures play an essential role in approximating the behavior of structures arising from the study of discrete random variables.

It is standard practice to denote random variables by capital letters, such as *X*, *Y*, and *Z*, and values they assume by lower case letters. In terms of his random variable *Z*, the statement Gus wins $1 is described by $Z = 1$.

> A random variable is sometimes described, less precisely, as a variable which takes on a value determined by chance. This is the case in the sense that when the process is repeated, some sample point occurs (determined by chance, if you will), and the value of the random variable that is determined is the one assigned to that sample point.

To determine if 40¢ per game is the fair price for Gus to pay Dan prior to a toss of Dan's die, let us envision the game being played a large number of times, say *N*. Then it makes sense to define the fair price per game that Gus should pay Dan prior to a play of the game to be the average amount per game that Gus can be expected to win from Dan in *N* tosses of the die. Since the random variable *Z* describes the amount Gus wins from Dan on a toss of the die, the fair price is the average value of *Z* over *N* throws. This is the sum of the values of *Z* in *N* throws divided by *N*. To determine this, let *k* denote the number of times the event $Z = -4$ occurs (that is, the number of times 1 shows) in *N* throws of the die. Then $N - k$ is the number of times the event $Z = 1$ occurs (that is, the number of times a face other than 1 shows). The sum of the values of *Z* in *N* throws is:

$$(-4)k + (1)(N - k)$$

Therefore the average value of *Z* is:

$$\frac{(-4)k + (1)(N - k)}{N} = (-4)\frac{k}{N} + (1)\frac{(N - k)}{N}$$

Now $\frac{k}{N}$ is the relative frequency of occurrence of the event $Z = -4$ in *N* throws of the die, and $\frac{N-k}{N}$ is the relative frequency of occurrence of the event $Z = 1$ in *N* throws of the die. By virtue of the relative frequency interpretation of probability., if N is large $\frac{k}{N}$ is approximately $P(Z=-4)$ and $\frac{N-k}{N}$ is approximately $P(Z=1)$. Therefore, if N is large,

$$(-4)\frac{k}{N}+(1)\frac{(N-k)}{N} \approx (-4)P(Z=-4)+(1)P(Z=1),$$

which is called the **expected value or mean value or expectation of** Z. Since $P(Z=-4)=P(1)=1/6$ and $P(Z=1)=P(2,3,4,5,6)=5/6$, we have,

$$(-4)P(Z=-4)+(1)P(Z=1)=(-4)\frac{1}{6}+(1)\frac{5}{6}=\frac{1}{6},$$

which means that over the long run the average amount that Gus can expect to win per game from Dan is approximately 1/6 of a dollar ($16\frac{2}{3}$ cents). Thus 1/6 of a dollar per game and not 40 cents per game is the fair price for Gus to pay Dan prior to each play of the game.

Note that on any toss of the die Gus will never win 1/6 of a dollar; he will either win \$1 or lose \$4; 1/6 of a dollar is Gus's average gain per toss of the die over a large number of tosses of the die.

> More generally, if X is a finite random variable which assumes values $x_1, x_2, \ldots, x_n$, then the **expected** or **mean value** of X, denoted by $E(X)$ or μ, is defined by:
>
> $$\begin{aligned}\mu &= x_1P(X=x_1)+\cdots+x_nP(X=x_n)\\ &= \sum_{i=1}^{n} x_iP(X=x_i)\end{aligned}$$

If the random process in question can be repeated a large number of times, then the mean value of X is approximately equal to the average value of X with respect to a large number of repetitions of the process.

If the random process will only occur once and values of the defined random variable X describe pay-offs to competitors, then the expected value of X gives us a sense of the fairness of the situation to the competitors. If $\mu \neq 0$, then the competition field is not level and one competitor has an advantage indicated by the magnitude of μ. If $\mu = 0$, then the competition field is level or **fair** and there is no inherent advantage in the situation for any competitor.

A random variable helps to put a spotlight on events of interest, described by the values it assumes. It is also useful to define a function which describes the probabilities with which these values are assumed.

If X is a finite random variable and x is a value assumed by X, then the function $p(x)$ defined by

$$p(x) = P(X = x)$$

is called the **probability distribution** or **probability distribution function** of X.

In terms of the probability distribution $p(x)$ of X, the definition of the mean value of X takes the following form:

$$\mu = x_1 p(x_1) + \cdots + x_n p(x_n)$$
$$= \sum_{i=1}^{n} x_i p(x_i)$$

Gullible Gus's probability distribution $p(x)$ is defined by:

$$p(-4) = \frac{1}{6}, \quad p(1) = \frac{5}{6}$$

Example 1 Lottery

To raise expense money for a chess tour the Bayside Chess Club has decided to run a lottery. There is to be one first prize of \$500, two second prizes of \$200, and three third prizes of \$100. One drawing is to be held. It is envisioned that 2000 tickets at \$1.00 each will be sold. Find the expected net winnings of Jean Jones, who buys one ticket.

Let X denote the expected net winnings of Jean Jones. The possible values of X are 499, 199, 99, and -1. We assume that 2000 tickets are indeed sold and that the ticket drawn is drawn at random. The computation of μ is simplified by use of the column organization shown in Table 1. Place the values of X in the first column, the values of $p(x_i)$ in the second column, and calculate and list their corresponding products in the third column. The sum of the values in this last column gives us μ, which in this case is -0.40.

Table 1

x_i	$p(x_i)$	$x_i p(x_i)$
-1	1994/2000	-1994/2000
99	3/2000	297/2000
199	2/2000	398/2000
499	1/2000	499/2000

$$\sum_{i=1}^{4} x_i p(x_i) = -0.40$$

In relative frequency terms, if Jean Jones were to participate in many lotteries of this kind, she would lose on average about 40 cents per play. On any one play she would, of course, win \$500, or \$200, or \$100, or \$0. For any one particular play of such a lottery -0.40 serves as a measure of the unevenness of the playing field for any one individual player.

EXERCISES

1. Identify each random variable *X* as being discrete or continuous. For each case explain the basis for your conclusion.

   (a) *X* is the number of persons who register each week for a course in statistics at Ecap University.

   (b) *X* is the sum of the numbers showing when a pair of dice are tossed.

   (c) *X* is the length of time that it takes a computer program to run.

   (d) *X* is the diameter of a circular ring made in a production process.

2. Dan to Gus: I have a die whose behavior is described by the following probability model: $S = \{1, 2, 3, 4, 5, 6\}$, $P(1) = P(3) = P(5) = \frac{1}{9}$, $P(2) = P(4) = P(6) = \frac{2}{9}$

   Now, if an even number shows on a throw of this die, I pay you \$1.00; if an odd number shows, you pay me \$1.50.

   (a) Define a random variable *Z* which describes the amount that Gus wins on a throw of this die.

   (b) Determine the probability distribution of *Z*.

   (c) Determine the mean of *Z*.

   (d) Give an interpretation for the value obtained in answer to (c).

(e) Is the game fair? Explain. If it is not a fair game, how could it be made fair?

3. New offer from Gus to Dan: If a number greater than 3 shows on a throw of this die, I pay you $1.25; if a number less than or equal to 3 shows, you pay me $1.50. Same questions (a) through (e).

4. A random variable Z has the following distribution function $p(x)$:

x	-5	0	5	10	15	20
$p(x)$	0.2	0.1	0.3	0.1	0.2	0.1

Find the following probabilities:

(a) $P(Z \geq 5)$

(b) $P(Z < 14)$

(c) $P(0 \leq Z < 15)$

(d) $P(Z \geq 9)$

(e) $P(-5 < Z < 15)$

(f) $P(Z < 7)$

5. One of the games offered at the casino of the Palace Hotel involves the tossing of a pair of fair dice. When the sum of the numbers showing is even the house wins $2.00, and when the sum of the numbers showing is odd the house loses $3.00.

(a) Define a random variable Z which describes the amount that the house wins on a throw of these dice.

(b) Determine the probability distribution of Z.

(c) Determine the mean of Z.

(d) Give an interpretation for the value obtained in answer to (c).

(e) Is the game fair? Explain. If it is not a fair game, how could it be made fair?

6. At one of their tables the casino of the Palace Hotel decided to switch to a pair of loaded dice with the property that outcomes with an even sum (such as (1,1),

(1,3) etc.) are twice as likely to occur as outcomes with an odd sum (such as (1,2), (2,3), etc.). Same questions (a) through (e).

7. The Energy Efficient Home Renovation Company offers free estimates in bidding for jobs. For a window replacement job Energy Efficient estimates the cost of providing a free estimate to be $200. It stands to make a profit of $5000 if awarded the job. The probability of being awarded the job has been estimated as 0.4. Determine Energy Efficient's expected net profit.

8. According to a life insurance company, the probability that a 30-year-old man will survive 1 year is 0.97, and that he will die within a year is 0.03. The company offers to sell such a man a $1000 one-year term life policy for a premium of $15.

   (a) What is the company's expected gain?

   (b) Give an interpretation for the value obtained in (a).

   (c) What premium would the company have to charge for such a policy if it is to break even, exclusive of administrative costs and taxes?

9. A sample of 2 items is chosen at random from a lot containing 10 items, 2 of which are defective.

   (a) Find the expected number of defective items drawn.

   (b) Give an interpretation for the value obtained in (a).

10. Does the following define a possible probability distribution $p(x)$ for a random variable X? Explain.

X	1	3	5	7	9
$p(x)$	0.30	0.25	0.15	0.22	0.11

11. The probability distribution $p(x)$ of the random variable X describing the number of accidents on a major highway in New Jersey on weekends is shown below.

x_i	0	1	2	3	4	5
$p(x_i)$	0.25	0.40	0.20	0.10	0.04	0.01

(a) Find the expected number of accidents.

(b) Give an interpretation for the value obtained in (a).

7.2 ■ Variance and Standard Deviation

How are the values of a random variables distributed about its mean? The **variance** and **standard deviation** of a random variable provide us with numerical measures of this.

> If *X* is a finite random variable which takes values $x_1, \ldots, x_n$ and has mean μ and probability distribution *p*(*x*), then the **variance** of *X*, also called the **variance** of *p*(*x*), is defined by:
>
> $$\sigma^2 = \sum_{i=1}^{n} (x_i - \mu)^2 \, p(x_i)$$
>
> The square root of the variance is called the **standard deviation** of *X*, and also the **standard deviation** of *p*(*x*).

The variance of Gullible Gus's random variable *Z*, which takes values -4 and 1 with probabilities 1/6 and 5/6, respectively, and has mean 1/6, is:

$$\begin{aligned}\sigma^2 &= \left(-4 - \frac{1}{6}\right)^2 \cdot \frac{1}{6} + \left(1 - \frac{1}{6}\right)^2 \cdot \frac{5}{6} \\ &= 3.47\end{aligned}$$

The standard deviation σ equals $\sqrt{3.47} \approx 1.86$.

The following computation theorem, which often simplifies the computation of σ^2, follows from the definition of variance.

> **Computation Theorem** for σ^2:
>
> $$\sigma^2 = \sum_{i=1}^{n} x_i^2 \, p(x_i) - \mu^2$$

The efficient use of this theorem is aided by a column arrangement of the needed ingredients, as illustrated in Table 2, for the calculation of variance for Gullible Gus's random variable *Z*.

Table 2

x_i	$p(x_i)$	$x_i^2 p(x_i)$
–4	$\frac{1}{6}$	16·—
1	—	·—
	1	$\sum x_i^2 p(x_i) = \frac{21}{6}$

Since $\mu = 1/6$, we have:

$$\sigma^2 = \frac{21}{6} - \left(\frac{1}{6}\right)^2 = 3.47$$

Example 1 Accident Occurrence

The probability distribution $p(x)$ of the random variable X describing the number of accidents on a major highway in a northeastern state on weekends is shown in Table 3. Find the variance and standard deviation of $p(x)$.

Table 3

x_i	0	1	2	3	4	5
$p(x_i)$	0.25	0.40	0.20	0.10	0.04	0.01

The column arrangement of Table 4 helps us to organize our work in a straightforward, simple manner.

Table 4

x_i	$p(x_i)$	$x_i p(x_i)$	$x_i^2 p(x_i)$
0	0.25	0	0
1	0.40	0.40	0.40
2	0.20	0.40	0.80
3	0.10	0.30	0.90
4	0.04	0.16	0.64
5	0.01	0.05	0.25
	1	1.31	2.99

From the sum given by column 3 we have that $\mu = 1.31$. Subtracting $(1.31)^2 = 1.72$ from the sum of column 4 yields $\sigma^2 = 2.99 - 1.72 = 1.27$. The standard deviation $\sigma = \sqrt{1.27} = 1.13$.

EXERCISES

1. Find the variance and standard deviation of the random variables introduced in Example 1 of Section 7.1.

2. (a)—(i) Find the variance and standard deviation of the random variables introduced in Exercises 2-9 and 11 of Section 7.1.

3. The number of inaccurate statements made by senatorial candidate J. J. Roberts per 15-minutes of speaking time, denoted by *X*, has been observed over a long period of time and found to have a mean of 16 with a standard deviation of 2. Estimate the probability that *X* will be between 12 and 20 during any 15 minute segment of candidate Roberts's speeches.

7.3 ■ Bernoulli Trial Random Variables

One of the most important examples of a finite random variable is the Bernoulli trial random variable *X* which describes the number of successes in *n* Bernoulli trials.

> From Section 6.8, if *X* is the number of successes in *n* Bernoulli trials, where *E* denotes the event whose occurrence or nonoccurrence on a trial is the focus of our interest, *p* is the probability that *E* occurs on a trial, and $q = 1 - p$, then the probability distribution of *X*, $p(x) = P(X = x)$, is defined by:
>
> $$p(x) = C(n, x)p^x q^{n-x}, \quad x = 0, 1, \ldots, n$$
>
> This **Bernoulli trial probability distribution** is called the **binomial distribution.** It can be shown that it has mean, variance and standard deviation given by:
>
> $$\mu = np, \quad \sigma^2 = npq, \quad \sigma = \sqrt{npq}$$

From the Bernoulli trial model developed for Larry's Apparel Shop in Example 3 of Section 6.8 (page 185), where $n = 100$, *E* is the event that a shopper makes a purchase, and $p = 0.20$, the random variable *X* describing the number of shoppers who make a purchase has probability distribution,

$$p(x) = C(100, x)(0.20)^x (0.800^{100-x}$$

with mean $\mu = 100(0.20) = 20$, variance $\sigma^2 = 20(0.80) = 16$, and standard deviation 4.

EXERCISES

In Exercises 1-8 define X, the number of successes in terms of the context provided, calculate the probability distribution $p(x)$, and the mean, variance and standard deviation of X.

1. Example 2, Section 6.8 (page 185).

2. Exercise 1, Section 6.8 (page 187).

3. Exercise 2, Section 6.8.

4. Exercise 3, Section 6.8.

5. Exercise 4, Section 6.8.

6. Exercise 5, Section 6.8.

7. Exercise 6, Section 6.8.

8. Exercise 7, Section 6.8.

8

The Remarkable Normal Curves

8.1 ■ The Family of Normal Curves

The normal curves are bell-shaped, with a single peak, and symmetric with respect to the line that is drawn through the peak. All normal curves lie above the horizontal axis; as we move along the horizontal axis away from the value at which a curve peaks, the curve gets closer and closer to the horizontal axis, but does not intersect it. The value at which a curve peaks is called its **mean.** Figure 1 shows three normal curves with the same mean, denoted by μ. These curves differ in a characteristic, called the **standard deviation** of the curve, which is a measure of the concentration of a curve about its mean; the smaller the standard deviation, the higher the peak of the curve, and the more concentrated is the curve about its mean. The standard deviations σ_1, σ_2, and σ_3 of the three normal curves shown in Figure 1 satisfy $\sigma_1 < \sigma_2 < \sigma_3$.

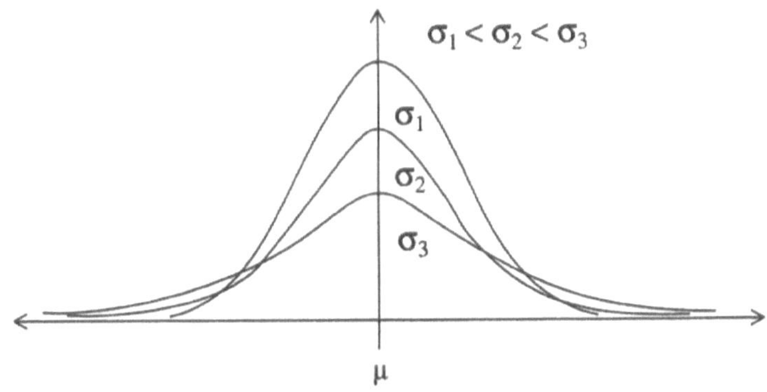

Figure 1

A normal curve is completely determined by its mean and standard deviation. The general equation of the normal curve with mean μ and standard deviation σ is

$$y = \frac{1}{\sigma\sqrt{2\pi}} e^{-(x-\mu)^2/2\sigma^2}$$

where π, the ratio of the circumference of a circle to its diameter, is approximately 3.14159, and *e* is a constant whose approximate value is 2.71828. When values for μ and σ are given, a specific normal curve is determined; for that normal curve values of *y* can be determined by substituting values for *x*.

The role played by μ and σ for the normal curves is analogous to the role played by the parameters *m* and *b* for the lines defined by $y = mx + b$. When values for *m* and *b* are given, a specific line is determined; for that line values of *y* can be determined by substituting values for *x*.

Fortunately, our work with normal curves does not require us to plot a curve from its equation; a quickly drawn sketch of a normal curve which shows its general character and mean will suffice. Our concern is with determining areas of regions which have a normal curve as their upper boundary. At first sight it might seem that this would require that we determine normal curve areas for each μ and that arise in practice. Once again we can breath a sigh of relief that this is not the case. We can restate an area determination problem that arises for any normal curve in terms of an area determination problem for a comparable region of a normal curve adopted as a standard. This **standard normal curve**, as it is called, has mean $\mu = 0$ and standard deviation $\sigma = 1$.

Given any normal curve with mean μ and standard deviation we transfer the scene to the standard normal curve by carrying out a change of scale by means of the formula:

$$z = \frac{x-\mu}{\sigma}$$

In this new *z*-scale (see Figure 2) *z* tells us in terms of standard deviation units how many standard deviations the *x*-value lies above or below its mean μ. For $z = 1$, for example, we have:

$$1 = \frac{x-\mu}{\sigma}$$

$$\sigma = x - \mu$$

$$x = \mu + \sigma$$

That is, $z = 1$ corresponds to x being one standard deviation unit above the mean μ in the x-scale, which is what Figure 2 shows.

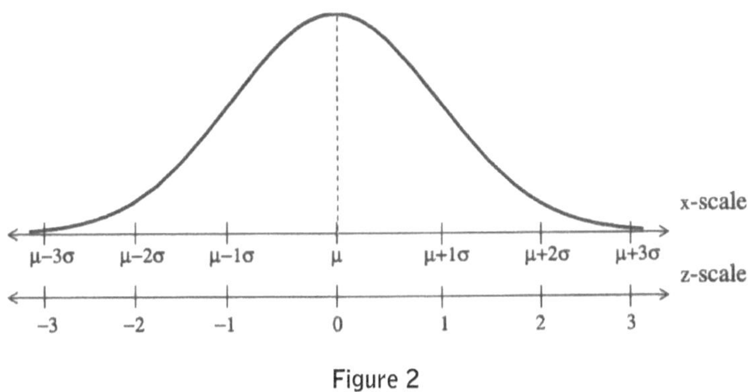

Figure 2

Areas of Regions Determined by the Standard Normal Curve

We first show how areas of regions determined by the standard normal curve can be obtained and then show how area determination problems with respect to other normal curves can be shifted to the standard normal curve.

Areas of regions defined by the standard normal curve have been tabulated and Table A (page 652) gives area values to four places. A short version of this table is reproduced here as Table 1. Table A specifies the standard normal curve area of the shaded region shown in Figure 3 for values of z from 0.00 to 3.09. The z values in our abbreviated Table 1 range from 0.00 to 1.19. The values in both tables give the area, to four places, of the region between the mean $\mu = 0$ and the given value of z. To obtain the area of the region between 0 and 0.63, for example (see Figure 4), we move across the row labeled 0.6 until we reach the column headed by 3, the last digit of 0.63. This yields 0.2357 as the area of our region, correct to four places.

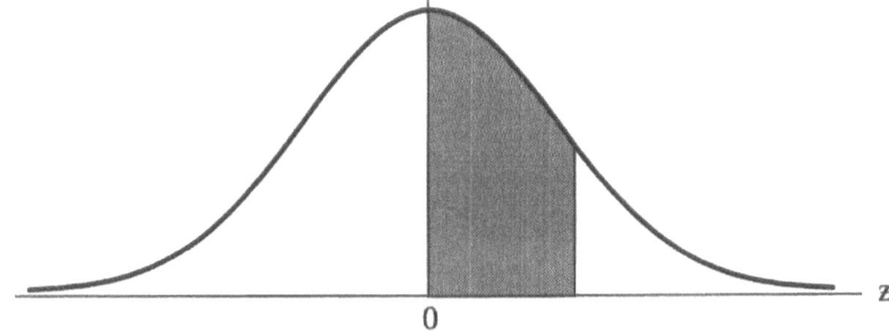

Figure 3

Table 1

z	.00	.01	.02	.03	.04	.05	.06	.07	.08	.09
0.0	.0000	.0040	.0080	.0120	.0160	.0199	.0239	.0279	.0319	.0359
0.1	.0398	.0438	.0478	.0517	.0557	.0596	.0636	.0675	.0714	.0753
0.2	.0793	.0832	.0871	.0910	.0948	.0987	.1026	.1064	.1103	.1141
0.3	.1179	.1217	.1255	.1293	.1331	.1368	.1406	.1443	.1480	.1517
0.4	.1554	.1591	.1628	.1664	.1700	.1736	.1772	.1808	.1844	.1879
0.5	.1915	.1950	.1985	.2019	.2054	.2088	.2123	.2157	.2190	.2224
0.6	.2257	.2291	.2324	.2357	.2389	.2422	.2454	.2486	.2517	.2549
0.7	.2580	.2611	.2642	.2673	.2704	.2734	.2764	.2794	.2823	.2852
0.8	.2881	.2910	.2939	.2967	.2995	.3023	.3051	.3078	.3106	.3133
0.9	.3159	.3186	.3212	.3238	.3264	.3289	.3315	.3340	.3365	.3389
1.0	.3413	.3438	.3461	.3485	.3508	.3531	.3554	.3577	.3599	.3621
1.1	.3643	.3665	.3686	.3708	.3729	.3749	.3770	.3790	.3810	.3830

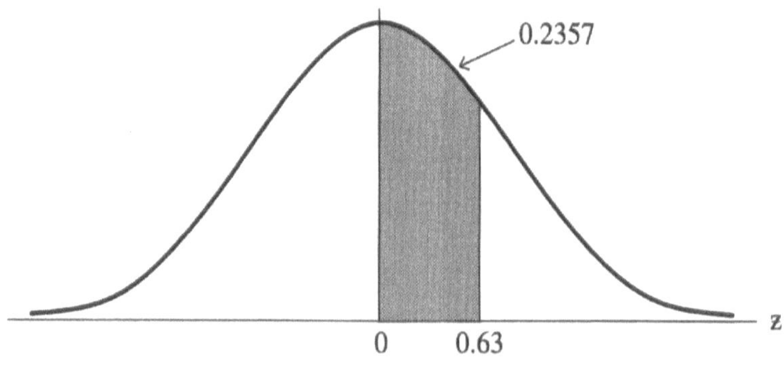

Figure 4

To obtain the area value corresponding to a negative value of z, $z=-1.18$, for example, we make use of the fact that the area of the region between 0 and -1.18 is the same as the area of the region between 0 and 1.18 because of the symmetry of the curve, and look up the area corresponding to $z=1.18$. We thus find that the area of our region, shown in Figure 5, is 0.3810.

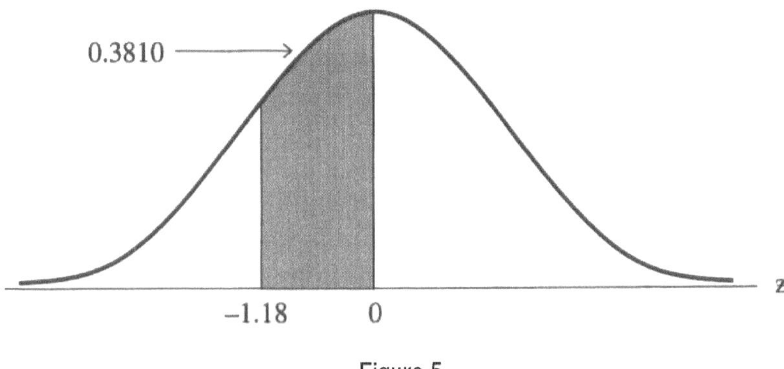

Figure 5

The area of the region under each of the normal curves is 1. In terms of the standard normal curve, the area of the region to the right of the line of symmetry, determined by $\mu = 0$, is 0.5000 as is the area of the region to the left of $\mu = 0$. To obtain the area of a region to the right of a positive value of z, $z = 0.94$, for example (see Figure 6) we subtract the area of the unwanted region between 0 and 0.94, 0.3264, from 0.5000 to obtain 0.1736.

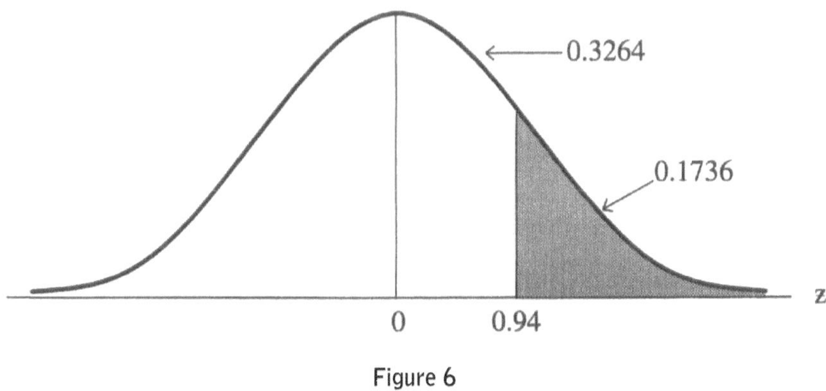

Figure 6

To obtain the area of a region to the left of a positive value of z, $z = 0.89$, say (see Figure 7), simply add 0.5000 to 0.3133, the tabular value for $z = 0.89$. This yields 0.8133 as the area of the region to the left of $z = 0.89$.

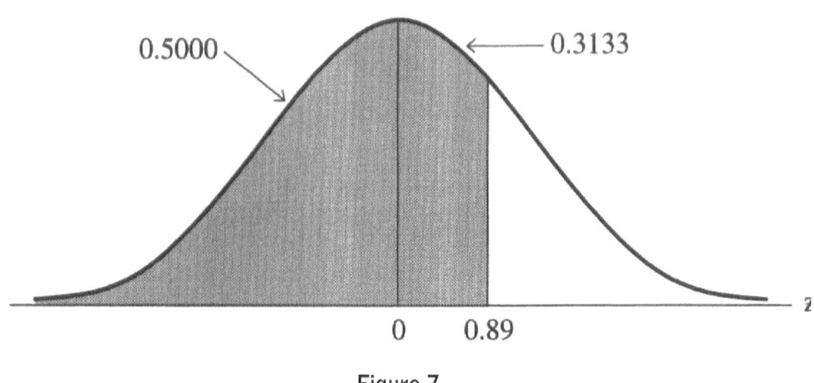

Figure 7

Also of importance are problems which lead to the normal curve area of a region between two given values of *z*. If both *z* values are on the same side of the mean, that is, both are positive or both are negative, then the area of the region between them is the difference in their tabular values. For example, the area of the region shown in Figure 8 determined by $z_1 = 0.72$ and $z_2 = 1.13$, is $0.3708 - 0.2642 = 0.1066$.

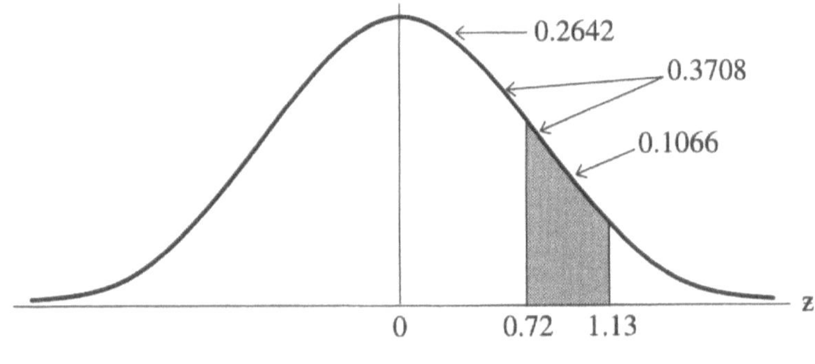

Figure 8

If the two *z* values are on opposite sides of the mean, that is, one is positive and one is negative, then the area of the region between them is the sum of their tabular values. For example, the area of the region shown in Figure 9, determined by $z_1 = -0.64$ and $z_2 = 1.18$, is $0.2389 + 0.3810 = 0.6199$.

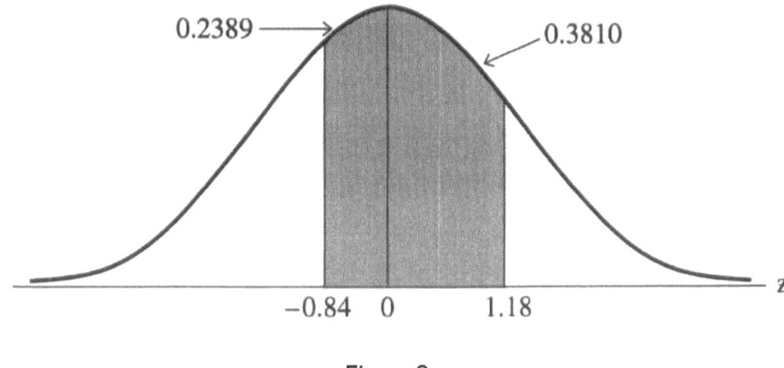

Figure 9

With all such area determination problems it is a good idea to make a diagram showing the region of concern. It then becomes much clearer how to proceed.

Areas of Regions Determined by Normal Curves

More generally, consider the general normal curve with mean μ and standard deviation and let us suppose that we are required to find the area of the region bounded by x_1 and x_2 as shown in Figure 10(a). To do this we shift the scene to the standard normal curve, determine the z_1 and z_2 values corresponding to x_1 and x_2

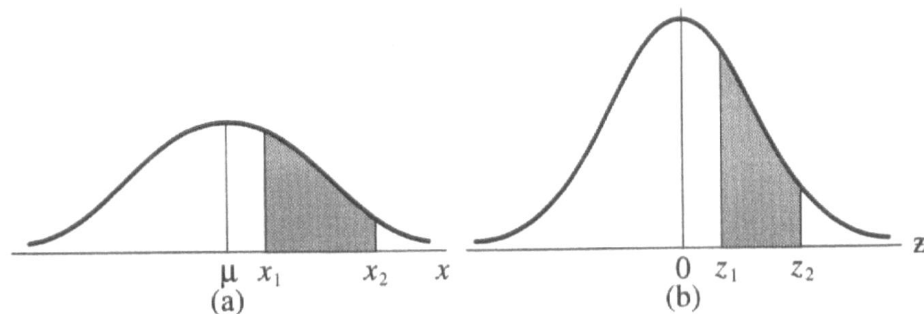

Figure 10

from the relations

$$z_1 = \frac{x_1 - \mu}{\sigma}, \quad z_2 = \frac{x_2 - \mu}{\sigma},$$

and find the area of this comparable region bounded by z_1 and z_2 (see Figure 10(b)). The area of this standard normal curve region is equal to the area of the original region (see Figure 10(a)).

Example 1

For the normal curve with mean $\mu = 4$ and standard deviation $\sigma = \sqrt{2}$, find the area of the region bounded by $x_1 = 4.5$ and $x_2 = 5.5$, shown in Figure 11.

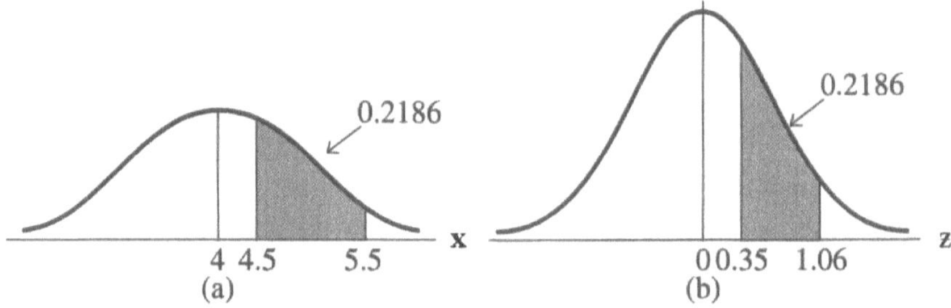

Figure 11

Corresponding to $x_1 = 4.5$ and $x_2 = 5.5$ we have:

$$z_1 = \frac{4.5 - 4}{\sqrt{2}} = \frac{0.5}{1.414} = 0.35$$

$$z_2 = \frac{5.5 - 4}{\sqrt{2}} = \frac{1.5}{1.414} = 1.06$$

The region under the standard normal curve between these z values, shown in Figure 11(b), has area $0.3554 - 0.1368 = 0.2186$. Thus the region under the given normal curve between $x_1 = 4.5$ and $x_2 = 5.5$ has area 0.2186.

Reversing Direction

Sometimes we are led to reverse direction in the sense that we start with an area value and seek to find the z-value, or come as close to it as we can, that yields the given area value. If, for example, we begin with the area value 0.3554, then by looking in the body of the standard normal curve table we find it exactly and read off the z-value which gives it as $z = 1.06$. If we begin with the area value 0.1955, then we see that the value closest to it in the standard normal curve table is 0.1950 with z-value 0.51. The next area value is 0.1985 so that 0.1955 is much closer to 0.1950 than 0.1985. Without going through an interpolation process, it is reasonable to take $z = 0.51$, to three places as the z-value for 0.1955.

EXERCISES

Find the area of the region under the standard normal curve that is defined by the following bounds.

1. Between 0 and $z = 2.38$.

2. Between 0 and $z = -1.46$.

3. To the right of $z = 1.56$.

4. To the left of $z = 2.04$.

5. To the right of $z = -1.20$.

6. Between $z_1 = -0.84$ and $z_2 = 1.58$.

7. Between $z_1 = -0.82$ and $z_2 = 1.98$.

8. Between $z_1 = -1.96$ and $z_2 = 1.90$.

9. To the left of $z = -1.24$.

10. To the right of $z = -0.50$.

11. Between $z_1 = -2.02$ and $z_2 = -0.44$.

12. To the left of $z = -1.08$.

Find the area of the region under the normal curve defined by the following values of μ and σ with respect to the given bounds.

13. $\mu = 12$, $\sigma = 4$, between $x_1 = 13$ and $x_2 = 16$.

14. $\mu = 20$, $\sigma = 6$, to the left of $x = 23$.

15. $\mu = 52$, $\sigma = 10$, between $x_1 = 50$ and $x_2 = 55$.

16. $\mu = 110$, $\sigma = 5$, to the right of $x = 115$.

17. $\mu = 64$, $\sigma = 2$, between $x_1 = 66$ and $x_2 = 68$.

18. $\mu = 82$, $\sigma = 6$, between $x_1 = 80.2$ and $x_2 = 83.3$.

19. $\mu = 100$, $\sigma = 4.28$, to the right of $x = 97.3$.

20. $\mu = 32.2$, $\sigma = 2.46$, to the left of $x = 30.1$.

21. $\mu = 4$, $\sigma = \sqrt{2}$, to the right of $x = 4.5$.

Find the *z*-values which yield the following area values.

22. 0.1879 23. 0.4750

24. 0.4900 25. 0.4500

26. 0.3830 27. 0.4950

28. Show that the area of the region under the normal curve with mean μ and standard deviation σ

(a) between $x_1 = \mu - \sigma$ and $x_2 = \mu + \sigma$ is 0.6826;

(b) between $x_1 = \mu - 2\sigma$ and $x_2 = \mu + 2\sigma$ is 0.9544;

(c) between $x_1 = \mu - 3\sigma$ and $x_3 = \mu + 3\sigma$ is 0.9972.

Draw diagrams.

In summary, these results say that for any normal distribution roughly 68% of the data fall within one standard deviation of the mean, roughly 95% of the data fall within two standard deviations of the mean, and roughly 99.7% of the data fall within three standard deviations of the mean.

8.2 ■ Normal Curve Estimates for Bernoulli Trial Probabilities

The normal curves exhibit many dimensions. One dimension involves approximating probability distributions of discrete random variables.

To explore this dimension we first return to the Bernoulli trial model with $n=8$ trials and probability $p=1/2$ of success on each trial. If we let X denote the number of successes in 8 trials, then the probability distribution $p(x)$ of X is:

$$p(x) \quad = \quad P(X=x) \quad = \quad C(8,x)\left(\frac{1}{2}\right)^x\left(\frac{1}{2}\right)^{8-x}$$

For $x=0,1,\ldots,8$, we obtain:

$$\begin{array}{ll} p(0)=0.0039 & p(5)=0.2188 \\ p(1)=0.0312 & p(6)=0.1094 \\ p(2)=0.1094 & p(7)=0.0312 \\ p(3)=0.2188 & p(8)=0.0039 \\ p(4)=0.2734 & \end{array}$$

The mean $\mu=np$ and standard deviation $\sigma=\sqrt{npq}$ of X (see Section 7.3) are:

$$\mu=4, \; \sigma=\sqrt{2}$$

These probabilities can be described geometrically as areas by means of a probability histogram by locating the possible values of X on the x-axis and then, centered on each of these 8 values, drawing a rectangle whose base has length 1 and whose height is the probability of the value on which the rectangle is centered. We obtain the histogram shown in Figure 12. The rectangle is centered at 0 has area 0.0039, which is the probability that $X=0$; the rectangle centered at 1 has area 0.0312, which is the probability that $X=1$, and so on.

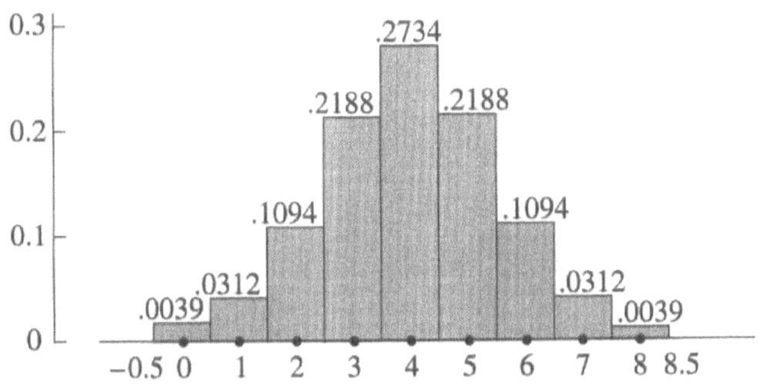

Figure 12

If we superimpose on this probability histogram the normal curve determined by mean $\mu = np = 4$ and standard deviation $\sigma = \sqrt{npq} = \sqrt{2}$, we obtain a remarkably good fit shown in Figure 13. A few comparisons of Bernoulli trial probability distribution values with their corresponding normal curve region areas are most suggestive. The probability of 5 successes, $p(5) = 0.2188$, expressed geometrically by the area of the rectangle centered at 5 whose base extends from 4.5 to 5.5, differs by 0.0002 from the area of the normal curve region lying above the same base, 0.2186 (see Section 8.1, Example 1, page 178). The probability of 5 or more successes, $p(5) + p(6) + p(7) + p(8) = 0.3633$, expressed geometrically by the sum of the areas of the rectangles centered at 5, 6, 7, and 8, whose base extends from 4.5 to 8.5, differs by 0.0001 from the area of the normal curve region lying on the same base, or equivalently, on the base which extends to the right of 4.5, 0.3632 (see Section 8.1, Exercise 21).

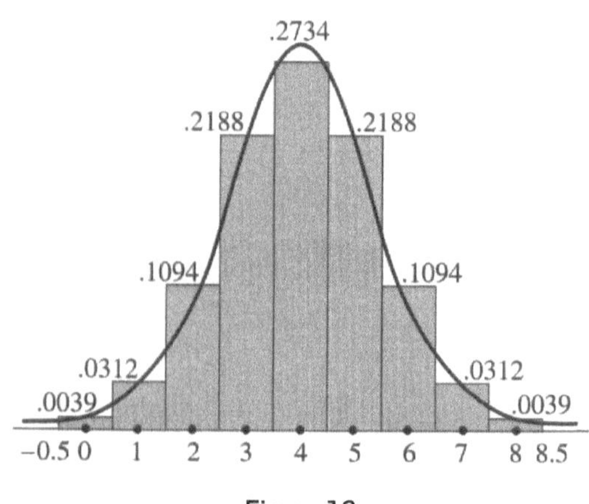

Figure 13

The accuracy of these approximations suggests that normal curves defined by $\mu = np$ and $\sigma = \sqrt{npq}$, under suitable conditions, can be used to approximate Bernoulli trial probabilities. This is indeed the case.

More generally, consider a Bernoulli trial model with n trials, an event E whose occurrence or nonoccurrence on each trial (called success and failure, respectively) is the focus of our interest, where the probability of success on each trial is p. Let X, the Bernoulli trial random variable, denote the number of successes in n trials. The probability distribution $p(x)$ of X is defined by:

$$p(x) \quad = \quad P(X = X) \quad = \quad C(n,x)p^x q^{n-x}, q = 1 - p, \text{ for } x = 0, 1, \ldots, n.$$

We shall more formally state a theorem in a moment, but the basic idea is the following: Suppose that we wish to estimate $p(x)$ for some value of x, 70 let us say, and that conditions are such that we have a green light to proceed. In geometric terms $p(70)$ is the area of a rectangle with height $p(70)$ whose base is centered at 70 and extends from 69.5 to 70.5 (see Figure 14(a)). Our problem is to approximate the area of this rectangle, which is shown in Figure 14(b) along with the "appropriate" normal curve arc which appears over the base from 69.5 to 70.5. The region defined by the normal curve differs from the rectangle in that it includes A, which is not part of the rectangle and does not include B, which is part of the rectangle. If the areas of A and B are approximately equal, and under suitable conditions they are, then the gain is balanced by the loss, and the area of the normal curve region is approximately that of the area of the rectangle, which is $p(70)$, the probability of 70 successes.

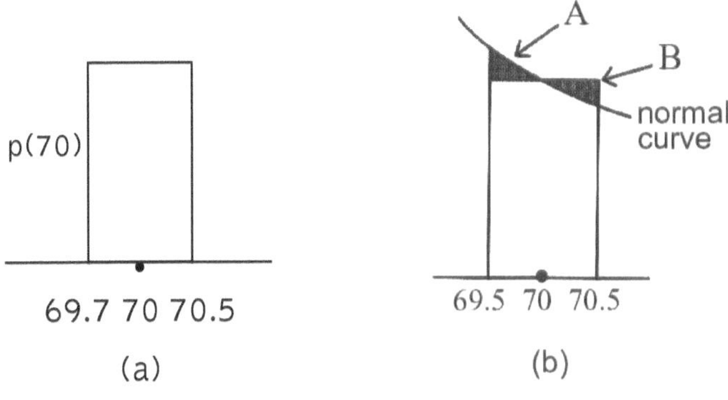

Figure 14

If instead of 70 successes we talk more generally about c successes, then the probability rectangle whose area is be approximated is centered at c with base extending from $c-1/2$ to $c+1/2$.

To take another example if we wish to approximate $P(c \le X \le d)$, the probability that the number of successes is between c and d, inclusive, then we are looking at a sequence of rectangles, the first centered at c and the last centered at d, the sum of whose areas is to be approximated. To include the areas of the first and last of these rectangles in our estimate we must take as our base the interval bounded by $c-1/2$ and $d+1/2$ (see Figure 15).

There are a number of variations on this basic theme which these examples hopefully indicate. More formally, we have the following theorem which was in part established by Abraham de Moivre and refined by Pierre Simon de Laplace.

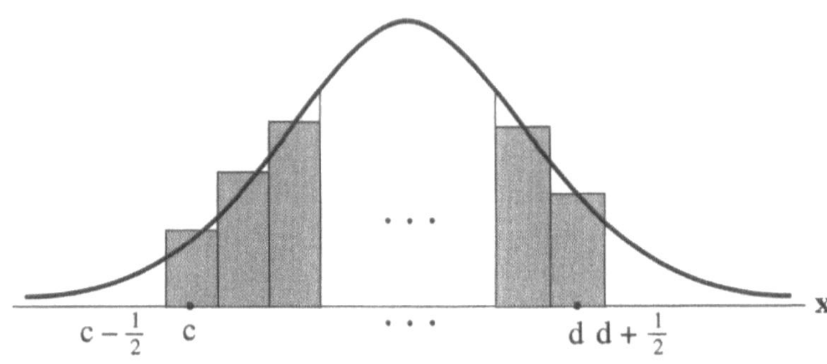

Figure 15

de Moivre—Laplace Theorem

If *np* and *nq* are both at least 5 (our green light), then the probability distribution *p*(*x*) of the Bernoulli trial random variable *X*, describing the number of successes in *n* Bernoulli trials, and Bernoulli trial probabilities in general, may be **approximated** by areas of normal curve regions, the normal curve being determined by mean μ and standard deviation σ defined by:

$$\mu = np, \quad \sigma = \sqrt{npq}$$

If, for example, we seek to estimate $P(c \leq X \leq d)$, the normal curve region is bounded by $c - 1/2$ and $d + 1/2$. If we seek to estimate $P(x \leq c)$, the normal curve region is the one to the left of $c + 1/2$, and so on.

Normal curve approximations of Bernoulli trial probabilities are best when *p* and *q* are close to 1/2. If *p* or *q* is very small, then *n* has to be sufficiently large to overcome the smallness of *p* or *q* to allow use of the normal curve approximation. This is what the guideline conditions $np \geq 5$, $nq \geq 5$ give us. If one of *np* and *nq* is less than 5, the normal curve approximation should ordinarily not be used. In such cases the Poisson distribution, which we shall not take up here, often provides excellent approximations.

Example 1 Return to Larry's Apparel Shop

Records kept by Larry's Apparel Shop indicate that 20% of the shoppers who come into the store to browse actually make a purchase. What is the probability that among 100 shoppers who examine Larry's merchandise at least 30 will make a purchase?

Our first task is to set up a probability model for this process. In examining this situation in Example 3 of Section 6.8 (page 142) we were led to take for this process a Bernoulli trial model with $n = 100$, E = the event that a shopper makes a purchase, $p = 0.20$. If we let X denote the number of purchases made in 100 trials, then the problem is to determine $P(X \geq 30)$.

To determine the feasibility of a normal curve approximation of $P(X \geq 30)$, we must examine np and nq.

$$np = 100(0.20) = 20, \qquad nq = 100(0.80) = 80$$

Since both values are greater than or equal to 5, we have a green light and may proceed.

The normal curve that we bring to this situation has mean $\mu = np = 20$ and standard deviation $\sigma = \sqrt{npq} = 4$. Our problem is to determine the area of the region to the right of $x_1 = 29.5$ (see Figure 16(a)). The z value corresponding to $x_1 = 29.5$ is $z_1 = 2.38$ (see Figure 16(b)). The area of the unwanted standard normal curve region between 0 and 2.38 is 0.4913, so that the area of the region to the right of 2.38 is $0.5000 - 0.4913 = 0.0087$.

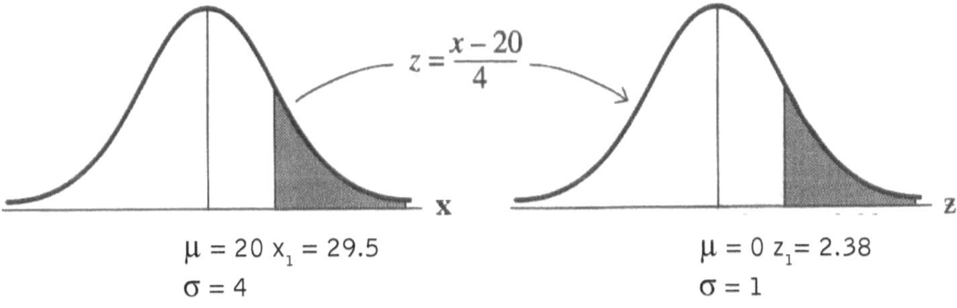

Figure 16(a) Figure 16(b)

Thus, the probability, in terms of our Bernoulli trial model, that among 100 shoppers who examine Larry's merchandise at least 30 will make a purchase is 0.0087. If we think in terms of groups of 100 shoppers, then at least 30 out of 100 will make a purchase, for many groups of 100 (the long run in this case), approximately 0.87% of the time.

Reminder

As precise as this may seem to be, let us keep in mind that it goes back to an **assumption** on which the Bernoulli trial model adopted for this situation is founded, namely, that however they interact shoppers have little impact on each other concerning the making of a purchase decision. This underlies taking the same value, p = 0.20, as the probability that a shopper will make a purchase. If the **realism** of this assumption is seriously open to question, then the Bernoulli trial model becomes seriously open to question, as does the 0.0087 probability value and its interpretation.

EXERCISES

1. Let X denote the number of successes in connection with a Bernoulli trial model with $n=100$ and $p=2/5$. Find (a) $P(X=30)$, (b) $P(X=25)$, (c) $P(X\leq 50)$, (d) $P(X>30)$, (e) $P(20\leq X\leq 40)$.

2. Let X denote the number of successes in connection with a Bernoulli trial model with $n=200$ and $p=0.30$. Find (a) $P(X=60)$, (b) $P(X=70)$, (c) $P(X\geq 60)$, (d) $P(40\leq X\leq 50)$, (e) $P(X>40)$.

In Exercises 3-9 state the probability model employed and the assumptions made.

3. A recently developed antibiotic for a spectrum of strep infections has been found to cause an allergic reaction in 5% of the people who take it. Find the probability that of 10,000 people given the antibiotic:

   (a) At least 551 allergic reactions will occur.

   (b) At least 490 allergic reactions will occur.

   (c) Between 480 and 510 allergic reactions will occur.

4. In the mass production of shafts for engine mounts at the MacNeil Company the probability that a defective shaft is produced is 0.02. For a production run of 12,000 shafts find the probability that:

   (a) No more than 200 defectives are produced.

   (b) At least 220 defectives are produced.

   (c) Between 230 and 260 defectives are produced.

5. Find the probability that a student will get at least 70 correct answers on a 100 question true—false test in economics if he answers each question by flipping a balanced coin and recording T if head shows and F if tail shows.

6. Dubin Outlets estimates that 2% of its accounts receivable cannot be collected. Find the probability that of 600 accounts receivable:

   (a) No more than 10 will not be collectable.

   (b) No more than 15 will not be collectable.

   (c) At least 20 will not be collectable.

7. A multicomponent flu vaccine has been found to be effective on 99% of the people inoculated. Of 500,000 who are inoculated, find the probability that:

   (a) Fewer than 2000 come down with the flu.

   (b) More than 4900 come down with the flu.

   (c) Between 4950 and 5100 come down with the flu.

8. An advertising agency claims that 22% of television viewers watch the new hit program *Happy Days and Foolish Nights.* Assuming this to be the case, if 300 television viewers are selected, what is the probability that:

   (a) At least 65 watch the program?

   (b) At most 80 watch the program?

   (c) Between 75 and 100 watch the program?

9. Studies have led medical authorities to conclude that 1 out of 500 teenagers in Southchester has the measles.

   (a) Find the probability that out of 10,000 teenagers selected at most 25 will have the measles.

   (b) Interpret the result you obtained.

   (c) Are the numbers obtained in answer to (a) and (b) credible? Explain

8.3 ▯ Normally Distributed Random Variables

Historically, the normal curves arose during the first third of the eighteenth century in response to the problem of approximating probabilities of a discrete random variable—the Bernoulli trial random variable.

Another dimension of the normal curves concerning continuous random variables, which we initiate a discussion of here, emerged during the nineteenth and twentieth centuries.

> A continuous random variable X is said to be **normally distributed** if its probability distribution is defined by the normal curve
>
> $$p(x) = \frac{1}{\sigma\sqrt{2\pi}} e^{-(x-\mu)^2/2\sigma^2}$$
>
> for suitable values of $\sigma > 0$ and μ.

The probability that X takes any specific value is 0, but the probability that X takes a value between two values, b and c, let us say, is given by the area of the region under $p(x)$ defined by b and c (see Figure 17(a)). The probability that X takes a value less than d, or less than or equal to d, is given by the area of the region under $p(x)$ to the left of d (see Figure 17(b)), and so on.

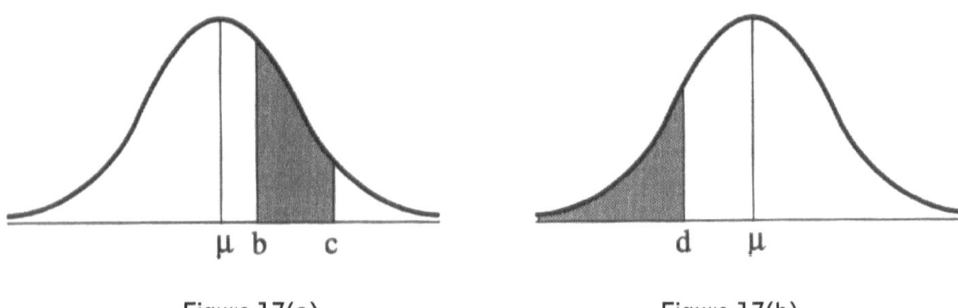

Figure 17(a) Figure 17(b)

The condition that a random variable be normally distributed is a mathematical ideal needed as a standard and for proving theorems, but this ideal is not realized "exactly" in practice. The question that emerges from applications is concerned with whether a probability distribution that arises in practice is close enough to being a normal distribution that we can treat it as such.

Example 1 Probability of Rejection

The Andrew J. Weil Company produces a certain kind of machine part. A deviation of more than ±0.01 millimeters from the desired length will result in rejection of the part. An analysis of the production process where the part is made led analysts to the hypothesis that the probability distribution approximating the probability behavior of the random variable describing the deviation from the desired length was the normal curve with $\mu = 0$ and $\sigma = 0.01$. The problem is to find the probability that a part will have to be rejected and interpret the result.

The region whose area we must first determine, bounded by $x_1 = -0.01$ and $x_2 = 0.01$, gives the probability that a part meets specifications (see Figure 18(a)). Transferring the scene to the standard normal curve, we obtain the corresponding *z*-values $z_1 = -1.00$ and $z_2 = 1.00$ (see Figure 18(b)). The area of the region defined by the *z*-bounds is 0.6826, which is therefore, the area defined by the *x*-bounds.

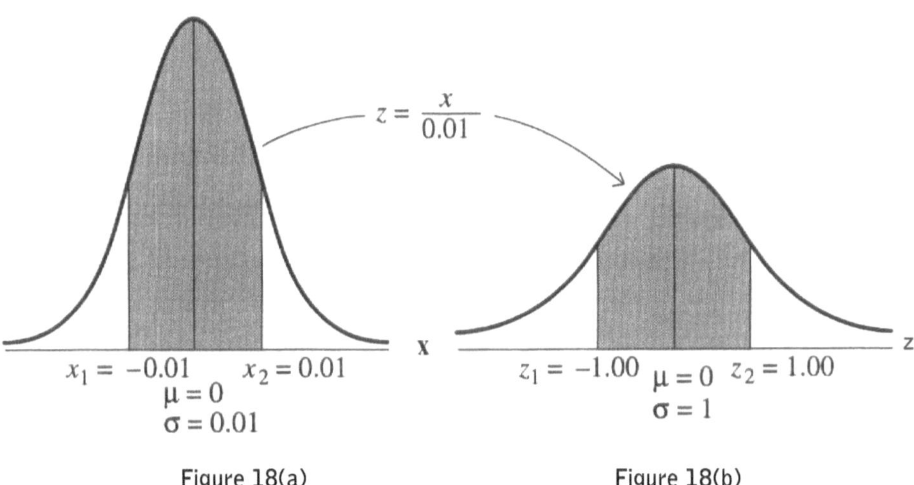

Figure 18(a)

Figure 18(b)

The probability that a part produced meets specifications is 0.6826, which implies that the probability that a part produced does not meet specifications and must be rejected is 0.3174.

In relative frequency terms, this means that in producing this part under current plant conditions approximately 31.7% of the output would have to be rejected.

This rejection rate was comparable to what the company was experiencing and management called for an overhaul of the plant's production machinery.

EXERCISES

1. The Andrew J. Weil Company (see Example 1) had its production machinery overhauled in light of the huge rejection rate that it was experiencing. After the overhaul was completed studies showed that the random variable describing the deviation from the specified length norm is approximately normal with $\mu = 0$ and $\sigma = 0.02$. As noted in Example 1, a deviation of more than ±0.01 millimeters from the desired length will result in rejection of the part. Find the probability that a part will be rejected and interpret the result.

2. The filling machines of Ramunės Gourmet Coffee are set to fill the 8-ounce jars of their special blend with 8 ounces of coffee. The actual amount that a filling machine puts into a jar varies and is described by a random variable which is approximately normal with $\mu = 8$ and $\sigma = 0.12$ ounces.

   (a) Find the probability that a jar contains between 7.8 and 8 ounces. Interpret the result obtained.

   (b) If at most 1% of the jars are to contain less than 8 ounces of coffee, what mean fill value should the filling machines be set for?

3. The commuting time from Dreamland to Huxley College is described by an approximately normally distributed random variable with $\mu = 40$ minutes and $\sigma = 10$ minutes. In terms of the mean commuting time, find the probability that a commuter will be:

   (a) Between 5 and 10 minutes late.

   (b) More than 5 minutes late.

   (c) Between 6 and 12 minutes early.

4. In Example 1 a conclusion reached was that the probability that a part produced does not meet specifications and must be rejected is 0.3174. The relative frequency interpretation of this result is that in the mass production of the machine part under current conditions approximately 31.7% of the output would have to be rejected. Suppose it turned out that the company was experiencing a 48% rejection rate. How are we to account for the discrepancy between the actual 48% rejection rate and the predicted 31.7% rejection rate?

5. In computing areas of normal curve regions how do we know when to move the side boundary (or boundaries) half a unit and when not do so?

When Is a Random Variable Normally Distributed?

One answer to this question is provided by what has come to be termed the hypothesis of elementary errors. To obtain a handle on the hypothesis of elementary errors, suppose that a measurement of a quantity of interest is to be taken; it might be the length of a rod, diameter of a ball-bearing, or distance of a star from some fixed point of reference, to take a few examples. A reading is taken and a value is obtained; another reading is taken and another value is obtained; a number of readings are taken, and a number of distinct values are obtained. An error, call it E, occurs each time a reading is taken.

E may be viewed as a random variable and one might inquire as to how its probability behavior can be described. This question occupied the attention of a number of scholars in the late eighteenth and early nineteenth centuries, one of whom was the astronomer Friedrich Wilhelm Bessel (1784-1846) who introduced a version of the hypothesis of elementary errors.

The Hypothesis of Elementary Errors views an error E arising in a measurement process as the sum of a large number of component elementary errors $e_1, e_2, \ldots, e_n$

$$E = e_1 + e_2 + \ldots + e_n,$$

which come from different, independent sources, each one of which is negligible in comparison to the sum E.

On the basis of his hypothesis of elementary errors Bessel [1], writing in 1838, derived as an approximation for the probability distribution describing the probability behavior of E the normal distribution in the form:

$$f(x) = \frac{1}{m\sqrt{2\pi}} e^{-x^2/2m^2}$$

Bessel's m, we should note, is our .

He further observed that:

> Cases in which an error of observation does not result from the coincidence of many mutually independent causes are doubtless extremely

rare. Even in seemingly very simple types of observation it is often possible to demonstrate numerous causes of their errors.

Bessel continued, by way of illustration:

To illustrate this by an example I shall assume that a number of distances of a fixed star from the zenith or the pole has been observed by means of a meridian circle and shall attempt to enumerate the causes of the errors the presence of which is apparent from a survey of the results.

First of all, the instrument must be focused on the star, and this focusing procedure may entail errors due to various causes including (1) the fact that the power of the telescope has limits within which its focusing direction remains arbitrary, (2) the fact that the point in the appearance of the star which one wishes to get into the line of sight can be arbitrary within certain limits, these being doubtless further apart in the case of large and bright stars than in that of smaller and less luminous ones, with the result that different points are selected by night and day or with a clearer and less clear sky, and (3) the fact that the star presents itself rarely or never at rest but in a tremorous movement resulting from the lack of equilibrium in the air, so that a choice must be made within the range of the extreme limits of this movement. These causes of error are joined by others which are totally unrelated to the focusing procedure, i.e., for instance (4) an influence of the elasticity of the metal of the instrument—an influence which (as the result of accidental external circumstances) may assume different values at different points in time, with the result that the direction of the telescope at the time when the observation is recorded need no longer be the same as when the instrument was focused, (5) an uncertainty in the reading of the circle resulting from small differences between the distances separating its lines of graduation as well as the lines of graduation of the verniers, which when a given observation is repeated—the lines of graduation of the scales that are made to coincide are generally not the same, (6) the uncertainty resulting from the restricted power of resolution of the optical device by means of which readings are effected, and (7) the errors resulting from the fact that the interpretation of the vernier reading can never be carried, for instance, beyond one half the smallest distance of 2' recorded by them, with the result that all the observations recorded on the basis of the four verniers of these instruments will close with a full second or with one fourth, one half, or three quarters of a second but never with any other fraction. Then there are other, external

> factors, such as for instance (8) the influence of the body warmth of the observer on the circle or other components of the instrument and (9) the influence of the generally present difference in temperature between the lower and upper edges of the circle, resulting in tension in the metal of the circle and hence in the changes in its shape. Furthermore, (10) a chance error is implied in the assumption that the level of the alidade is in all readings in a state of unimpaired equilibrium, and (11) a similar error is introduced by the assumption that the instrument remains in precisely the same condition from one observation to the next with which the first is to be compared, although the findings is by no means rare that the instrument undergoes changes within shorter or longer periods of time. The so-called error of observation is further increased by (12) the influence exerted by the faulty assumption that the state of the atmosphere determining the refraction of rays at a particular moment is precisely the state suggested by the barometric and thermometric readings and (13) the influence of small imperfections in the elements of reduction of the observations. I have doubtless overlooked several points that should have been included in this list of causes which result through their interaction in what appears to be an error of observation—aside from the fact that I intentionally made no mention of the accidental negligence entering into the execution of the various component parts of an observation, nor of the advantageous or unsteady illumination of the threads and lines of graduation, of the influence of low temperatures on the instrument itself, and of other items of this kind. In any event, this enumeration of causes of error is quite sufficient to achieve the purpose it was intended to serve, which is to make clear that even the simplest kind of observation must entail a total error arising from numerous causes each one of which is independent of others . . . [1, pp. 398-400] Translated by Dr. Alexander Gode.

Bessel also called attention to the fact that the normal distribution cannot be regarded as universally applicable by obtaining

$$f(x) = \frac{1}{\pi} \cdot \frac{1}{\sqrt{a^2 - x^2}}$$

as the probability distribution for a case arising in angle measurement in which one source of error dominated the others. He regarded as unlikely that the probability distributions governing the elementary errors were normal and viewed the normal distribution as arising through the interaction of a large number of sources of elementary errors.

Example 2 Probability of a Measurement Error

In carrying out the measurement of a parameter an error having the structure described by the hypothesis of elementary errors arises. The probability distribution approximating the behavior of the random variable describing the error is the normal curve with $\mu = 0$ and $\sigma = 0.391$ units. (The units are in terms of some prescribed unit of measurement.) Find the probability that the error is between -0.1 and 0.1 units. How is this result to be interpreted?

The probability value we seek is given by the area of the region under the normal curve determined by $\mu = 0$ and $\sigma = 0.391$ between $x_1 = -0.1$ and $x_2 = 0.1$ (see Figure 19(a)). To calculate this area we shift the scene to the standard normal curve, shown in Figure 19(b), by calculating the z_1 and z_2 values corresponding to $x_1 = -0.1$ and $x_2 = 0.1$.

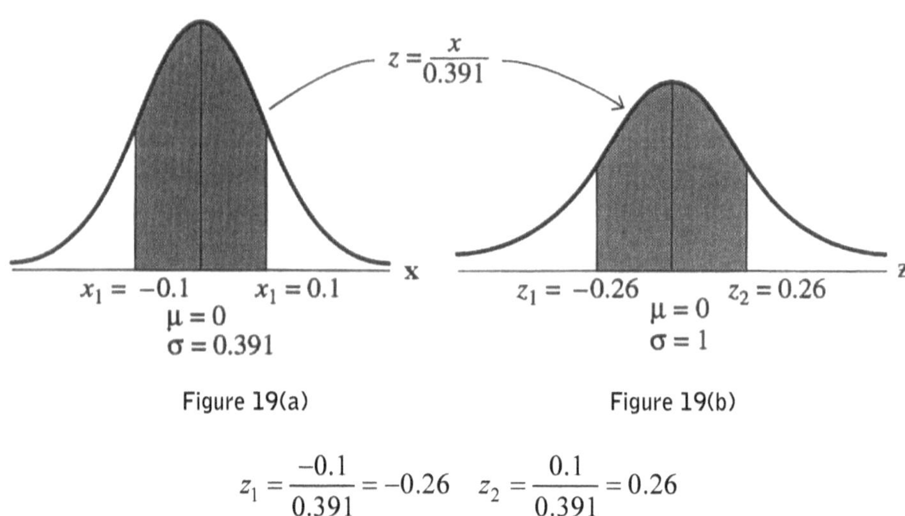

Figure 19(a) Figure 19(b)

$$z_1 = \frac{-0.1}{0.391} = -0.26 \quad z_2 = \frac{0.1}{0.391} = 0.26$$

The area of the region between 0 and 0.26 is 0.1026, as is the area of the region between 0 and -0.26. Their sum, 0.2052, is the area of the region bounded by -0.26 and 0.26, which is equal to the area of the region bounded by -0.1 and 0.1 in terms of the original normal curve.

The probability that the error arising in the measurement of the parameter in question lies between -0.1 an 0.1 units is, therefore, 0.2052. In relative frequency terms, if a large number of observations of the parameter in question are made under the same conditions, then approximately 20.5% of the errors arising from the process will lie between -0.1 and 0.1 units.

Processes Governed by Normal Curves

A measurement process giving rise to an observation error having the structure described by the hypothesis of elementary errors is one example of a process which is the additive result of a large number of independently-acting random factors, each having a negligibly small influence on the process as a whole. There are many others, and the normal curves are fundamental to the study of the probabilistic behavior of all.

Another interesting illustration is provided by mass-production processes in industry. Consider an item with given dimensions or specifications which is to be mass produced. Each item will deviate slightly from the specified norm and this deviation is often the cumulative result of a large number of independently-acting factors, each of which contributes minutely to the cumulative effect. Contributing factors may include temperature variation in the mass production process, vibration effects, slight fluctuations in the homogeneity of the raw material used (weight and shape, for example), human errors due to slight changes in the physical or psychological state of the person tending the machinery.

As another illustration, consider artillery fire or projectile launching in general. The aiming devices are set to attain a certain flight distance. But no two projectiles ever attain the same flight distance. Deviations, which in the case of very careful firing are slight, are always observed. Such deviations can often be viewed as the cumulative result of a large number of independently-acting random factors, each contributing a very slight amount to the total effect. Among these contributing factors are vibrations in the gun barrel, temperature variations, slight variations in instrument settings, fluctuations in atmospheric conditions, variations in the weight and shape of the projectile fired.

Still another interesting example comes from the study of growth processes. The growth of organs of a living organism, animal or plant, is dependent on numerous, independently acting nutritional and environmental factors, each of which has a small influence on the growth of organs, but which tend to increase or decrease their size.

REFERENCE

1. F.W. Bessel, "Untersuchungen über der Wahscheinlichkeit der Beobachtungsfehler" (Investigations on the Probability of Errors of Observation), Astronomishce Nachricchten, Vol. 15, No. 358-359(1838), 368-404.

EXERCISES

6. In connection with Example 2, find the probability that:

   (a) The error is between -0.2 and 0.2 units.

   (b) The error exceeds 1.5 units.

   (c) The error is less than 1 unit.

   (d) State the relative frequency interpretation of the results obtained in answer to (a), (b), and (c).

7. In carrying out the measurement of a parameter an error having the structure described by the hypothesis of elementary error arises. The probability distribution of the random variable describing the error is normal with $\mu = 0$ and $\sigma = 0.12$ units. Find the probability that:

   (a) The error is between -0.05 and 0.05 units.

   (b) The error is between -0.01 and 0.1 units.

   (c) The error exceeds 0.2 units.

   (d) State the relative frequency interpretation of the results obtained in answer to (a), (b), and (c).

8.4 ■ Normally Distributed Populations

The family of normal curves considered to this point, defined by

$$y = \frac{1}{\sigma\sqrt{2\pi}} e^{-(x-\mu)^2/2\sigma^2} \qquad (1)$$

has the property that the area of the region under each curve is 1. These curves, as we have seen, lend themselves to describing and estimating probability values under suitable conditions.

The more general family of normal curves

$$y = \frac{K}{\sigma\sqrt{2\pi}} e^{-(x-\mu)^2/2\sigma^2} \tag{2}$$

obtained by multiplying the right side of (1) by a positive constant *K* can, in many instances, be used to approximate the histogram of a frequency distribution of a population or sample.

Normal Curve Approximations of Frequency Distributions

If the underlying population has a large number of data, *N*, the width of the frequency distribution is *c*, and μ and are the mean and standard deviation of the distribution, then the approximating normal curve is defined by:

$$y = \frac{Nc}{\sigma\sqrt{2\pi}} e^{-(x-\mu)^2/2\sigma^2} \tag{3}$$

Usually, the smaller *c*, the better the approximation.

Example 1 Return to Ecap University's Statistics Text Situation

As you recall, a recent edition of *Basic Statistics* was the focus of a wrist strength test at Ecap University. Table 2 gives a frequency distribution for the time lengths, in minutes, that the population of 156 students at Ecap University who used the text were able to hold it in one sitting before their wrists gave out (see Sec. 3.1, Example 1, page 85). Figure 20 shows the histogram for this distribution (see Sec. 3.2, Figure 2, page 96). This histogram strongly suggests that the underlying population is approximately normal. Since $N = 156$, $c = 1$, $\mu = 4.6$, and $\sigma = 1.6$ (see Sections 5.2 and 5.3, Examples 1 and 1, page 127 and page 138), the normal curve fit for this distribution in terms of its histogram is defined by (3) for $Nc = 156$, $\mu = 4.6$ and $\sigma = 1.6$.

A rough graph of this curve shown against the underlying histogram is given in Figure 21. To increase the accuracy of the normal curve approximation for the population we would have to refine the frequency distribution by taking more classes so that *c* decreases.

Table 2

Time Interval	f_i
0.0-0.9	2
1.0-1.9	7
2.0-2.9	13
3.0-3.9	31
4.0-4.9	42
5.0-5.9	32
6.0-6.9	16
7.0-7.9	10
8.0-8.9	3
	156

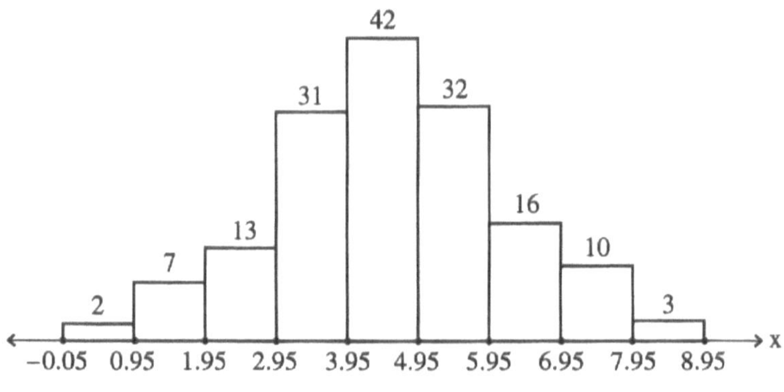

Figure 20

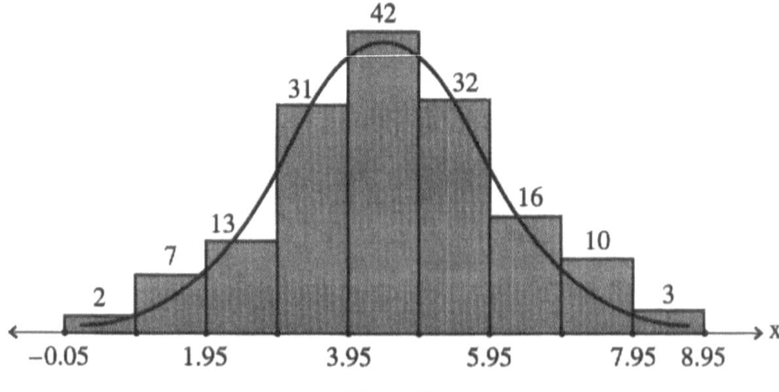

Figure 21

Example 2 Arm Lengths of Bougainville Tribesmen

In his book *Statistics* (Holden-Day, 1973) Roger Carlson refers to the frequency distribution of arm lengths, measured to the nearest centimeter, of 385 tribesman of southern Bougainville. The mean length is 75.1 cm., and the standard deviation is 3.8 cm. The histogram for this frequency distribution and its normal curve fit [Carlson, p. 148] are shown in Figure 22.

How can we determine whether a population is approximately normally distributed? One somewhat rough way is to construct a histogram for the population in terms of its frequency distribution where there are many classes so that c is small. One can often obtain a good sense from the histogram whether the population is approximately normal. Then there are more formal hypothesis tests which may be used to test for normality. These are considered in Sections 16.4 and 17.3.

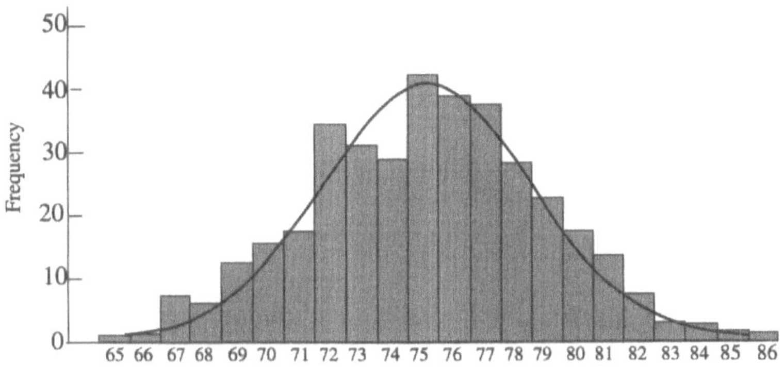

Figure 22

Normally Distributed Populations

A normally distributed population is one whose distribution of values can be described by a normal curve. As is the case with normally distributed random variables, the normally distributed population is a mathematical ideal which serves as a foundation for proving theorems, but this ideal is not realized "exactly" in practice. The question that emerges from applied situations concerns whether a population that arises in practice is close enough to being normally distributed that we may treat it as such.

Populations which arise from random variables that are approximately normal are, by virtue of their origins, approximately normal. Thus, the population of error values of an error term whose structure is described by the hypothesis of elementary errors is approximately normal. More generally, the population of values of any quantity which may be viewed as the additive result of a large number of independently-acting random factors, each having a negligible influence on the quantity as a whole, is approximately normal. As we saw in Section 8.3, mass production processes involving deviation from a specified norm (error, if you will), projectile launching at a specified target, and growth phenomena satisfying the aforenoted generalized elementary error structure lead to populations that are approximately normal. Populations of anthropometric data (such as heights, weights, I.Q. scores, exam scores) are often found to be approximately normal.

Hypothesis tests for normality are discussed in Sections 16.4 and 17.3.

Normal Distributions Imposed from Above

Normal distributions are sometimes imposed on populations from on high. An example of this is the practice generally referred to as "grading on the curve." An instructor confronted with a population of exam grades or term averages must determine the percentage of A, B, C, D, and F grades to give. Some instructors argue that grades or averages require no massaging. Grades, they argue, should reflect individual effort, not comparative states of achievement. They should reflect what is there, no more and no less, and not what we believe should be there or want to be there.

The practice of grading on the (normal) curve reflects various points of view, and sometimes a combination of points of view. The immense authority of Sir Francis Galton (1822-1911), who strongly favored the normal curve to describe anthropometric data, looms large. The neo-Galtonians would impose a normal curve on anthropometric data because of a view that to do so is to conform to nature as it is. Belief in the pervasiveness of the normal curve had become so strong in the late nineteenth century that the story began to circulate that the natural scientists believed in the normal curve because they thought it had been established by the mathematicians and that the mathematicians believed in the normal curve because they thought it had been established by the natural scientists.

Then there are those who never heard of Galton and have not been influenced by the neo-Galtonians, but are reluctant to accept what they view as excesses. Too many high grades may suggest declining standards and too many low grades may suggest that the exam was unfair and lead to trouble or an uprising, in this view. The answer, use a curve to avoid such excesses. Some departments and schools mandate grading on a curve.

Some industrial enterprises whose output is based on piecework use a normal curve to determine, comparatively speaking, who their most productive workers are and who has to measure up.

Example 3

An instructor has decided to assign grades based on the normal curve with mean μ and standard deviation σ to scores based on the following mechanism:

(a) A if the score exceeds $\mu + 1.75\sigma$;

(b) B if the score exceeds $\mu + 0.4\sigma$ and does not exceed $\mu + 1.75\sigma$;

(c) C if the score exceeds $\mu - 0.6\sigma$ and does not exceed $\mu + 0.4\sigma$;

(d) D if the score exceeds $\mu - 1.75\sigma$ and does not exceed $\mu - 0.6\sigma$;

(e) F if the score does not exceed $\mu - 1.75\sigma$.

The problem is to determine the percentage of each grade to be given.

The percentage of A's to be given is equal to the area of the tail of the normal curve with mean μ and standard deviation σ to the right of $\mu + 1.75\sigma$ as shown in Figure 23(a). In terms of the standard normal curve, the *z*-value corresponding to $\mu + 1.75\sigma$ is 1.75 (see Figure 23(b)). The area of the tail to the right of 1.75 is $0.5000 - 0.4599 = 0.0401$. Thus, 4% of the assigned grades are A's. Similarly, there are 30% B's, 36% C's, 23% D's, and 4% F's (see Exercise 1). For this to work the underlying population must, of course, be approximately normal with mean μ and standard deviation σ.

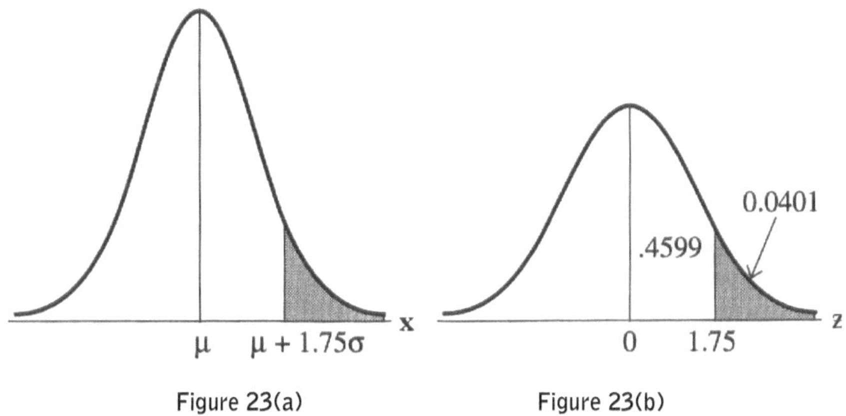

Figure 23(a) Figure 23(b)

The percentage of grades for each group can be adjusted by modifying the values of the multipliers c_1 and c_2 on $\mu + c_1\sigma$ and $\mu - c_2\sigma$.

EXERCISES

1. In connection with Example 3 show that there are 30% B's, 38% C's, 23% D's, and 4% F's.

2. Suppose that grades based on the normal curve with mean μ and standard deviation σ are to be assigned according to the following mechanism:

   (a) A if the score exceeds $\mu + 1.5\sigma$.

   (b) B if the score exceeds $\mu + 0.5\sigma$ and does not exceed $\mu + 1.5\sigma$;

   (c) C if the score exceeds $\mu - 0.5\sigma$ and does not exceed $\mu + 0.5\sigma$;

   (d) D if the score exceeds $\mu - 1.5\sigma$ and does not exceed $\mu - 0.5\sigma$;

   (e) F if the score does not exceed $\mu - 1.5\sigma$.

   Find the percentage of A's, B's, C's, D's and F's to be given.

3. Under the system described in Exercise 2:

   (a) Is it possible to have a term average of 85 (of 100) and receive F in the course? Explain.

   (b) Is it possible to have a term average of 45 (of 100) and receive A in the course? Explain.

4. The frequency distribution of the grades on an economics exam taken by 85 students at Huxley College is given in Table 3.

   (a) Draw the histogram for this distribution.

   (b) Does a normal curve seem like a good fit to the grade distribution? Explain.

   (c) Suppose grades were assigned by the following mechanism: A: 90-99, B: 80-89, C: 70-79, D: 60-69, F: below 60. What would the percentages of A's, B's, C's D's and F's be?

(d) If grades were assigned by the mechanism described in Exercise 2, what bounds would define grades of A, B, C, D, and F?

(e) In this case how would the grade categories compare with those defined in (c)?

Table 3

Grade	f_i
20-29	2
30-39	6
40-49	5
50-59	7
60-69	8
70-79	27
80-89	19
90-99	11
	85

(f) What grades would the following scores receive?

(i) 95 (ii) 45 (iii) 63 (iv) 79

(g) Something doesn't seem to be right here? What is your view and explanation?

(h) Which grading system makes most sense to you? Explain.

5. Assume that I.Q.'s are approximately normally distributed with $\mu = 110$ and $\sigma = 20$. The High I.Q. Society admits as members those who are in the upper 5 percent of the I.Q. scale. The Low I.Q. Society admits as members those who are in the lower 10 percent of the I.Q. scale.

(a) What is the lowest I.Q. that will allow one to be admitted to H.I.Q.S.?

(b) What is the highest I.Q. that will allow one to be admitted to L.I.Q.S.?

9

Sampling Distributions

9.1 ■ Sampling Distributions of Sample Statistics

Statistical inference is concerned with the development of a methodology for reaching and supporting conclusions about population parameters based on sample evidence obtained from the population. Probability theory is used to provide us with a sense of reliability for the conclusions reached through use of statistical inference.

One foundation stone for statistical inference is provided by the concept of sampling distribution of a sample statistic which is "close" to the population parameter that we seek to estimate. To initiate a development of this idea we look at one of the most important cases, the sampling distribution of the sample mean $\overline{x}$, first in miniature and then in a larger framework.

9.2 ■ The Sampling Distribution of the Sample Mean

To develop the basic idea of sampling distribution of $\overline{x}$ consider the population $Q = \{1,3,5,7,9\}$. By direct computation we obtain the mean μ and standard deviation σ of Q to be:

$$\mu = \frac{1+3+5+7+9}{5} = 5$$

$$^2 = -[(1-5)^2 + \quad + (9-5)^2] = 8$$

so that $\sigma = \sqrt{8} = 2\sqrt{2}$.

Due to the small size of Q it is a simple matter to obtain its mean and standard deviation. With populations running into millions, billions, and more values, whose content in many cases is not even clear, the problem of determining population means and standard deviations by direct computation ranges from extraordinarily difficult to impossible.

We get around these difficulties by taking a random sample from the population under study, determining its mean and standard deviation, and using these values to get a handle on their population counterparts μ and σ. To construct this handle we must examine how the random variable $\overline{x}$, which assigns to all samples of a selected size their mean, behaves in general.

To explore this behavior in the context of our population Q, consider the process of choosing a sample of size 2 at random from Q. We take as our sample space *S* for this process the set of all unordered samples of 2 which can be chosen from Q.

$$S = \{(1,3),\ (1,5),\ (1,7),\ (1,9),\ (3,5),\ (3,7),\ (3,9),\ (5,7),\ (5,9),\ (7,9)\}$$

The number of samples of 2 in *S* equals the number of ways of choosing 2 objects (numbers in this case) from 5 without regard to order, which is $C(5,2) = 10$. Checking back we see that we have 10 sample points in *S*, so that in compiling our list we didn't miss any.

The randomness of the selection of a sample is best reflected by the probability function *P* which assigns the same value to each sample point in *S*. We have:

$$P(1,3) \quad = \quad P(1,5) \quad = \quad \cdots \quad = \quad P(7,9) \quad = \quad \frac{1}{10}$$

We define on *S* the random variable $\overline{x}$ which assigns to each sample point in *S*, that is, sample of 2 from *S*, the mean of the sample drawn. This gives us:

$$\begin{array}{ll} \overline{x}(1,3) = 2 & \overline{x}(3,7) = 5 \\ \overline{x}(1,5) = 3 & \overline{x}(3,9) = 6 \\ \overline{x}(1,7) = 4 & \overline{x}(5,7) = 6 \\ \overline{x}(1,9) = 5 & \overline{x}(5,9) = 7 \\ \overline{x}(3,5) = 4 & \overline{x}(7,9) = 8 \end{array}$$

The probability distribution $p(x)$ of $\overline{x}$, defined by $p(x) = P(\overline{x} = x)$, is called the **sampling distribution of** $\overline{x}$. It is defined as follows in this case:

$$p(2) \quad = \quad P(\overline{x} = 2) \quad = \quad P(1,3) \quad = \quad \frac{1}{10}$$

$$p(3)=\frac{1}{10}\quad p(6)=\frac{2}{10}$$
$$p(4)=\frac{2}{10}\quad p(7)=\frac{1}{10}$$
$$p(5)=\frac{2}{10}\quad p(8)=\frac{1}{10}$$

Our next task is to compute the mean and standard deviation of the random variable $\overline{x}$, but first there is a question of how to denote these values. A useful notation for the mean of $\overline{x}$ is μ with the subscript $\overline{x}$, $\mu_{\overline{x}}$, which tells us at a glance in the μ part that we are talking about a mean and in the subscript part that the random variable is $\overline{x}$. We shall follow this practice in general, so that if our interest is in the mean of the random variable which is the difference of two means, $\overline{x}_1-\overline{x}_2$, the notation would be $\mu_{\overline{x}_1-\overline{x}_2}$. If our interest is in the variance or standard deviation of the random variable $\overline{x}$, the notation would be $\sigma^2_{\overline{x}}$ and $\sigma_{\overline{x}}$, respectively.

The mean $\mu_{\overline{x}}$ of the random variable $\overline{x}$ under consideration is given by:

$$\begin{aligned}\mu_{\overline{x}} &= 2p(2)+3p(3)+4p(4)+5p(5)+6p(6)+7p(7)+8p(8)\\ &= 2\cdot\frac{1}{10}+3\cdot\frac{1}{10}+4\cdot\frac{2}{10}+5\cdot\frac{2}{10}+6\cdot\frac{2}{10}+7\cdot\frac{1}{10}+8\cdot\frac{1}{10}\\ &= 5\end{aligned}$$

The variance and standard deviation of $\overline{}$ are given by:

$$\begin{aligned}\sigma^2_{\overline{x}} &= (2-5)^2\,p(2)+(3-5)^2\,p(3)+\cdots+(8-5)^2\,p(5)\\ &= (-3)^2\cdot\frac{1}{10}+(-2)^2\cdot\frac{1}{10}+\cdots+(3)^2\cdot\frac{1}{10}\\ &= 3\end{aligned}$$

Thus, $\sigma_{\overline{x}}=\sqrt{3}$.

In summary, we have:

Population parameters: $\mu=5,\quad \sigma=2\sqrt{2}$,

Sampling distribution of $\overline{x}$ values: $\mu_{\overline{x}}=5,\quad \sigma_{\overline{x}}=\sqrt{3}$

The first natural question is, what relationships exist between the population values and the sampling distribution of $\overline{x}$ values? It is immediately apparent that the means are equal:

$$\mu_{\overline{x}} = \mu = 5$$

As to the standard deviations, there is a sharp, clear relationship, but its nature is not immediately apparent. To see its nature consider the formula:

$$\frac{\sigma}{\sqrt{n}}\sqrt{\frac{N-n}{N-1}},$$

where σ is the population standard deviation, *n* is the sample size, and *N* is the population size. For population Q, $\sigma = 2\sqrt{2}$, $N = 5$ and the sample size is $n = 2$. Substituting these values into

$$\frac{\sigma}{\sqrt{n}}\sqrt{\frac{N-n}{N-1}}$$

yields:

$$\frac{2\sqrt{2}}{\sqrt{2}} \cdot \sqrt{\frac{3}{4}} = \frac{2\sqrt{2}}{\sqrt{2}} \cdot \frac{\sqrt{3}}{\sqrt{4}}$$
$$= \sqrt{3},$$

which is the value of $\sigma_{\overline{x}}$, the standard deviation of the sampling distribution of $\overline{x}$.

> We therefore have the following relationships between population parameters and sampling distribution of $\overline{x}$ values in this special case:
>
> $$\mu_{\overline{x}} = \mu$$
>
> $$\sigma_{\overline{x}} = \frac{\sigma}{\sqrt{n}}\sqrt{\frac{N-n}{N-1}}$$

The natural follow-up question is, do these relationships hold in general? As we shall see, the answer is yes.

The More General Framework

μ More generally, consider a population $Q = \{x_1, x_2, \ldots, x_N\}$ of N values with mean μ and standard deviation . Also consider the process of choosing a sample of size n at random from Q. We take as our sample space S for this process the set of all unordered samples of size n which can be chosen from Q. The number of unordered samples of size n which can be chosen from Q is given by $C(N, n)$, the number of combinations of N objects taken n at a time. At a glance we have the situation shown in Figure 1.

The randomness of the sample selection leads us to the probability function P which assigns the same value, $1/C(N, n)$, to each sample of size n. (Note that for even modest values of N and n, $C(N, n)$ is rather large. For $N = 52$ and $n = 5$, $C(N,n) = 2,598,960$.)

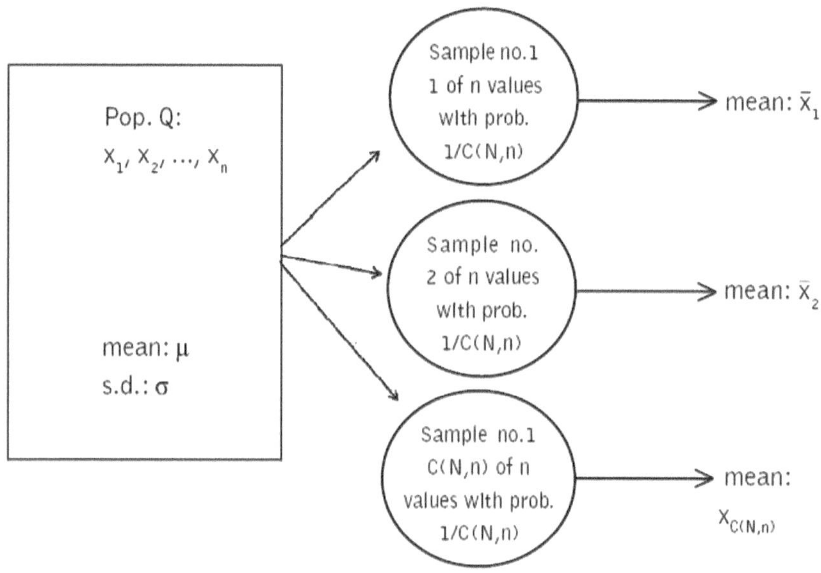

Figure 1

We define on S the random variable $\bar{x}$ which assigns to each sample point in S the mean of the sample drawn. The probability distribution of $\bar{x}$, defined by

$$p(x) = P(\bar{x} = x),$$

is called the **sampling distribution of** $\bar{x}$.

As suggested by our consideration of the earlier miniature example, it can be established that the mean $\mu_{\bar{x}}$ and standard deviation $\sigma_{\bar{x}}$ of the sampling distribution of $\bar{x}$ are related to the population Q parameters μ and σ in the following way:

$$\mu_{\bar{x}} = \mu, \quad \sigma_{\bar{x}} = \frac{\sigma}{\sqrt{n}}\sqrt{\frac{N-n}{N-1}}$$

The expression under the preceding square root sign may be written:

$$\frac{N-n}{N-1} = \frac{N}{N-1} - \frac{n}{N-1} = \frac{1}{\left(1-\frac{1}{N}\right)} - \frac{n}{N-1}$$

For N large in its own right and N large in comparison to sample size n, both $1/N$ and $n/(N-1)$ are close to zero, so that

$$\frac{N-n}{N-1} \text{ is close to 1.}$$

$$\sqrt{\frac{N-n}{N-1}} \text{ is then even closer to 1, and}$$

$$-\quad \frac{\quad}{\sqrt{\quad}}\sqrt{\frac{N \quad n}{\quad}} \text{ becomes, to a close approximation, } \frac{\sigma}{\sqrt{n}}.$$

Summary Result

The mean $\mu_{\bar{x}}$ and standard deviation $\sigma_{\bar{x}}$ of the sampling distribution of $\bar{x}$ are given by

$$\mu_{\bar{x}} = \mu, \quad \sigma_{\bar{x}} = \frac{\sigma}{\sqrt{n}},$$

where the underlying population Q is large and the sample size n is small in comparison to the population size. The sample size n is considered small in comparison to the population size N if n does not exceed 5% of N, that is, if the population size is at least 20 times that of the sample size.

$\sigma_{\bar{x}} = \frac{\sigma}{\sqrt{n}}$ is called the **standard error of the mean.**

It serves as a measure of variation of the sample means about $\mu_{\overline{x}}$. Since $\mu_{\overline{x}}$ equals the population mean μ, $\sigma_{\overline{x}}$ serves as a measure of the variation of the sample means about the population mean. It is to be expected that as the sample size increases the sample means will be more tightly concentrated about the population mean, but these results describing $\sigma_{\overline{x}}$ tell us that the measure of concentration is inversely proportional to the square root of the sample size.

EXERCISES

1. Population $Q = \{2, 4, 6, 8, 10\}$. Consider the process of drawing a sample of size 2 at random from Q.

   (a) Define a probability model for this process.

   (b) Define the random variable $\overline{x}$ on the sample space of the model given in answer to (a).

   (c) Determine the sampling distribution of $\overline{x}$.

   (d) Find the mean of this sampling distribution.

   (e) Find the standard deviation of this sampling distribution.

   (f) Find: (i) $P(5 \le \overline{x} \le 7)$, (ii) $P(\overline{x} > 6)$

2. Population $Q = \{5, 6, 7, 8, 9, 10\}$. Consider the process of drawing a sample of size 2 at random from Q. Same questions (a) through (f) as stated for Exercise 1.

3. Assuming the underlying population is very large, what happens to the standard error of the mean if we quadruple the sample size?

4. Assuming the underlying population is very large, what would we have to do to the sample size if we wanted to reduce the standard error of the mean by a factor of 3?

5. A population Q, consisting of 30 values, has mean 10 and standard deviation $\sqrt{5}$. Consider the process of drawing a sample of size 3 at random from Q.

(a) In defining a probability function P on the collection of unordered samples of 3 that can be drawn from Q, what value would P assign to each of these sample points?

(b) What is the basis for defining P as described in answer to (a)?

(c) Find the mean of the sampling distribution of the mean.

(d) Find the standard deviation of the sampling distribution of the mean.

6. A population Q, consisting of 52 values, has mean 22 and standard deviation $\sqrt{20}$. Consider the process of drawing a sample of size 5 at random from Q.

(a) In defining a probability function P on the collection of unordered samples of 5 that can be drawn from Q, what value would P assign to each of these sample points?

(b) What is the basis for defining P as described in answer to (a)?

(c) Find the mean of the sampling distribution of the mean.

(d) Find the standard deviation of the sampling distribution of the mean.

9.3 ■ A Central Limit Theorem

It is feasible to determine the sampling distribution of $\overline{x}$, $p(x) = P(\overline{x} = x)$, by direct means when the size N of the population Q being sampled from and the size n of the sample taken at random from Q are both small. When N and n are even of moderate size the direct determination of $p(x)$ is no longer feasible and we are led to seek tools that would allow us to give close approximations of probabilities which arise from $p(x)$. Fortunately, there is a central limit theorem that comes to our rescue.

Central Limit Theorem for $\overline{x}$: If a sample of size n is chosen at random from a "large" population Q, then as n gets larger and larger without bound, the sampling distribution of $\overline{x}$, $p(x) = P(\overline{x} = x)$ gets closer and closer to the normal curve with mean $\mu_{\overline{x}}$ and standard deviation $\sigma_{\overline{x}}$ defined by

$$\mu_{\overline{x}} = \mu, \quad \sigma_{\overline{x}} = \frac{\sigma}{\sqrt{n}},$$

where μ and σ are the mean and standard deviation of population Q.

This raises an important question: If we wish to approximate the sampling distribution of $\overline{x}$ by the normal curve with mean μ and standard deviation $\sigma/\sqrt{n}$, how large must *n* be for the approximation to be "good enough" for statistical applications?

Theory does not give us an answer to this question, but from statistical practice we have $n=30$ as a guideline value; this defines what is termed **large sample size.**

In terms of statistical practice, the following result geared to statistical applications emerges.

Central Limit Theorem for Statistical Applications. If a sample of size $n \geq 30$ (large sample size) is chosen at random from a "large" population Q with mean μ and standard deviation , then the sampling distribution of $\overline{x}$ may be approximated by the normal curve with mean

$\mu_{\overline{x}} = \mu$ and standard deviation $\sigma_{\overline{x}} = \dfrac{\sigma}{\sqrt{n}}$.

How "good" the normal curve approximation is for the sampling distribution of $\overline{x}$ depends on the nature of the population Q from which the sample is taken.

(1) If Q is itself a normally distributed population, then the sampling distribution of $\overline{x}$ is normally distributed for all values of *n*, both small and large.

(2) If Q is mound shaped or approximately normal (see Figure 2), then the sampling distribution of $\overline{x}$ is approximately normal for large sample size ($n \geq 30$).

(3) If Q is strikingly non-normal (as illustrated by Figures 3, 4, and 5), then a sample size of $n=30$ may not be large enough to yield a "good" normal curve approximation for the sampling distribution of $\overline{x}$.

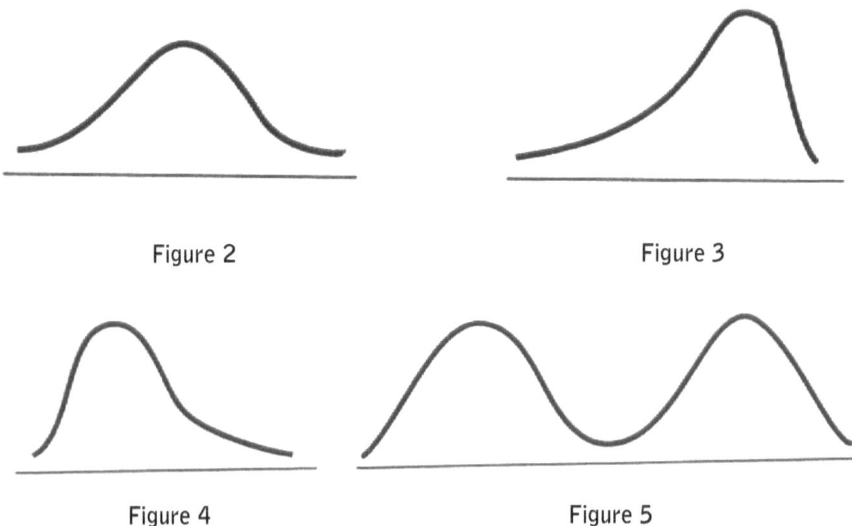

Figure 2

Figure 3

Figure 4

Figure 5

A grade distribution dominated by very low grades or an income distribution dominated by very low incomes would be approximated by the kind of curve shown in Figure 4, which is said to be sharply **skewed to the left.**

A grade distribution dominated by very high grades or an income distribution dominated by very high incomes would be approximated by the kind of curve shown in Figure 3, which is sharply **skewed to the right.** A grade or income distribution dominated by low and high grades or low and high incomes, with little in the middle, has the shape of a **bimodal curve** shown in Figure 5.

> In brief, if the population Q being sampled from is "large" and does not deviate "significantly" from normality (by being, for example, "significantly" skewed to the left or right or by being bimodal), then a large sample size of $n \geq 30$ will yield a "good" normal curve approximation for the sampling distribution of $\overline{x}$.

By a "large" population Q we mean one that is large compared with n, the size of the sample being chosen from Q, in that n is less than 5 percent of the size of Q when n takes such values as 30, 50, 100, 150 and the like.

As to the terms "significantly" and "good," they appear in quotation marks in the preceding because their lack of precision leaves them open to different interpretations. It is, however, the best that can be done at this point.

> If a population Q deviates "significantly" from normality, then sample size n has to be taken larger than 30 (how much larger, we cannot say in advance) if the normal curve approximation for the sampling distribution of $\overline{x}$ is to be a "good" one.

Example 1 Probability that the Sample Mean is Between Given Bounds

A sample of size 49 is to be chosen at random from a large population whose mean and standard deviation are $\mu = 20$ and $\sigma = 14$. Find the probability that the sample mean $\overline{x}$ will be between 18 and 24.

Since $n = 49 \geq 30$, the central limit theorem for $\overline{x}$ is applicable here. The normal curve that approximates the sampling distribution of $\overline{x}$ has mean $\mu_{\overline{x}} = \mu = 20$ and standard deviation $\sigma_{\overline{x}} = \sigma / \sqrt{n} = 14 / \sqrt{49} = 2$ (see Figure 6).

The problem reduces to finding the area of the region under the normal curve with $\mu_{\overline{x}} = 20$ and $\sigma_{\overline{x}} = 2$ between $x_1 = 18$ and $x_2 = 24$. To do this we find the area of the comparable region under the standard normal curve. The z_1 and z_2 values corresponding to $x_1 = 18$ and $x_2 = 24$ are:

$$z_1 = \frac{18 - 20}{2} = -1.00, \; z_2 = \frac{24 - 20}{2} = 2.00$$

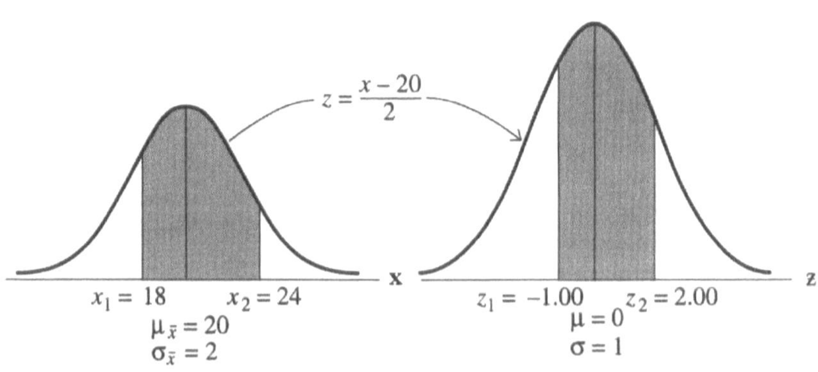

Figure 6 Figure 7

The corresponding standard normal curve region is shown in Figure 7. From the standard normal curve table (Table A, page 652) we have that the areas of the regions determined by $z_1 = -1.00$ and $z_2 = 2.00$ are 0.3413 and 0.4772, respectively.

Thus, the normal curve approximation for the probability that $\overline{x}$ will be between 18 and 24 is 0.8185.

EXERCISES

1. In connection with Example 1, find:

   (a) $P(\overline{x} < 16)$
   (b) $P(\overline{x} \geq 19)$
   (c) $P(19 < \overline{x} < 23)$?
   (d) $P(18.5 \leq \overline{x} < 25)$

2. A sample of size 36 is to be chosen at random from a large population whose mean and standard deviation are $\mu = 12$ and $\sigma = 18$. Find:

   (a) $P(\overline{x} \geq 14)$
   (b) $P(11 \leq \overline{x} \leq 15)$
   (c) $P(\overline{x} < 15.5)$
   (d) $P(13 < \overline{x} < 16)$

3. A sample of size 40 is to be chosen at random from a large population. The mean and standard deviation of Q are 25 and $\sigma = \sqrt{10}$.

   (a) Find $\sigma_{\overline{x}}$
   (b) Find $P(24.3 < \overline{x} < 25.6)$
   (c) Find $P(\overline{x} \geq 26.2)$
   (d) Find $P(25.2 < \overline{x} < 26)$

4. A sample of size 49 is to be chosen at random from a large population with mean $\mu = 10$ and standard deviation $\sigma = 21$. The normal curve approximating the sampling distribution of $\overline{x}$ is shown in Figure 8.

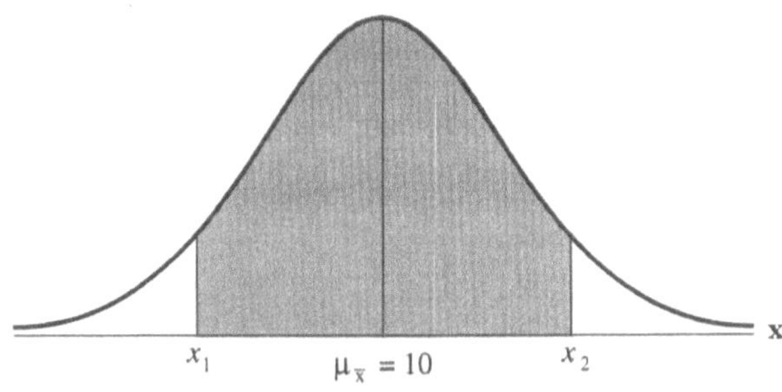

Figure 8

The values x_1 and x_2 denote equally spaced bounds on both sides of the mean such that the area under the curve between them is 0.95.

   (a) Find x_1 and x_2.

   (b) Find x_1 and x_2 if the area under the curve between them is 0.98.

SELF-TESTS FOR PART TWO

Allow 90 or so minutes for each self-test. Go over each one before going on to the next.

Self-Test I

NOTE: Combinatorial expressions or products which may arise need not be multiplied out.

1. The Science Department of Ecap University has 80 faculty, 30 of which are biology faculty, 25 of which are chemistry faculty, 15 of which are physics faculty, and 10 of which are geology faculty. A committee of 8 is to be chosen from the Science Department to study curriculum changes.

   (a) Set up a probability model for the selection process.

   (b) State the assumption on which your probability model is based.

   (c) Find the probability that the committee chosen contains 3 biology faculty.

   (d) Find the probability that the committee chosen contains Robert Weiss and Janet Fox of the Science Department.

   (e) Find the probability that the committee chosen contains 3 biology faculty and 2 chemistry faculty.

2. In a certain mass production process the probability that a defective item is produced is 0.01. A run of 10,100 items is produced.

   (a) Set up a probability model for this production process.

   (b) On what assumption is your probability model based?

   (c) Let X denote the number of defectives produced. Use a normal curve approximation to find $P(X \leq 105)$. As a point of information note that $\sqrt{99.99} = 10$.

3. The following is a probability model for the process of tossing a certain die:

$S=\{1,2,3,4,5,6\}$; $P(1)=\frac{2}{13}$, $P(2)=\frac{1}{13}$, $P(3)=\frac{3}{13}$, $P(4)=\frac{2}{13}$, $P(5)=\frac{3}{13}$, $P(6)=\frac{2}{13}$

Let A denote the event that an odd number shows.

(a) Find $P(A)$.

(b) State the relative frequency interpretation of the conclusion reached in (a).

(c) In tossing the die described by the above model 1300 times an odd number was observed to show 795 times. Does this evidence establish that the conclusion reached in (a) is valid? Explain.

(d) Is the conclusion reached in (a), interpreted in relative frequency terms, true? Explain.

(e) Are the events "an odd number shows" and "a number greater than 3 shows" independent? Explain.

4. In connection with a random process it is stated that the probability of A is 0.10. State the subjective-probability interpretation of this statement.

5. Let X denote the number of successes in connection with a Bernoulli trial model with $n=625$ and $p=1/5$. Use a normal curve approximation to find $P(124 \le X \le 130)$.

Self-Test 2

1. Population $Q=\{6,8,10,12\}$. Consider the process of choosing a sample of size 2 at random from Q.

(a) Define a probability model for this process.

(b) What assumption underlies your model?

(c) Define the sampling distribution of the mean $\overline{x}$

(d) Determine the mean of the sampling distribution of $\overline{x}$.

(e) Determine the standard deviation of the sampling distribution of $\overline{x}$.

(f) Find $P(8 \leq \overline{x} \leq 10)$.

2. Dan to Gus: "I have a die whose behavior is described by the probability model with sample space $S = \{1, 2, 3, 4, 5, 6\}$ and probability function:

$$P(1) = P(2) = \frac{2}{12}, \quad P(3) = P(4) = \frac{3}{12}, \quad P(5) = P(6) = \frac{1}{12}$$

I suggest the following: If a number less than 4 shows on a throw of this die, you pay me \$3.00; if a number greater than or equal to 4 shows, I pay you \$2.00."

(a) Define a random variable Z which describes the amount that Gus wins on a throw of this die.

(b) Determine the mean of Z.

(c) How is the mean of Z to be interpreted?

(d) Is the game fair? Explain.

(e) If the game is not fair, what would have to be done to make it fair?

(f) Determine the standard deviation of Z.

3. State the hypothesis of elementary errors.

4. Carbon steel bolts are made by the Axel Manufacturing Company. The bolt diameters being produced in one of the plants of the Axel Company have mean 0.420 inches and standard deviation 0.02 inches. The distribution of the bolt diameters is assumed to be approximately normal. Bolts with a diameter greater than 0.430 inches or less than 0.410 inches do not meet quality control specifications and must be rejected.

(a) What percentage of the output will have to be rejected?

(b) Below what diameter value will we find the lowest 20% of the output?

5. The Philosophy Club of Ecap University has 20 members. In how many ways can the positions of president, vice-president, treasurer, and secretary be filled if no person is to hold more than one position?

6. In LOTTO you must pick 6 numbers out of 44. The order in which the numbers are chosen is not significant. In how many ways can this be done?

Self-Test 3

1. The following is a probability model for tossing one of Dapper Dan's dice. $S = \{1, 2, 3, 4, 5, 6\}$

$$P(1) = P(2) = \frac{3}{16}, \quad P(3) = \frac{4}{16}, \quad P(4) = \frac{1}{16}, \quad P(5) = \frac{2}{16}, \quad P(6) = \frac{3}{16}$$

(a) Find the probability that a number greater than 3 shows.

(b) State the relative frequency interpretation of the result obtained in answer to (a).

(c) Are the events "a number greater than 2 shows" and "an even number shows" independent? Explain.

Dan offered his friend Gus the following deal: "If a number greater than 3 shows on a toss of this die, I pay you \$1.00; if a number less than or equal to 3 shows, you pay me \$1.00."

(d) Define a random variable Z which describes the amount Gus wins on a throw of this die.

(e) Determine the mean of Z.

(f) How is the mean of Z to be interpreted?

(g) Is the game fair? Explain.

(h) Determine the standard deviation of Z.

(i) In tossing Dan's die 800 times a number greater than 3 showed 496 times. Dan, who expected to be winning around $200 at this point, found himself $192 in the red. He could not understand what had gone wrong and posed the following two questions which should be answered in appropriate detail.

(i) Does this evidence mean that the probability value obtained in (a) for a number greater than 3 showing isn't valid?

(ii) If mathematical reasoning is as precise as it is reputed to be, how could I end up with $192 in the red instead of $200 in the black? Explain.

2. A measles vaccine has been found to be ineffective on 1 out of 1000 people who are given it. Plans call for administering the vaccine to 60,000 people in Bell City.

(a) Describe a probability model for this process.

(b) On what assumption(s) is your model based?

(c) Is your assumption realistic? Explain.

(d) In terms of your model state (but do not compute) the probability that there are at most 70 cases where the measles vaccine is ineffective.

(e) Is it possible to approximate the probability of at most 70 vaccine failures by means of a normal curve estimate? Explain.

(f) Assuming that the answer to (e) is yes, determine a normal curve estimate for the probability of at most 70 vaccine failures.

3. Population $Q = \{2, 6, 10, 14, 16\}$. Consider the process of choosing a sample of size 2 at random from Q.

(a) Define a probability model for this process.

(b) What assumptions underlies your model?

(c) Define the sampling distribution of the mean $\overline{x}$.

(d) Determine the mean of the sampling distribution of $\overline{x}$.

(e) Determine the standard deviation of the sampling distribution of $\overline{x}$.

(f) Find $P(8 \leq \overline{x} \leq 11)$.

PART THREE

INTRODUCTION TO STATISTICAL INFERENCE

10

Estimation

10.1 ■ Estimation Problems

Numerous situations lead to problems which require the estimation of population parameters such as means, proportions, and variances. The following three examples illustrate.

Ecap University administers a mathematics placement exam to all of its students prior to their taking a mathematics course at the university. A student's score on this 50-question exam along with information about his intended major area of study are used by the mathematics department to determine the student's placement into a mathematics course. The mathematics department is interested in estimating the mean score on the placement exam.

A medical society wishes to estimate the proportion of its membership that favors a Canadian type medical insurance system over other alternatives.

Ramunė's Gourmet Coffee is negotiating a contract with a processing company to fill 10 ounce jars with a new coffee blend to be introduced into the market. One concern is with the variability of the fill weights, which must be reasonably small, and Ramunė's Gourmet Coffee desires to estimate the variance of the coffee fills.

Inherent in the problem of obtaining estimates for such parameters is the problem of establishing a measure of reliability for the estimates provided. Our approach to this problem is based on the idea of defining a sample statistic which is closely related to the parameter to be estimated and whose probability behavior can be described. We first turn to population means.

10.2 ■ Estimation of Means: Large Sample Case

Interval Estimates

Consider the problem of estimating the value of a population mean μ, where the population is understood to be large with standard deviation σ. Each sample of size n chosen at random from the population yields a sample mean $\bar{x}$; different samples yield, in general, different sample means. The Central Limit Theorem for $^{-}$ tells us that, under mild conditions, for large sample size n the sampling distribution of $\bar{x}$ may be approximated by a normal distribution with mean,

$$\mu_{\overline{x}} = \mu$$

and standard deviation:

$$\sigma_{\overline{x}} = \frac{\sigma}{\sqrt{n}}$$

The guideline used by applied statisticians for "large sample size n" is that n must be at least size 30: $n \geq 30$. Also, we require that the sample size n not exceed 5% of the size of the population from which the sample is chosen.

To take a specific probability level to work with, let us agree to obtain an estimate for μ which is to hold with probability 0.95. With a small modification in our analysis we could easily change the probability level to 0.98 or any level that we wish.

The normal curve describing the sampling distribution of $\bar{x}$ is shown in Figure 1. Its standard normal curve counterpart is shown in Figure 2.

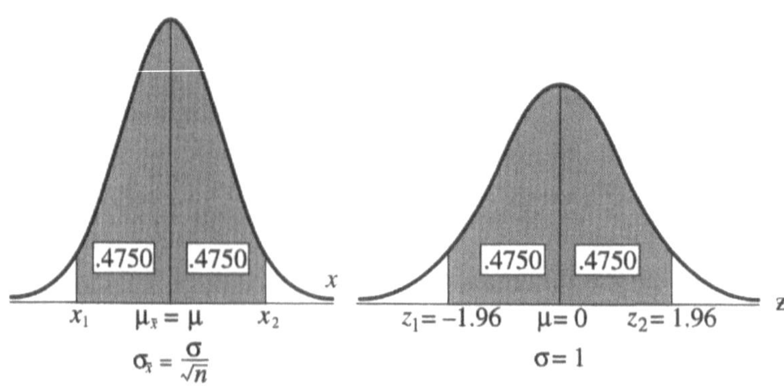

Figure 1 Figure 2

Let x_1 and x_2 denote equally spaced bounds on both sides of the mean $\mu_{\bar{x}} = \mu$ (Figure 1) so that the area under the normal curve between x_1 and x_2 is 0.95. The area of each piece between $\mu_{\bar{x}}$ and x_1, and $\mu_{\bar{x}}$ and x_2, is 0.4750. In terms of the standard normal curve, the counterparts of x_1 and x_2 are z_1 and z_2, as shown in Figure 2. The area under the standard normal curve between 0 and z_1, and 0 and z_2 is also 0.4750.

What are z_1 and z_2? Pinpointing 0.4750 in the body of the standard normal curve table (Table A, page 652) and reading off the *z*-value which yields it gives us $z_1 = -1.96$, $z_2 = 1.96$.

In terms of the *z*-value of $\bar{x}$, the statement that $\bar{x}$ is between x_1 and x_2 with probability 0.95 is equivalent to saying that the *z*-value of $\bar{x}$ is between -1.96 and 1.96, with probability 0.95. That is:

$$-1.96 < \frac{\bar{x} - \mu}{\sigma / \sqrt{n}} < 1.96 \text{ with prob. 0.95.}$$

To obtain our desired estimate for μ we must isolate μ from the preceding double inequality. This requires the application of basic algebra with patience and fearlessness.

We begin by multiplying the preceding double inequality through by $\sigma / \sqrt{n}$, which is positive. This yields:

$$-1.96\frac{\sigma}{\sqrt{n}} < \bar{x} - \mu < 1.96\frac{\sigma}{\sqrt{n}} \text{ with prob. 0.95}$$

Adding $-\bar{x}$ to all three components yields:

$$-\bar{x} - 1.96\frac{\sigma}{\sqrt{n}} < -\mu < -\bar{x} + 1.96\frac{\sigma}{\sqrt{n}} \text{ with prob. 0.95}$$

Multiplying through by -1 (which gives us μ in the middle and reverses the direction of the inequalities) gives us

$$\bar{x} + 1.96\frac{\sigma}{\sqrt{n}} > \mu > \bar{x} - 1.96\frac{\sigma}{\sqrt{n}} \text{ with prob. 0.95,}$$

which is equivalent to:

$$\bar{x} - 1.96\frac{\sigma}{\sqrt{n}} < \mu < \bar{x} + 1.96\frac{\sigma}{\sqrt{n}} \text{ with prob. 0.95}$$

In geometric terms this double inequality describes an interval centered at $\bar{x}$ with radius $1.96\frac{\sigma}{\sqrt{n}}$, shown in Figure 3.

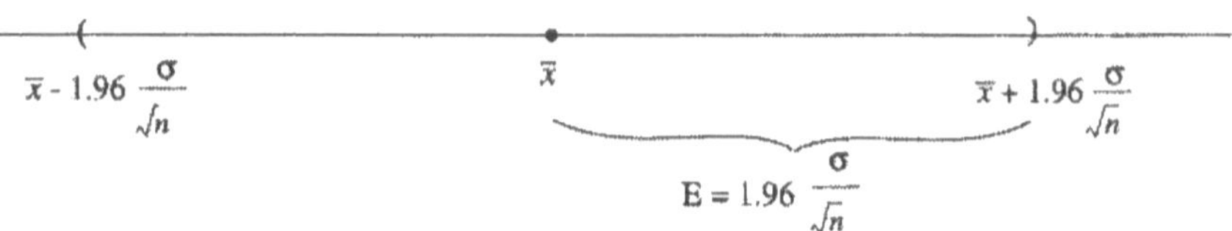

Figure 3

> The lower and upper bounds $\bar{x}-1.96\sigma/\sqrt{n}$ and $\bar{x}+1.96\sigma/\sqrt{n}$ which contain μ are called **95% confidence limits** for μ. The interval between these confidence limits is called the **95% confidence interval**, and the 0.95 probability value is called the **confidence level** or **degree of confidence.**

Construction of a 95% confidence interval for μ requires the population standard deviation σ, which is most often unknown.

As noted in the definition of sample variance (See 4.4) the definition chosen gives us, in general, a better estimate of population variance than the alternative open to us. The string attached is that the sample size be sufficiently large. A sample size of at least 30 qualifies as sufficiently large. (See Sec. 10.9 for discussion of estimators and their properties.

> If for $n \geq 30$, we estimate σ by *s*, the 95% confidence interval for μ takes the form:
>
> $$\bar{x}-1.96\frac{s}{\sqrt{n}} < \mu < \bar{x}+1.96\frac{s}{\sqrt{n}}$$
>
> The lower and upper 95% confidence limits take the form $\bar{x}-1.96\frac{s}{\sqrt{n}}$ and $\bar{x}+1.96\frac{s}{\sqrt{n}}$.
>
> We may interpret 95% confidence intervals in relative frequency terms as follows: If we calculate 95% confidence intervals for the many, many population mean estimation problems that arise, then in approximately 95% of these cases the population mean μ will lie in the 95% confidence interval constructed for it. That is, over the long run 95% confidence intervals contain the population mean μ in approximately 95% of the cases.

Figure 4 shows ten 95% confidence intervals for μ based on ten randomly drawn samples of the same size. The confidence intervals in 1 and 7 do not contain μ. Over the long run we can expect this to happen approximately 5% of the time.

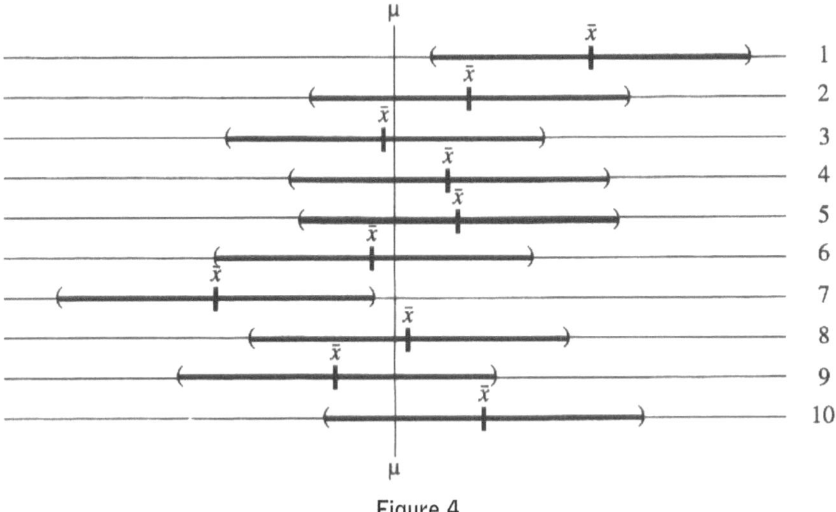

Figure 4

Example 1 Ecap University

Ecap University administers a mathematics placement exam to all of its students prior to their taking a mathematics course at the university. The exam, for which there are several equivalent versions, consists of 50 multiple choice questions.

The problem is to construct a 95% confidence interval for the mean test score μ. A randomly chosen sample of 50 test scores had a mean of 38 points (out of 50) with a standard deviation of 5 points.

The 95% confidence limits for μ are given by

$$\overline{x}+1.96\frac{\sigma}{\sqrt{n}} \quad \Box \quad 38\pm\frac{1.96(5)}{\sqrt{50}}=38\pm1.39$$
$$=36.6 \text{ and } 39.4,$$

where $s=5$ is used as an estimate for σ. This is allowable, in terms of standard statistical practice, since the sample size $n=50\geq30$.

Thus:

$$36.6<\mu<39.4 \text{ with prob. } 0.95$$

Does μ lie in this interval or not? We would have to say that we're not sure. The 0.95 probability value is a measure of the reliability of 95% confidence intervals; over the long run they do their job approximately 95% of the time. In terms of odds (Section

6.10, page 152), it would be appropriate to give odds of 95 to 5, or 19 to 1, that this interval does contain μ, but we cannot be absolutely sure that it does.

A reader of this analysis might argue that we would like to be more than 95% certain of results obtained. Let's increase the reliability level to 98%, for example.

For the 98% confidence interval the only change in the analysis is to replace $z_1 = -1.96$ and $z_2 = 1.96$ by values such that the area under the standard normal curve between 0 and z_1, and between 0 and z_2, is half of 0.98, or 0.4900 (instead of 0.4750 for the 0.95 confidence level). Getting as close as possible to 0.4900 in the standard normal curve table (Table A, page 652) and reading off its corresponding z value gives us $z_1 = -2.33$ and $z_2 = 2.33$. The lower and upper 98% confidence limits for μ are

$$\bar{x} \pm 2.33 \frac{\sigma}{\sqrt{n}}$$

which, in terms of standard statistical practice, may be approximated by

$$\bar{x} \pm 2.33 \frac{s}{\sqrt{n}}$$

when $n \geq 30$.

The 99% confidence limits for μ are

$$\bar{x} \pm 2.58 \frac{\sigma}{\sqrt{n}}$$

which may be approximated by

$$\bar{x} \pm 2.58 \frac{s}{\sqrt{n}}$$

when $n \geq 30$ (see Exercise 3).

The 98% confidence limits for the mean placement test score at Ecap University are 36.4 and 39.7 (see Exercise 2), whereas the 99% confidence limits are 36.2 and 39.8 (see Exercise 4).

What this makes clear, which we can also see from 1.96 increasing to 2.33 and 2.58 as we increase the confidence level from 95% to 98% and 99%, is that there is no free statistical lunch. The more confident we want to be that μ is in the confidence interval constructed for it, the wider will that interval have to be.

If we want to tighten up on the reliability of our estimate of μ (from 95% to 98% or 99% for example), and maintain the same sample size n, then the give will have to come from the precision of the confidence interval as measured by its radius, and vice versa.

Point Estimates

The other side of the confidence interval coin involves using $\overline{x}$ as an estimate, called a **point estimate**, for μ. The amount by which the estimator $\overline{x}$ differs from the estimated μ, $|\mu-\overline{x}|$, is the error committed in using $\overline{x}$ as an estimate for μ. If μ is in the confidence interval constructed for it, then $|\mu-\overline{x}|$ cannot exceed the radius of the confidence interval, For the 95% confidence interval the radius is $1.96\sigma/\sqrt{n}$ (see Figure 5). For other degrees of can confidence the 1.96 coefficient in $1.96\sigma/\sqrt{n}$ changes accordingly.

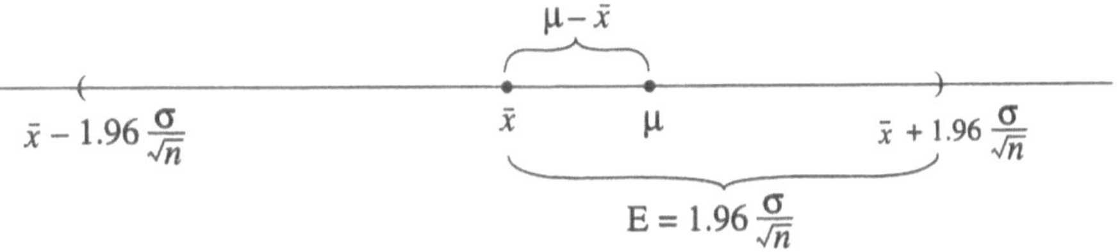

Figure 5

In summary we have the following:

> If the mean $\overline{x}$ of a randomly chosen sample of size $n \geq 30$ is used as a point estimate for μ, then with probability 0.95 the error in the estimate, $|\mu-x|$ does not exceed $E=1.96\sigma/\sqrt{n}$, which is the largest error that might arise.
>
> $E=1.96\sigma/\sqrt{n}$ is formally defined as the **error** inherent in the estimate at the 0.95 confidence level. For $n \geq 30$ E may be approximated by $1.96s/\sqrt{n}$.
>
> If the 0.95 confidence degree is changed, then the confidence level coefficient 1.96 changes accordingly.

Example 2 Return to Ecap University

If we use $\overline{x} = 38$ as a point estimate for the mean test score μ, then with probability 0.95 the error inherent in this estimate is:

$$E = 1.96\frac{\sigma}{\sqrt{n}} \approx \frac{1.96(5)}{\sqrt{50}} = 1.4$$

Is it possible that the difference between μ and $\overline{x}$, $|\mu - \overline{x}|$, exceeds 1.4? Yes, but 0.95 serves as a measure of our degree of belief that $|\mu - \overline{x}|$ does not exceed 1.4. Appropriate odds that $|\mu - \overline{x}|$ does not exceed 1.4 are 19 to 1.

A String Attached

Once a computation machinery is put in place and we become geared up to produce confidence intervals and point estimates on a mass scale, the fact that there is a string attached to the size of the sample n to be taken tends to fade into the background. Let us be sure to keep in mind the following observation from Section 9.3.

> If a population deviates "significantly" from normality, then sample size n has to be taken larger than 30 (how much larger we cannot say in advance) if the normal curve approximation for the sampling distribution of $\overline{x}$ is to be a good one.

Another dimension to sample size is considered in the next subsection.

A Question of Sample Size

In our discussion of Ecap University's mean test score estimation problem we began our analysis with a decision having been taken as to what size sample should be used. Here it was $n = 50$.

Taking a step back to before this decision was taken, let us consider the question of what size should be taken to attain a desired degree of accuracy. Suppose, for example, that we want to obtain an estimate of the mean test score μ and, with probability 0.95, restrict the error in the estimate to not exceed 1 point. How large a sample do we need at the 0.95 confidence level to attain this degree of accuracy?

We know that if $\bar{x}$ is used as an estimate for μ, then at the 0.95 confidence level the error inherent in the estimate is:

$$E = 1.96\frac{\sigma}{\sqrt{n}}$$

If the error in the estimate is not to exceed 1 point, then taking $E = 1$ gives us:

$$1 = 1.96\frac{\sigma}{\sqrt{n}}$$

The population standard deviation σ is unknown and there is no sample standard deviation *s* which might serve as an estimate for σ since a sample has not been taken. One option is to look at previous studies of a "similar kind" and, "if reasonable", base our estimate for σ on the results of such studies. The problem with this is in the vague nature of "similar kind" and "if reasonable." Reasonable to one might not be reasonable to another. But with the pressure to get the job done, a previously done study might more and more look as being of a "similar kind" to the one being undertaken and "if reasonable" might end up translated as "just do it."

Another option is to take a preliminary sample of size 30 or more to obtain a sample standard deviation value *s* that could be used as an estimate of σ. We would then proceed with the analysis to determine how much we would have to add on to this preliminary sample to obtain the precision desired.

Let us suppose that doing this, one way or the other, leads us to estimate σ as 5.5 points. This gives us:

$$1 = \frac{1.96(5.5)}{\sqrt{n}}$$

The application of basic algebra yields:

$$\sqrt{n} = 10.78$$
$$n = 116.2$$

Rounding up to give us a viable integer value for *n* yields $n = 117$. (Why not round down to 116?)

Thus, a sample of size $n = 117$ must be chosen at random from the placement test scores if the error committed in estimating μ by $\bar{x}$ is not to exceed 1 point at the 0.95 confidence level.

More generally, the error inherent in estimating μ by $\bar{x}$ at the 95% confidence level is:

$$E = 1.96\frac{\sigma}{\sqrt{n}}$$

Solving for *n* yields:

$$E\sqrt{n} = 1.96\sigma$$

$$\sqrt{n} = \frac{1.96\sigma}{E}$$

$$n = \left(\frac{1.96\sigma}{E}\right)^2$$

> If a desired degree of accuracy is prescribed, replace *E* by that prescribed value (1 in the previous test score problem). For us to determine *n*, the population standard deviation σ must be known or we must be able to estimate it.
>
> With these values for *E* and σ we solve the above for *n* and round up to obtain a viable integer value consistent with the 0.95 confidence level.
>
> If the 0.95 confidence level is changed, then the 1.96 confidence level coefficient changes accordingly.

EXERCISES

1. Construct a 95% confidence interval for the means of the populations from which each of the following samples was chosen. It is understood that the sample is small compared to the population. Can we be certain that the population mean is in the confidence interval obtained for it? Explain.

   (a) $n = 36, \bar{x} = 20, s = 2$.

   (b) $n = 40, \bar{x} = 50, s = 3.1$.

   (c) $n = 50, \bar{x} = 1500, s = 25$.

2. Determine the 98% confidence limits for Ecap University's mean placement test score based on a randomly chosen sample of 50 test scores with mean $\bar{x} = 38$ and $s = 5$.

3. Show that the confidence level coefficient for the 99% degree of confidence is 2.58, where the sample size *n* is large.

4. Determine the 99% confidence limits for Ecap University's mean placement test score as per Exercise 2.

5. For the situations given in Exercise 1, the sample mean is taken as a point estimate of the population mean. Determine the error inherent in the estimate at the 95% confidence level. Can we be certain that the error inherent in the estimate cannot be exceeded? Explain.

6. Determine the lower and upper confidence limits for the following degrees of confidence, where the sample size n is large.

   (a) 97% (c) 85%

   (b) 39% (d) 92%

   (e) Why must we require that the sample size n be large?

7. (a) What is the relative frequency interpretation of the 98% confidence interval for the population mean?

   (b) What would appropriate odds be that a population mean lies in the 98% confidence interval constructed for it?

8. **MERT:** The City Council of Bell City wishes to obtain a sense of how quickly the city's Medical Emergency Response Team (MERT) was responding to calls. A response time is the time elapsed between the time a call is received and the arrival time at the destination, in minutes, rounded off. A random sample of 36 response times chosen from the response time records of the past year is given in Table 1.

Table 1

12	13	17	9	10	15
6	8	20	10	8	5
10	5	8	8	11	19
12	7	16	11	8	10
9	10	21	18	5	7
6	9	10	4	6	4

   (a) Determine 95% confidence limits for MERT's mean response time.

(b) How is the answer obtained for (a) to be interpreted?

(c) What sample size would be needed if the City Council wanted to estimate MERT's mean response time with the sample mean so that the error inherent in the estimate does not exceed 1 minute with probability 0.95?

9. Dr. Janet Terence, Dean of the School of Education of Ecap University, wants to estimate the mean yearly salary of master's degree graduates of the education school three years after graduation. A randomly chosen sample of 36 graduates had a mean salary of $35,200 three years after graduation with a standard deviation of $5000.

(a) Determine 98% confidence limits for the mean salary of all recent graduates of the school of education three years after graduation.

(b) Dean Terence was asked whether she could guarantee that the mean she was seeking to estimate was between the confidence limits established in answer to (a). State an appropriate reply with explanation.

(c) Discuss the significance of the 0.98 degree of confidence.

(d) If $35,200 is taken as an estimate for the mean salary of all recent master's graduates of the school of education three years after graduation, what is the error inherent in the estimate at the 0.95 confidence level?

(e) Is it possible for the error determined in answer to (d) to be exceeded in giving $35,200 as an estimate for the population mean? Explain.

(f) What sample size would be needed if Dean Terence desired to estimate the mean salary so as to be able to assert with probability 0.98 that the error in the estimate does not exceed $1000?

10. In a discussion about sampling it was argued by one of the participants that "irrespective of the confidence level adopted, a random sample of 100 chosen from a population of 100,000 will give as precise an estimate for the population mean as a random sample of 100 chosen from a population of 100 million, provided that the standard deviations of the populations are the same." Would you agree or disagree with this assertion? Explain.

11. The Dean of the Nursing School of State University wants to estimate the mean SAT score of students in the nursing school so that the error inherent in the estimate does not exceed 10 points with probability 0.98. The problem is to determine the sample size that should be taken to carry out the study. The population standard deviation σ is needed, but it is not known and there are no reliable estimates for σ from other studies.

    Is there any way to obtain an estimate for σ?

12. Based on a randomly chosen sample of 49 plumbers who work for Larsen Associates, a large chain of plumbing companies, the following 92% confidence interval was obtained for the mean dollar hourly wage μ: $18 < \mu < 22$.

    (a) Can you say without reservation that μ lies in this interval? Explain.

    (b) One interpretation of this confidence interval that has been offered is that if many samples were chosen at random from the plumbers working for Larsen Associates, μ would fall between 18 and 22 approximately 92% of the time. Would you agree with this interpretation? Explain.

    (c) What interpretation would you give?

    (d) What would be appropriate odds that μ is between 18 and 22?

    (e) Determine $\overline{x}$ and s.

    (f) If we took $\overline{x}$ as an estimate for μ, what would be the error inherent in the estimate at the 92% confidence level?

    (g) Construct the 95% confidence interval for μ.

    (h) What are the odds that μ lies in the 95% confidence interval constructed for it?

    (i) What sample size would be needed if the management of Larsen Associates wanted to be 99% sure that the error in taking $\overline{x}$ as an estimate for μ did not exceed $1.50?

13. The Water Quality Control Division of Cameron City's Health Department is responsible for monitoring the chlorine levels of the city's reservoir system. A randomly drawn sample of 34 readings of chlorine levels from the main reservoir on a specified day is shown in Table 2. The readings describe parts per million chlorine per ten gallons of water.

(a) Determine 98% confidence limits for the mean chlorine level of the reservoir on the day in question.

Table 2

4.8	5.3	5.2	4.8	5.0	6.1
5.1	5.4	4.9	6.2	5.8	5.2
5.0	4.9	6.1	6.2	5.7	5.6
6.1	6.0	5.7	4.8	4.9	5.2
5.5	5.2	4.9	5.7	5.6	5.3
6.0	5.1	5.3	5.1		

(b) How is the result obtained in answer to (a) to be interpreted?

(c) What is the probability that the mean chlorine level of the reservoir does not lie within the bounds established in answer to (a)?

(d) How is the result obtained in answer to (c) to be interpreted?

14. Alexis Nurseries, which does a large mail order business, wishes to estimate the mean dollar size of orders received during the recent spring seasons. A randomly selected sample of 50 orders yielded, $\Sigma x_i = 1350$ and $\Sigma x_i^2 = 47{,}050$, where x_i denotes the dollar amount of the ith order in the sample, the sum being taken from 1 to 50.

(a) Determine 85% confidence limits for the mean dollar amount ordered during the recent spring seasons.

(b) Discuss the significance of the 0.85 level of confidence.

(c) If the sample mean is taken as an estimate for the population mean, what is the error inherent in the estimate at the 0.92 confidence level?

(d) Is it possible that the error exceeds the value given in answer to (c)? Explain.

(e) What sample size would be needed if the management of Alexis Nurseries wanted to estimate the mean order size with the sample mean so that the error inherent in the estimate does not exceed $2 with probability 0.92?

15. As part of a health awareness program the Keith City Health Services Administration determined the cholesterol level of a random sample of 45 individuals chosen from a large pool of Keith City residents who had volunteered to be part of the study. The sample mean was 180 mg/dl (milligrams per deciliter) with a standard deviation of 12 mg/dl.

(a) Find 98% confidence limits for the population mean.

(b) How is the result obtained in answer to (a) to be interpreted?

(c) A cholesterol reading above 200 mg/dl is considered high. Does the result obtained in answer to (a) imply that the mean cholesterol level of Keith City residents is satisfactory? Explain.

(d) The Keith City Health Services Administration desires to formulate conclusions about the cholesterol level of all residents of Keith City. Does the way in which the sample was chosen allow them to do this? Explain.

10.3 ■ The Small Sample Case and the t Distributions

The mechanism regulating blood coagulation is a complex one which involves the interaction of a number of factors in blood plasma, some of which are enormously difficult to identify and study. The bleeder's disease hemophilia, to take an example, comes in different forms which depend on which blood factor is missing. The advancement of our awareness and understanding of these blood factors depends in large measure on our finding people who lack them, which sometimes happens by accident, as was the discovery of Factor XII or Hageman factor, after the person who was first found to have Factor XII deficient blood.

Factor XII deficient blood is rare and difficult to find. The physiologist Leo Vroman, who spent much of his career studying the mechanism of blood coagulation, noted that in the seven years after he took blood from a Factor XII deficient patient, only one other person lacking Factor XII has been found in New York City. [5, p.32]

The study of materials which are rare or difficult to locate, difficult to prepare for study, or expensive lead us to take small samples. As was first noted by William Sealy Gosset (1876-1937), the central limit theorem and the normal distributions do not, at least in some circumstances, provide sufficiently accurate approximations when the sample size is small. Gosset, employed as a scientist at Guinness Brewery, found himself working with a substance which some would consider even more precious than blood, namely beer. Gosset's work at the Experimental Brewery was concerned with the analysis of malt and hops and the behavior of beer, which limited him to small samples. Company policy did not allow employees to publish research under their own names, but Gosset was allowed to publish his work under a pseudonym, for which he chose Student. Gosset's 1908 paper [4] marks the emergence of one of the most important classes of distributions, the *t* distributions or Student *t* distributions, as they have also come to be called.

The *t* distributions make up a family of probability distributions which are similar in appearance to the standard normal distribution. They are centered at 0, bell shaped, symmetric with respect to $t = 0$, lie above the horizontal axis, and get closer and closer to the horizontal axis as we take *t* values further and further away from 0. *t* distributions depend on the sample size through a parameter called the **degrees of freedom**, which is denoted by ν (the Greek letter nu) or d.f., depending on the context of the discussion. Figure 6 shows the standard normal curve and *t* distributions determined by 3 and 7 degrees of freedom. The *t* curves are less peaked than the standard normal curve and higher in the tails, which carry more of the area. As the number of degrees of freedom becomes larger and larger the *t* curves approach the standard normal curve.

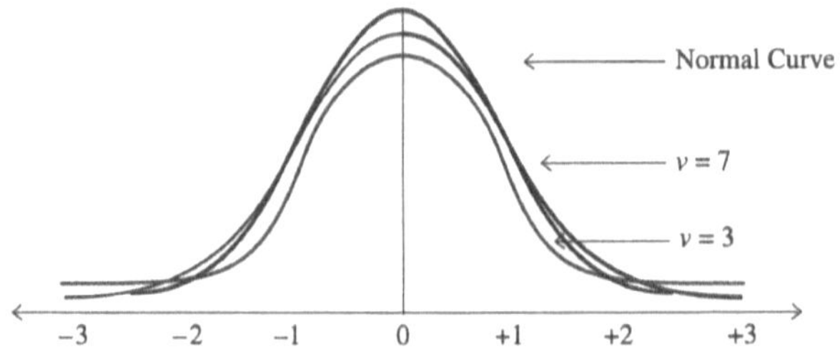

Figure 6 Standard Normal Curve and *t* Distributions

The t curves are defined by

$$y = c\left(1+\frac{t^2}{v}\right)^{-\frac{v+1}{2}},$$

where c is a constant depending on ν and is chosen in such a way that the total area under the curve is equal to 1. As with the normal curves, our work will not require us to plot a t curve from its equation. Our concern will be with areas of regions determined by t curves.

A table of t curve values for 1 through 29 degrees of freedom is given in Table B (page 653). A short version of this table is reproduced here as Table 3.

Table 3

d.f.	$t_{.025}$	$t_{.010}$	$t_{.005}$	d.f.
1	12.706	31.821	63.657	1
2	4.303	6.965	9.925	2
3	3.182	4.541	5.841	3
4	2.776	3.747	4.604	4

The left and right hand columns of Table 3 list degrees of freedom. The column labeled $t_{.025}$, for example, describes values for the degrees of freedom cited such that the area of the region to the right of $t_{.025}$ (tail) is 0.025 (see Figure 7). Thus, $t_{.025}$ for 3 degrees of freedom is 3.182.

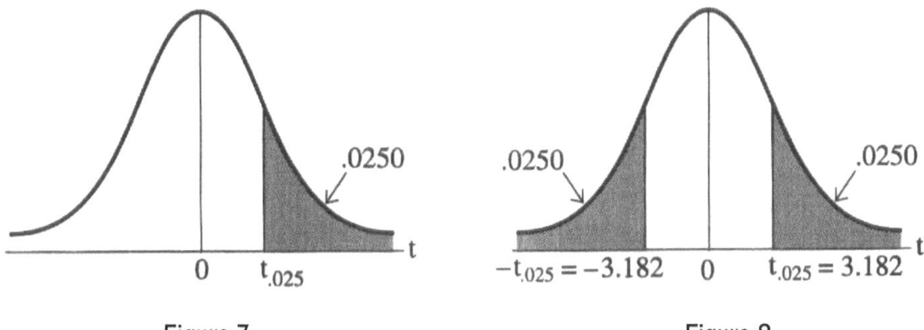

Figure 7

Figure 8

Since the t curves are symmetric about 0, $-t_{.025}$ is -3.182 (see Figure 8).

Confidence intervals for μ, where the sample size n is small, the sample is chosen at random from a **population which is normally distributed** or approximately so and the population standard deviation σ is not known, are based on the probability behavior of the sample statistic

$$t = \frac{\bar{x} - \mu}{s / \sqrt{n}}$$

where $\bar{x}$ is the sample mean. The variability of t is due to $\bar{x}$ and s.

Sampling Distribution of t. Under the afore conditions the sampling distribution of

$$t = \frac{\bar{x} - \mu}{s / \sqrt{n}}$$

is, to a close approximation, the t distribution with $n-1$ degrees of freedom.

To obtain a 95% confidence interval for μ, let us observe that 95% of the area under a t curve is between the bounds $-t_{.025}$ and $t_{.025}$ (see Figure 9). Thus, with probability 0.95, $t = \frac{\bar{x} - \mu}{s / \sqrt{n}}$ falls between $-t_{.025}$ and $t_{.025}$; that is,

$$-t_{.025} < \frac{\bar{x} - \mu}{s / \sqrt{n}} < t_{.025} \text{ with prob. 0.95.}$$

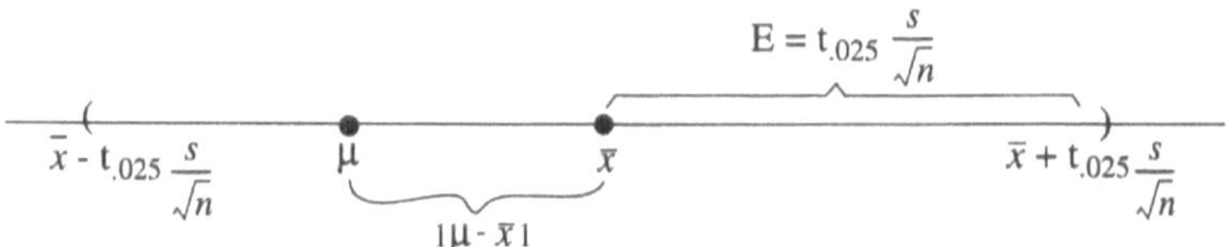

Figure 9

This double inequality has the same structure as the one which emerged in the large sample situation, with s appearing instead of σ, and $-t_{.025}$ and $t_{.025}$ which depend on the sample size through the number of degrees of freedom, appearing instead of -1.96 and 1.96.

The same follow through yields this 95% confidence interval for μ:

$$\bar{x} - t_{.025} \frac{s}{\sqrt{n}} < \mu < \bar{x} + t_{.025} \frac{s}{\sqrt{n}} \text{ with prob. 0.95}$$

An approximate value for $t_{.025}$ is to be looked up in Table B (page 653) for d.f. $= n-1$.

As is the case with large sample confidence intervals, the other side of the confidence interval coin involves using $\overline{x}$ as a point estimate for μ. The amount by which the estimator $\overline{x}$ differs from the estimated μ, $|\mu - x|$, is the error committed in using $\overline{x}$ as an estimate for μ. If μ is in the 95% confidence interval constructed for it, then cannot exceed the radius $E = t_{0.025}\frac{s}{\sqrt{n}}$ of the confidence interval (see Figure 10).

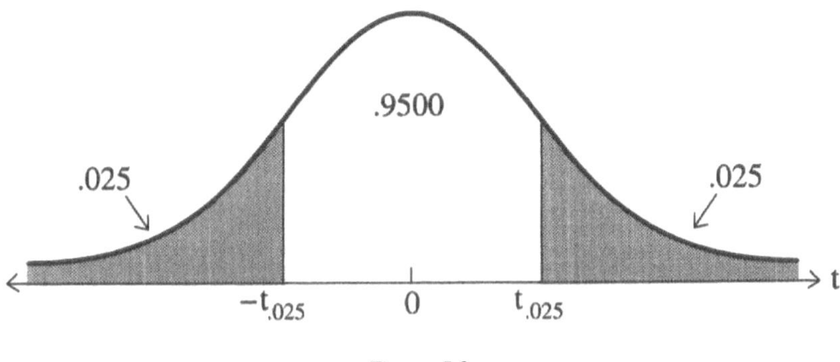

Figure 10

$E = t_{.025}\frac{s}{\sqrt{n}}$ is formally defined as the **error** inherent in the estimate at the 0.95 confidence level, where n is a small sample size ($n < 30$).

If the 0.95 confidence level is changed, then the confidence level coefficient $t_{.025}$ changes accordingly.

Strings Attached

Two strings attached for the application of the small sample confidence interval and error term are that the population being sampled from is normally distributed or approximately so, and that the population standard deviation σ is not known. Roughly put, the first string means that a mound shaped population like that shown in Figure 11 may be sampled from, whereas highly skewed populations like those shown in Figures 12 and 13 do not satisfy the required condition.

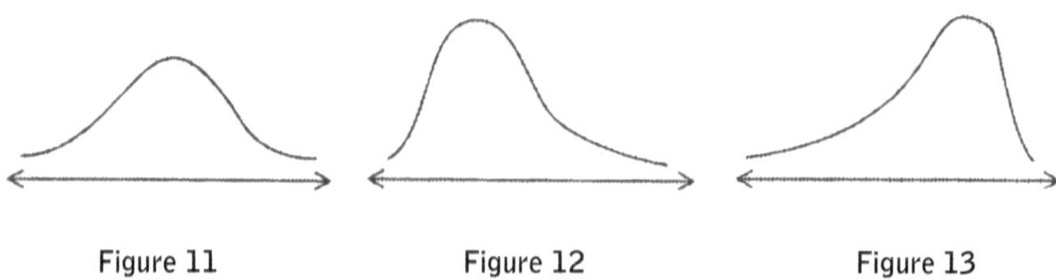

Figure 11 Figure 12 Figure 13

> Why all this fuss about strings attached? Strings attached are like the small print in a legal document which we tend not to look at closely, often much to our subsequent dismay. When the strings attached to a theorem, test, or procedure are violated, the results promised are usually illusionary.

The 95% small sample confidence interval for μ is strikingly similar to its large sample counterpart after *s* has been used as an estimate for σ, namely:

$$\bar{x} - 1.96\frac{s}{\sqrt{n}} < \mu < \bar{x} + 1.96\frac{s}{\sqrt{n}}$$

While the end-of-the-trail computation procedures are strikingly similar, for a proper perspective on this situation it should be kept in mind that the trails leading to these results are very different.

In the large sample case the sample statistic that started us going is $\bar{x}$. A central limit theorem brought a normal distribution into play independent of the sample size. The sample standard deviation *s* was later used as an estimate for σ when σ was unknown.

In the small sample case the sample statistic that started us going is

$$t = \frac{\bar{x} - \mu}{s/\sqrt{n}}.$$

This is a package deal, so-to-speak, with *s* **not serving as an estimate for** σ, but as an integral part of the package. A *t* distribution determined by the sample size through the number of degrees of freedom ($\text{d.f.} = n-1$ in this case) comes into play.

To construct 98% and 99% small sample confidence intervals replace 2.33 and 2.58 in the large sample case by $t_{.01}$ and $t_{.005}$, respectively.

Example 1 Friendly Furry Creatures

Robert Franks, an experimental psychologist, seeks an estimate of the time it takes a species of friendly furry creature he is studying to run through a maze. A randomly chosen sample of 10 friendly furry creatures traversed the maze in the following times, in seconds: 52, 48, 59, 60, 55, 51, 51, 50, 49, 58.

(a) Determine the sample mean.

(b) If the sample mean is taken as an estimate of the population mean, what is the error inherent in the estimate at the 0.95 confidence level?

(c) How is the result obtained in answer to (b) to be interpreted?

(d) What is the population whose mean Robert Franks is seeking to estimate?

(e) Are there conditions on this population which Mr. Franks should be concerned with? Explain.

(f) If these conditions are not met, then what should be said about the reliability of the error estimate given in answer to (b) when the sample mean is taken as an estimate for the population mean? Explain.

(a) $\sum x = 533$; thus $\overline{x} = 53$

(b) $\sum x^2 = 28,581$, and $s^2 = \dfrac{10(28,581) - (533)^2}{90} = 19.1$; $s = 4.4$

$t_{.025} = 2.262$ for $\text{d.f.} = 9$

$E = 2.262\dfrac{(4.4)}{\sqrt{10}} = 3.1$ sec. with prob. 0.95

(c) 0.95 expresses the reliability of the estimate. It is a numerical measure of our degree of belief that the error does not exceed 3.1 seconds if μ is estimated as 53 seconds.

(d) The population of times that it takes for the members of the species of friendly furry creature to run through the maze.

(e) The population must be normally distributed or approximately so.

(f) The error term would have to be considered unreliable in the sense that the 0.95, which expresses the reliability of the estimate of μ by $\overline{x} = 53$, would be open to question.

The towering figure of modern statistical theory and practice, Sir Ronald Fisher (1890-1962), observed that "The study of the exact distributions of statistics commences in 1908 with Student's paper *The Probable Error of a Mean.* Once the true nature of the problem was indicated, a large number of sampling problems were within reach of mathematical solution" [2; 14th ed., p.23] Gosset's work was not quickly appreciated, and it was through Fisher's efforts and work that it gained the recognition it deserved. Gosset himself, as Joan Fisher Box puts it, "was an extraordinarily appealing individual, generous to a fault, humble, enthusiastic in the pursuit of his varied interests, and helpful. He had a flair for grasping the heart of a problem that made his contributions to statistics important out of all proportion to their number." [1;p.61]

REFERENCES

1. J.F. Box, "Gosset, Fisher, and the *t* Distribution," *The American Statistician,* 35, 2 (May 1981), 61-66.

2. R.A. Fisher, *Statistical Methods for Research Workers* (Edinburgh: Oliver and Boyd, 1925); 14th ed., 1970.

3. J.O. Irwin, "William Sealy Gosset," *International Encyclopedia of the Social Sciences,* D.L. Sills, ed. (New York: The MacMillan Co. & The Free Press, 1968), 6, 211-214.

4. Student (W.S. Gosset), "The Probable Error of a Mean," *Biometrika,* 6, 1 (1908).

5. L. Vroman, *Blood* (Garden City, New York: The Natural History Press, 1967).

EXERCISES

1. Find the value of t. Sketch a diagram for each case.

   (a) $t_{.100}$, d.f. = 15 (c) $t_{.025}$, d.f. = 29

   (b) $t_{.010}$, d.f. = 8 (d) $t_{.005}$, d.f. = 5

2. Construct a 95% confidence interval for the means of the populations from which each of the following samples was chosen. What strings attached should

we be aware of? Why? Can we be certain that the population mean is in the confidence interval constructed for it? Explain.

(a) $n = 12,\ \overline{x} = 32,\ s = 3$.

(b) $n = 7,\ \overline{x} = 16,\ s = 2$.

(c) $n = 5,\ \overline{x} = 26,\ s = 4$.

3. For the situations given in Exercise 2, the sample mean is taken as a point estimate of the population mean. Determine the error inherent in the estimate at the 98% confidence level. Can we be certain that the error inherent in the estimate cannot be exceeded? Explain.

4. For the small sample case, what is the relative frequency interpretation of the 95% confidence interval for the population mean?

5. A sample of 12 values chosen at random from a normally distributed population is found to have mean 14.1 and standard deviation 2.3.

(a) Find 95% confidence limits for the population mean.

(b) How is the result obtained in answer to (a) to be interpreted?

(c) If 14.1 is taken as an estimate for the population mean, what is the error inherent in the estimate at the 0.98 probability level?

(d) How is the result obtained in answer to (c) to be interpreted?

6. Jennifer Clarke, a quality control manager for an auto manufacturer, wants to get an estimate for the mean dollar damage to the new Turbo model when it is driven into heavier autos at 30 miles per hour. She has been authorized to conduct tests with 5 Turbo model cars. A random sample of 5 cars that were tested yielded a mean damage amount of $18,250, with a standard deviation of $3500.

(a) Find 98% confidence limits for the population mean. How is your result to be interpreted?

(b) Evidence available for models similar to the Turbo suggest that the population of dollar damage values is U shaped, with values concentrated at the low and high ends, indicating that collision costs tended to be minimal or very high. Would this information prompt you to question the reliability of the confidence limits obtained in answer to (a)? Explain.

7. A test of the breaking strengths of 15 safety belts chosen at random from the production line of the Tobin Company yielded $\Sigma x_i = 1430$ pounds and $\Sigma x_i^2 = 137{,}200$ pounds, where x_i is the breaking strength, in pounds, of the ith belt in the sample, the sum being taken from 1 to 15.

   (a) Find 98% confidence limits for the mean breaking strength of the belts made by the Tobin Company.

   (b) How is the result obtained in answer to (a) to be interpreted?

   (c) If the sample mean is taken as an estimate for the population mean, what is the error inherent in the estimate at the 0.99 confidence level?

   (d) How is the answer obtained in answer to (c) to be interpreted?

   (e) What assumption underlies the confidence limits and error term obtained in answer to (a) and (c)?

   (f) Suppose it were later found that the population from which the sample was chosen deviates considerably from the assumption given in answer to (e); where does this leave us in connection with the results obtained in response to (a) through (d)? Explain.

8. To help trim its budget, the Water Quality Control Division of Cameron City's Health Department (see Exercise 13, Sec. 10.2) has been instructed to monitor the chlorine level of the City's reservoir system by means of small sample tests, rather than the large sample tests which they had been using. The readings, in parts per million chlorine per ten gallons of water, from a randomly drawn sample of 8 containers of water, obtained on a specified day, were: 4.9, 5.2, 5.2, 5.3, 5.8, 5.6, 5.1, 5.5.

   (a) Find 90% confidence limits for the mean chlorine level of the City's reservoir system on the day in question.

(b) How is the result obtained in answer to (a) to be interpreted?

(c) What assumption underlies the confidence limits obtained in answer to (a)?

(d) Suppose it was later found that the population from which the sample was chosen deviates markedly from the assumption given in answer to (c); where does this leave the confidence interval obtained in answer to (a) and interpretation given in answer to (b)? Explain.

9. As has been observed, statistics texts have not only become more expensive in recent years, but bigger and heavier as well. A recent edition of *Basic Statistics* was the focus of a wrist strength test at Ecap University. Table 4 gives the time lengths, in minutes, that the population of 156 students at Ecap University who used the text were able to hold it in reading position in one sitting before their wrists gave out.

Table 4

1.7	4.3	6.0	4.7	1.7	7.2	3.1	3.9	3.3	7.0	2.5	4.9	5.2
2.3	4.5	6.9	3.4	3.8	6.6	3.6	3.5	5.5	5.4	5.9	6.1	4.4
7.9	4.7	5.1	6.0	5.9	8.4	5.1	6.3	5.2	4.4	5.2	3.0	4.4
4.3	3.3	5.0	3.8	7.0	2.9	3.0	4.1	4.5	5.2	3.3	4.2	2.5
5.1	2.1	4.6	2.9	4.1	5.9	4.7	4.8	4.1	5.5	5.0	2.6	4.4
6.1	5.0	6.3	4.6	8.1	4.6	6.1	6.6	3.2	5.9	1.8	3.2	1.5
5.7	7.3	4.3	3.2	3.4	4.1	5.0	1.7	4.3	4.2	3.5	4.1	4.5
4.8	3.7	3.0	4.9	7.2	4.5	4.9	6.5	3.1	5.2	4.8	3.9	4.9
7.5	3.8	4.5	6.8	4.1	0.3	5.9	2.4	3.6	3.4	1.9	3.4	1.6
3.5	3.4	3.6	2.6	5.9	5.1	4.9	6.2	5.0	3.9	601	5.6	4.6
7.5	6.7	4.4	5.8	4.5	5.1	5.3	4.4	5.8	4.7	2.5	3.5	7.6
4.3	4.7	5.1	5.8	3.1	2.3	5.7	8.2	0.8	2.5	7.2	2.6	6.7

(a) By using the procedures discussed in Section 2.3 on using the table of random numbers (page 51), take a random sample of 5 times. See Example 1 of Section 2.3.

(b) Find 90% confidence limits for the population mean.

(c) How is the result obtained in answer to (b) to be interpreted?

(d) Must any assumptions be made about the population you are sampling from? Explain.?

(e) If so, are the assumptions realistic? Explain.

(f) Is the population mean definitely contained in the confidence interval constructed in answer to question (b)?

(g) Would it not be better to obtain the 95% interval for μ as opposed to the 90% interval since the 95% level gives us a higher level of reliability than the 90% level? Explain.

10.4 ■ What Should Be Considered in Constructing Confidence Intervals?

Statistical life in the classroom and textbook is in many respects easier than that in the corner of the "real statistical world" which we happen to inhabit. In the classroom and textbook we can freely say such things as "take a random sample" or "a sample of size 32 yields mean and standard deviation . . ." and by a wave of a magic statistical wand it is done. When, in fact, someone has to do such things it is far from being the simple matter it so innocently sounds like, as we have seen in connection with random sampling. The same might be said about the problem of determining the sample size that we should take in a particular study. Why take a sample of size 32, or whatever, as opposed to other options? Why this or that confidence degree? We have discussed some concerns and we now look at a more complete picture.

Large Versus Small Samples

One advantage of large samples, $n \geq 30$, is that large samples free us, to a "large extent," from concerns about the nature of the distribution of values in the population being sampled from. Small sample analysis. $n < 30$, requires that the population being sampled from be normally distributed or, at least, mound shaped. This condition must be looked into before a small sample can be taken. If the population differs from the assumption we require of it and we go ahead with a small sample analysis, through convenience, defiance or ignorance, our conclusions, though impressive sounding, are not mathematically valid and may be nonsense about the subject matter in question.

Mathematical voodoo is more deadly than its real world cousin. With the real world thing you have a better idea of where you stand. Mathematical voodoo—deriving mathematical conclusions without a sufficient basis—is more subtle in leading us to believe we have something of substance which in fact may be nonsense. **Always watch the assumptions.**

As to our working guideline of what distinguishes large samples, $n \geq 30$, from small samples, $n < 30$, it should be kept in mind that 30 is a guideline which has come out of statistical practice. We need a specific guideline value and this one will do. The central limit theorem on the sampling distribution of the sample mean $\overline{x}$ tells us that the sampling distribution of $\overline{x}$ converges to a normal distribution as the sample size n increases. It says nothing about 30 being a defining value which distinguishes large from small samples. If we look at 95% confidence limits, to take a specific confidence level, for $n = 29$ (small) and $n = 30$ (large) we see little difference:

$$\overline{x} \pm 2.045 \frac{s}{\sqrt{n}}, \quad n = 29$$

$$\overline{x} \pm 1.960 \frac{s}{\sqrt{n}}, \quad n = 30, \ s \text{ approximates } \sigma$$

The value 2.045 differs from 1.960 by 0.085, so that there is no cataclysmic change in practice in going from $n = 29$ to $n = 30$. Still, the guideline 30 value distinguishing small from large samples should be adhered to. It does not happen in practice that one is faced by a big decision whether to take $n = 29$ or $n = 30$. "Small samples" taken in practice are small and "large samples" taken in practice are large.

Trade-Offs: Sample Size, Reliability and Precision

Sample size n, reliability of $\overline{x}$ as an estimate for μ—measured by the confidence level, and the precision of the estimate for μ—measured by the radius of the confidence interval, are intimately linked so that you cannot change one without affecting at least one of the other two. To take specific cases, the 95%, 98%, and 99% large sample confidence limits for μ and their accompanying maximum error terms E are the following:

95% $\quad \overline{x} \pm 1.960 \frac{\sigma}{\sqrt{n}}, \quad E = 1.960 \frac{\sigma}{\sqrt{n}}$

98% $\quad \overline{x} \pm 2.33 \frac{\sigma}{\sqrt{n}}, \quad E = 2.33 \frac{\sigma}{\sqrt{n}}$

99% $\bar{x} \pm 2.58\frac{\sigma}{\sqrt{n}}$, $E = 2.58\frac{\sigma}{\sqrt{n}}$

In geometric terms the situation is shown in Figure 14. For the same sample size *n*, if we seek to strengthen the reliability of our estimate, then it will have be achieved at the expense of precision, and vice versa.

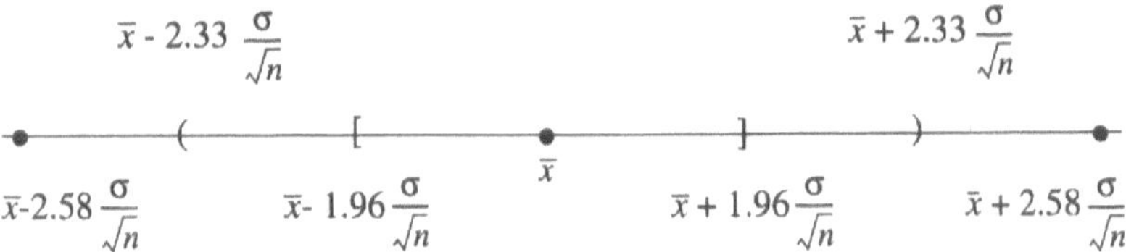

Figure 14

How reliable must an estimate be? 95%, which is equivalent to odds of 19 to 1, 98%, which is equivalent to odds of 49 to 1, 99%, which is equivalent to odds of 99 to 1, or something larger or smaller? Statisticians are not definitive about this, and the answer will have to come out of the context of the situation under study through use of judgment. What is clear is that to tighten up on reliability there will have to be movement in terms of precision or sample size or both. These observations hold, even more strongly, for small sample confidence intervals.

Small Sample Size and Precision

We pay a high premium for small sample size in terms of precision, which is expressed by the error *E* inherent in a confidence interval estimate of μ, and defined by the radius of the confidence interval.

Let us consider, for example, sample size 9 versus sample size 49, other factors being the same. Suppose the confidence level is 95%, and $s = 5$, just to make things concrete. We further assume that the underlying population is approximately normal.

Large sample: $E = 1.96\frac{s}{\sqrt{n}} = 1.96\frac{(5)}{\sqrt{49}} = 1.4$, with probability 0.95

Small sample: $E = t_{.025}\frac{s}{\sqrt{n}} = 2.306\frac{(5)}{\sqrt{9}} = 3.8$, with probability 0.95

In this case E for the small sample is 2.7 times E for the large sample. In general, E for the small sample exceeds E for the large sample because the t-value multiplying $s/\sqrt{n}$ exceeds the z-value multiplying this factor and $\sqrt{n}$ for small n is smaller than $\sqrt{n}$ for large n.

Small sample size is costly in terms of the precision of a confidence interval estimate of μ.

Restricted Resources

As we have seen, working with rare or difficult to obtain materials places a severe constraint on sample size. Time is another resource which is often in short supply. If obtaining the sample must be accomplished within a tight time frame, this constraint will have an impact on sample size.

And then there is the matter of cost of sample collection, which includes the fixed cost of setting up the project, overhead, and related costs, irrespective of the sample size taken, and a variable cost which describes the cost per data value in the sample.

If the project consists of choosing a sample of screws at random from a production line and measuring their lengths, it might cost something like \$5000 to set up the project and \$2 per measurement. The sampling cost function $C(n)$, where n is the sample size, would be:

$$C(n) = 5000 + 2n$$

If the project involves obtaining a mean dollar damage estimate for an automobile model when it is driven into heavier autos at 30 miles per hour (see Section 10.3, Exercise 6), it might cost something like \$5000 to set up the project and \$8000 per measurement. The sampling cost function would be:

$$C(n) = 5000 + 8000n$$

From a sampling cost point of view we would be much less hesitant about increasing sample size in the former case than the latter.

When budget and other resource limitations restrict sample size n the consequences are usually borne by the precision E of the estimate. Certain minimal levels of confidence are expected and there is not much room to maneuver with here. On the other hand, the value of the information provided by a sample depends on the precision with which it is given.

> Thus, the determination of sample size must also take into account the balance between the value of the information to be obtained from the sample and the limitations on resources for obtaining that information.

EXERCISES

1. With the summer season about the arrive, the health department of Ocean City will begin testing the beach area water to monitor the concentration of E. Coli bacteria, which at sufficiently high levels pose a health hazard. Mary Smith has been put in charge of the project of determining a confidence interval for the mean E. Coli count for each day. Random samples of the water are to be taken daily and tested, but there are questions about what sample size, confidence level, and precision would be appropriate within a context provided by health, budget and mathematical considerations.

   Ms. Smith has asked you to help her reach a decision on these matters by writing a summary of the factors she should be considering about confidence intervals. Please oblige her.

2. Arthur Jackson, quality control manager for a truck manufacturer, has been given the responsibility of determining the mileage traveled by the new Xcel model before a transmission overhaul is required. A random sample of this model is to be taken and a confidence interval constructed from the results obtained. It has been determined that the cost function $C(n) = 20,000 + 10,000n$ provides a satisfactory estimate, in dollars, of the cost of conducting tests on n Xcel trucks. A preliminary budget of \$150,000 has been set for the project.

   Mr. Jackson has questions about the sample size, confidence level, and precision that would be appropriate, considering the context provided by the situation and the mathematics involved. Advise Mr. Jackson on the factors he should be considering in setting up a confidence interval.

3. Suppose it were determined that the population Mr. Jackson is sampling from is highly skewed to the right, J shaped. How would this affect the advice you would give him? Explain.

4. (a) For large sample size, what affect would quadrupling the sample size have on a confidence interval?

   (b) More generally, what affect would increasing a given sample size by a factor of k have on a confidence interval?

5. (a) For large sample size, what affect would going from a 99% to 80% confidence level have on a confidence interval?

   (b) Why not adopt the 80% confidence level as a standard for confidence interval applications? Explain.

10.5 ■ Estimation of Proportions

Television executives are interested in the proportion of viewers who are tuned into their network, political establishments are interested in the proportion of voters who favor their candidate, marketing personnel are interested in the proportion of consumers who prefer their product, medical doctors are interested in the proportion of those suffering from an illness who respond to a proposed therapy.

In all of these situations there is a population and an option or characteristic, call it A, which is favored or preferred by some proportion of the population as opposed to not-A. It is a two-sided situation, option A versus not-A. We shall denote the **population proportion** which favors option A by the Greek letter π (pi).

Confidence Interval Estimates for the Population Proportion

The problem we set ourselves is that of developing confidence interval estimates for π. To do this we make use of the sample proportion $p = x/n$, where x is the number of members of the sample who favor option A, and n is the size of a sample randomly chosen from the population. If, for example, a sample of size $n = 100$ were chosen from a population and 40 favored option A (a specific network, candidate for office, particular product, etc.), the sample proportion in favor of option A would be $40/100 = 0.40$ or 40%. The key to using $p = x/n$ as our sample statistic to obtain a confidence interval estimate for π is the following central limit theorem on the probability behavior of $p = x/n$.

> **Central Limit Theorem for $p = x/n$.** If a sufficiently large sample of size n is chosen at random from a population which is large, the sampling distribution of $p = x/n$ may be approximated by the normal curve with mean μ_p and standard deviation σ_p given by,
>
> $$\mu_p = \pi, \quad \sigma_p = \sqrt{\frac{\pi(1-\pi)}{n}},$$
>
> where π is the population proportion, n is the size of a sample drawn at random from the population, and x is the number of members of the sample who favor option A.

To implement this theorem we require that np and $n(1-p)$ be at least 5.

To be specific let us take 0.95 as our working confidence level. The normal curve describing the sampling distribution of $p = x/n$ is shown in Figure 15 and its standard normal curve counterpart is shown in Figure 16.

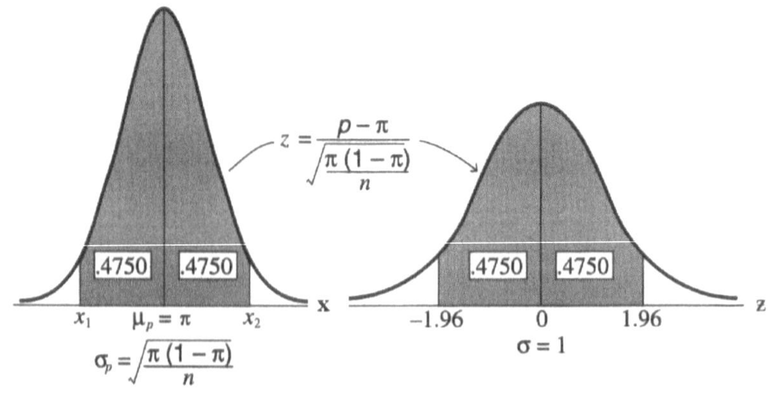

Figure 15 Figure 16

If we let x_1 and x_2 denote equally spaced bounds on both sides of $\mu_p = \pi$ (Figure 15) so that the area under the normal curve between them is 0.95, then the standard normal curve counterparts of x_1 and x_2 are -1.96 and 1.96, respectively (Figure 16). To say that $p = x/n$ is between x_1 and x_2 with probability 0.95 is equivalent to saying that the *z*-value of $p = x/n$ is between -1.96 and 1.96 with probability 0.95. That is:

$$-1.96 < \frac{p - \pi}{\sqrt{\frac{\pi(1-\pi)}{n}}} < 1.96 \text{ with prob. 0.95}$$

To obtain our desired estimate for π we must isolate π from the preceding double inequality. This requires the same sort of application of basic algebra which we employed in Section 10.2 to isolate the population mean from a double inequality. Doing so (see Exercise 9) yields the following 95% confidence interval for π:

$$p - 1.96\sqrt{\frac{\pi(1-\pi)}{n}} < \pi < p + 1.96\sqrt{\frac{\pi(1-\pi)}{n}} \text{ with prob. 0.95}$$

In geometric terms this double inequality is an interval centered at the sample proportion $p = x/n$ with radius $1.96\sqrt{(\pi(1-\pi))/n}$, shown in Figure 17.

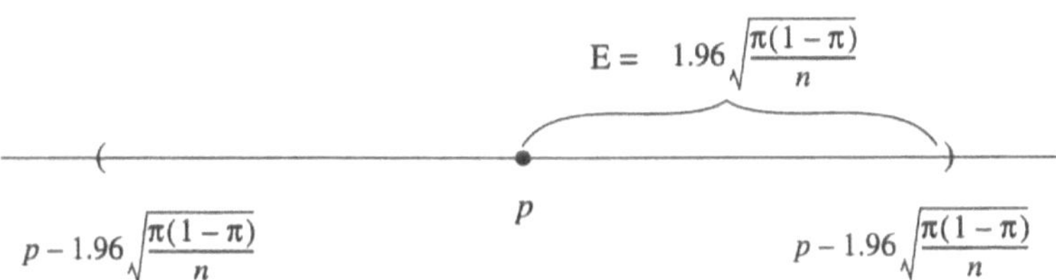

Figure 17

As in the previous cases considered, the values

$$p - 1.96\sqrt{\frac{\pi(1-\pi)}{n}} \text{ and } p + 1.96\sqrt{\frac{\pi(1-\pi)}{n}}$$

are called **95% confidence limits for π.**

The problem with this confidence interval is that its radius involves π itself, which is to be estimated in terms of the confidence interval. We ran into a similar situation in developing confidence intervals for the population mean μ, large sample case, where σ was unknown. Our solution there was to take *s* as an estimate for σ. This solution is viable as long as the sample size is at least 30.

Sample Size

Our solution here is to take the sample proportion $p = x/n$ as an estimate for π. For this solution to be viable the sample size n must be sufficiently large, or else p would not be a close enough estimate for π and the accuracy of the confidence interval estimate for π would be illusionary. There is no generally accepted benchmark value like 30 in this situation, but it is reasonable to say that any sample size less than 200 would not yield a value of p which would serve as a sufficiently accurate estimate for π.

The problem of estimating π with a "crude" estimate $p = x/n$ to obtain a more "exact" confidence interval estimate arises when the sample size is to be initially stated and the error of the estimate subsequently determined. As we shall see, the problem does not arise if the desired error term is initially stated with the sample size to be determined.

> Replacing π by $p = x/n$, $n \geq 200$, in the radius of the 95% confidence interval for π gives us the following working 95% confidence interval for π:
>
> $$p - 1.96\sqrt{\frac{p(1-p)}{n}} < \pi < p + 1.96\sqrt{\frac{p(1-p)}{n}} \text{ with prob. 0.95.}$$
>
> Confidence intervals based on other degrees of confidence require that the coefficient 1.96 be replaced by the appropriate value. For 98% and 99% confidence intervals replace 1.96 by 2.33 and 2.58, respectively.

Applicability

> While other applications of confidence interval analysis considered in this chapter require that the data scale be at least interval, confidence interval analysis of proportions is applicable to nominal and ordinal data. In Example 1 the issue centers on whether a tube produced is defective (D) or non-defective (N), which entails an ordinal scale; $p = x/n$ records the proportion of defectives in a sample of size n. In Example 2 the issue centers on whether a vote cast was for Leonard Truck (Y) or not (N), which entails a nominal scale; $p = x/n$ records the proportion of votes for Truck in a sample of size n,

Example 1 Quality Control

The Glascow Company produces cathode ray picture tubes for television sets. Part of the company's quality control maintenance procedures is to monitor the proportion of defectives produced by random sampling from the production line. In a random sample of 320 units, 6 were found to be defective. Find 98% confidence limits for the proportion of defectives produced by the company.

Since $n = 320 > 200$, and $np = 6$, $n(1-p) = 314$, both of which exceed 5, the required guidelines are satisfied and we may proceed. The 98% confidence limits for the population proportion of defectives are:

$$\begin{aligned} p \pm 2.33\sqrt{\frac{p(1-p)}{n}} &= 0.0188 \pm 2.33\sqrt{\frac{0.188(0.9812)}{320}} \\ &= 0.0188 \pm 0.0177 \\ &= 0.0011 \text{ and } 0.0365 \end{aligned}$$

With probability 0.98, the population proportion, or percentage, of defectives is between.11% and 3.65%. In terms of odds, the odds are 49 to 1 that the proportion of defectives produced is between.11% and 3.65%

Point Estimates

As we have seen, the other side of the coin of a confidence interval estimate is a point estimate wherein the sample value is used as an estimate for the population value with a statement of the error inherent in the estimate.

> **Theorem:** If for a randomly drawn, sufficiently large sample, the sample proportion $p = x/n$ is used as an estimate for π, the error in the estimate, $|\pi - p|$, is less than
>
> $$E = 1.96\sqrt{\frac{\pi(1-\pi)}{n}}$$
>
> with probability 0.95 (see Figure 18).

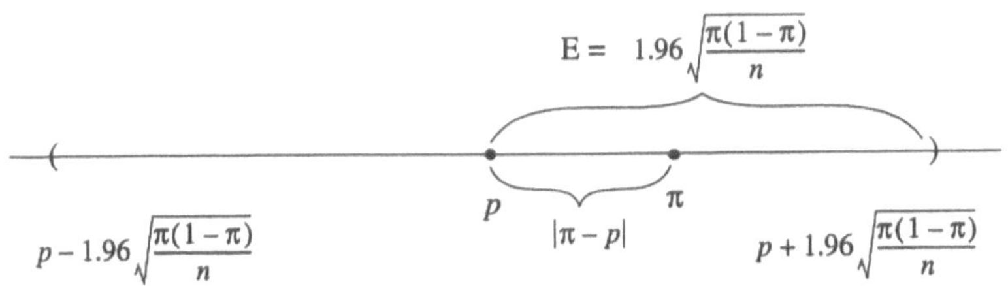

Figure 18

The radius of the confidence interval E, called the **error** inherent in the estimate at the 95% confidence level, is the largest error that might occur at the 0.95 probability level in estimating π by p.

If we change the degree of confidence from 0.95 to some other value, the coefficient 1.96 changes accordingly. For the 98% and 99% confidence levels 1.96 is replaced by 2.33 and 2.58, respectively.

> For n sufficiently large (200 is our suggested guideline minimum), we may estimate
>
> $$E = 1.96\sqrt{\frac{\pi(1-\pi)}{n}} \text{ by } 1.96\sqrt{\frac{p(1-p)}{n}}$$
>
> at the 0.95 confidence level. For other confidence levels the coefficient 1.96 is replaced accordingly.

Example 2 Political Preference

Leonard Truck is seeking the Democratic Party's nomination to run for governor of a state in the southwest. It's been a tough fight against strong opposition, and on the eve of the primary a poll was taken to establish where the contenders stood. In a random sample of 500 party members 275 expressed a preference for Truck over the combined opposition.

If $\mathrm{p} = 275/500 = 0.55$ is taken as an estimate for the population proportion of party members who favor Truck, what is the error inherent in the estimate at the 95% confidence level?

The sample size 500 is large and both $np = 500 \cdot (0.55)$ and $n(1-p) = 500 \cdot (0.45)$ clearly exceed our guideline value of 5. If we take $p = 0.55$ as our estimate for π, the error inherent in the estimate at the 95% confidence level is:

$$E = 1.96\sqrt{\frac{(0.55)(0.45)}{500}} = 0.0436$$

In terms of odds, the odds are 19 to 1 that if 55% is taken as an estimate of the percentage of party members who favor Truck as their candidate, the error inherent in the estimate does not exceed 4.36%.

Leonard Truck was fully confident of victory in the primary, but when it came to pass he was soundly defeated, receiving only 25% of the vote. Mr. Truck was not a happy candidate and his humor was not improved by an aide's philosophical observation that people are more difficult to deal with than television cathode ray tubes. His reaction was that the pollsters were a bunch of bums who shouldn't be allowed to practice, and he demanded an explanation of what had gone wrong. In this connection see Section 2.5 and Exercise 5.

More on Sample Size

In the two examples considered a sample size was chosen and the error term obtained emerged as a consequence of sample size and, of course, the chosen degree of confidence. Suppose, on the other hand, that we prescribe the degree of confidence, 95% let us say, and require that the error inherent in the estimate, *E*, not exceed a certain level. How large a sample *n* should be taken?

To answer this question we begin with the error term inherent in the 95% confidence level,

$$E = 1.96\sqrt{\frac{\pi(1-\pi)}{n}},$$

which we must solve for *n*. Squaring both sides and following up with basic algebra yields:

$$E^2 = (1.96)^2\left[\frac{\pi(1-\pi)}{n}\right]$$

$$nE^2 = (1.96)^2\pi(1-\pi)$$

$$n = \pi(1-\pi)\left[\frac{1.96}{E}\right]^2$$

The problem with this expression for *n* is that it involves π, which is what we are seeking to estimate. Two possibilities arise: If there are previous studies of a "similar kind", it might be "reasonable" to use π determined there as a working estimate for π in the situation at hand; maybe, and then maybe not. The danger in this practice is that it makes the estimate of *n* dependent on an assumption which might be seriously open to question.

The second option is to replace $y = \pi(1-\pi) = \pi - \pi^2$ by the largest value it could have. This is a conservative approach which would yield a larger value for *n* than we would need, but it would work and does not require any assumptions about π.

To find the largest value that $y = \pi - \pi^2$, $0 < \pi < 1$, could have it is useful to examine the graph of this function of π. This function is a quadratic function and plotting a few points yields the graph shown in Figure 19.

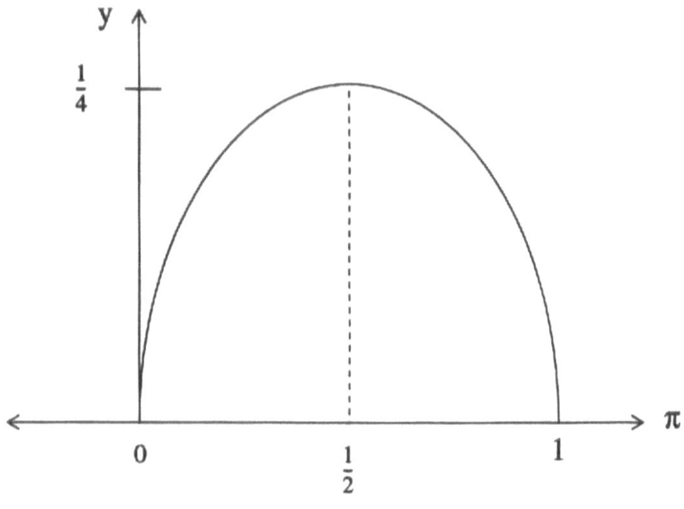

Figure 19

$\pi = 1/2$ yields the largest value of 1/4 for $y = \pi(1-\pi)$.

> As a conservative estimate for *n* we obtain,
>
> $$n = \frac{1}{4}\left[\frac{1.96}{E}\right]^2$$
>
> where *n* is rounded up to the nearest integer value, for the 95% confidence degree, and where *E* is specified.
>
> For other confidence levels, 1.96 is replaced accordingly.

Example 3 Sample Size for 95% Degree of Confidence

If Leonard Trunk's campaign manager had insisted that the poll be taken with 95% degree of confidence and with a sampling error which did not exceed 2%, then the sampling size required would have been:

$$n = \frac{1}{4}\left[\frac{1.96}{0.02}\right]^2$$
$$= 2401$$

EXERCISES

1. Computer World Security, Inc. produces disk drive locks to prevent unauthorized disk drive use. To monitor the proportion of defectives produced random samples are periodically taken from the production line and inspected. In a random sample of 500 units 5 were found to be defective. The problem is to obtain confidence limits for the population proportion of defectives.

   (a) What guidelines must be satisfied before we may proceed? Are they satisfied?

   (b) Find 95% confidence limits for the population proportion of defectives produced.

   (c) How is the result obtained in answer to (a) to be interpreted?

   (d) If the population proportion of defectives is estimated at 1%, what is the error inherent in the estimate at the 98% confidence level?

   (e) How is the result obtained in answer to (d) to be interpreted?

2. Ecap University's faculty council wished to determine the faculty's attitude toward student evaluations of Ecap's faculty, which play a role in determining salary increases, promotion, and tenure. A random sample of 200 of the University's 4200 faculty showed that 81 were not in favor of student evaluations of faculty which have an impact on salary increases, promotion and tenure.

   (a) Find 95% confidence limits for the proportion of faculty that do not favor student evaluations.

   (b) Did you check the conditions which must be satisfied for part (a) to be undertaken? Looking into conditions which must be satisfied should be considered standing operating procedure. What results did you obtain?

   (c) How are the results obtained in answer to (a) to be interpreted?

   (d) If the sample proportion is taken as an estimate for the population proportion, what is the error inherent in the estimate at the 98% confidence level?

(e) If the faculty council wished to restrict the error inherent in the estimate to 3% at the 98% confidence level, what sample size would have to be taken?

(f) What percentage of the population is the sample size obtained in answer to (e)?

(g) A random sample of size obtained in answer to (e) showed that 981 faculty were not in favor of student evaluations of faculty. Find 98% confidence limits for the population proportion.

3. Ecap University's faculty council also wished to determine the faculty's attitude toward faculty evaluations of academic administration, which plays a role in determining salary increases, promotion and tenure. A random sample of 200 of Ecap's faculty showed that 70 were in favor of faculty evaluations of academic administration.

   (a) Find 90% confidence limits for the proportion of Ecap's faculty that favor faculty evaluations of academic administration.

   (b) How are the results obtained in answer to (a) to be interpreted?

   (c) If the sample proportion is taken as an estimate for the population proportion, what is the error inherent in the estimate at the 95% confidence level?

   (d) If the faculty council wished to restrict the error inherent in the estimate to 4% at the 95% confidence level, what sample size would have to be taken?

   (e) A random sample of the size obtained in answer to (d) showed that 476 were in favor of faculty evaluations of academic administration. Find 90% confidence limits for the population proportion.

4. What is the relative frequency interpretation of the 98% confidence interval for the population proportion π?

5. As noted in the aftermath of Example 2, Leonard Truck was most upset at his decisive loss in the primary when the results of a poll held just prior to the

primary indicated that he was strongly favored to win the nomination. He has come to you for an explanation of what might have gone wrong. Please oblige him.

6. Alan Chow, head of the consumer research division of a leading pharmaceutical manufacturer, has been asked to determine the proportion of lower back pain sufferers who prefer RELIEF, a medication manufactured by his company, to aspirin for relief of lower back pain. It occurred to him that the approach developed in this section would be suitable for addressing this problem. Do you agree or disagree? Explain.

7. A campaign task force set up to work for the reelection of the President was charged with determining the President's approval rating. One poll of 1200 randomly selected voters had 438 approving of his performance. Another poll of 1500 randomly selected voters taken one month later had 510 approving of his performance.

   (a) For each of these results construct 98% confidence limits for the population approval proportions for the President's performance.

   (b) The sample proportion based on the second poll is less than that based on the first poll. May we conclude that the President's approval rating has dropped? Explain.

8. The City Transportation Authority wanted to determine the attitudes of riders using the subway system on whether the system provided them with reliable transportation during the rush hour periods in the early morning and late afternoon. A random sample of 500 people using the system during these hours was selected. They were asked to rate the system as satisfactory or unsatisfactory and comment on any particular problems that they experienced. Three hundred ten of these interviewed rated the system as unsatisfactory.

   (a) Find 98% confidence limits for the proportion of the ridership which views the subway system as unsatisfactory during rush hour periods.

   (b) How is the result obtained in answer to (a) to be interpreted?

   (c) Changes were made in the subway system and three years later another poll was taken. In a random sample of 600 interviewed 354 rated the system as unsatisfactory. Find 98% confidence limits for the proportion of the ridership which views the subway system as unsatisfactory in terms of this second poll.

(d) May we conclude on the basis of these results that the unsatisfactory rating has gone down and the ridership feels that the reliability of the subway system has gone up? Explain.

9. Starting with the double inequality,

$$-1.96 < \frac{p-\pi}{\sqrt{\frac{\pi(1-\pi)}{n}}} < 1.96 \text{ with prob. } 0.95,$$

derive the 95% confidence interval for π:

$$p-1.96\sqrt{\frac{\pi(1-\pi)}{n}} < \pi < p+1.96\sqrt{\frac{\pi(1-\pi)}{n}}.$$

10.6 ■ The Central Limit Theorems

Roughly put, a central limit theorem is any mathematical result which gives us conditions (strings attached, if you will) under which a probability distribution may be approximated by a normal distribution. The emergence of the central limit theorem is one of the great triumphs of probability and, as we have already begun to see, these results are at the center of a good deal of statistical analysis.

The de Moivre-Laplace theorem considered in Section 8.3 was the first established of the central limit theorems. Its initial formulation by Abraham de Moivre (1667-1754) is found in a privately printed pamphlet dated November 12, 1733 [6]. The objective of de Moivre's investigations was to strengthen the foundation laid by Jacob Bernoulli in his *Ars Conjectandi* (The Art of Conjecture) [2] for the estimation of probabilities by means of empirical data. The remaining two-thirds of the eighteenth century saw the further development of the use of areas of normal curve regions as an approximation tool had become sufficiently important that Laplace was led to suggest the construction of a table of areas of normal curve regions.

A concept of probability distribution of errors that arise from a measurement process, due to Thomas Simpson, first appeared in 1755, the year following de Moivre's death, but during the last part of the eighteenth century what we now call the normal distribution was neither thought of as a probability distribution of errors nor as a probability distribution in its own right. This changed at the end of the eighteenth century with Gauss' first proof of the method of least squares, formulated around 1798 and made generally available through the publication of his *Theory of Motions...* [4] in 1809. The method of least squares is a procedure for combining the results of measurement of a quantity so as to minimize the error committed.

The second major central limit theorem to appear is due to Laplace. It made its initial appearance in a memoir published in 1810 and two years later in Laplace's great work, *Théorie analytique des probabilités*, which contains his work on probability that appeared over a period of almost forty years. Laplace's central limit theorem is the culmination of three problems which he gave many years of thought to: the problem of determining the probability that the arithmetic mean of the errors of observation arising from a measurement process,

$$\overline{x} = \frac{e_1 + e_2 + \ldots + e_n}{n},$$

is contained within given bounds, the problem of determining the probability that the mean inclination of the orbits of a specified number of comets to a given plane is between given bounds, and the problem of justifying the method of least squares.

Laplace's central limit theorem reads roughly as follows when put into modern form: Suppose that a sample of size *n*, $(x_1 + x_2, \ldots, x_n)$, is chosen at random from a large population. The probability distribution of the sample mean $\overline{x}$, a random variable defined by

$$\overline{x} = \frac{x_1 + x_2 + \ldots + x_n}{n},$$

is, for n large, approximately normal. This theorem describes the probability distribution of a sample statistic and in doing so it, and others like it, provide the major pillar on which statistical inference rests. The foundation for employing this theorem is developed in the next chapter. Its role in statistical inference is taken up in Chapters 10, 11, and elsewhere.

If the seed of The Central Limit Theorem was sown by Jacob Bernoulli and its germination was initiated by Abraham de Moivre, then it is fair to say that the flowering of this incredible mathematical orchid is due to Pierre Simon de Laplace. Laplace's methods for establishing his central limit theorem were not only not mathematically rigorous by modern standards, but generated a good deal of skepticism in his own time.

A number of attempts were made to simplify Laplace's analysis, but within the context provided by the standard of mathematical rigor of the time.

The emergence of an "abstract" central limit theorem based on carefully stated conditions and a modern level of rigor is due to Pafnuti Lvovich Chebyshev (1821-1894) in 1887. Most important refinements and extensions of his work were carried out by his students Andrei A. Markov (1856-1922) and Alexander M. Lyapunov (1857-1918) in the last part of the nineteenth and early twentieth centuries. The problem of determining necessary and sufficient conditions for the Central Limit Theorem, the last word on the subject, so-to-speak, was not decided until 1937 by William Feller.

REFERENCES

1. W.J. Adams, *The Life and Times of the Central Limit Theorem* (New York: Kaedmon Pub. Co., 1974).

2. J. Bernoulli, *Ars Conjectandi*, Basel: 1713.

3. W. Feller, "The Fundamental Limit Theorems in Probability," *Bulletin of the American Mathematical Society*, 51, 11 (Nov. 1945), 800-832.

4. C.F. Gauss, *Theoria Motus Corporum Coelestium in Sectionibus Conicis Arbientium* (1808). English translation: *Theory of the Motions of the Heavenly Bodies Moving About the Sun in Conic Sections*, 1857, (New York: Dover, 1963).

5. P.S. de Laplace, *Theorié analytique des probabilités* (Paris, Courcier, 1812, 1814 (2nd ed.), 1820 (3rd ed.)).

6. A. de Moivre, *Approximatio ad summam terminorum Binomii* $(a+b)^n$ *in seriem expansi (A method of approximating the sum of the Terms of the Binomial* $(a+b)^n$ *expanded into a Series*), 1733.

10.7 ■ The Chi-Square Distributions

This will serve to introduce a third major family of probability distributions of statistics called the chi-square distributions. To denote these distributions it is standard practice to use the Greek letter χ (called chi, pronounced ky to rhyme with my) with the exponent 2, so that we have the notation χ^2 for chi-square. The χ^2 distributions or curves are a family of curves which, as with the t distributions, depend on a quantity called the degrees of freedom denoted by ν or d.f., whose value depends on the situation under study. Distributions for 1, 3, 5, and 15 degrees of freedom are shown in Figure 20.

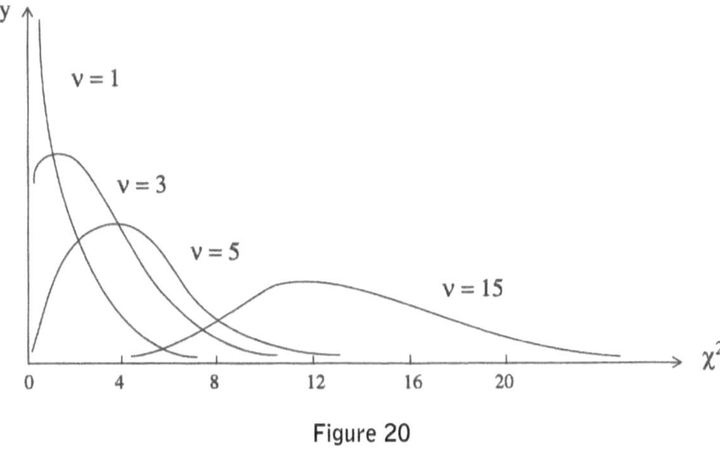

Figure 20

Properties of the χ^2 Distributions

1. Note from Figure 20 that χ^2 values are represented on the horizontal axis and that y values corresponding to χ^2 values on the vertical or y-axis.

2. χ^2 values are positive, unlike the x and t values on which the normal and t distributions are defined.

3. The χ^2 curves are not symmetric about any vertical line, unlike the normal and t curves.

4. As χ^2 increases, the χ^2 curves approach the horizontal axis but do not intersect it.

5. For those who wanted to know but were afraid to ask, the χ^2 distributions are defined by

$$y = c(\chi^2)^{\frac{\nu-2}{2}} e^{\frac{-\chi^2}{2}},$$

where e is a constant whose approximate value is 2.71828, ν is the number of degrees of freedom, and c is a constant depending on ν and determined in such a way that the area under the curve determined by ν is equal to 1.

As with normal and *t* curves, we will not have to plot χ^2 curves with great precision. Our interest centers on areas of regions defined by χ^2 curves and vertical bounds.

Areas of Regions Defined by χ^2 Curves

A table of χ^2 curve values for 1 through 30 degrees of freedom is given in Table C (page 654). A short version of this table is reproduced here as Table 5. The left and right hand columns of Table 5 list degrees of freedom. The column labeled $\chi^2_{.025}$, for example describes values for the degrees of freedom cited such that the area of the region to the right of $\chi^2_{.025}$ is 0.025 (see Figure 21). Thus, $\chi^2_{.025}$ for 5 degrees of freedom is 12.832; the area of the tail to the right of 12.832 is 0.025. $\chi^2_{.975}$ for 4 degrees of freedom is 0.484 (see Figure 22); the area of the region to the right of 0.484 is 0.975, or equivalently, the area of the tail to the left of 0.484 is 0.025.

Table 5

d.f.	$\chi^2_{.99}$	$\chi^2_{.975}$	$\chi^2_{.025}$	$\chi^2_{.01}$	d.f.
.	.	.	.	.	.
.	.	.	.	.	.
.	.	.	.	.	.
3	0.115	0.216	9.348	11.345	3
4	0.297	0.484	11.143	13.277	4
5	0.554	0.831	12.832	15.086	5
.	.	.	.	.	.
.	.	.	.	.	.
.	.	.	.	.	.

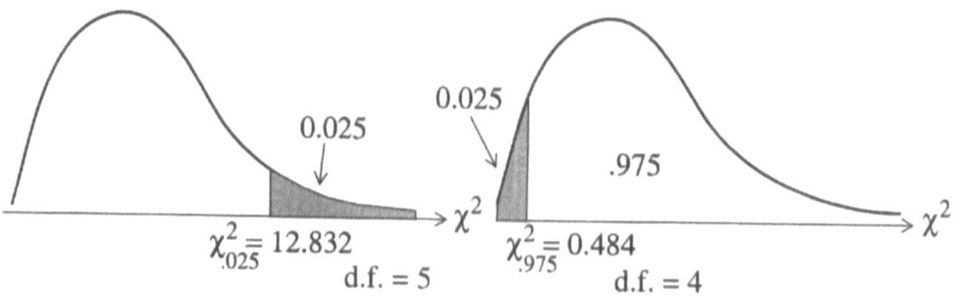

Figure 21 Figure 22

EXERCISES

1. Find the value of χ^2. Sketch a diagram for each case.

(a) $\chi^2_{.025}$, d.f. = 10
(b) $\chi^2_{.01}$, d.f. = 15
(c) $\chi^2_{.99}$, d.f. = 12
(d) $\chi^2_{.995}$, d.f. = 7

2. The area under the χ^2 curve between the two bounds given is 0.95. Find the bounds. Sketch a diagram for each case.

(a) $\chi^2_{.975}$, $\chi^2_{.025}$, d.f. = 17
(b) $\chi^2_{.975}$, $\chi^2_{.025}$, d.f. = 5
(c) $\chi^2_{.975}$, $\chi^2_{.025}$, d.f. = 3
(d) $\chi^2_{.975}$, $\chi^2_{.025}$, d.f. = 9

3. The area under the χ^2 curve between the two bounds given is 0.98. Find the bounds. Sketch a diagram for each case.

(a) $\chi^2_{.99}$, $\chi^2_{.01}$, d.f. = 10
(b) $\chi^2_{.99}$, $\chi^2_{.01}$, d.f. 25
(c) $\chi^2_{.99}$, $\chi^2_{.01}$, d.f. = 7
(d) $\chi^2_{.99}$, $\chi^2_{.01}$, d.f. = 23

10.8 ■ Estimation of Variance and Standard Deviation

Confidence intervals for and σ, where the sample of size n is chosen at random from a population which is normally distributed, or "very close" to it, is based on the probability behavior of the sample statistic

$$\chi^2 = \frac{(n-1)s^2}{\sigma^2},$$

where is the sample variance and is the population variance. This sample statistic is denoted by χ^2 because its sampling distribution is a chi-square distribution.

Sampling Distribution of χ^2: Under the aforenoted conditions the sampling distribution of

$$\chi^2 = \frac{(n-1)s^2}{\sigma^2}$$

is, to a close approximation, the χ^2 distribution with $n-1$ degrees of freedom.

To obtain a 95% confidence interval for σ^2 let us observe that 95% of the area under χ^2 curve is between the bounds $\chi^2_{.975}$ and $\chi^2_{.025}$ (see Figure 23).

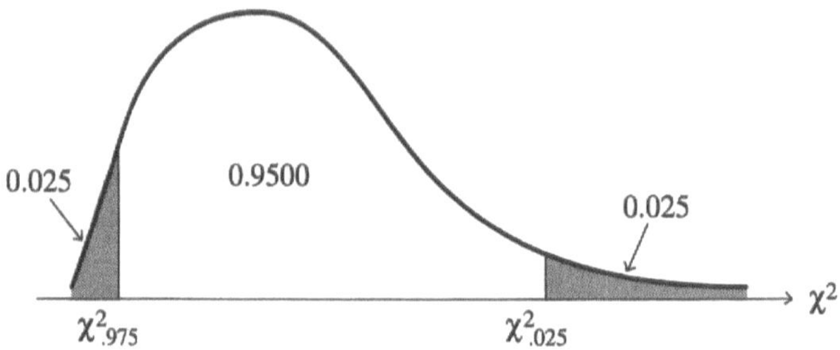

Figure 23

Thus, with probability 0.95, $\chi^2 = \frac{(n-1)s^2}{\sigma^2}$ is between $\chi^2_{.975}$ and $\chi^2_{.025}$; that is

$$\chi^2_{.975} < \frac{(n-1)s^2}{\sigma^2} < \chi^2_{.025} \text{ with prob. 0.95}$$

The task before us is to isolate σ^2 from this double inequality. Application of basic algebra (see Exercise 10) yields:

$$\frac{(n-1)s^2}{\chi^2_{.025}} < \sigma^2 < \frac{(n-1)s^2}{\chi^2_{.975}} \text{ with prob. 0.95}$$

Taking square roots of the terms in this double inequality yields a 95% confidence interval for σ.

If the confidence level changes, $\chi^2_{.975}$ and $\chi^2_{.025}$ change accordingly. For the 98% confidence level $\chi^2_{.975}$ and $\chi^2_{.025}$ change to $\chi^2_{.99}$ and $\chi^2_{.01}$, respectively; for the 99% confidence level $\chi^2_{.975}$ and $\chi^2_{.025}$ are replaced by $\chi^2_{.995}$ and $\chi^2_{.005}$, respectively.

Example 1 Return to Ramunė's Gourmet Coffee Problem

In Section 10.1 we noted that Ramunė's Gourmet Coffee is negotiating a contract with a processing company to fill 10-ounce jars with a new coffee blend to be introduced into the market. Ramunė's is concerned with the variability of the fill weights, which must be reasonably small, and the consulting firm they hired wants to obtain 98% confidence intervals for the variance and standard deviation of the coffee fills.

A randomly chosen sample of 9 10-ounce jars yielded a mean fill weight of 10.1 ounces and a variance of 0.42 ounces. It was established that the population of fill weights is approximately normally distributed.

Since $\text{d.f.} = 8$, from Table C (page 654) we obtain $\chi^2_{.01} = 20.090$ and $\chi^2_{.99} = 1.646$. This yields:

$$\frac{8(0.42)}{20.090} < \sigma^2 < \frac{8(0.42)}{1.646} \text{ with prob. } 0.98$$

$$0.167 < \sigma^2 < 2.041 \text{ with prob. } 0.98$$

Taking square roots gives us:

$$0.41 < \sigma < 1.43 \text{ with prob. } 0.98$$

The odds are 49 to 1 that the standard deviation of the coffee fills is between 0.41 ounces and 1.43 ounces. Ramunė's management will have to decide whether this is satisfactory or to insist that the processing company reset its filling machines.

The confidence intervals just described are usually used when small samples are involved, but may also be used for large samples. Their application requires that the population being sampled from be normally distributed or "very close" to it.

EXERCISES

1. (a) State 98% confidence intervals for σ^2 and σ. Sketch a diagram indicating the location of the bounds defining these confidence intervals.

   (b) What are the bounds for: (i) $n = 5$, (ii) $n = 12$, (iii) $n = 15$, (iv) $n = 22$?

2. (a) State 99% confidence intervals for σ^2 and σ. Sketch a diagram indicating the location of the bounds defining these confidence intervals.

   (b) What are the bounds for: (i) $n = 7$, (ii) $n = 15$, (iii) $n = 20$, (iv) $n = 25$?

3. In reviewing the expenses charged to Ramunė's by its consulting firm the accountant noted a charge for work done to establish that the population of fill weights is normally distributed. He could not understand the necessity for this

work and concluded that the consulting firm was adding some unnecessary work to the project in order to inflate the bill. If he came to you for an explanation of why this work was necessary, what would you tell him?

4. The management of the Russell Coal Company is interested in obtaining an estimate of the variation in the heat producing capacity of coal from a newly opened mine (in millions of calories per ton). A random sample of 6 specimens from the mine yielded a standard deviation of 192 million-calories per ton. Determine the 95% confidence interval for the standard deviation of the mine's heat producing capacity.

5. Janet Long, standards control inspector for the State Standards Maintenance Department, is concerned about variation in tensile strength (measured in thousand pounds per square inch) of steel bars to be used in a highway construction project. She instructed that a random sample of 8 bars be chosen from a shipment of steel bars to be used in the project and that their tensile strength be measured. It was determined that $\Sigma x_i = 336$ and $\Sigma x_i^2 = 15,276$, where x_i denotes the tensile strength of the ith steel bar in the sample, the sum being taken from 1 to 8.

   (a) Determine the 98% confidence interval for the standard deviation of the tensile strength of the steel bar population.

   (b) How is the result obtained in answer to (a) to be interpreted?

6. In Example 1 of Section 10.2 (page 261) we noted that Ecap University administers a mathematics placement exam to students prior to their taking a mathematics course at the university. In a study of student performance on the exam a randomly chosen sample of 50 test scores was found to have a mean of 38 points with a standard deviation of 5 points.

   (a) Determine the 95% confidence interval for the standard deviation of the placement test scores. As an approximation take $\chi^2_{.025} = 71$ and $\chi^2_{.975} = 32$.

   (b) How is the result obtained in answer to (a) to be interpreted?

7. Andrew Adams has decided on which model 27 inch television set to purchase. Being a person who wishes to get value for his money, Andrew investigated

price variation for the model he had selected by choosing at random 5 retail outlets in the metropolitan area that sell television sets and contacting them for price information. He was given the following prices: \$450, \$575, \$500, \$550, \$500.

(a) Find a 90% confidence interval for the standard deviation in prices for this model for retail outlets in the metropolitan area.

(b) How is the result obtained in answer to (a) to be interpreted?

8. Now that the computation dust has settled a bit, you have been appointed Statistical Inspector General charged with examining the work carried out in connection with Exercises 4, 5, 6 and 7. Are there any concerns that you would express to the individuals or institutions involved? Explain.

9. In looking at confidence intervals for μ and π we saw that in geometric terms they were intervals centered at the sample statistics $\bar{x}$ and $p = x/n$, respectively. Can we give a similar geometric interpretation for a confidence interval for σ^2? Explain.

10. From the double inequality

$$\chi^2_{.975} < \frac{(n-1)s^2}{\sigma^2} < \chi^2_{.025} \quad \text{with prob. 0.95,}$$

derive the 95% confidence interval for σ^2:

$$\frac{(n-1)s^2}{\chi^2_{.025}} < \sigma^2 < \frac{(n-1)s^2}{\chi^2_{.975}}$$

10.9 ■ Estimators and Their Properties

As we have seen in the preceding sections, statistical estimation is concerned with deriving an approximate value for a population parameter on the basis of a sample statistic, within a context of reliability provided by the confidence level.

To obtain a general perspective on statistical estimation it is useful to introduce the terms estimator and estimate for a population parameter and consider properties that make some estimators better than others. To take a specific illustration, the sample mean $\bar{x}$, a random variable whose value depends on the sample chosen, was used as an estimator of the population parameter μ. When a sample was chosen and $\bar{x}$ took on the specific value of the sample mean of this sample, also denoted by $\bar{x}$, this value served as an estimate of μ.

More generally, let θ (Greek letter theta) denote a population parameter. A random variable θ_1, whose value depends on the sample chosen, is said to be an **estimator** of θ if for specific samples that are drawn the value of θ_1 is used as an **estimate** for θ. Thus, $p = x/n$ is an estimator of the population proportion π, s^2 is an estimator of the population variance σ^2, and *s* is an estimator of the population standard deviation σ.

Three desirable properties of an estimator, which we now examine, are unbiasedness, consistency, and relative efficiency.

Unbiased Estimators

The estimator θ_1 is said to be an **unbiased estimator** of parameter θ if the expected value or mean of θ_1 is equal to θ, that is $E(\theta_1) = \theta$. This is an appealing property of an estimator because, in less precise terms, it says that on average, over the long run, an unbiased estimator gives us the correct value of θ; that is, if we were to take many samples, calculate θ_1 for each sample, and then average these values of θ_1, that average would closely approximate θ.

The sample mean $\bar{x}$ is an unbiased estimator of μ since $E(\bar{x}) = \mu$. The sample variance s^2 defined for a sample of size *n* by taking the sum of the squared deviations from the mean, $\Sigma(x_i - \bar{x})^2$, and dividing by $n-1$, not *n*, is an unbiased estimator of σ^2 since $E(s^2) = \sigma^2$. If we had defined s^2 by dividing $\Sigma(x_i - \bar{x})^2$ by *n*, this would have resulted in a biased estimator of σ^2, one that produced on average an s^2 that was smaller than σ^2. The sample standard deviation *s* is a biased estimator of σ since $E(s) \neq \sigma$. The sample proportion $p = x/n$ is an unbiased estimator of π.

Consistent Estimators

We intuitively expect an estimator θ_1 to, in some sense, provide better estimates for θ for larger as opposed to smaller sample size *n*. An estimator θ_1 of parameter θ is said to be **consistent** if the probability with which θ_1 differs from θ grows smaller, that is, approaches zero, as the sample size increases. It is in this sense that a consistent estimator provides better estimates of θ for larger and larger sample size *n*. When probability is involved one cannot flat out guarantee that every time you increase the sample size your estimator will, without reservation, provide you with a better estimate of θ.

The sample mean $\bar{x}$, sample proportion $p = x/n$, and sample variance s^2, are consistent estimators of μ, π, and σ^2, respectively.

Unbiasness implies that for any sample size errors in estimates average to zero, although individually they may be large. Consistency implies that errors in estimates approach zero, in a probability sense, as the sample size increases.

Efficient Estimators

The variance of an estimator θ_1 of θ describes a sense of concentration of the estimates of θ about the mean of θ_1, which is θ itself if θ_1 is an unbiased estimator. The unbiased estimator θ_1 of θ with smallest variance is said to be **efficient.**

The sample mean $\overline{x}$, sample proportion $p = x/n$, and sample variance s^2 are unbiased, consistent, and efficient estimators of μ, π, and σ^2, respectively.

The Maximum Likelihood Estimate

To estimate θ a random sample of size n is taken which yields values $x_1, x_2, \ldots, x_n$. These are the values actually obtained, and the **maximum likelihood** estimate of θ is defined as that estimate for which the probability of obtaining the sample data $x_1, x_2, \ldots, x_n$ actually obtained is largest.

Maximum likelihood estimators are widely used and have a number of advantages, but they are not necessarily unbiased. As an example of interest, s^2 defined by dividing the sum of the squared deviations from the mean, $\Sigma(x_i - \overline{x})^2$, by n, rather than $n-1$, is a maximum likelihood estimator of σ^2 which is, however, a biased estimator. Defining s^2 by dividing the sum of the squared deviations from the mean by $n-1$, as we have chosen to do, gives us an unbiased estimator of σ^2, but not a maximum likelihood estimator.

It is a question of trade-offs and some statisticians and writers on statistics prefer to define s^2 by dividing the sum of the squared deviations from the mean by n, giving up unbiasedness for the maximum likelihood property.

Reader beware; when you pick up a statistics book be sure that you know the author's definition of s^2. It makes a difference in the consequences.

11

Hypothesis Testing

11.1 ■ Trial by Statistics

Hypothesis testing as would seem reasonable from the title is about testing hypotheses, that is, statistical hypotheses. A **statistical hypothesis** differs from most hypotheses that we encounter in that it is an assertion made about a parameter of a population ($\mu = 10$, $\sigma = 2$, etc.) or parameters of populations ($\mu_1 = \mu_2$, $\sigma_1 = \sigma_2$ etc.) The following four examples illustrate statistical hypotheses.

Ramunė's Gourmet Coffee stands behind its claim that the mean weight of the 8 ounce jars of its Deluxe Blend is 8 ounces.

The manufacturer of Deadly Lights cigarettes claims that this brand contains an average of 0.6 milligrams of nicotine per cigarette.

The campaign headquarters of John Exon asserts that 60% of the voters in his district support him for the state senate.

The quality control manager of Lee Tires claims that there is no difference in the average life of the tires made at its Wells and Jason City plants.

The last of these examples involves a comparison of means of two populations, the mean lifetime of the tires made at the Wells plant compared with the mean lifetime of those made at the Jason City plant, which requires that a sample be drawn from each of the two populations. Tests of this sort are a special case of multi-sample tests which we initiate a discussion of in Section 11.7.

The first three of these examples have the same structure in that a population and some population parameter θ emerge. It is claimed or hypothesized that θ is equal to some specific value θ_1. Ramunė's claims that μ, the mean of its population of coffee weight fills, is 8 ounces; Deadly Lights cigarettes claims that μ, the mean nicotine content of its cigarettes, is 0.6 milligrams; John Exon's campaign headquarters claims that π, the proportion of voters who support him, is 0.60.

Generally speaking, we emerge with an hypothesis about a population parameter of the following kind:

$$H_o : \theta = \theta_1$$

H_o is the hypothesis to be tested. It asserts, claims, alleges that the population parameter θ is equal to a specific value θ_1 which comes out of the context provided by the situation being considered. H_o is called the **null hypothesis**, the term null meaning no difference; there is no difference between the value of the population parameter θ and the value θ_1 that it is claimed to be. The tests that are developed are directed toward the null hypothesis.

Once the null hypothesis H_o is formulated we must formulate an **alternative or alternate hypothesis** H_a so that if we are led to reject the null hypothesis H_o we then accept the alternative hypothesis H_a. Depending on circumstances, the alternative hypothesis may take one of three forms:

Form 1: $H_a : \theta \neq \theta_1$
Form 2: $H_a : \theta < \theta_1$
Form 3: $H_a : \theta > \theta_1$

The null hypothesis that emerges from Ramunė's Gourmet Coffee setting is:

$$H_o : \mu = 8$$

Ramunė's is concerned with both overfills $(\mu > 8)$, which are costly because more of the product is being given away than claimed, and underfills $(\mu < 8)$ which means that the company's product is not living up to expectations and the consumer is being short-changed. This situation may draw very negative reactions from consumers and consumer protection groups and, needless-to-say, is not good for the company's image and business. Ramunė's would take as its alternative hypothesis:

$$H_a : \mu \neq 8$$

If a consumer protection agency were investigating Ramunė's claim, then the alternative hypothesis,

$$H_a : \mu < 8,$$

would make sense from a consumer protection point of view. A higher than claimed mean fill of 8 ounces would not be against consumer interests, but a lower than claimed mean fill of 8 ounces would be against consumer interests and cause for concern.

> If we think of the null hypothesis $H_o : \theta = \theta_1$ as a defendant to go to trial, then we may find the defendant guilty and conclude that $\theta = \theta_1$ as charged or we may reserve judgment, meaning that we are hesitant about declaring the defendant guilty, but we cannot conclude that H_o is innocent based on the evidence obtained, or we find H_o innocent and take as our conclusion H_a, the alternative hypothesis that was set up.

As to the mechanism for conducting the trial, if the population were small, say $Q = \{2, 4, 6, 8\}$, and it were claimed that $\mu = 6$, it would be a simple matter to determine directly that $\mu = 5$ and establish that the null hypothesis is false. But situations of interest which arise from applied settings are far more complex and do not lend themselves to such a direct approach, and so we must conduct the trial by taking a sample from the population and evaluating the results obtained—trial by statistical methods.

The basic idea is simple. Take a sample at random from the population under consideration and determine the appropriate sample statistic. If the value obtained is, in some sense, close enough to what is to be expected when the null hypothesis is true, then this is evidence in support of the null hypothesis and it cannot be rejected. On the other hand, if the value obtained is far removed from what is to be expected when the null hypothesis is true, then this is evidence against it. We then reject the null hypothesis and accept the alternative hypothesis that had been formulated.

> Sample based evidence is by its very nature incomplete and there is always the possibility of rejecting a true null hypothesis, called a **type I error**, or accepting a false null hypothesis, called a **type II error**. Hypothesis testing theory provides us with a statistical framework for deciding whether to accept, reserve judgment on, or reject a null hypothesis. It does not eliminate the possibility of error, but it does provide us with procedures through which the probability of each kind of error can be evaluated and controlled.

EXERCISES

1. State and give a rationale for an alternative hypothesis for $H_o: \mu = 0.6$, the null hypothesis that arose from the Deadly Lights cigarette claim.

2. State and give a rationale for an alternative hypothesis for $H_o: \pi = 0.60$, the null hypothesis that arose from John Exon's campaign headquarters.

In Exercises 3-10 formulate null and alternative hypotheses for the situation described. State a rationale for your alternative hypothesis.

3. The manufacturer of Kool Kola claims that the mean weight of their 32 ounce bottles of Kola is 32 ounces, as advertised.

4. Dapper Dan, known from Chapter 7 for his extensive collection of dice, claims that the coin he has been using in various coin tossing games is fair.

5. The manufacturer of Deadly Lights cigarettes claims that Deadly Lights contains an average of 8 milligrams of tar per cigarette.

6. President G. Marx of Huxley College claims that the mean combined SAT scores of Huxley's students is at least 1400.

7. Jason Farnham, candidate for the office of Lord High Executioner, claims that at least 55% of the voters support him for this office.

8. Lee Tires claims that its new super-ride steel belted radials have an average life of 60,000 miles.

9. A group of labor economists has put out a report claiming that the mean income of factory workers in the midwest does not exceed $25,000.

10. The anthropologist Alice Williams claims that recently discovered humanoid fossil remains in Central America are one million years old.

11.2 ■ A Case Study: Tests Concerning Means

To develop the procedures of hypothesis testing we turn first to what is perhaps the most straightforward of the hypothesis testing situations for developing basic ideas, namely the one-sample test to determine whether sample evidence refutes or tends to support some hypothetical value μ_0 of the population mean μ.

Example 1 Ramunė's Gourmet Coffee

Ramunė's Gourmet Coffee stands behind its claim that the mean weight of the 8 ounce jars of its Deluxe Blend is 8 ounces.

The problem we set ourselves is to develop procedures that would not only enable us to test the hypothesis which emerges from this situation, but others like it which emerge from not too different situations. This leads us to the following sequence of steps.

1. State the null hypothesis H_o.

In this case we have,

$$H_o : \mu = 8,$$

where μ is the mean of the population of coffee weights of its 8 ounce jars and 8 is its claimed value.

2. Formulate the alternative hypothesis H_a.

As we shall discuss further, null hypotheses are always stated in equality form, but there may be more to it than that. The alternative hypothesis should reflect the overall nature of the situation and not just the equality condition in the null hypothesis.

Ramunė's is concerned with both overfills $(\mu > 8)$ and underfills $(\mu < 8)$ so that the appropriate alternative hypothesis which takes into account deviation in both directions, is:

$$H_a : \mu \neq 8$$

3. Give thought to the sample size and sample statistic to be used.

As we have seen in our discussions of estimation, there are a number of factors which affect the decision on the sample size to be taken (see Sections. 10.3 and 10.4). If a small sample size $(n < 30)$ is suggested, then it is imperative that we be sensitive to conditions on the underlying population that may be required.

For this situation, let us suppose that it has been decided to take a large sample $(n \geq 30)$ of size 49 at random from the population. The sample statistic is $\overline{x}$ and we shall be looking at the z-value of $\overline{x}$ with s being used as an estimate for σ.

4. Specify the level of significance at which the test is to be carried out.

The idea is that if the sample statistic obtained is "sufficiently close" to the population parameter value we are testing then this is evidence in support of the null hypothesis. If the sample statistic obtained is "far removed" from the population parameter we are testing, then this is evidence which favors rejecting the null hypothesis and accepting the alternative hypothesis. The **level of significance**, denoted by the Greek letter α (alpha), is a probability value which makes precise, in probability language, the terms "sufficiently close" and "far removed".

In this situation we must make precise what it means to say that $\overline{x}$ is sufficiently close to 8, from both directions, and $\overline{x}$ is far removed from 8, from both directions. We know that the sampling distribution of $\overline{x}$ is approximately normal with mean $\mu_{\overline{x}} = \mu$, which by H_o is 8, and standard deviation $\sigma_{\overline{x}} = \sigma/(\sqrt{49})$ which we shall approximate by $s/\sqrt{49}$. We need bounds x_1 and x_2 to serve as measures of the closeness of the sample mean $\overline{x}$ to 8 (see Figure 1(a)).

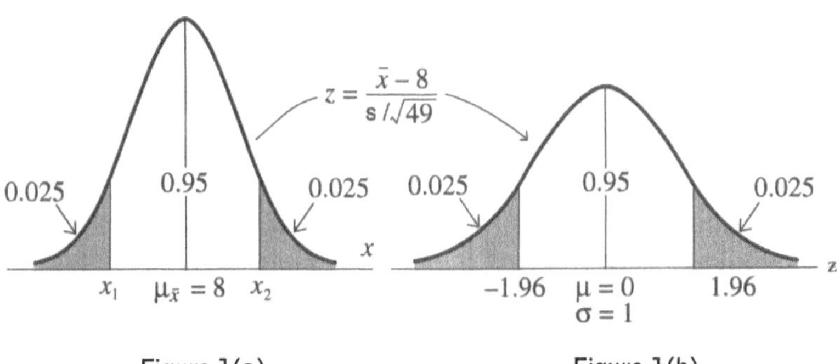

Figure 1(a) Figure 1(b)

Suppose we take our level of significance to be $\alpha = 0.05$. Since there are two bounds x_1 and x_2 symmetrically placed on both sides of $\mu_{\overline{x}} = 8$, $\alpha = 0.05$ determines x_1 and x_2

as those values for which the area of the region between x_1 and x_2 is 0.95. This means that the area of each of the two tails to the left of x_1 and to the right of x_2, is 0.025.

In terms of the standard normal curve the *z*-bounds corresponding to x_1 and x_2 are -1.96 and 1.96 respectively (see Figure 1(b)), and the last stone of our acceptance/rejection machinery is now in place.

5.	Review the acceptance/rejection machinery implicit in the level of significance chosen.

For this case, consider the *z*-value of the mean $\overline{x}$ of the sample chosen.

If -1.96 ≤ *z*-value of $\overline{x} \le 1.96$, accept H_o or reserve judgment; if the *z*-value of $\overline{x} > 1.96$ or the *z*-value of $\overline{x} \le -1.96$, reject H_o and accept H_a (see Figure 2).

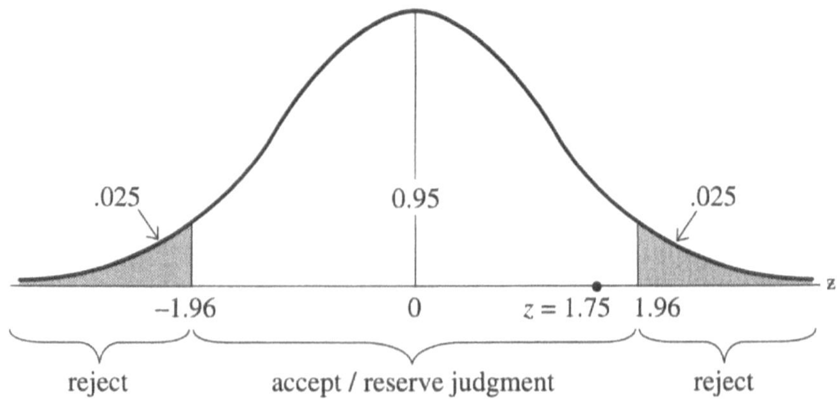

Figure 2

6.	Take the sample at random from the population and determine the sample statistic.

For this case, let us suppose that the sample of size 49 yields $\overline{x} = 8.2$ ounces with $s = 0.8$ ounces. Then:

$$z = \frac{8.2 - 8}{0.8/\sqrt{49}} = 1.75$$

7.	Make a decision on H_o.

The individual or group with the responsibility for making a decision on H_o is sometimes, for convenience, called the **decision maker**. In this case, since $z = 1.75$ the decision maker cannot reject H_o. H_o is either accepted or judgment is reserved.

Observations

> When the test statistic falls within the accept/reserve judgment region we say that the result of the test is **not statistically significant.** It is not statistically significant in the sense that the difference between what we expect and what we actually obtain from the sample is sufficiently small, as made precise by our chosen level of significance, that it may be attributed to chance rather than a basic underlying difference between the population parameter and what it is claimed to be: When the test statistic falls within the reject region we say that the result is **statistically significant.** In terms of our chosen level of significance, the difference is viewed as reflecting a difference between the population parameter and what it is claimed to be that cannot be attributed to chance.
>
> Hypothesis tests are often termed **tests of significance.**

In the case at hand $z = 1.75$ is not statistically significant since it falls within the accept/reserve judgment region established by the bounds -1.96 and 1.96 in terms of the 0.05 level of significance.

Since $z = 1.75$ is not statistically significant, doesn't this prove that the null hypothesis $\mu = 8$ is true? To put it in legal terms, doesn't this establish that μ has been found guilty of being equal to 8? Trial by statistical hypothesis testing is great from the defendant's point of view because it is very conservative about reaching the conclusion that the null hypothesis is guilty as charged. After all, μ might actually be equal to some value that is very close to 8 (there are many, many such values), which would be very difficult, if not impossible, to detect.

Still the value 8 has been thrust upon us by circumstances and we must deal with it. Unless a decision must be taken, the preferred option is to reserve judgment on a null hypothesis rather than accepting it outright when the test statistic is not statistically significant. Ramunė's Gourmet Coffee, however, is not in a position to reserve judgment. The question before its decision makers (management) is, do we continue with the current settings on the filling machines or do we stop production and reset the machines? Since $z = 1.75$ is not statistically significant, there is no compelling evidence that the null hypothesis $\mu = 8$ is untenable and that Ramunė's should stop production and reset the filling machines. Ramunė's decision makers would opt to accept the null hypothesis and continue production.

In accepting the null hypothesis $\mu = 8$ is it not possible that Ramunė's decision makers have committed a type II error by accepting a null hypothesis which is false? The answer, of course, is yes.

> The probability of committing a type II error by accepting a null hypothesis which is false is denoted by the Greek letter β (beta). The value of β depends on what the value of the population parameter actually is as opposed to what the null hypothesis claims it to be.

We shall have more to say about type II errors and their probabilities in Section 11.3.

Suppose $z = 2.12$ or some such value in the reject region had been obtained as the test statistic. The decision maker would then reject the null hypothesis $\mu = 8$ and accept the alternative hypothesis $\mu \neq 8$. In rejecting the null hypothesis it is possible that the decision maker has committed a type I error by rejecting a null hypothesis which is true.

> The probability of committing a type I error, denoted by α, is equal to the probability that the test statistic falls in the reject region, which is equal to the level of significance (also denoted by α) of the test.
>
> Establishing the accept/reserve judgment and reject regions by setting the level of significance α also establishes the probability of a type I error α.

In this case $\alpha = 0.05$.

At first sight it would seem desirable to take α as small as possible so as to minimize the probability of a type I error. Unfortunately, it is no as simple as that. The size of α determines the accept/reserve judgment and reject regions and has an affect on the size of β, the probability of committing a type II error. It is a matter of trade-offs.

If we take $\alpha = 0.01$, for example, then the accept/reserve judgment and reject regions are defined by the *z*-bounds -2.58 and 2.58 (see Figure 3).

In this case the *z*-statistic $z = 2.18$ would not be statistically significant and thus would not lead to rejection of the null hypothesis $\mu = 8$.

Decision makers engaged in quality control or maintenance of standards studies concerned with processes that would have to be shut down if the test statistic fell into the reject region, such as Ramunė's coffee fill process, favor taking a very small α to keep the accept region wide and the probability of making a type I error very small. Stopping a production process, resetting the machinery, and restarting the process is time consuming and expensive. It should only be done in cases of extreme necessity in the judgment of decision makers concerned with such processes. Keep α low, very low, is their motto. The trade-off is in a higher β. An escape valve of sorts in this balancing act is provided by the option to increase the sample size.

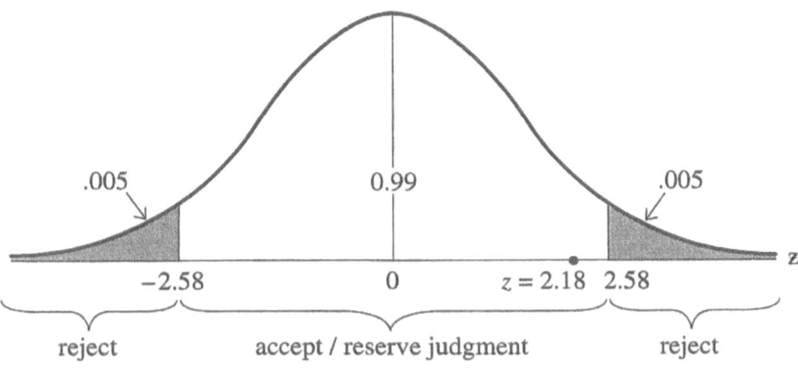

Figure 3

In studies other than the afore mentioned the standard values taken for α tend to be 0.05, 0.02 and 0.01.

As we have seen, it is possible that the decision maker would have to reject a null hypothesis at one level, 0.05 for example but not necessarily at another; 0.01 comes to mind. In carrying out hypothesis testing procedures it is important that the level of significance of the test be set prior to a sample being taken to avoid the temptation of manipulating the test to yield desired preferences. (See Sec. 11.12.)

The management of Ramunė's Gourmet coffee was concerned with both overfills ($\mu > 8$) and underfills ($\mu < 8$) and was led to the alternative hypothesis $\mu \neq 8$ to take into account deviation in both directions. This required that we set up two bounds to measure variation in both directions from the hypothesized mean. Tests of this sort are called **two-sided tests.** With two-sided tests the significance level value α is distributed equally to each of the two tails emerging from the two bounds. As we saw, with $\alpha = 0.05$, each of the two tails had an area value of 0.025.

If the alternative hypothesis is one-sided, $\mu < 8$ or $\mu > 8$, for example, then one bound to measure deviation in the one direction is required. Tests of this sort are called **one-sided tests.** With one-sided tests the significance level value α is distributed in total to the one tail that emerges from the one bound.

The following example illustrates a one-sided test.

Example 2 Ramunė's Gourmet Coffee: The Consumer's View

A consumer protection agency desires to test Ramunė's assertion that the mean weight of the 8 ounce jars of its Deluxe Blend is 8 ounces.

$$H_o : \mu = 8$$

A higher than claimed mean fill of 8 ounces would not be counter to consumer interests, but a lower than claimed mean fill would be cause for complaints and possible action. Accordingly, the consumer protection agency took

$$H_a : \mu < 8$$

as its alternative hypothesis.

Let us assume that a large sample size is feasible from a cost point of view. It is also advantageous in that the consumer protection agency's analysts need not be concerned about whether the population of coffee weights from the 8 ounce jars is normally distributed. The analysts decide on a sample size of 60, let us assume.

Significance level: $\alpha = 0.02$.

Acceptance/rejection machinery:

This is a one-sided test since the alternative hypothesis is one-sided. We need one lower bound to make precise what it means to say that the *z*-value of the sample mean is significantly small at the 0.02 level. The lower *z*-bound corresponding to the one-sided 0.02 level is -2.05 (see Figure 4).

A sample of size $n = 60$ chosen at random from the population had mean $\bar{x} = 7.8$ ounces and standard deviation $s = 0.9$, let us say.

Since

$$z = \frac{7.8 - 8}{0.9/\sqrt{60}} = -1.72$$

is not less than the lower bound -2.05, the null hypothesis $\mu = 8$ must either be accepted or judgment reserved. Either way the consumer protection agency has no basis for taking issue with Ramunė's claim that the 8 ounce jars of its Deluxe Blend have a mean fill of 8 ounces.

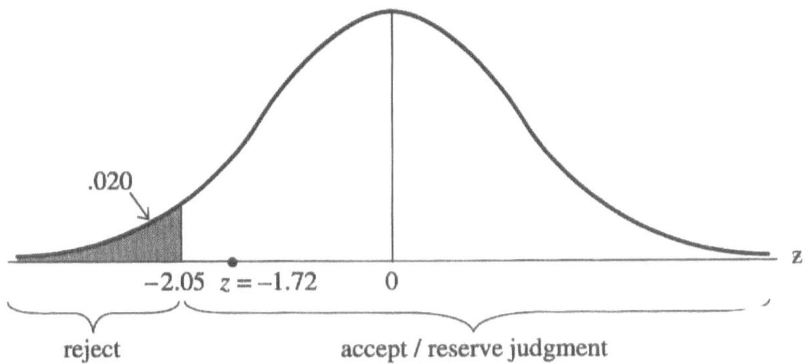

Figure 4

EXERCISES

Null and alternative hypotheses are stated for Exercises 1-4. Also given are the sample size and level of significance. State the test statistic to be examined, the accept/reserve judgment and reject regions, and the probability of committing a type I error. Draw a diagram in each case.

1. $H_o: \mu = 300, H_a: \mu \neq 300, n = 49, \alpha = 0.05$.
2. $H_o: \mu = 100, H_a: \mu > 100, n = 64, \alpha = 0.05$.
3. $H_o: \mu = 420, H_a: \mu \neq 420, n = 30, \alpha = 0.02$.
4. $H_o: \mu = 1000, H_a: \mu < 1000, n = 32, \alpha = 0.01$.

Example 3 Ramunė's Gourmet Coffee: Small Sample Size

Let us suppose that in Example 1 Ramunė's management decided to carry out the test using a small sample size, $n = 15$. Their analysts argued that it is realistic to view the coffee fills of the 8 ounce jars from the point of view of the hypothesis of elementary errors (see Section 8.3, page 222), that is, as determined by a large number of independently acting random factors arising in the filling process, each having a negligible influence on the fill weight as a whole. As such, they pointed out, the population of fill weights is approximately normally distributed and a small sample test is feasible. What is the same and what changes?

We have the same H_o, H_a, and α.

$$H_o: \mu = 8$$
$$H_a: \mu \neq 8$$
$$\alpha = 0.05$$

The sample statistic changes from z to:

$$t = \frac{\overline{x} - 8}{s / \sqrt{15}}$$

The z-bounds -1.96 and 1.96 defining the acceptance/rejection region for $\alpha = 0.05$ change to t-bounds $-t_{.025} = -2.145$ and $t_{.025} = 2.145$ for $\text{d.f.} = 14$ (see Figure 5).

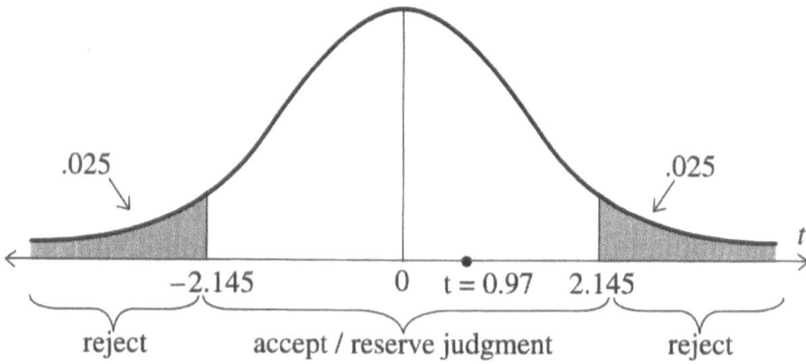

Figure 5

Let us suppose that the sample of 15, chosen at random from the population of fill weights, yielded the same mean and standard deviation: $\overline{x} = 8.2$ ounces, $s = 0.8$ ounces. Then,

$$t = \frac{8.2 - 8}{0.8 / \sqrt{15}} = 0.97,$$

which falls in the accept/reserve judgment region. This value is not statistically significant and Ramunė's decision makers would accept the null hypothesis that $\mu = 8$ and continue production.

"At Least" and "At Most" Claims

Example 4 Huxley College's SAT Scores

President G. Marx of Huxley College claims that the mean combined SAT scores of Huxley's students is at least 1400. A Higher Education Board (HEB) set up to monitor claims made by colleges and universities to curtail misleading or fraudulent advertising has decided to investigate President Marx's claim.

The first task of HEB's analysts is to set up a null hypothesis. This situation is unlike the others considered in that it is not being claimed that the mean is exactly 1400, but at least 1400, which translates to greater than or equal to 1400. Let us denote the claim by H_c, so that we have:

$$H_c : \mu \geq 1400$$

We must go from H_c to H_o which must be stated in equality form if we are to emerge with an unequivocal test statistic and probability of a type I error for a specified level of significance. Such are necessary if we are to be able to apply the hypothesis testing procedures that have been developed.

We handle the "at least" condition leading to $H_c : \mu \geq 1400$ in the following way in setting up H_o and H_a.

$$H_o : \mu = 1400$$
$$H_a : \mu < 1400$$

The H_a part is easy; we are led to $\mu < 1400$ as the alternative to the "at least" condition in President Marx's claim that the mean SAT score is at least 1400.

Sample size: $n = 70$.
Level of significance: $\alpha = 0.01$.

Accept/reserve judgment and reject regions: There is one lower bound, $z = -2.33$, to make precise what it means to say that the z-value of $\overline{x}$ is significantly small at the 0.01 level of significance. See Figure 6.

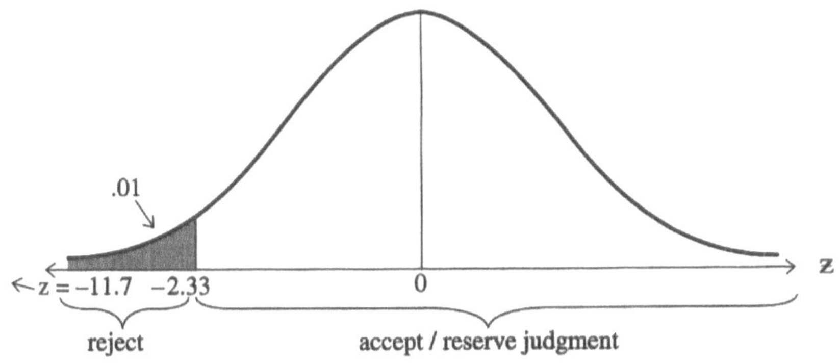

Figure 6

The sample of size 70 chosen at random from the population of SAT scores yielded $\bar{x} = 980$ points with $s = 300$ points. The z-value of $\bar{x}$ is

$$z = \frac{980 - 1400}{300/\sqrt{70}} = -11.7,$$

which falls in the reject region.

Pointing to the result obtained, HEB took issue with President Marx's assertion that the mean combined SAT scores of Huxley's students is at least 1400.

> An "at least" claim that a population mean μ is at least equal to a value μ_0 can be summarized by the condition
>
> $$H_c : \mu \geq \mu_0$$
>
> which in turn may be formulated in terms of the null and alternative hypotheses as follows:
>
> $$H_o : \mu = \mu_0$$
> $$H_a : \mu < \mu_0$$
>
> An "at most" claim that a population mean μ is at most equal to a value μ_0 can be summarized by the condition
>
> $$H_c : \mu \leq \mu_0$$
>
> which in turn may be formulated in terms of the null and alternative hypotheses in the following way:
>
> $$H_o : \mu = \mu_0$$
> $$H_a : \mu > \mu_0$$

Summary of Hypothesis Testing Procedures

1. State the null hypothesis H_o.

2. Formulate the alternative hypothesis H_a.

3. Give thought to the sample size and sample statistic to be used.

4. Specify the level of significance at which the test is to be carried out.

5. Based on H_a, set up the acceptance/reserve judgment/rejection machinery implicit in the level of significance α chosen.
   H_a of the unequal to form leads to two bounds and a two-sided test.
   H_a of the less than or greater than form leads to one bound and a one-sided test.

6. Take a sample at random from the population and determine the sample statistic.

7. Make a decision on H_o.

EXERCISES

Null and alternative hypotheses are stated for Exercises 5-10. Also given are the sample size and level of significance envisioned. State the test statistic to be examined, the accept/reserve judgment and reject regions, and the probability of committing a type I error. Draw a diagram for each case.

5. $H_o : \mu = 500,\ H_a : \mu \neq 500,\ n = 60,\ \alpha = 0.05$.

6. $H_o : \mu = 120,\ H_a : \mu < 120,\ n = 12,\ \alpha = 0.01$.

7. $H_o : \mu = 1050,\ H_a : \mu > 1050,\ n = 36,\ \alpha = 0.02$.

8. $H_o : \mu = 5,\ H_a : \mu \neq 5,\ n = 10,\ \alpha = 0.01$.

9. $H_o : \mu = 30,\ H_a : \mu > 30,\ n = 8,\ \alpha = 0.05$.

10. $H_o : \mu = 4000,\ H_a : \mu < 4000,\ n = 70,\ \alpha = 0.05$.

11. In Exercises 5-10 are there any concerns about the population being sampled from that should be kept in mind? Explain.

A claim is stated about a population mean in Exercises 12-17. Also given are the sample size and level of significance envisioned. Formulate null and alternative hypotheses, state the test statistic to be examined, state the accept/reserve judgment and reject regions, and draw a diagram for each case.

12. The mean is at least 12; $n = 50, \alpha = 0.05$.

13. The mean is at least 25, $n = 15, \alpha = 0.01$.

14. The mean is at most 50, $n = 42, \alpha = 0.02$.

15. The mean is at most 120, $n = 16, \alpha = 0.05$.

16. The mean is at least 2000, $n = 55, \alpha = 0.01$.

17. The mean is at most 600, $n = 9, \alpha = 0.01$.

18. In Exercises 12-17 are there any concerns about the population being sampled from that should be kept in mind? Explain.

19. The manufacturer of Kool Kola stands behind the claim that the mean weight of its 32 ounce bottles of Kola is 32 ounces, as advertised. As a member of the manufacturer's in-house quality control department you have been asked to conduct a statistical test to evaluate the accuracy of this claim.

   (a) Formulate null and alternative hypotheses for this claim.

   (b) What is the basis for your alternative hypothesis? Explain.

   (c) At what level of significance would you conduct the test? Explain. Use this level of significance in the follow-up analysis. Let us suppose that $\alpha = 0.01$ is chosen.

   (d) Your next step is to choose a sample of Kool Kola at random from the output and determine the sample mean weight and standard deviation. The question is, should a large or small sample be chosen? What factors would you consider in making this decision? What sample size would you take if it were up to you? Let us take $n = 50$

(e) What accept/reserve judgment and reject regions emerge? Draw a diagram.

(f) Let us further suppose that a randomly chosen sample of 50 bottles yields a mean weight of 31.2 ounces of Kool Kola with a standard deviation of 1.5 ounces. Would this tend to support or contradict the manufacturer's claim? Explain.

(g) Let us now suppose that a randomly chosen sample of 15 bottles yields $\overline{x} = 31.2$ ounces with $s = 1.5$ ounces. Would this tend to support or contradict the manufacturer's claim? Explain.

20. The manufacturer of Deadly Lights cigarettes claims that Deadly Lights contains an average of 8 milligrams of tar per cigarette. As an analyst for the Regional Health Protection Society you have been asked to test this claim at the 0.02 level of significance.

(a) Formulate null and alternative hypotheses for this claim.

(b) What is the basis for your alternative hypothesis? Explain.

(c) Now, on to choosing a sample. What factors would you consider in whether to choose a large or small sample? Let us suppose that a large sample is chosen.

(d) Describe the accept/reserve judgment and reject regions that emerge. Draw a diagram.

(e) Let us suppose that a randomly chosen sample of 55 cigarettes yielded $\overline{x} = 8.5$ milligrams with $s = 1.6$ milligrams. What would you conclude?

(f) It came to pass that before you issued your report an individual claiming to be a member of the 0.01 Significance Level Society urged you to change the significance level of your test from 0.02 to 0.01, arguing that with the 0.01 level you cut the probability of making a type I error in half and also that the 0.01 level based on 1 is more satisfying than the 0.02 level based on 2 on esthetic grounds. What would your reply be?

(g) Do you think the 0.01 significance level enthusiast might have had some other reason for wanting you to change the significance level from 0.02 to 0.01?

21. The anthropologist Alice Williams claims that recently discovered humanoid fossil remains in Central America are 1 million years old. You have been asked by the Anthropologists Union to conduct a statistical test on this claim. You have agreed to undertake the job and have decided to use the 0.05 level of significance.

(a) Formulate null and alternative hypotheses.

(b) What is the basis for your alternative hypothesis? Explain.

(c) The number of humanoid fossil remains is not very large to begin with and a small sample will have to be used. A sample of 4 remains chosen randomly from the available group yielded the following dates, in millions of years: 0.8, 0.75, 0.9, 1.1. What conclusion do you reach from these data? Explain.

(d) Do you have any concerns or reservations about the conclusion you have reached? Explain.

(e) How would you report your results to the Anthropologists Union?

22. A group of labor economists has put out a report claiming that the mean income of factory workers in the midwest does not exceed $25,000. The Midwest Economic Development Association (MEDA) desires to test this claim. The 0.01 level of significance is to be used.

(a) Formulate null and alternative hypotheses.

(b) What is the basis for your null and alternative hypotheses? Explain.

(c) It has been suggested that a sample of size 10 be chosen because it is relatively easy to work with, does not consume an inordinate amount of time, and allows us to stay within budget. What is your reaction to this suggestion? Explain.

(d) Suppose that a sample of size 10 were to be used. What would the accept/reserve judgment and reject regions be? Draw a diagram.

(e) The analysis team ultimately decided on a sample size of 35. Could the same accept/reserve judgment and reject region you described in answer to (d) be used in this case? Explain.

(f) If you answered no to (e), then what accept/reserve judgment and reject regions would you recommend? Explain. Draw a diagram.

(g) Let us suppose that a random sample of size 35 yielded $\bar{x} = \$25,300$ with $s = \$800$. What conclusion do you reach from these data? Explain.

(h) What does your finding mean to MEDA in terms of possible follow-up action?

23. In connection with Example 1, determine the bounds x_1 and x_2 (Figure 1 (a)) which have -1.96 and 1.96 as their counterparts in terms of the standard normal curve (Figure 1 (b)).

24. Oscar Wilmot has been asked by a car rental company to test the claim that private passenger cars not driven for business purposes are driven, on average, at most 12,000 miles per year. It is estimated that $\sigma = 2500$ miles. Oscar set up the null and alternative hypotheses $H_o : \mu = 12,000, H_a : \mu > 12,000$ and he plans to take a random sample of 100 such cars. He will accept H_o if the average mileage on the 100 cars were at most 12,750 miles and reject H_o if it exceeded 12,750 miles. Under these conditions, what is α, the probability of making a type I error?

11.3 ■ Balancing Type I and Type II Errors

Automaker, Inc. specializes in making 300 horsepower engines. As we know, not every engine will put out exactly 300 h.p., but if the manufacturing process is working properly then on average the engines will put out 300 h.p.

The quality control engineers believe it realistic to view the horsepower rating of an engine as determined by a large number of independently acting factors arising in the production process, each having a negligible influence on the horsepower level as a whole. Thus, it follows that the population of engine horsepower values is normally distributed (see Section 8.3) and it is feasible to run small sample quality control checks. This is what the quality control team does: A random sample of 9 engines is taken periodically from the production line and their mean horsepower $\bar{x}$ is calculated. Based on the value of $\bar{x}$ a decision has to be reached as to whether the manufacturing process is working properly, that is, whether engines with an average output of 300 h.p. are being produced, or not. If it is the former, then the process is continued; if it is the latter, then the process will have to be stopped and the machinery adjusted. We should note that the standard deviation of the population of engine horsepower values σ has been determined as 6; $\sigma = 6$ h.p.

The quality control team is working with the following null and alternative hypotheses:

$$H_o : \mu = 300$$
$$H_a : \mu \neq 300$$

The question is, what accept/reject criterion should be used to determine whether to accept H_o and continue the production process or reject H_o and stop the process and adjust the machinery? We shall look at some possibilities and the trade-offs that emerge between the probabilities of committing type I and type II errors, α and β. The accept/reject bounds can be set by stating the level of significance α, or equivalently the probability α of the type I error at which the test is to be conducted, or by stating the bounds directly from which α follows. In this case, for analysis purposes, it is convenient to do the latter.

Let us assume that the following decision rule has been put into operation:

If $294 \leq \bar{x} \leq 306$, then accept H_o.
If $\bar{x} < 294$ or $\bar{x} > 306$, then reject H_o.

Based on the null hypothesis $\mu = 300$, the normality of the population the quality control engineers are sampling from, and the information that $n = 9$ and $\sigma = 6$, it follows that the sampling distribution of $\bar{x}$ is normal with mean $\mu_{\bar{x}} = 300$ and $\sigma_{\bar{x}} = 6/\sqrt{9} = 2$, as shown in Figure 7(a). To obtain α, which is the sum of the areas of the two shaded tail regions to the left of $x_1 = 294$ and to the right of $x_2 = 306$ (Figure 7(a)) we shift the scene to the standard normal curve (Figure 7(b)), determine the corresponding z_1 and z_2 values and use them to obtain the areas sought.

$$z_1 = \frac{294 - 300}{2} = -3.00, \quad z_2 = \frac{306 - 300}{2} = 3.00$$

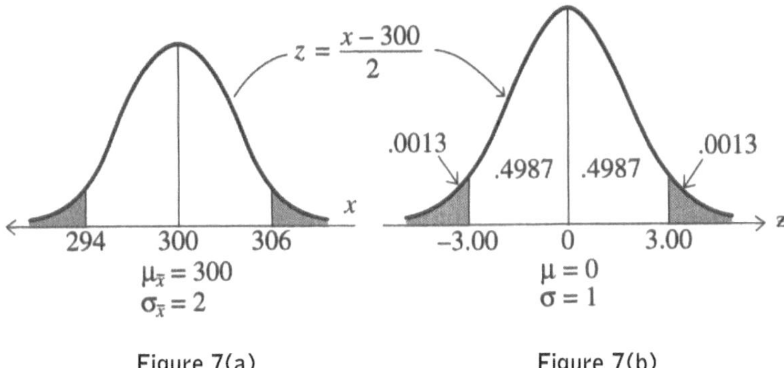

Figure 7(a) Figure 7(b)

Since the area of each of the regions determined by $z_1 = 3.00$ and $z_2 = 3.00$ is 0.4987, each tail has area 0.0013. Thus, the probability of a type I error for the bounds 294 and 306 is given by:

$$\alpha = 0.0026$$

A type II error is committed when a false null hypothesis is accepted, and its probability β depends on the actual value of μ as opposed to what the null hypothesis claims it to be, which is 300 in this case. Let us suppose that the production process has suffered deterioration and that the actual mean horsepower value has slipped to 295. This being the case, the sampling distribution of $\bar{x}$ is normal with mean $\mu_{\bar{x}} = 295$ (instead of 300) and $\sigma_{\bar{x}} = 2$ (as before), as shown in Figure 8 (a). The acceptance bounds 294 and 306 have

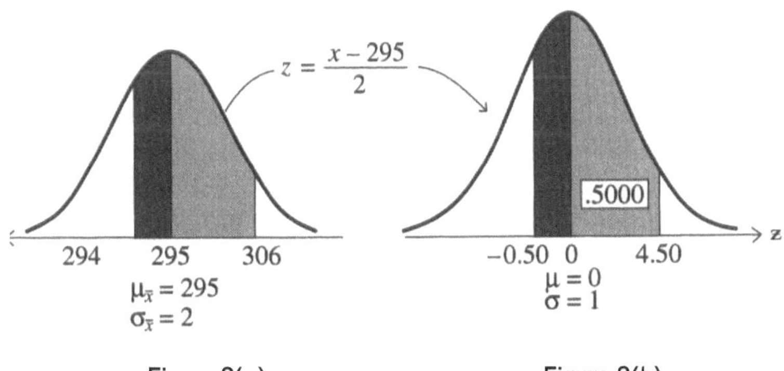

Figure 8(a) Figure 8(b)

not been changed and we will accept the false null hypothesis if $\bar{x}$ is between these bounds, inclusive of them. The probability of a type II error β is, therefore, equal to the area of the region determined by $x_1 = 294$ and $x_2 = 306$ under the normal curve with mean 295 (Figure 8(a)). To obtain the area of this region we shift the scene to the standard normal curve (Figure 8(b)), determine the corresponding z_1 and z_2 values, and use them to obtain the area sought.

$$z_1 = \frac{294-295}{2} = -0.50, z_2 = \frac{306-295}{2} = 5.50$$

Since the areas of the regions determined by $z_1 = -0.50$ and $z_2 = 5.50$ are 0.1915 and 0.5000, respectively, the area the region bounded by $x_1 = 294$ and $x_2 = 306$ is 0.6915.

Thus, for the acceptance bounds 294 and 306,

$$\alpha = 0.0026, \quad \beta = 0.6915,$$

where the actual value of μ is assumed to be 295. In summary, we have Figures 9(a) and 9(b).

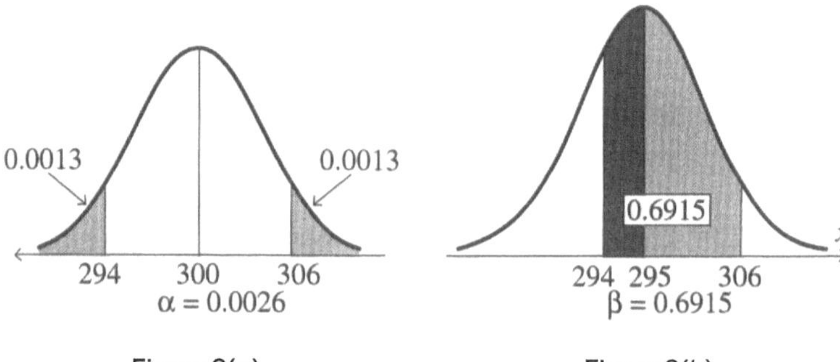

Figure 9(a) Figure 9(b)

If we narrow the acceptance bounds further by raising 294 to 296 and reducing 306 to 304, then the same analysis shows that α increases to 0.0456 and β decreases to 0.3085 (Exercise 1). In summary, we have Figures 10(a) and 10(b).

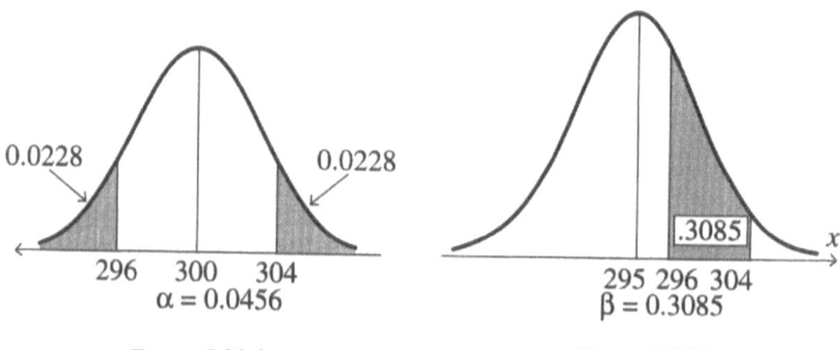

Figure 10(a) Figure 10(b)

If we narrow the acceptance bounds further by raising 296 to 298 and lowering 304 to 302, then α increases to 0.3174 and β decreases 0.0668 (Exercise 2). In summary, we have Figures 11(a) and 11(b).

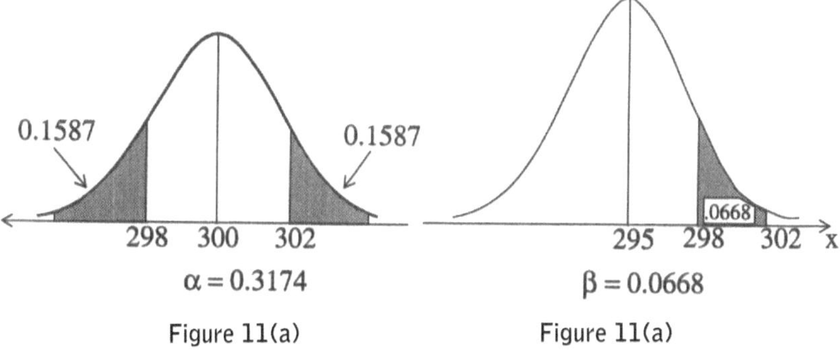

Figure 11(a) Figure 11(a)

> Geometrically speaking, we see from Figures 9(a)-11(b) that as we narrow the acceptance bounds, the tail regions, which define α, increase in area while the region described by the acceptance bounds, which defines β, decrease in area.

The management of Automaker, Inc. would like to set the accept/reject bounds so that α is small because it wants to keep to a minimum the number of disruptive and costly situations where the manufacturing process is stopped unnecessarily when the 300 h.p. mean rating is being met. From this point of view, setting the bounds at 294 and 306 is most satisfactory in that the small $\alpha = 0.0026$ emerges. The cost is in the high $\beta = 0.6915$, when the manufacturing process has deteriorated to the point that the mean horsepower rating is 295.

There has been an almost 2% drop in the mean horsepower level from the advertised norm with a 69% probability that the quality control standard established by the accept/reject bounds will not pick this up. Management cannot tolerate such a high β. As we have seen, we decrease β by adjusting the bounds so as to increase α. Management will have to decide on the proper balance between a low α and not too high β and set its accept/reject bounds accordingly. Some relief can be obtained by increasing the sample size from 9 to some higher value, but there is a cost dimension to this option which would have to be factored into the analysis.

EXERCISES

1. Show that if the accept/reject bounds at Automaker, Inc. are set at $x_1 = 296$ and $x_2 = 304$, then $\alpha = 0.0456$ and $\beta = 0.3085$, where $H_o : \mu = 300$, but $\mu = 295$.

2. Show that if the accept/reject bounds at Automaker, Inc. are set at $x_1 = 298$ and $x_2 = 302$, then $\alpha = 0.3174$ and $\beta = 0.0668$, where $H_o: \mu = 300$, but $\mu = 295$.

3. Suppose that the accept/reject bounds at Automaker, Inc. were set at $x_1 = 299$ and $x_2 = 301$, where $H_o: \mu = 300$, but $\mu = 295$.

   (a) Find α,

   (b) Find β.

4. Suppose that the accept/reject bounds at Automaker, Inc. were set at $x_1 = 299$ and $x_2 = 301$, where $H_o: \mu = 300$, but $\mu = 299$.

   (a) Find α,

   (b) Find β.

   (c) As part of the management team of Automaker, Inc., which would disturb you more, the high α or the high β? Explain.

5. The management of Automaker, Inc. has decided that it does not pay to stop the production process if sample data indicates that μ is no longer 300, but has moved higher. We have $H_o: \mu = 300$, $H_a: \mu < 300$. The accept/reject criterion has been modified as follows: If $\bar{x} \geq 296$, accept H_o; if $\bar{x} < 296$, reject H_o. As before, $\sigma_{\bar{x}} = 2$.

   (a) Find α.

   (b) Find β, where $\mu = 295$.

   (c) Find β, where $\mu = 297$.

   (d) Find β, where $\mu = 299$.

11.4 ■ Power and Operating Characteristic Curves

As we have discussed, the probability of accepting $H_o: \mu = \mu_0$ when it is false because $\mu \neq \mu_0$, but is equal to another numerical value μ_1, is designated by β. Often the decision maker focuses on the probability of not making a type II error for a specific numerical value μ_1 other than the hypothesized one μ_0. The probability of this, $1-\beta$, is called the **power** of the hypothesis test. For Automaker, Inc., for example with $\mu_0 = 300$ and $\mu_1 = 295$, we obtained $\beta = 0.6915$. Thus, the power of the hypothesis test for $\mu_1 = 295$ is $1-0.6915 = 0.3085$. Since there is only about a 31% probability of not accepting the false H_o the test is not very powerful under these circumstances. When the acceptance interval of the decision rule was shortened and β decreased to 0.3085 the hypothesis test for $\mu_1 = 295$ became more powerful since $1-\beta$ increased to about 69%.

If the various powers associated with different possible values of μ_1 are calculated and a graph is constructed showing how power varies with μ_1 values, the resulting curve is called the **power curve** of the hypothesis test. Table 1 shows the values of β and $1-\beta$ corresponding to values of μ_1 ranging from 300 h.p. to 300 ± 12 h.p. The acceptance bounds are $x_1 = 294$ and $x_2 = 306$, so that $\alpha = 0.0026$.

Table 1

μ_1	β	$1-\beta$	μ_1	β	$1-\beta$
300 ± 0	0	1	300 ± 7	.3085	.6915
300 ± 1	.9938	.0062	300 ± 8	.1587	.8413
300 ± 2	.9772	.0228	300 ± 9	.0668	.9332
300 ± 3	.9332	.0668	300 ± 10	.0228	.9772
300 ± 4	.8413	.1587	300 ± 11	.0062	.9938
300 ± 5	.6915	.3085	300 ± 12	.0013	.9987
300 ± 6	.5000	.5000			

Figure 12 shows the power curve of the Automaker, Inc. hypothesis test.

By observing the shape of the power curve a decision maker can evaluate the effectiveness of the current decision rule with respect to the power of the test over (i) those possible values of μ_1 for which a high degree of power is desired and (ii) those possible values of μ_1 for which a lower degree of power is desirable. For example, looking at Figure 12 we see that for values of μ_1 between 293 and 307 the test is not very powerful. It begins to get somewhat powerful for values of μ_1 equal to 292 or 308 and achieves a high degree of power for values μ_1 of 291 or less or 309 or more. Whether or not this is satisfactory to the quality control team of the company is up to them. If it is, they will either change the rule or the sample size so that the test becomes more powerful.

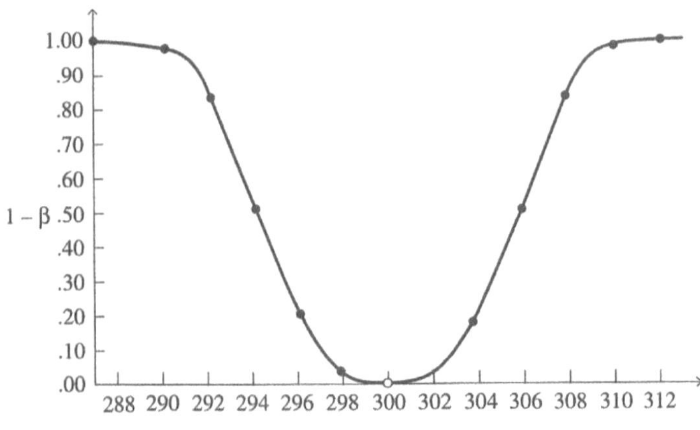

Figure 12

Operating Characteristic (OC) Curves

The complement of a power curve is called an **operating characteristic (OC) curve**. It is a plot of the values of β corresponding to various values of μ_1 in contrast to the power curve which plotted values of $1-\beta$ corresponding to various of μ_1. Figure 13 shows the OC curve for the Automaker. Inc. problem. We see from Figure 13 that the OC curve shows the same information as the power curve except with respect to a different frame of reference. That is, rather than indicate how powerful in terms of $1-\beta$ the hypothesis test is for various values of μ_1, it shows the probability β that a false hypothesis will be accepted for possible values of μ_1.

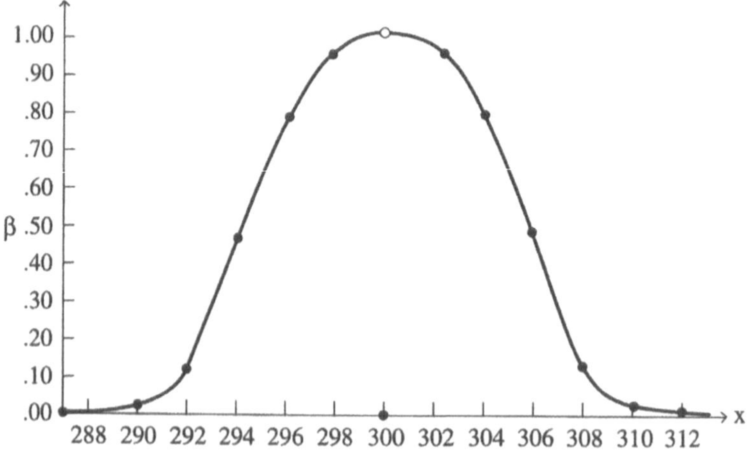

Figure 13

EXERCISES

1. Concerning Automaker, Inc., suppose that the acceptance bounds were changed to $x_1 = 296$ and $x_2 = 304$.

   (a) Determine β for $\mu_1 = 300 \pm k$, $k = 1, 2, \ldots, 12$

   (b) Sketch the power curve.

   (c) Sketch the OC curve.

2. Consider Ramunė's Gourmet Coffee (Sec. 11.2, Example 1), where $H_o : \mu = 8$, $H_a : \mu \neq 8$, $\alpha = 0.05$, $n = 49$ and σ is approximated by $s = 0.8$.

   (a) Determine β for $\mu_1 = 8 \pm k$, $k = 0.1, 0.2, \ldots, 0.5$.

   (b) Sketch the power curve.

   (c) Sketch the OC curve.

3. In connection with the production of a product, as illustrated by Automaker, Inc. and Ramunė's Gourmet Coffee, in what sense does α express the producer's risk and β the consumer's risk?

11.5 ■ Tests Concerning Proportions

The hypothesis testing framework developed in Section 11.2 with the focus on population means is generally applicable to one-sample test situations where one sample is chosen at random from a population under study. As the population parameter of interest changes, the theorem describing the sampling distribution of the "appropriate" sample statistic changes accordingly, but the hypothesis testing framework remains the same.

When the focus shits to a population proportion π, the appropriate sample statistic is the sample proportion $p = x / n$ whose sampling distribution we recall from Section 10.5.

Central Limit Theorem for $p = x / n$.

If a sufficiently large sample of size n is chosen at random from a large population, the sampling distribution of $p = x/n$ may be approximated by the normal curve with mean μ_p and σ_p given by,

$$\mu_p = \pi, \quad \sigma_p = \sqrt{\frac{\pi(1-\pi)}{n}},$$

where π is the population proportion, n is the sample size, and x is the number of members of the sample who favor the option or situation in question.

Let us keep in mind that hypothesis tests concerning proportions are applicable to situations where the underlying data are on a **nominal scale.** The following example illustrates.

Example 1 The Case of Dapper Dan's Coin

Dapper Dan, known for his extensive collection of dice, claims that the coin he has been using in various coin tossing games is fair. Some of Dan's associates in these games would like to test this claim and propose using the 0.05 significance level.

For a fair coin, which does not favor head showing over tail showing, it is reasonable to take as our null hypothesis $\pi = 0.50$, where π is the population proportion, or probability, of head showing. Since significant variation in either direction would refute this null hypothesis, it makes sense to take as the alternative hypothesis $\pi \neq 0.50$.

Thus we have:

$$H_o : \pi = 0.50$$
$$H_a : \pi \neq 0.50$$
$$\alpha = 0.05$$

A sample size of 150 has been agreed on. The test is two-sided since the alternative hypothesis is two-sided. For the 0.05 significance level the decision bounds are ± 1.96, as shown in Figure 14.

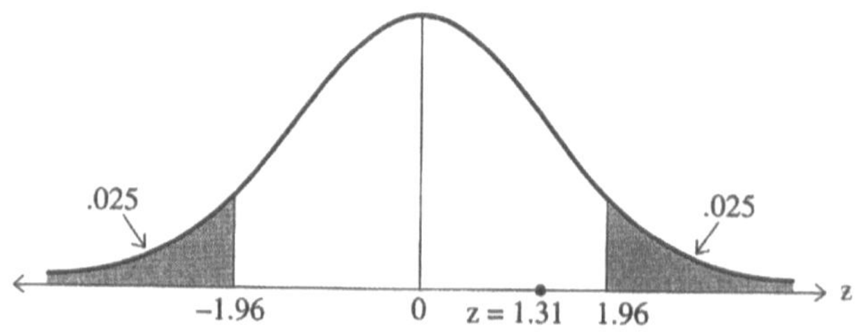

Figure 14

Let us suppose that the sequence of 150 tosses of Dan's coin yielded 83 heads; then $x/n = 83/150 = 0.5533$, and

$$z = \frac{x/n - \pi}{\sqrt{\frac{\pi(1-\pi)}{n}}} = \frac{0.5533 - 0.50}{\sqrt{\frac{0.50(0.50)}{150}}} = 1.31$$

which falls within the accept/reserve judgment region.

Thus, Dan's associates must either accept Dan's claim that his coin is fair or reserve judgment. They have no basis for a counter-claim that Dan has been cheating them. If they decide to engage in further activities with Dan's coin, then this is tantamount to accepting the null hypothesis.

EXERCISES

1. A medication which has become standard for the treatment of rheumatoid arthritis is effective in 75% of the cases treated. It is claimed that a newly developed, less expensive medication for this condition is at least as effective as the standard.

   (a) Formulate null and alternative hypotheses for this claim.

   (b) What is the basis for your alternative hypothesis? Explain.

   (c) It has been decided to test the claim at the 0.02 level of significance. What accept/reserve judgment and reject regions emerge? Draw a diagram.

   (d) The new medication proved effective on 72 of 100 patients who were chosen at random and administered the medication. What conclusion can be drawn from this finding?

   (e) What follow-up in terms of the new medication might acceptance of the null hypothesis have as opposed to reserving judgment

   (f) If such follow-up were undertaken, what would be the consequences of a type II error having been committed?

2. The campaign headquarters of John Exon asserts that 60% of the voters in his district support him for state senator. The claim is to be tested at the 0.05 significance level.

(a) Formulate null and alternative hypotheses for this claim?

(b) What is the basis for the alternative hypothesis? Explain.

(c) What accept/reserve judgment and reject regions emerge? Draw a diagram.

(d) Suppose that 110 of 200 voters chosen at random stated that they support Exon. What conclusion can be drawn from this finding?

(e) How could a type II error occur and what would be its consequences?

3. Dapper Dan has a die for which the probability, or true proportion, of an even number showing is 0.67, he claims (see Sec. 7.1, Exercise 1). Dan's friend Gullible Gus would like to test this claim at the 0.01 level of significance.

(a) How should Gus formulate null and alternative hypotheses?

(b) What is the basis for the alternative hypothesis?

(c) What accept/reserve judgment and reject regions emerge? Draw a diagram.

(d) Suppose that in 150 tosses of Dan's die an even number showed 84 times. What conclusion can be drawn from this result?

(e) How could a type I error arise and what would be its consequences?

(f) What are the odds of committing a type I error?

4. The marketing division of the Rasa Company claims that 45% of the perfume market has been captured by its new perfume Essence of Rasa. The marketing director of a competitor wants to test this claim by comparing consumer preference for Essence of Rasa against his own company's highly popular Desire.

(a) Is there any "difficulty" in using the methods developed in this section to compare Essence of Rasa to Desire? Explain.

(b) If your answer to (a) is yes, are there conditions under which the difficulty is minimized or eliminated? Explain.

(c) Formulate null and alternative hypotheses for this claim.

(d) What is the basis for the alternative hypothesis? Explain.

(e) The 0.05 level of significance is to be used. What accept/reserve judgment and reject regions emerge? Draw a diagram.

(f) Suppose that of 120 consumers chosen at random and asked their preference, 60 expressed a preference for Essence of Rasa over Desire. Can we conclude that the Rasa Company's claim has been sustained? Explain.

(g) Since the sample proportion of 50% is larger than the 45% claimed by the Rasa Company can we conclude that the Rasa's Company's share of the market is actually larger than what it claims? Explain.

5. Mendelian inheritance theory tells us that certain crosses of peas yield 75% yellow and 25% green peas. The basic question is whether deviations from theory obtained in practice are significant at the 1% level.

(a) Formulate null and alternative hypotheses.

(b) What is the basis for the alternative hypothesis? Explain.

(c) What accept/reserve judgment and reject regions emerge? Draw a diagram.

(d) Let us suppose that a sample of 200 pea crossings yielded 140 yellow and 60 green peas. What conclusion can be drawn?

(e) Does the result obtained prove that Mendelian inheritance theory is true? Explain.

6. A television producer claims that at least 25% of television viewers in the northwest regularly watch his program *Outrageous Acts*, which had been introduced to the northwest on a trial basis. The principal advertiser for *Outrageous Acts* wishes to test this claim at the 2% level.

(a) Formulate null and alternative hypotheses.

(b) What is the basis for the alternative hypothesis? Explain.

(c) What accept/reserve judgment and reject regions emerge? Draw a diagram.

(d) Suppose that of a randomly chosen sample of 150 television viewers in the northwest who were asked whether they watch *Outrageous Acts* on a regular basis, 24 replied that they do. What conclusion can be drawn?

(e) How could a type II error arise and would be its most likely consequences?

11.6 ■ Variance and Standard Deviation

In many situations it is important to consider the variability in values of a population. In the manufacture of a product, such as razor blades, for example, the thickness of the blade is of major concern and cannot vary greatly from blade to blade. There are standards for the time it takes to complete various tasks and variation from such standards is of concern. In considering air pollution levels of a pollutant, sulfur dioxide, for example, one is concerned with variation in the daily emission levels in addition to their mean level.

Situations of such a nature lead to estimation and hypothesis testing problems involving σ and σ^2. We turn here to hypothesis testing problems, for which our major tool is the sampling distribution of the sample statistic:

$$\chi^2 = \frac{(n-1)s^2}{\sigma_0^2}$$

with respect to the null hypothesis that the population standard deviation σ equals some value σ_0.

> The sampling distribution of
>
> $$\chi^2 = \frac{(n-1)s^2}{\sigma_0^2}$$
>
> is, to a close approximation, the chi-square distribution with $n-1$ degrees of freedom, **assuming that the sample of size n is drawn at random from a normally distributed population.** s^2 is the sample variance.

Example 1 Variation in Width

The Villars Company makes razor blades. Their stainless-steel-plus Super Shave blade is to have a mean width of 0.650 inches with a standard deviation of 0.015 inches. The quality control department of the Villars Company wants to test the null hypothesis $\sigma = 0.015$ at the 0.01 level of significance. Since too large a variation in σ is what is troublesome, the appropriate alternative hypothesis is $\sigma > 0.015$. We have:

$$H_o : \sigma = 0.015$$
$$H_a : \sigma > 0.015$$
$$\alpha = 0.01$$

An important underlying condition for this test is that the population of blade widths be normally distributed. In support of this condition the quality control department argued that it is realistic to view blade width as determined by a large number of independently acting random factors arising in the production process each having a negligible influence on the blade width as a whole. From this it follows that the population of blade widths is approximately normally distributed.

Sample size: A sample size of 20 has been decided on, let us assume, so that the number of degrees of freedom is 19.

Acceptance/rejection machinery: This is a one-sided test since the alternative hypothesis is one-sided. We need one upper bound to make precise what it means to say that the value of the sample χ^2 statistic is significantly large at the 0.01 level. At the 0.01 level the bound for $\text{d.f.} = 19$ is $\chi^2_{0.1} = 36.191$ (see Figure 15).

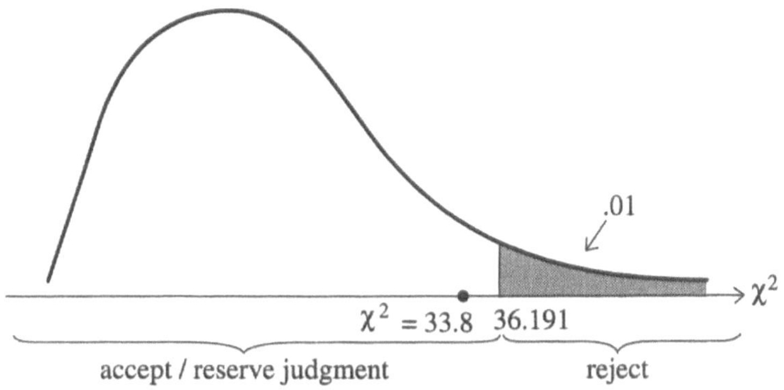

Figure 15

A randomly chosen sample of 20 blades yielded $s^2 = 0.0004$, so that the sample χ^2-value is:

$$\chi^2 = \frac{19(0.0004)}{(0.015)^2} = 33.8$$

Thus, the quality control department has no basis for rejecting the null hypothesis. They accept the null hypothesis and continue the production process as it is currently operating.

EXERCISES

1. The manufacturer of Deadly Lights cigarettes claims that Deadly Lights contains an average of 8 milligrams of tar per cigarette (considered in Section 11.2, Exercise 20) and that the standard deviation does not exceed 0.85 milligrams. The Regional Health Protection Society wishes to test this claim at the 0.025 level.

   (a) Formulate the null and alternative hypotheses for this claim.

   (b) What is the basis for your alternative hypothesis? Explain.

   (c) A random sample of 15 cigarettes is to be chosen and analyzed. Describe the accept/reserve judgment and reject regions that emerge. Draw a diagram.

   (d) Do you have any concerns or reservations about the hypothesis test that is being undertaken? Explain.

   (e) Let us suppose that the randomly chosen sample of 15 cigarettes had a standard deviation of 1.1 milligrams. What would you conclude?

   (f) What does your finding mean to the manufacturer of Happy Lights?

2. The Yang Company makes ball bearings for machine tools. The diameters of the ball bearings vary to some extent, but if they are to be used this variation must be small. It is required that the standard deviation of their diameters not exceed 0.009 millimeters. The quality control department wishes to test the null hypothesis $\sigma = 0.009$ at the 0.01 level.

(a) Formulate an alternative hypothesis for this claim.

(b) What is the basis for your alternative hypothesis? Explain.

(c) The quality control department believes that the population of ball bearing diameters is approximately normally distributed. What point of view about the manufacturing process would justify this belief? Explain.

(d) A random sample of 20 ball bearings is to be chosen from the production line and analyzed. What accept/reserve judgment and reject regions emerge? Draw a diagram.

(e) Let us suppose that the randomly chosen sample of 20 ball bearings had a standard deviation of 0.012 millimeters. What would you conclude?

(f) What does your finding mean to the Yang Company?

(g) The quality control department plans to repeat the aforenoted sampling procedure a large number of times at regular intervals. What interpretation does setting α at 0.01 have in this connection?

(h) Let us suppose that the production process has deteriorated and that α has shifted upward so that $\beta = 0.4100$.

3. The Tasty Company has begun manufacturing a frozen yogurt called Tasty-but-Healthy. It is to have a mean fat content of 2 grams per half-gallon with a standard deviation of 0.25 grams. The quality control department desires to test the null hypothesis $\sigma = 0.25$ at the 0.05 level. If σ significantly exceeds 0.25, then the department is concerned about high inconsistency in the fat content of the product.

(a) Formulate an alternative hypothesis.

(b) What is the basis for your alternative hypothesis? Explain.

(c) A random sample of 12 frozen yogurt scoops is to be chosen from the production line and analyzed. What accept/reserve judgment and reject regions emerge? Draw a diagram.

(d) Let us suppose that the randomly chosen sample of 12 frozen yogurt scoops had a standard deviation of 0.37 grams. What would you conclude?

(e) What does your finding mean to the Tasty Company?

(f) Before issuing its report the quality control department was visited by our undaunted friend from the 0.01 Significance Level Society (first encountered in Section 10.2, Exercise 20(f) (page 266), who urged the department to mend its ways before its too late and change the significance level to 0.01. What would an appropriate reply be?

(g) Suppose that 0.01 had been initially chosen as the significance level. What would your chi-square statistic mean to the Tasty Company in this case?

(h) Suppose it were later found that the population of sample standard deviations was highly skewed to the right and could not be considered approximately normally distributed. What effect would this have on the result of the analysis? Explain.

4. Ecap University administers a placement exam to all of its students prior to their taking a mathematics course at the university. The time currently allowed for the exam is 60 minutes, but variation in the time it takes a student to complete the exam is troublesome. The mathematics department has decided that it will have to modify the exam or the time allowed to take it if σ exceeds 6 minutes. The department wishes to test the null hypothesis $\sigma = 6$ at the 0.05 level. Six randomly chosen students who sat for the exam were observed to complete it in the following times (in minutes): 45, 53, 62, 65, 50, 54.

(a) Formulate an alternative hypothesis.

(b) What is the basis for your alternative hypothesis? Explain.

(c) What accept/reserve judgment and reject regions emerge? Draw a diagram.

(d) Before reaching a conclusion, are there any factors which must be given thought? Explain.

(e) Assuming the conditions of (d) are satisfactorily met, what do you conclude?

(f) What does your finding mean to the mathematics department?

11.7 ■ Difference Between Means

Our focus thus far has been on hypotheses made about one population parameter. This led to consideration of a variety of one-sample tests where one sample is drawn from the population in question and a conclusion concerning H_o is reached based on that one sample.

Many situations of interest lead us to compare parameters of two or more populations. When two populations are involved we end up choosing a sample from each one, for which the term **two-sample test** is appropriate. More generally, the term **multi-sample test** is appropriate for hypothesis testing situations involving two or more populations where a sample is chosen from each one. Some multi-sample test situations involving more than two populations are considered in Chapter 17.

Two Populations: Large Samples, Independently Chosen

To take an example, in Section 11.1 we noted that the quality control manager of Lee Tires claims there is no difference in the average life of the tires made at its Wells and Jason City plants. Two populations emerge, the population of tire lifetimes of the tires made at the Wells plant and the population of the tire lifetimes of the tires made at the Jason City plant, with means μ_1 and μ_2. The quality control manager believes that $\mu_1 - \mu_2 = 0$. or equivalently, $\mu_1 = \mu_2$. To test this hypothesis samples would have to be drawn at random from the two plants and their means $\bar{x}_1$ and $\bar{x}_2$ calculated. If their difference were significantly small, then the null hypothesis of no difference in the population means would be accepted or judgment reserved; if their difference were significantly large, then the null hypothesis of no difference would be rejected and a suitable alternative hypothesis accepted.

More generally, consider two populations, P_1 with mean μ_1 and standard deviation σ_1, and P_2 with mean μ_2 and standard deviation σ_2. Suppose that large samples of sizes n_1 and n_2 $(n_1 \geq 30, n_2 \geq 30)$ are drawn at random and independently from populations P_1 and P_2, that is, the drawing of either sample does not affect the composition of the other. From these samples we obtain their means and standard deviations.

In summary, we have Table 2.

Table 2

	Pop. P_1	Pop. P_2
Pop. mean	μ_1	μ_2
Pop. s.d.	σ_1	σ_2
Sample size	n_1	n_2
Sample mean	$\overline{x}_1$	$\overline{x}_2$
Sample s.d.	s_1	s_2

The null hypothesis of no difference is

$$H_o : \mu_1 - \mu_2 = 0, \quad \text{or} \quad \mu_1 = \mu_2$$

against an alternative hypothesis H_a which may take one of three forms:

(1) $H_a : \mu_1 \neq \mu_2$

(2) $H_a : \mu_1 > \mu_2$

(3) $H_a : \mu_1 < \mu_2$

The first alternative hypothesis leads to a two-sided or two-tailed test since a difference in the sample means in either direction might be significant. The other two lead to one-sided, or one-tailed, tests since a difference in the sample means in only one direction might be significant.

The key sample statistic which gets us off the ground in situations of this sort is the difference in sample means $\overline{x}_1 - \overline{x}_2$. Its sampling distribution is described by the following central limit theorem:

Central Limit Theorem for $\overline{x}_1 - \overline{x}_2$

The basic background and notation are summarized in Table 2. Consider large samples $(n_1 \geq 30, n_2 \geq 30)$ drawn at random and independently from two populations P_1 and P_2. The sampling distribution of $\overline{x}_1 - \overline{x}_2$ is approximately normal with mean $\mu_{\overline{x}_1 - \overline{x}_2}$ and standard deviation $\sigma_{\overline{x}_1 - \overline{x}_2}$ given by:

$$\mu_{\bar{x}_1-\bar{x}_2} = \mu_1 - \mu_2, \quad \sigma_{\bar{x}_1-\bar{x}_2} = \sqrt{\frac{\sigma_1^2}{n_1} + \frac{\sigma_2^2}{n_2}}$$

With respect to the null hypothesis $H_o : \mu_1 - \mu_2 = 0, \mu_{\bar{x}_1-\bar{x}_2}$ is 0. If the population variances σ_1^2 and σ_2^2 are unknown, they may, in accordance with standard statistical practice, be approximated by the sample variances s_1^2 and s_2^2 since n_1 and n_2 are large.

Example 1 Lee Tires

The quality control manager of Lee Tires decided to test his null hypothesis that there is no difference in the mean tire lifetimes of the tires made at the two plants against the alternative hypothesis that there is a difference at the 0.05 level. We have:

$$H_o : \mu_1 - \mu_2 = 0$$
$$H_a : \mu_1 - \mu_2 \neq 0$$
$$\alpha = 0.05$$

The decision bounds are ± 1.96, as shown in Figure 16.

Randomly chosen samples of 36 and 42 tires taken from the Wells and Jason City plants, respectively, yielded mean lifetimes (in thousands of miles) and variances shown in Table 3.

The population variances σ_1^2 and σ_2^2 are unknown, but since the sample sizes are large we may approximate them by the sample variances $s_1^2 = 5.4$ and $s_2^2 = 4.8$, respectively.

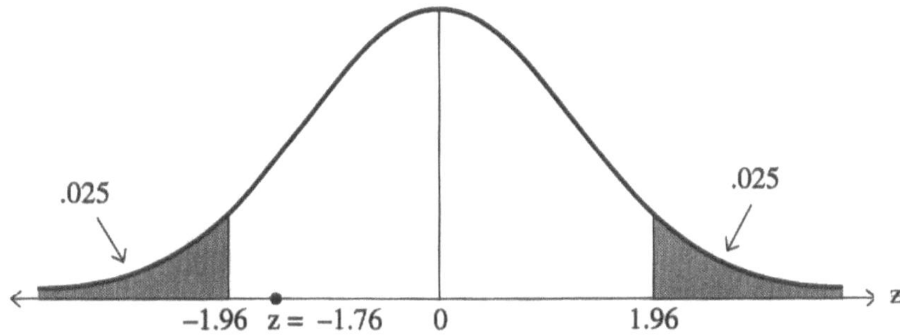

Figure 16

Table 3

	Wells	Jason City
Pop. mean	μ_1	μ_2
Pop. var.	σ_1^2	σ_2^2
Sample size	$n_1 = 36$	$n_2 = 42$
Sample mean	$\bar{x}_1 = 25.3$	$\bar{x}_2 = 26.2$
Sample var.	$s_1^2 = 5.4$	$s_2^2 = 4.8$

The z-statistic has mean $\mu_{\bar{x}_1 - \bar{x}_2} = \mu_1 - \mu_2 = 0$, by the null hypothesis, and standard deviation:

$$\sigma_{\bar{x}_1 - \bar{x}_2} = \sqrt{\frac{5.4}{36} + \frac{4.8}{42}} = 0.51$$

Thus, the z-value of $\bar{x}_1 - \bar{x}_2$ is

$$z = \frac{\bar{X}_1 - \bar{X}_2}{\sqrt{\frac{s_1^2}{n_1} + \frac{s_2^2}{n_2}}} = \frac{25.3 - 26.2}{0.51} \quad -1.76 \,,$$

which is within the accept/reserve judgment bounds ± 1.96 (see Figure 16). The quality control manager has no basis for rejecting the null hypothesis and must either accept it or reserve judgment. Either way, there is no basis for taking action at either plant because of signs that one is not operating as effectively as the other.

Example 2 Another Point of View on Lee Tires

Let us suppose that the quality control manager of Lee Tires had been led to suspect that the production process had deteriorated at the Wells plant and that the mean lifetime of the tires being produced there was less than that of the tires being produced at the Jason City plant. What hypothesis test formulation would this lead to?

The null hypothesis would remain the same,

$$H_o : \mu_1 - \mu_2 = 0$$

but the alternative hypothesis would now take the form:

$$H_a : \mu_1 - \mu_2 < 0, \quad \text{or} \quad \mu_1 < \mu_2$$

The test is now one-sided and we need one lower bound to make precise what it means to say that μ_1 is significantly smaller than μ_2. For the same level of significance as previously used, $\alpha = 0.05$, the *z*-bound is -1.65, as shown in Figure 17.

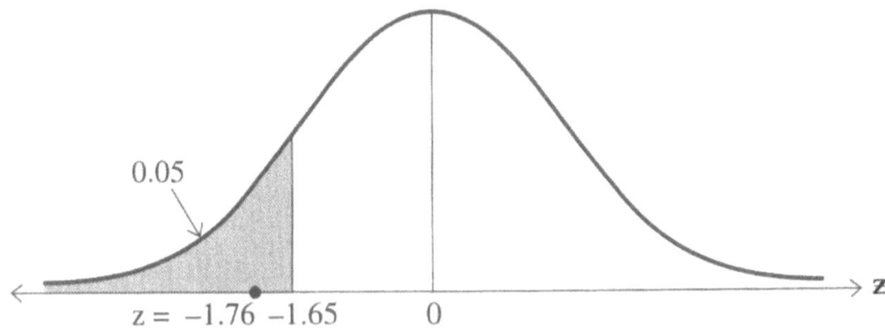

Figure 17

For samples yielding the same results, $z = -1.76$, so that in this situation the quality control manager would reject the null hypothesis and accept the alternative hypothesis that $\mu_1 < \mu_2$.

This outcome suggests that inferior tires are being made at the Wells plant, which might lead to adjustments being made in the plant's production machinery.

Small Samples, Independently Chosen

As previously noted (Section 10.3), the study of materials which are rare or difficult to locate, difficult to prepare for study, or expensive lead us to consider taking small samples. Even at Lee Tires the testing procedure involving samples of the order of 36 and 42 tires chosen from the two plants and tested can become expensive if repeated on a regular basis.

Tests of differences between means where small samples n_1 and n_2 $(n_1 < 30, n_2 < 30)$ are drawn from populations P_1 and P_2 come with strong statistical strings attached.

1. It is **required** that the populations P_1 and P_2 be normally distributed, or approximately so.

2. The populations P_1 and P_2 must have the **same standard deviations**; $\sigma_1 = \sigma_2$.

3. The samples **must be randomly and independently drawn** from P_1 and P_2. That is, the drawing of either sample must not affect the composition of the other.

The test statistic that emerges in this case is the *t*-statistic

$$t = \frac{\overline{x}_1 - \overline{x}_2}{\sqrt{\dfrac{(n_1 - 1)s_1^2 + (n_2 - 1)s_2^2}{n_1 + n_2 - 2} \cdot \left(\dfrac{1}{n_1} + \dfrac{1}{n_2}\right)}}$$

or equivalently,

$$t = \frac{\overline{x}_1 - \overline{x}_2}{\sqrt{\dfrac{\sum (x_1 - \overline{x}_1)^2 + \sum (x_2 - \overline{x}_2)^2}{n_1 + n_2 - 2} \cdot \left(\dfrac{1}{n_1} + \dfrac{1}{n_2}\right)}}$$

with $n_1 + n_2 - 2$ degrees of freedom.

Here $\sum (x_1 - \overline{x}_1)^2$ is the sum of the squared deviations from the mean $\overline{x}_1$ of the sample values chosen from population P_1 and $\sum (x_2 - \overline{x}_2)^2$ is the sum of the squared deviations from the mean $\overline{x}_2$ of the sample values chosen from population P_2.

The second cited version of t follows from the first from the relations

$$\sum (x_1 - \overline{x}_1)^2 = (n_1 - 1)s_1^2, \quad \sum (x_2 - \overline{x}_2)^2 = (n_2 - 1)s_2^2,$$

which are immediate consequences of the definition of sample variance.

The second cited version of t would be used when working directly from sample data, whereas the first is preferable if the sample variances have already been calculated or have been given to us.

Although this *t*-statistic is a mouthful, its appearance is more intimidating than its reality. Even with a primitive calculator one can handle the calculation of t with relative ease. The more challenging part is to keep in mind the statistical strings which underlie its use. When it comes to determining the accept/reserve judgment and reject regions we obtain *t*-bound(s) in much the same way that we obtain *z*-bound(s), keeping in mind the degrees of freedom factor $n_1 + n_2 - 2$.

Example 3 Return to Lee Tires

The quality control manager of Lee Tires would like to test the null hypothesis $H_o : \mu_1 - \mu_2 = 0$ against the alternative hypothesis $H_a : \mu_1 - \mu_2 \neq 0$ at the 0.05 level on a regular basis. Testing on a regular basis by use of large samples becomes expensive and its cost could quickly exceed the amount budgeted for this purpose. To help keep costs within budget the quality control manager proposes to use small samples of size 11 chosen randomly and independently from the Wells and Jason City plants. The questions that arise are: What concerns must be addressed? How should we proceed?

Strings Attached

The concerns that must be addressed are centered on the statistical strings attached which underlie the application of the small sample test of no difference in the population means. Are the populations of tire lifetimes of the tires made at the Wells and Jason City plants approximately normal? The quality control manager believes that it is realistic to view the tire lifetimes at both plants as determined by a large number of independently acting random factors arising from the production process, each having a negligible influence on the tire lifetime as a whole. From this generalized hypothesis of elementary errors it follows that these populations of tire lifetimes are approximately normal (see Section 8.3). Hypothesis tests for normality are discussed in Sections 16.4 and 17.3.

Are the population standard deviations σ_1 and σ_2 equal? There is a hypothesis test for equality of population standard deviations which we take up in Section 11.10 and apply to this situation (Example 1). For now we shall proceed, subject to an affirmative answer. If it should turn out that the answer is no, then the foundation for this hypothesis test would be seriously undermined and its conclusion made doubtful.

We tentatively proceed by setting up our accept/reserve judgment and reject regions in terms of the 0.05 level. The test is two-sided since $H_a : \mu_1 - \mu_2 \neq 0$ is two-sided. Since samples of size $n_1 = 11$ and $n_2 = 11$ are to be drawn, $\text{d.f.} = 11 + 11 - 2 = 20$, and $_{.025}$ 2.086.

In summary, we have H_o, H_a, α and accept/reserve judgment and reject bounds shown in Figure 18.

$$H_o : \mu_1 - \mu_2 = 0$$
$$H_a : \mu_1 - \mu_2 \neq 0$$
$$\alpha = 0.05$$

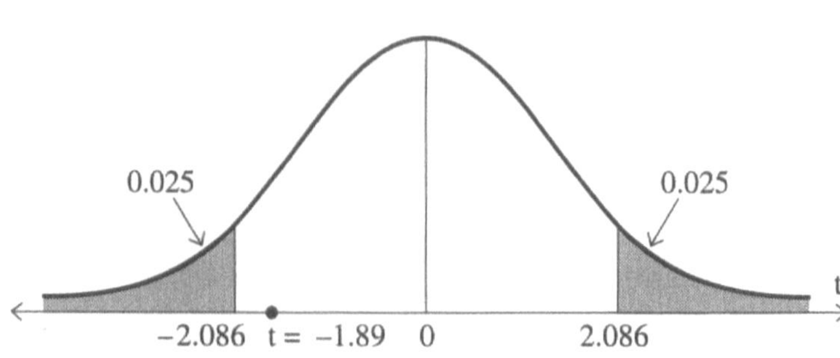

Figure 18

Randomly and independently chosen samples of 11 tires from the Wells and Jason City plants yielded the results summarized in Table 4.

The sample variances obtained support the condition $\sigma_1 = \sigma_2$ (see Section 11.11, Example 1, page 380), which gives us a green light to proceed with this hypothesis test.

Table 4

	Wells	Jason City
Pop. mean	μ_1	μ_2
Pop. var.	σ_1^2	σ_2^2
Sample size	$n_1 = 11$	$n_2 = 11$
Sample mean	$\overline{x}_1 = 24.6$	$\overline{x}_2 = 26.4$
Sample var.	$s_1^2 = 5.8$	$s_2^2 = 4.2$

The t-statistic takes the value

$$t = \frac{24.6 - 26.4}{\sqrt{\frac{10(5.8)+10(4.2)}{20}\left(\frac{2}{11}\right)}} = -1.89,$$

which falls within the accept/reserve judgment region determined by ± 2.086 (see Figure 18).

Either way, the quality control manager of Lee Tires has no support for the view that inferior tires, in terms of mean lifetime, are being produced at one of the plants.

11.8 ■ Which Math Model is Better Suited to the Problem?

Example 1 Carl's Fertilizer Problem

To determine whether a new fertilizer is more effective than one in current use for growing corn, Carl Cairns selected 5 test plots on his farm at random which he treated with the fertilizer in current use and another 5 plots at random which he treated with the new fertilizer. The 0.05 level was to be used in carrying out the analysis. The following results (in bushels per acre) were obtained.

Current fertilizer: 70.5, 80.2, 90.3, 60.2, 85.4
New Fertilizer: 73.2, 82.1, 93.1, 63.6, 87.1

(a) Formulate null and alternative hypotheses.

(b) What is the basis for your alternative hypothesis?

(c) What accept/reserve judgment and reject regions emerge?

(d) Are there special concerns that must be addressed?

(e) Assuming that the concerns noted in (d) are satisfied, what would you conclude if you were Carl?

(f) Instead of proceeding as described, would it be preferable to generate paired yields obtained by selecting 5 plots at random, treating half of each plot with the currently used fertilizer and the other half with the new fertilizer, and conduct a hypothesis test on the equality of means arising from data obtained in this way?

(g) If there is an advantage to pairing as described in (f), would the methods of this section be applicable to a hypothesis test on the equality of means?

Let μ_1 and μ_2 denote the average corn yields of plots treated with the fertilizer in current use and the new fertilizer, respectively.

(a) We have:

$$H_o : \mu_1 - \mu_2 = 0$$
$$H_a : \mu_1 - \mu_2 < 0, \text{ or } \mu_1 < \mu_2$$
$$\alpha = 0.05$$

(b) H_a states that use of the new fertilizer results in higher average corn yields (μ_2) than use of the fertilizer currently employed (μ_1).

(c) Since H_a is one of the less than form we need one lower bound to make precise what significantly less than means in a statistical sense. The statistic to be employed is,

$$t = \frac{\overline{x}_1 - \overline{x}_2}{\sqrt{\left(\frac{4s_1^2 + 4s_2^2}{8}\right)\left(\frac{1}{5} + \frac{1}{5}\right)}}$$

The accept/reserve judgment and reject regions emerge from the lower bound of the t-curve defined by 8 degrees of freedom ($t_{.05} = 1.860$ for $\text{d.f.} = 8$) shown in Figure 19. Accept H_o or reserve judgment if $t \geq -1.86$; reject H_o if $t < -1.86$.

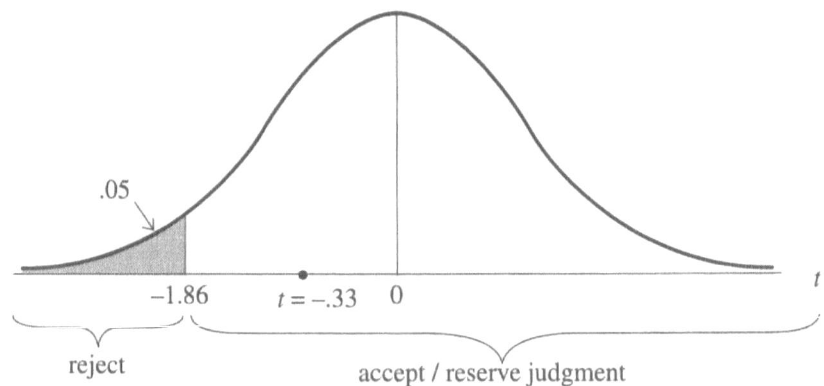

Figure 19

(d) Three concerns emerge:

(1) **Normality.** The populations of corn yields obtained through use of the fertilizer currently employed and the new fertilizer must be normal, or approximately normal. It is shown that these conditions are satisfied in Example 2 and Exercise 1 of Section 16.4.

(2) **Equality of population standard deviations.** This is shown by the hypothesis test to be carried out in Exercise 11 of Section 11.10.

(3) **Randomness and independence of the samples chosen.** From the way in which the samples are chosen it is clear that the intent is to satisfy these conditions.

Since the conditions required for carrying out this hypothesis test are satisfied, Carl has a green light to proceed. Whether or not this is the best test to employ is discussed in part (f).

(e) For the fertilizer in current use we have:

$$\sum x_1 = 386.6, \quad \sum x_1^2 = 30{,}474, \quad \bar{x}_1 = 77.3, \quad s_1^2 = 145.4$$

For the new fertilizer we have:

$$\sum x_2 = 399.1, \quad \sum x_2^2 = 32{,}398, \quad \bar{x}_2 = 79.8, \quad s_2^2 = 135.4$$

This yields:

$$t = \frac{77.3 - 79.8}{\sqrt{\left(\frac{4(145.4) + 4(135.4)}{8}\right)\left(\frac{2}{5}\right)}} = -0.33,$$

which falls in the accept/reserve judgment region (see Figure 19).

What this result means to Carl is that there is no advantage to changing fertilizers in terms of increasing average corn yields.

(f) Yes; the problem is that the procedure employed—choose 5 test plots at random and employ the fertilizer in current use, choose another 5 plots at random and employ the new fertilizer, and then compare results—is open to criticism because it does not effectively screen out extraneous factors (called **confounding factors**) which also play a role in determining crop yields. Such confounding factors include soil conditions, moisture, and weather conditions, which could vary considerably on the plots chosen (even though they are chosen at random) and mask the true effects of fertilizers on corn yields.

The effectiveness of a statistical test in revealing accurate information about the factors that are the focus of the study (effectiveness of the fertilizers in enhancing average corn yields, in this case) depends on how well the test screens out confounding factors. The technical term is **experimental design**, and this test is not built on an appropriate experimental design.

An experimental (math/statistical model) design that would more effectively screen out the effects of confounding factors calls for choosing a number of plots at random, 5 let us say, divide each into approximately equal pieces, and on one piece of each pair employ the fertilizer in current use and on the other the new fertilizer.

(g) The method employed in the analysis carried out would not be applicable to paired-plots chosen as described because the independence requirement—the drawing of each sample must not affect the composition of the other—is not satisfied.

Another statistical modeling approach is needed to handle paired samples and such a modeling approach is developed in the next section. When applied to Carl's problem it yields quite different results.

EXERCISES

1. A basic underlying condition for applying the methods developed here is that the samples are independently drawn. Are the samples arising from the following situations independent? Explain.

   (a) To study the effectiveness of a new sleeping pill the sleeping patterns of a number of participants in a study are monitored before taking the pill and then after taking the pill. Their mean daily number of hours of sleep is determined before and after taking the pill.

   (b) A standard final examination in financial accounting is administered to all students in both the day and evening sessions at the Kaunas School of Business. To study whether there is a difference in the mean scores on the exam for the day and evening session students the chairperson of the accounting department plans to take samples of the day and evening session test scores and study the results.

(c) A psychology project is being planned to study whatever differences exist between the IQ's of husbands and wives. Samples of men and their wives are to be chosen and their IQ levels determined and compared.

(d) The Proctor Company makes light bulbs at their Waverly and Turin plants. The quality control manager wants to determine whether there is a difference in the mean lifetimes of the bulbs made by the two plants and plans to take samples from the output of both and analyze the results.

2. The quality control manager of Lee Tires decided to test the null hypothesis $H_o : \mu_1 - \mu_2 = 0$ against the alternative hypothesis $H_a : \mu_1 - \mu_2 > 0$, where μ_1 and μ_2 are the mean lifetimes of the tires produced at the Wells and Jason City plants, respectively. The test is to be carried out at the 0.02 level of significance.

(a) Randomly chosen samples of 40 tires are to be chosen from the Wells and Jason City plants and their mean lifetimes determined. Describe the accept/reserve judgment and reject regions that emerge. Draw a diagram.

(b) The samples yielded $\overline{x}_1 = 25.7$ (in thousands of miles), $\overline{x}_2 = 24.2$, $s_1^2 = 5.8$, $s_2^2 = 6.1$. What do you conclude?

(c) What does your finding mean to Lee Tires?

3. The quality control manager of Lee Tires decided to test the null hypothesis $H_o : \mu_1 - \mu_2 = 0$ against the alternative hypothesis $H_a : \mu_1 - \mu_2 > 0$ on a regular basis. To help keep costs within budget he proposes to use samples of size 10 chosen randomly and independently from the Wells and Jason City plants. The testing is to be carried out at the 0.01 level.

(a) In order to proceed, what concerns must be addressed?

(b) Describe the accept/reserve judgment and reject regions that emerge. Draw a diagram.

(c) The samples yielded $\overline{x}_1 = 26.1$, $\overline{x}_2 = 24.1$, $s_1^2 = 5.9$, $s_2^2 = 6.3$. What do you conclude?

(d) What does your finding mean to Lee Tires?

4. Huxley College's mathematics placement exam is administered to freshman and transfer students at both its Dale and Richfield campuses. The mathematics department wants to test the null hypothesis that there is no difference between the mean placement exam scores on the two campuses against the alternative hypothesis that there is a difference at the 0.01 level. Samples of 36 test scores are to be chosen at random from the files of each of the campuses and analyzed.

   (a) Formulate null and alternative hypotheses.

   (b) What accept/reserve judgment and reject regions emerge? Draw a diagram.

   (c) Are there special concerns that must be addressed? Explain.

   (d) The samples drawn yielded the following results. Dale campus: $\overline{x}_1 = 74.6$, $s_1^2 = 15.2$; Richfield campus: $\overline{x}_2 = 70.2$, $s_2^2 = 12.3$. What do you conclude?

   (e) What does your finding mean to Huxley College?

5. A fierce competition has developed between the manufacturers of the Longlife and Neverdie batteries. The Xxon Company, which makes Longlife, was planning to run a series of television commercials claiming that the life of Longlife was unsurpassed by that of Neverdie. To be sure that is was on firm ground it planned to carry out a hypothesis test on the difference between the mean lifetimes of Longlife and Neverdie, $\mu_1 - \mu_2$. The test was to be carried out at the 0.01 level and samples of 13 batteries were to be chosen at random from stocks of Longlife and Neverdie batteries and analyzed.

   (a) Set up null and alternative hypotheses.

   (b) What is the basis for your alternative hypothesis? Explain.

   (c) What accept/reserve judgment and reject regions emerge? Draw a diagram.

   (d) The samples drawn yielded the following results. Longlife: $\overline{x}_1 = 42$ hours, $s_1^2 = 3.4$ hours; Neverdie: $\overline{x}_2 = 44$ hours, $s_2^2 = 14.2$ hours. What do you conclude?

(e) What does your finding mean to Xxon Company?

(f) The Xxon Company went ahead with its advertising campaign and the manufacturer of Neverdie brought false advertising charges against Xxon to the Fairness in Advertising Association (FAA).

A hearing was subsequently scheduled at which the following exchange took place between an FAA official and Xxon Company representative.

FAA: Since the mean life of Neverdie exceeds the mean life of Longlife in your test, I conclude that your commercial misrepresents the facts.

Xxon: No; this result does not imply that our claim is untrue. At the 0.01 level of significance the difference in sample means is not statistically significant. The significance level 0.01 is so small that the odds are 99 to 1 that our claim is true.

FAA: In carrying out your hypothesis test, isn't it required that the populations you are sampling from be normally distributed?

Xxon: No; it follows from the central limit theorem that the probability behavior of the difference in sample means does not depend on whether the populations being sampled from are normal.

FAA: I have been informed that a hypothesis test established that the standard deviations of the population you sampled from are unequal. What are the implications of this result for your study?

Xxon: It's not relevant to the study and conclusion reached.

Does the FAA official have cause to take issue with any of the Xxon representative's replies? If yes, what should be said?

6. A study of wage rates paid to student graduate assistants in public and private colleges in the northeast asserts that there is no difference in the mean hourly rate paid to graduate assistants. The test was conducted at the 0.05 level. The samples, which were chosen at random and independently, yielded the following data.

Public colleges: $n_1 = 36$, $\bar{x}_1 = \$15.75$, $s_1^2 = 11.20$.

Private colleges: $n_2 = 32$, $\bar{x}_2 = \$14.25$, $s_2^2 = 12.50$.

(a) Does the data obtained support the conclusion of the wage study? Explain.

(b) Suppose the alternative hypothesis had been that the mean hourly rate paid at the public colleges exceeded that paid by the private colleges; would the data obtained have supported this hypothesis? Explain.

(c) Could we now conclude that the mean hourly rate paid by the public colleges exceeds that paid by the private colleges? Explain.

(d) Knowing what we now know, could we now change the alternative hypothesis to the mean rate paid by the public colleges exceeds that paid by the private colleges and conclude this result? Explain.

(e) On hearing about the wage study and its conclusion a graduate assistant at Huxley College, a private school, noted that Huxley pays its graduate assistants \$18.50 per hour and asserted that this fact contradicts the conclusion reached in the study. Would you agree with this conclusion? Explain.

7. The Statistical Inspector General (SIG), a mathematical and statistical conservative, called into question the afore study of wage rates paid to student graduate assistants because of its unqualified use of sample variances as estimates for population variances even though the samples are large. How would you, as a member of the team of analysts who conducted the study, address the SIG's concerns? Explain.

8. Aaron Henry, who has long had an interest in baseball and statistics, decided to apply statistics to test the hypothesis that the end-season mean number of runs scored per game in the National League is the same as that for the American League at the 0.05 level. Samples of 6 and 7 games chosen at random and independently from the records of the leagues yielded the following values:

National League: 8, 12, 2, 5, 15, 18.

American League: 14, 8, 1, 9, 6, 17, 6.

(a) Formulate null and alternative hypotheses.

(b) What is the basis for your alternative hypothesis? Explain.

(c) What accept/reserve judgment and reject regions emerge? Draw a diagram.

(d) Are there special concerns that must be addressed? Explain.

(e) If the special concerns noted in (d) are satisfied, what would you conclude if you were Aaron?

(f) If the special concerns noted in (d) are not satisfied, what would you conclude if you were Aaron?

9. True to its name, Neverdie (Exercise 5) has launched a counterattack with extensive ads claiming that the life of Neverdie is longer than that of Longlife. To test this claim an independent testing agency chose 6 items at random from a list of battery operated items of different models and makes and operated them by using Neverdie batteries. Another 6 randomly chosen items were operated by using Longlife batteries. The following lifetime data (in hours) were obtained:

Neverdie: 53, 48.2, 38.6, 40.1, 54.2, 42.6

Longlife: 52.2, 47.4, 37.8, 39.4, 53.5, 41.8

(a) Formulate null and alternative hypotheses.

(b) Are there special concerns that must be addressed? Explain.

(c) Assuming that the concerns noted in (b) are satisfied, what conclusion do you reach after carrying out the test at the 0.05 level?

(d) What is the significance of your results to Neverdie?

10. The small sample hypothesis testing framework developed in this section requires that the standard deviations of the two populations in question be equal. In a number of situations it is realistic to assume that one population $P_2 = \{y_1, y_2, \ldots, y_n\}$ arises from the other $P_1 = \{x_1, x_2, \ldots, x_n\}$ by means of a "treatment" whose effect is to change each of the values in P_1 by a fixed amount c; that is, $y_1 = x_1 + c, \ldots, y_n = x_n + c$

. The value c is constant, but it may be positive, negative, or zero. Thus, for example, a chemical added to gasoline may raise by a fixed amount the mileage per gallon obtained on a given model car; material of a certain kind added as part of a blend to a fabric may add to its strength by a fixed amount; fertilizer added to farm land may enhance the yield obtained by a fixed amount.

(a) Show that if μ_1 and μ_2 are the means of P_1 and P_2, respectively, then $\mu_2 = \mu_1 + c$.

(b) Show that if σ_1^2 and σ_2^2 are the variances of P_1 and P_2, respectively, then $\sigma_1^2 = \sigma_2^2$.

That is, if "treatments" have additive effects, they change the means by an amount described by the additive effect, but the equality of variances is undisturbed.

11.9 ■ Matched Pairs' Mean Difference

With some tests of mean equality situations it is preferable to pair each value of one sample with a particular value of the other, as opposed to choosing the samples independently. As an illustration, consider the situation described in Example 4 of the preceding section.

Example 1 A Return to Carl's Fertilizer Problem

To determine whether a new fertilizer is more effective than the one in current use for growing corn Carl Cairns selected 5 test plots on his farm at random which he treated with the fertilizer in current use and another 5 plots at random which he treated with the new fertilizer. The following results (in bushels per acre) were obtained.

Current fertilizer: 70.5, 80.2, 90.3, 60.2, 85.4
New fertilizer: 73.2, 82.1, 93.1, 63.6, 87.1

The underlying significance level was taken at 0.05.

With respect to this framework the data yield $t = -0.33$, which is within the accept/reserve judgment region. We cannot reject the null hypothesis and conclude that the new fertilizer is more effective than the currently used one.

But is this the best experimental design for extracting information about the effectiveness of the new fertilizer relative to the one in current use? Would it not be better to select 5 plots at random, treat half of each plot with the currently used fertilizer and the other half with the new, and consider the results obtained from the pairings? This approach would more effectively screen out the effects of such extraneous factors as soil conditions, moisture, and weather and focus on the effects of the fertilizer.

The methods of the preceding section are not applicable to paired samples because the condition that the samples be independently chosen is violated.

Test Statistic for the Matched Pairs' Mean Difference

Given the two underlying populations P_1 and P_2 from which the samples are drawn (the populations of corn yields obtained from using the current and new fertilizers, in our case) we can circumvent this difficulty by introducing the population of differences P_d obtained by subtracting values y_i of P_2 from the values of x_i of P_1 and examining the behavior of the *t*-statistic

$$t = \frac{\bar{x}_d - \mu_d}{s_d / \sqrt{n}}, \quad \text{d.f.} = n - 1$$

where $\bar{x}_d$, μ_d, s_d, and *n* are the mean of the sample drawn from P_d, mean of P_d, standard deviation of the sample of paired differences, and sample size of the sample drawn from P_d, respectively.

$$\bar{x}_d = \frac{\sum (x_i - y_i)}{n},$$

where $x_i - y_i$ is the difference between the paired sample values obtained from P_1 and P_2.

$$\mu_d = \mu_1 - \mu_2,$$

where μ_1 and μ_2 are the means of P_1 and P_2. With respect to the null hypothesis, $\mu_d = 0$.

$$s_d = \frac{\sum (x_i - y_i - \bar{x}_d)^2}{n - 1}$$

The **underlying condition** for the application of this result is that the population of differences be normal, or approximately so. If the populations P_1 and P_2 from which P_d is obtained are normal, or approximately so, then P_d will also be normal, or approximately so.

Example 1 (continued)

Suppose that the resulting yields obtained from the currently used and new fertilizers were obtained from paired plots as summarized in Table 5. What conclusion can be drawn at the 0.05 level of significance?

Table 5

Pair	Yield x_i (current fert.)	Yield y_i (new fert.)
1	70.5	73.2
2	80.2	82.1
3	90.3	93.1
4	60.2	63.6
5	85.4	87.1

Let μ_1 and μ_2 denote the population means of the corn yields (bushels per acre) obtained from employing the current and new fertilizers, respectively. We have:

$$H_o : \mu_d = 0$$
$$H_a : \mu_d < 0$$
$$\alpha = 0.05$$

The accept/reserve judgment and reject regions for $\text{d.f.} = 4$ are shown in Figure 20.

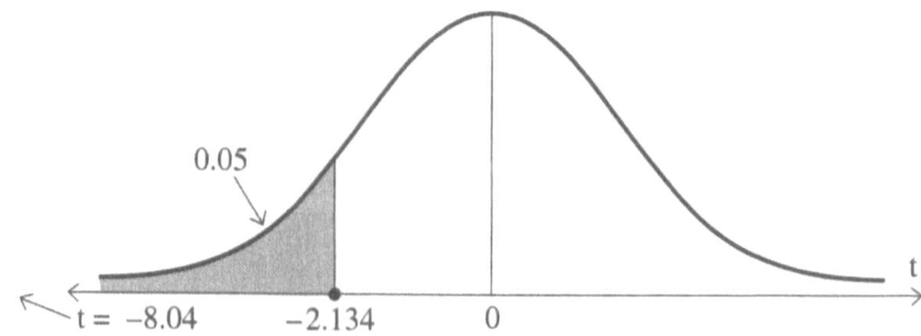

Figure 20

The computations for the value of the t-statistic come out of Table 6.

Table 6

x_i	y_i	$d_i = x_i - y_i$	$d_i - \bar{x}_d$ $(\bar{x}_d = -2.5)$	$(d_i + 2.5)^2$
70.5	73.2	-2.7	-0.2	0.04
80.2	82.1	-1.19	0.6	0.36
90.3	93.1	-2.8	-0.3	0.09
60.2	63.6	-3.4	-0.9	0.81
85.4	87.1	-1.7	0.8	0.64
		-12.5	0	1.94

$$\bar{x}_d = \frac{-12.5}{5} = -2.5, \quad s_d^2 = \frac{1.94}{4} = 0.484, \quad s_d = 0.696$$

$$t = \frac{\bar{x}_d - \mu_d}{s_d / \sqrt{n}} = \frac{-2.5 - 0}{0.696 / \sqrt{5}} = -8.04$$

Since $t = -8.04$ is less than the t-bound -2.134, we reject the null hypothesis of no difference and conclude that the new fertilizer is more effective than the one in current use for growing corn.

We should keep in mind that this analysis presupposes that the underlying populations of corn yields obtained by use of both fertilizers are approximately normal.

It is clear in this situation that by employing matched pair samples we were successful in blocking out extraneous factors and more sharply focusing on the effects of the fertilizers on crop yield.

When Should Matched Pair Samples Be Employed?

There is no sharp unequivocal answer to this question; it is a matter of judgment, but judgment refined through study and consideration of examples. The Lee Tires situation considered in the preceding section does not lend itself to matched pair samples. We have two independently operating plants and there is no clear advantage to the consideration of matched pairs or how such a matching would even be carried out.

Situations involving a comparison between one way of doing things (old fertilizer, for example) and a new way (new fertilizer) where we want to block out extraneous effects and focus on the two ways might be handled advantageously through use of matched pair samples, depending on circumstances.

Example 2 Mathematics Review Tapes

Ecap University has been giving thought to what could be done to help entering freshmen improve their mathematics level of preparation prior to their taking the required initial year of freshman mathematics. One suggestion being considered would have the students view a series of six one-hour mathematics review videotapes and work through the exercises in a workbook which accompanies the tapes. A faculty member would be available as a resource person. This suggestion met with a positive response, but its implementation on a large scale is formidable and expensive, with questions about its effectiveness.

Before implementing the program on a large scale it was decided to conduct a paired sample study. The admissions department determined a pool of entering freshmen who were willing to participate in the study. Twenty students were chosen at random from the pool and paired on the basis of similar academic profiles. One member of each pair took the videotape review program and the other did not. All took the same initial semester of mathematics with the same instructor. The differences in final exam test scores of the pairs was to be examined at the 0.05 level of significance.

Let μ_1 and μ_2 denote the means of the populations of final exam scores of the students who took the videotape review program prior to taking the semester of mathematics, and those who did not, respectively.

$$\mu_d = \mu_1 - \mu_2$$

The question of whether the videotape review program is of benefit in preparing students for their initial study of mathematics leads to the alternative hypothesis $H_a : H_d > 0$.

In summary, we have:

$$H_o : \mu_d = 0$$
$$H_a : \mu_d > 0$$
$$\alpha = 0.05$$

The accept/reserve judgment and reject regions for $\text{d.f.} = 9$ are shown in Figure 21.

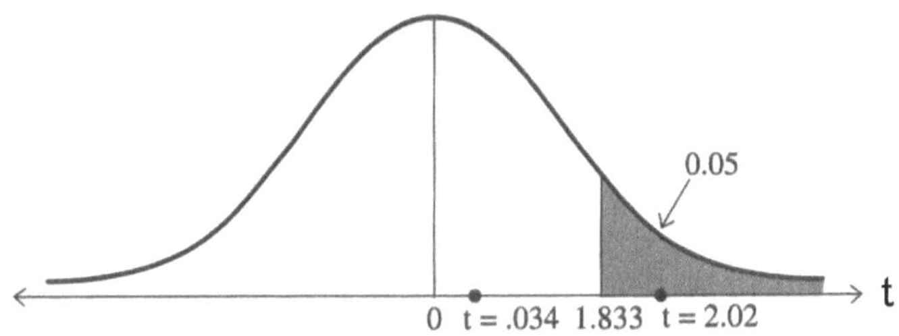

Figure 21

Table 7 shows the scores obtained by the pairs on the final exams and the computations needed to determine the *t*-statistic.

Table 7

Pair	Took video program x_i	Did not take video program y_i	$d_i = x_i - y_i$	$d_i - \bar{x}_d$ $(\bar{x}_d = 3)$	$(d_i - 3)^2$
1	85	76	9	6	36
2	72	64	8	5	25
3	66	67	-1	-4	16
4	59	61	-2	-5	25
5	67	65	2	-1	1
6	91	83	8	5	25
7	73	72	1	-2	4
8	72	74	-2	-5	25
9	76	77	-1	-5	16
10	88	80	8	5	25
			30	0	198

$$\bar{x}_d = \frac{30}{10} = 3, \quad s_d^2 = \frac{198}{9} = 22, \quad s_d = 4.69$$

$$t = \frac{\bar{x}_d - \mu_d}{s_d / \sqrt{n}} = \frac{3 - 0}{4.69 / \sqrt{10}} = 2.02$$

Since $t = 2.02$ exceeds the t-bound 1.833, we reject the null hypothesis of no difference and conclude that the videotape program is of benefit in preparing students for their initial study of mathematics.

We should keep in mind that this analysis **presupposes** that the **underlying populations of final exam grades are approximately normal.**

Example 3 Return to the Mathematics Review Situation

In Example 2 the criterion defining the effectiveness of the videotape review programs is

$$\mu_d = \mu_1 - \mu_2 > 0,$$

where μ_1 and μ_2 denote the means of the population of final exam scores of the students who took the videotape review programs prior to taking the semester of mathematics, and those who did not.

Suppose that it had been successfully argued that, considering the formidable problems inherent in implementing the videotape review program on a large scale and the expense involved, the program should be judged effective if the mean final exam score of those who took the review program was more than 2.5 points higher than those who did not take the program. What would the result have been if this had been the criterion for the videotape program's effectiveness?

The null and alternative hypotheses would have to be modified as follows:

$$H_o : \mu_d - 2.5 = 0 \text{ (or } \mu_d = 2.5\text{)}$$
$$H_a : \mu_d - 2.5 > 0 \text{ (or } \mu_d > 2.5\text{)}$$

The t-statistic is modified to

$$t = \frac{\overline{x}_d - 2.5 - \mu_d}{s_d / \sqrt{n}},$$

which in this case yields:

$$t = \frac{3 - 2.5 - 0}{4.69 / \sqrt{10}} = 0.34$$

Since $t = 0.34$ falls short of the t-bound 1.833 (see Figure 21) we cannot reject H_o and conclude that the videotape review program is of benefit in preparing students for their initial study of mathematics.

With respect to this criterion for the videotape review program's effectiveness, Ecap University would not have grounds for its implementation.

We should again keep in mind that this analysis presupposes that the underlying populations of final exam grades are approximately normal.

> If we have **reservations about the realism of this assumption**, then this approach is untenable. A non-parametric alternative which does not require population normality is discussed in Example 1 of Section 16.3.

EXERCISES

1. (a) In connection with the mathematics review situation described in Example 2, it was suggested that the statistical study of the effectiveness of the videotapes by conducted as follows: Choose 6 students at random from those who agreed to participate in the study, have them take the videotape review program, and place them into the first semester mathematics course. Independently, choose another 6 students at random from those who agreed to participate in the study and place them in the same course with the same instructor without having been exposed to the videotape review program. Conduct a hypothesis test on the null hypothesis $H_o: \mu_1 - \mu_2 = 0$ versus $H_a: \mu_1 - \mu_2 > 0$, where μ_1 and μ_2 are the mean final exam scores of the freshmen populations who had and did not have the videotape review, respectively.

   Which approach, this or the paired sample approach employed in Example 2, is preferable? Explain.

   (b) Suppose the approach described in (a) were employed to test $H_o: \mu_1 - \mu_2 = 0$ versus $H_a: \mu_1 - \mu_2 > 0$ with $\alpha = 0.05$ and the same scores were obtained by those who undertook the videotape review and those who did not. What conclusion would this approach yield? Explain.

   (c) What assumptions underlie the approach described in (a)?

(d) Are the assumptions underlying the approach described in (a) more demanding than those underlying the paired sample approach adopted in Example 2? Explain.

(e) It has been argued that the videotape review sessions may be viewed as a "treatment" whose effect is to change the final exam scores of all freshmen taking the initial course in mathematics by a fixed amount of c; it may not be clear whether c is positive, negative or zero, but it is constant. This being the case, it follows (see Section 11.7, Exercise 10) that the standard deviations of the populations of final exam grades of those who took the videotape review and those who did not are equal. Would you agree? Explain.

2. A new method for measuring the amount of lead in water has been developed. It promises to be easier and less costly to apply than the "standard method" in current use, but does it yield, on average, the same readings? To address this question eight samples were analyzed by the two methods. The following results (in parts per billion) were obtained:

New Method	25	4	12	6	3	42	10	1
Standard	24	3	11	8	3	41	8	3

(a) Set up null and alternative hypotheses.

(b) What is the basis for your alternative hypothesis?

(c) What accept/reserve judgment and reject regions emerge for the 0.05 significance level? Draw a diagram.

(d) Are there special conditions of concern in carrying out the test? Explain.

(e) Test H_o at the 0.05 level; what do you conclude?

(f) What does your finding mean for the new method?

3. Suppose the Neverdie versus Longlife test considered in Exercise 9 (page 298) of the preceding section were conducted by choosing 6 items at random from

a list of battery operated items and testing both batteries in each of the items. The same data, let us assume, are obtained for the battery lifetimes (hours), but now we have paired values per item as shown in Table 8.

Table 8

Item	1	2	3	4	5	6
Neverdie	53	48.2	38.6	40.1	54.2	42.6
Longlife	52.2	47.4	37.8	39.4	53.5	41.8

(a) Set up null and alternative hypotheses.

(b) Are there special conditions of concern in carrying out the test? Explain.

(c) Test H_o at the 0.05 level; what do you conclude?

(d) What does your finding mean to Neverdie?

(e) How do you explain the different conclusions obtained here and from Exercise 9 (page 363) of the preceding section?

(f) Which approach to the Neverdie versus Longlife test situation is preferable? Explain.

4. The productivity of eleven randomly chosen workers was measured for one week when they were under strict supervision and one week when they were not. The general belief is that workers generally work harder and produce more when they are under strict supervision. Table 9 gives the measures assigned to each worker for both approaches.

Table 9

Worker	John	Harry	Alice	Mary	Ralph	Mark	Lou	Bill	Andy	Jean	Paula
With sup.	46	83	59	76	51	58	66	87	64	54	53
Without sup.	47	75	54	82	37	47	68	79	60	57	49

At the 5% level of significance, does the data indicate that strict supervision results in a higher mean level of productivity?

5. To study the effect of temperature on skiing time fourteen randomly selected skiers were clocked on a course (in minutes) on two days, once when the temperature as 20°F and once when it was slightly above the freezing mark. Wind and cloud conditions were similar on the two days. The skiing times are shown in Table 10.

Table 10

Skiers	Esther	Joshua	May	Bruce	Carol	Bob	Darice	Barry	Ellen	Lou	Sharon	Heather	Rick	Donna
20°F	8.1	7.9	10	8.6	7.8	9.2	8.6	8.4	8.5	9.1	8.5	8.9	7.9	8.8
35°F	8.4	8.2	9.7	8.8	7.9	9.7	8.2	8.3	8.6	9.1	8.4	9.2	8.1	9.2

The feeling was that the increase in temperature might result in an increase in the mean skiing time. Test, at the 0.05 level of significance, the hypothesis that temperature has no affect on the mean skiing time against the alternative hypothesis that the warmer temperature adversely affects it.

6. Lister Hospital is considering two vendors, the Algis and Maria Companies, as sources for its medical supplies and drugs. Ten items were selected from a list of commonly used supplies and medications and their prices (in hundred dollars per unit) for the two vendors noted. The data are summarized in Table 11.

Table 11

Item	1	2	3	4	5	6	7	8	9	10
Algis	2	5.2	10	32	6	11	4	48	15	8
Maria	1.5	4.8	11.8	34	5.6	12.7	3.6	50.5	16.7	9

(a) Does that data indicate, at the 0.05 level, that the difference in prices charged by the two vendors is, on average, not the same? Explain.

(b) In deciding on which vendor to order supplies from what other factors would you take into account?

(c) Another approach to testing for mean price differences is to proceed as follows: Take a random sample of items of interest available from the Algis Company and determine their mean price $\overline{x}$; independently, take a random sample of items of interest available from the Maria Company and determine their mean price $\overline{y}$; by employing the statistic $\overline{x}-\overline{y}$ use the hypothesis testing procedure developed in Section 11.7 to test the null hypothesis of no difference between the price means of the two vendors against the alternative hypothesis that there is a difference.

Is this approach preferable to the paired-difference approach for obtaining relevant information? Explain.

(d) Suppose that the afore procedure were used and the same data, independently obtained, were determined for the Algis and Maria Companies.

Does the data indicate, at the 0.05 level, that there is a difference in the mean prices charged by the two vendors?

(e) Do the conclusions obtained from (a) and (d) agree? If not, which is more revealing? Explain.

(f) What **assumptions** underlie the tests carried out in (a) and (d)?

7. Ramunė's Gourmet Coffee launched an intensive advertising campaign for its Deluxe Blend and would like to evaluate the results. To do so sales figures (in thousands of dollars) were obtained from eight randomly chosen distributors for the company's product for the month preceding the campaign and the month following it. The data obtained are summarized in Table 12.

Table 12

Distributor	1	2	3	4	5	6	7	8
Before	40	82	72	92	101	68	55	92
After	55	100	98	126	136	90	77	118

For the advertising campaign to be judged a success the average sales of the distributors must have increased by more than $20,000.

(a) Does this approach to evaluating the advertising campaign's success require assumptions about the coffee market? Explain.

(b) Set up null and alternative hypotheses.

(c) At the 1% level of significance, would you conclude that the campaign was a success? Explain.

8. What **assumptions** underlie the tests carried out in Exercises 4, 5, and 7?

11.10 ■ The Family of *F* Distributions

This will introduce a fourth major family of probability distributions of statistics called *F* distributions.

The *F* distributions, or curves, are a two parameter family of curves depending on values assigned to parameters v_1 and v_2, called the numbers of degrees of freedom. The *F* curves are defined by

$$y = \frac{cF^{(v_1-2)/2}}{(v_2+v_1)^{(v_1+v^2)/2}} \tag{1}$$

where c is a constant depending on v_1 and v_2 appears only in the denominator of (1), it is usually referred to as the **denominator degrees of freedom**; v_1 is referred to as the **numerator degrees of freedom.**

The graphs of (1) vary considerably as v_1 and v_2 change. The graph of a "typical" *F* distribution where $v_1 > 2$, is shown in Figure 22.

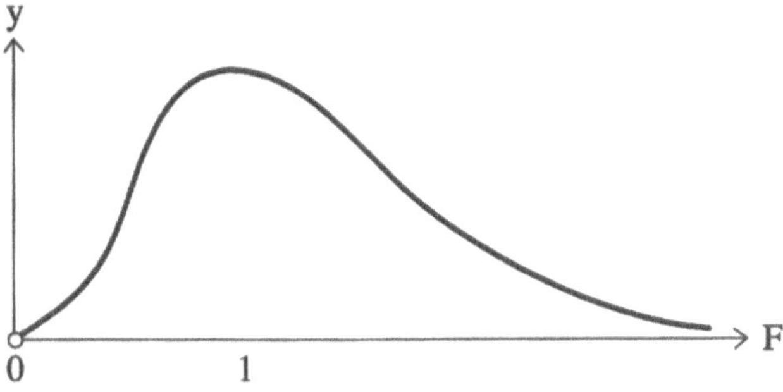

Figure 22

Properties of the F Distributions

1. Note from Figure 22 that F values are represented on the horizontal axis and y values are represented, as usual, by values on the vertical or y-axis.

2. The numbers of degrees of freedom v_1 and v_2 are positive integers, their values in an applied situation depending on the sample sizes in the problem.

3. F values are non-negative.

4. The F curves, like the χ^2 curves, are not symmetric about any vertical line.

5. As F increases, the F curves approach the horizontal axis, but do not intersect it.

6. For $v_1 > 2$, the F curves reach their highest point at:

$$F = \frac{v_1 - 2}{v_1} \cdot \frac{v_2}{v_2 + 2} \tag{2}$$

This value is less than 1 since both fractions in (2) are less than 1. As v_1 and v_2 increase, both fractions in (2) approach 1, so that for large v_1 and v_2 the maximum height of the *F* curves occurs around $F = 1$.

Before stating property 7, which involves the use of tables, we first turn to the use of tables to obtain area values.

Areas of Regions Defined by *F* Curves

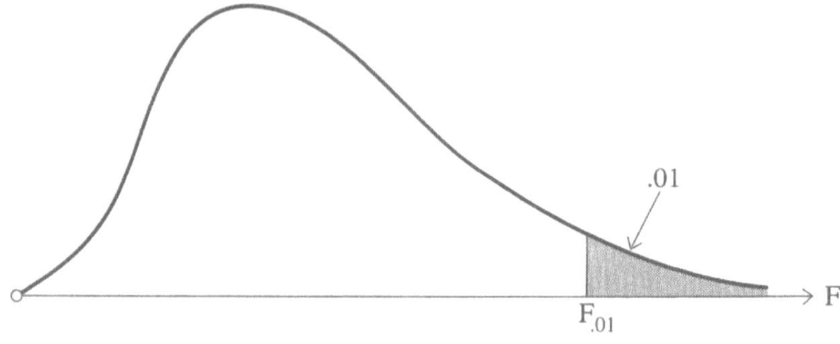

Figure 23

A short version of Table D (page 656) with $F_{.01}(v_1, v_2)$ values based on Figure 23 is given in Table 13. Thus, $F_{.01}(5,10) = 5.64$; for the *F* distribution determined by $v_1 = 5$ and $v_2 = 10$ the area of the tail to the right of $F_{0.01}(5,10) = 5.64$ is 0.01.

Table 13: $F_{.01}(v_1, v_2)$

$v_2 \backslash v_1$		Degrees of Freedom for Numerator												
		3	4	5	6	7	8	9	10	12	15	20	24	30
Degrees of Freedom for Denominator	4	16.7	16.0	15.5	15.2	15.0	14.8	14.7	14.5	14.2	14.2	14.0	13.9	13.8
	5	12.1	11.4	11.0	10.7	10.5	10.3	10.2	10.1	9.89	9.72	9.55	9.47	9.38
	6	9.78	9.15	8.75	8.47	8.26	8.10	7.98	7.87	7.72	7.56	7.40	7.31	7.23
	7	8.45	7.85	7.46	7.19	6.99	6.84	6.72	6.62	6.47	6.31	6.16	6.07	5.99
	8	7.59	7.01	6.63	6.37	6.18	6.03	5.91	5.81	5.67	5.52	5.36	5.28	5.20
	9	6.99	6.42	6.06	5.80	5.61	5.47	5.35	5.26	5.11	4.96	4.81	4.73	4.65
	10	6.55	5.99	5.64	5.39	5.20	5.06	4.94	4.85	4.71	4.56	4.41	4.33	4.25

For the comparable left tail with area 0.01, shown in Figure 24, the area of the region to the right of the defining bound in 0.99, so that the bound itself is denoted by $F_{.99}$. To obtain the area of the tail to the

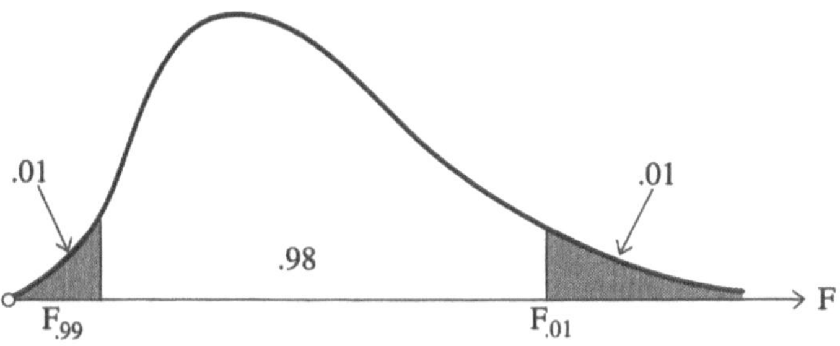

Figure 24

left of $F_{.99}$ we employ the following remarkable property 7.

7. $F_{1-\alpha}(v_1, v_2) = \dfrac{1}{F_\alpha(v_2, v_1)}$

Thus,

$$\begin{aligned} F_{.99}(5,10) &= \frac{1}{F_{.01}(10,5)} \\ &= \frac{1}{10.1} \\ &= 0.099 \end{aligned}$$

For the F distribution determined by $v_1 = 5$ and $v_2 = 10$ the area of the tail to the left of 0.099 is 0.01; equivalently, the area of the tail to the right 0.099 is 0.99.

We put the F distributions to work in the next section in developing a hypothesis test for the equality of variances.

EXERCISES

Values of $F_{.01}(v_1, v_2)$ and $F_{.05}(v_1, v_2)$ are given in Table D (page 656).

1. Find: (a) $F_{.01}(4,6)$, (b) $F_{.01}(6,4)$, (c) $F_{.01}(10,20)$, (d) $F_{.01}(20,10)$, (e) $F_{.99}(4,6)$, (f) $F_{.99}(6,4)$, (g) $F_{.99}(10,20)$

   For each situation sketch a diagram showing the region whose area is being obtained.

2. (a) Find: (i) $F_{.05}(8,12)$, (ii) $F_{.05}(12,15)$, (iii) $F_{.05}(24,10)$.

   (b) For each situation sketch a diagram of the region whose area is being determined.

   (c) Find (i) $F_{.95}(8,12)$, (ii) $F_{.95}(12,15)$, (iii) $F_{.95}(24,10)$. For each situation sketch a diagram of the region whose area is being determined.

11.11 ■ Equality of Variances and Standard Deviations

In considering hypothesis tests for equality of means where small samples are independently chosen from two populations (Section 11.7) we saw that one of the underlying conditions for carrying out such tests is that the standard deviations, or equivalently, variances, of the populations be equal. Other concerns lead us to testing for equality of variances as well.

Sampling Distribution of $F = \dfrac{s_1^2/\sigma_1^2}{s_2^2/\sigma_2^2}$

> It can be shown that if samples of size n_1 and n_2 are n_1 **randomly chosen from normally distributed populations** with variances σ_1^2 and σ_2^2, the sampling distribution of the variance ratio
>
> $$F = \frac{s_1^2/\sigma_1^2}{s_2^2/\sigma_2^2}$$
>
> may be approximated by the F distribution with $v_1 = n_1 - 1$ and $v_2 = n_2 - 1$ degrees of freedom.

As the notation itself suggests, s_1^2 and s_2^2 are the sample variances of the samples drawn from the populations with variances σ_1^2 and σ_2^2, respectively.

In testing hypotheses about the equality of variances or standard deviations the null hypothesis takes the form

$$H_o : \sigma_1^2 = \sigma_2^2, \quad \text{or} \quad \sigma_1^2 - \sigma_2^2 = 0,$$

with any of three possible alternative hypotheses:

$$H_a : \sigma_1^2 \neq \sigma_2^2$$
$$H_a : \sigma_1^2 > \sigma_2^2$$
$$H_a : \sigma_1^2 < \sigma_2^2$$

With respect to the null hypothesis of variance equality, the test statistic reduces to the ratio of the sample variances:

$$F = \frac{s_1^2}{s_2^2}, \quad v_1 = n_1 - 1, \ v_2 = n_2 - 1$$

Example 1 Another Look at Lee Tires

In Example 3 of Section 11.7 (page 353) we note that the quality control manager of Lee Tires wanted to carry out a small-sample test of equality of mean lifetimes of the tires made at the Wells and Jason City plants. One prerequisite for carrying out this test is that the population variances be equal. We undertake to test this hypothesis at the 0.10 level. We have,

$$H_o : \sigma_1^2 = \sigma_2^2$$
$$H_a : \sigma_1^2 \neq \sigma_2^2$$
$$\alpha = 0.10,$$

where σ_1^2 and σ_2^2 are the variances of the tire lifetimes of the tires made at the Wells and Jason City plants, respectively.

We will examine the behavior of $F = s_1^2 / s_2^2$ for the samples obtained, but we first set up the accept/reserve judgment and reject bounds for the 0.10 level. Eleven tires were selected at random from each of the plants, so that $v_1 = 10$ and $v_2 = 10$. Thus, we have:

$$F_{.05}(10,10) = 2.98, \qquad F_{.95}(10,10) = \frac{1}{2.98} = 0.34$$

This gives us the bounds shown in Figure 25.

From Example 3 of Section 11.7 we have that $s_1^2 = 5.8$ and $s_2^2 = 4.2$, so that:

$$F = \frac{s_1^2}{s_2^2} = \frac{5.8}{4.2} = 1.38$$

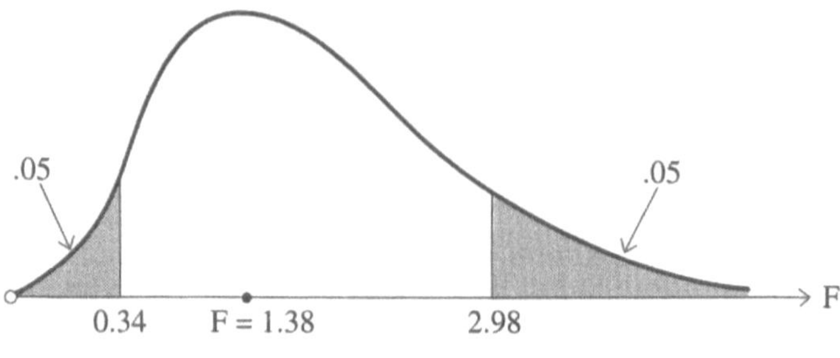

Figure 25

We have no basis for rejecting the null hypothesis of equality of variances, and we thus accept it since the quality control manager of Lee Tires needs an answer in terms of the equality of means testing situation of Example 3 of Section 11.7. The quality control manager has a green light to go ahead with that hypothesis test.

EXERCISES

1. Why not take $\sigma_1^2 < \sigma_2^2$ as the alternative hypothesis in Example 1?

2. In the competition between Longlife and Neverdie batteries described in Exercise 5 of Section 11.8 (page 360) randomly drawn samples of 13 Longlife and 13 Neverdie batteries yielded the sample variances $s_1^2 = 3.4$ and $s_2^2 = 14.2$, respectively. In the course of an exchange of views between a Fairness in Advertising Association official and Xxon Company representative (Xxon makes Longlife) it was asserted that the standard deviations of the populations being sampled from are unequal.

   (a) Test the hypothesis of standard deviation equality at the 0.10 level. Does your result support the aforenoted assertion? Explain.

   (b) Are any conditions required for this test to be carried out?

3. Eight ounce jars of the Deluxe Blend of Ramunė's Gourmet Coffee are filled at the company's Audre and Vytis processing plants. The variability in the fills from the two plants is of concern to the production manager. In a test random samples of 16 and 13 bottles drawn from the Audre and Vytis plants had variances $s_1^2 = 1.12$ and $s_2^2 = 1.41$, respectively.

   (a) Formulate null and alternative hypotheses.

(b) What is the basis for your alternative hypothesis?

(c) Test the null hypothesis at the 0.02 level.

(d) What are the implications of the test result for Ramunė's Gourmet Coffee?

(e) The production manager also wanted to carry out a test for equality of mean fills. Are the results of this test relevant to this undertaking? Explain.

(f) Are there any other concerns relevant to these tests? Explain.

4. At some point the production manager of Ramunė's Gourmet Coffee became concerned that the variability in fills at the Audre plant might significantly exceed that at the Vytis plant and decided to conduct a test.

(a) Formulate appropriate null and alternative hypotheses.

(b) Randomly drawn samples of 21 and 18 bottles from the Audre and Vytis plants had variances $s_1^2 = 1.14$ and $s_2^2 = 0.32$, respectively. What conclusion does this lead to at the 0.01 level?

(c) What are the implications of this test result for Ramuné's Gourmet Coffee? Explain.

(d) The production manager wanted to carry out a test for equality of means. Are the results of the F test relevant to this undertaking? Explain.

(e) What are the odds that we were wrong in rejecting the null hypothesis?

(f) Are there any other concerns relevant to these tests? Explain.

5. As we noted in Exercise 8 of Section 11.8 (page 362), Aaron Henry, who has long had an interest in baseball and statistics, decided to apply the latter to study the former. Is the variance of the end-season number of runs scored per game in the National League the same as that for the American League, is the question to be addressed at the 0.10 level. Samples of 6 and 7 games chosen at random from the records of the leagues yielded the following values:

National League: 8, 12, 2, 5, 15, 18

American League: 14, 8, 1, 9, 6, 17, 6

(a) Formulate null and alternative hypotheses.

(b) What is the basis for your alternative hypothesis?

(c) Test the null hypothesis at the 0.10 level.

(d) Assuming that the underlying run populations are normally distributed, what are the implications of the test result for Aaron Henry's desire to test the hypothesis that the end-season mean number of runs scored per game in the National League is the same as that for the American League, as per Exercise 8 of Section 11.7? Explain.

6. George and Amy have been in a friendly competition over their science grades for a number of years now. Randomly chosen samples of 13 of George's grades and 16 of Amy's grades had variances $s_1^2 = 25.2$ and $s_2^2 = 12.8$, respectively.

(a) Test the hypothesis that George's and Amy's science grades are equally consistent, that is, have the same variance, at the 0.02 level.

(b) It was later established that the population of George's science grades is U shaped and that the population of Amy's science grades is J shaped. Is this finding relevant to the conclusion obtained in (a)? Explain.

(c) George to Amy: "Since we also have the sample means of our science grades, let's test the hypothesis of no difference in the mean grades against the alternative hypothesis that the mean of my science grades exceeds yours." What would an appropriate (serious) reply be?

7. Albina Uris, a psychologist doing research at Huxley College's Behavioral Studies Institute, is planning to administer an IQ test to two populations who are similar in terms of mean age and age dispersion. It has been established that the ages of the two populations are approximately normal. The next step is to test the hypothesis of variance equality for the two populations of age values.

(a) Formulate appropriate null and alternative hypotheses.

(b) Randomly selected samples of 21 and 16 people from the population groups had variances $s_1^2 = 10.2$ and $s_2^2 = 13.1$, respectively. At the 0.10 level, does this support the null hypothesis of variance equality for the two populations? Explain.

(c) Professor Uris has calculated the mean ages of the two samples obtained. Is it feasible for her to carry out a hypothesis test of equal mean ages for the two populations in terms of the approach developed in Section 11.7? Explain.

8. Ecap University's Enrollment Planning Division has a large amount of promotion material printed. Price estimates are obtained for all printing jobs and the feeling has developed that the price variation among the larger printing establishments is greater than that among the small houses.

To put this view to the test random samples of 25 and 21 estimates on a standard promotion piece by the larger and smaller printing houses, respectively, were obtained from Enrollment Planning's files. These samples were found to have variances $s_1^2 = 152$ and $s_2^2 = 51$, respectively.

(a) Formulate appropriate null and alternative hypotheses.

(b) Test the null hypothesis at the 0.05 level.

(c) What are the implications of this test result for the Enrollment Planning Division?

(d) It was pointed out by the Director of Enrollment Planning that the difference in the mean price estimates for the large and small printing houses is of interest. This was followed by the suggestion that the hypothesis of no difference in the mean price estimates for the large and small houses be tested against the alternative hypothesis that the mean price estimate from the larger houses exceeds that of the smaller ones.

Is it feasible to act on this suggestion, **assuming** that the populations in question are **approximately normal**?

(e) Is it feasible to act on the aforenoted suggestion, assuming that the populations in question are J shaped and U shaped, respectively?

9. As described in Exercise 9 of Section 11.8 (page 363), the advertising war between the manufacturers of Longlife and Neverdie batteries continues. To test a claim made by Neverdie concerning the mean life of its batteries compared to that of Longlife we must test for equality of battery life variances. Randomly selected samples of 6 Neverdie and 6 Longlife batteries yielded the following lifetime values:

Neverdie: 53 48.2 38.6 40.1 54.2 42.6
Longlife: 52.2 47.4 37.8 39.4 53.5 41.8

(a) Formulate appropriate null and alternative hypotheses.

(b) A statistical observer argued that since the manufacturer of Neverdie is claiming that the life of Neverdie is longer than that of Longlife, the appropriate alternative hypothesis in this situation is $\sigma_1^2 > \sigma_2^2$, where σ_1^2 and σ_2^2 are the variances of the two populations of Neverdie and Longlife lifetimes, respectively. Do you agree with this point of view? Explain.

(c) Test the null hypothesis at the 0.02 level.

(d) **Assuming** that the underlying populations are **approximately normal**, what are the implications of your test result for the difference between means test proposed in Exercise 9 of Section 11.7?

10. Ball bearings are produced by the Henrik Company at its Riverside and Townsend plants. The rejection rate of the bearings from the Riverside plant has exceeded that of the Townsend plant in the last several tests that were conducted and the quality control staff suspects that the diameter variance of the bearings from the Riverside plant has come to exceed that from the Townsend plant. It was decided to conduct a hypothesis test to determine whether this suspicion had merit.

(a) Formulate appropriate null and alternative hypotheses.

(b) he quality control staff has decided to conduct the test at the 0.05 level. Randomly chosen samples of 16 and 19 bearings from the Riverside and Townsend plants had variances $s_1^2 = 0.00082$ and $s_2^2 = 0.00023$, respectively.

Assuming that the populations of ball bearing diameters are **approximately normal**, what conclusion do you draw from this data? Explain.

(c) What are the implications of the test results for the Henrik Company?

11. As we learned from Example 1 of Section 11.8 (page 355), Carl Cairns selected 5 test plots on his farm at random which he treated with the fertilizer currently used for growing corn. He selected another 5 plots at random which he treated with a new fertilizer. The question we address here is, are the corn yield variances the same for the two fertilizers? Carl wants to test the hypothesis of equal variances at the 0.10 level.

(a) Formulate appropriate null and alternative hypotheses.

(b) What is the basis for your alternative hypothesis?

(c) The following yields, in bushels per acre, were obtained:

Current fertilizer: 70.5, 80.2, 90.3, 60.2, 85.4

New fertilizer: 73.2, 82.1, 93.1, 63.6, 87.1

Assuming that the populations of corn yields for the two fertilizers are **approximately normal**, what conclusion do you draw from this data? Explain.

(d) What are the implications of this test result for Carl's situation as described in Example 4 of Section 11.8 (page 355)? Explain.

12. Consider two normally distributed populations with equal variances. Let us suppose that a hypothesis test for equality of variances is to be carried out versus the alternative hypothesis of unequal variances at the 0.02 level.

(a) Show that if samples of size 10 are chosen at random from the populations, one sample variance could be more than five times the other without leading to rejection of the null hypothesis of equal population variances.

(b) Show that if samples of sizes 5 are chosen at random from the populations, one sample variance could be almost sixteen times the other without leading to rejection of the null hypothesis of equal population variances.

(c) What are the counterparts of the aforenoted factors 5 and 16 for the 0.10 significance level?

(d) What do these results suggest concerning hypothesis testing for variance equality by means of small samples?

11.12 ■ *p*-values

In considering the case of Huxley College's SAT scores (Section 11.2, Example 4, page 322) we noted President Marx's claim that the mean combined SAT scores of Huxley's students is at least 1400. A Higher Education Board (HEB), set up to monitor claims made by colleges and universities, decided to investigate this claim, we observed. This led to the following:

$$H_o : \mu = 1400$$
$$H_a : \mu < 1400$$
$$\alpha = 0.01$$

Analysis, based on a large sample of SAT scores to be chosen at random from Huxley's files, leads to the standard normal curve and the accept/reserve judgment and reject bound -2.33 shown in Figure 26.

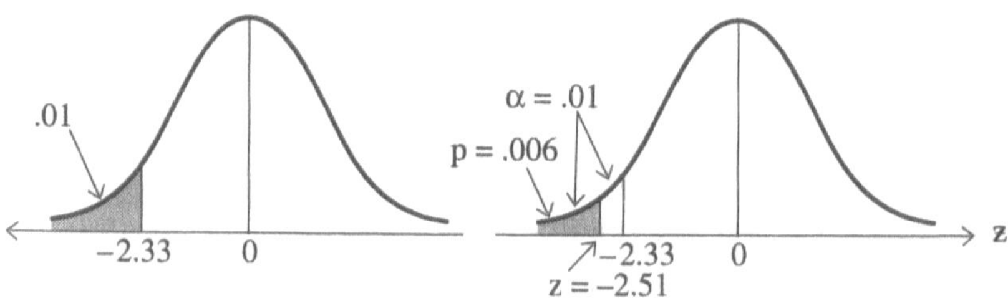

Figure 26 Figure 27

A large sample is chosen at random and the z-value of its mean $\bar{x}$ is found to be $z=-2.51$, let us say. The probability of obtaining a sample mean whose z-value is more extreme than -2.51, expressed by the area of the left tail defined by -2.51 (see Figure 27), is called the **p-value** of the observed sample statistic $\bar{x}$ or, more formally, the **significance probability of $\bar{x}$**. The area determined by $z=-2.51$ is 0.4940, so that $p=0.5000-0.4940=0.006$.

In terms of the level of significance $\alpha=0.01$ any p-value less than 0.01 would call for rejection of H_o since the z-value for such a p would be less than the rejection bound. The p-value, compared with the level of significance α, gives us a sense of how "decisively" the test statistic falls into the reject region. For $\bar{x}$ yielding $z=-2.34$, $p=0.0096$ which, compared to $\alpha=0.01$, puts us just over the edge into the reject region. For $\bar{x}$ yielding $z=-11.7$, the situation in which President Marx of Huxley College found himself, $p=0.0000$, which puts us deep into reject territory. For $\bar{x}$ yielding $z=-2.00$, $p=0.0228$, exceeding the level of significance $\alpha=0.01$, which puts us in the accept/reserve judgment region shown in Figure 28.

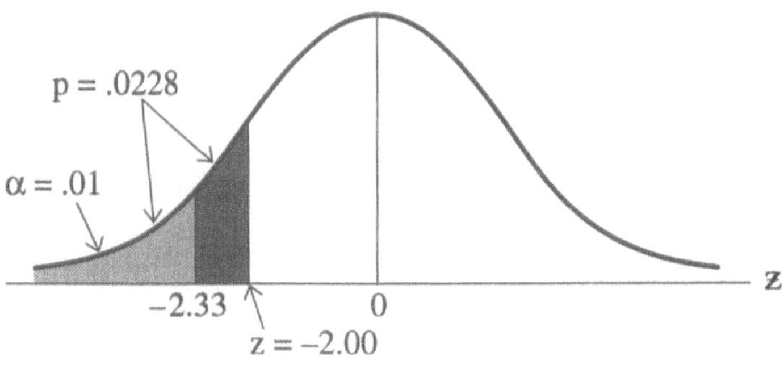

Figure 28

Example 1 Deadly Lights Cigarettes

The manufacturer of Deadly Lights cigarettes claims that Deadly Lights contains an average of 8 milligrams of tar per cigarette. As an analyst for the Regional Health Protection Society you have been asked to test this claim at the 0.02 level of significance. This situation, described in Exercise 20 of Section 11.2 (page 327), leads to the following:

$$H_o: \mu = 8$$
$$H_a: \mu > 8$$
$$\alpha = 0.02$$

A randomly chosen sample of 55 cigarettes yielded $\overline{x} = 8.5$ milligrams with $s = 1.6$ milligrams. The corresponding z-value is $z = 2.32$. We emerge with the situation shown in Figure 29.

The greater than type alternative hypothesis, $H_a : \mu > 8$, leads to a right tail with p-value given by the area of the region to the right of $z = 2.32$, which is 0.0102.

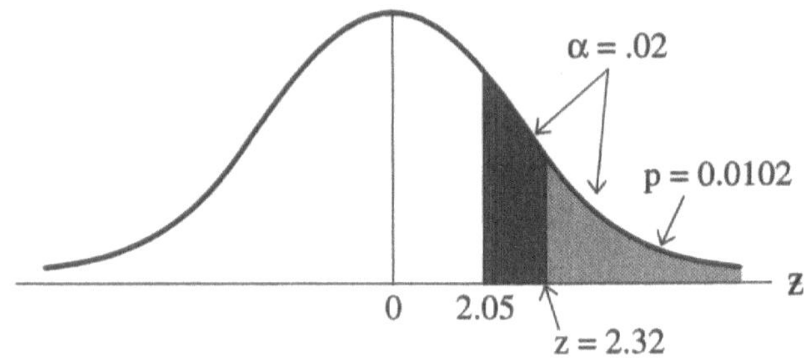

Figure 29

Example 2 Lee Tires

Lee Tires has for quite some time run advertisements claiming that its super-ride steel-belted radials have an average life of 60,000 miles. Just to be on the safe side management decided to carry out a test, which led to the following:

$$H_o : \mu = 60,000$$
$$H_a : \mu \neq 60,000$$
$$\alpha = 0.05$$

Analysis based on a large sample to be chosen at random from Lee Tires' stock leads to the standard normal curve and the accept/reserve judgment and reject bounds -1.96 and 1.96 shown in Figure 30.

When the alternative hypothesis is two-sided, as is the case here, a sufficiently large or sufficiently small value of the test statistic will lead to rejection of the null hypothesis. Before the test is carried out we do not know in which direction the test statistic will fall, and dividing the level of significance α between the two tails prepares us for either case. The p-value assigned to a situation reflects this state of affairs. With two-tailed tests half of the p-value is assigned to each tail or, put another way, the p-value is the sum of the halves assigned to the two tails.

In terms of the Lee Tires situation, if z_o is the *z*-value of the sample mean as shown in Figure 31, then the *p*-value for the test is given by the sum of the areas of the tails defined by z_o and $-z_o$; if $z_o = 2.14$, for example, then $p/2 = 0.0162$ and $p = 0.0324$.

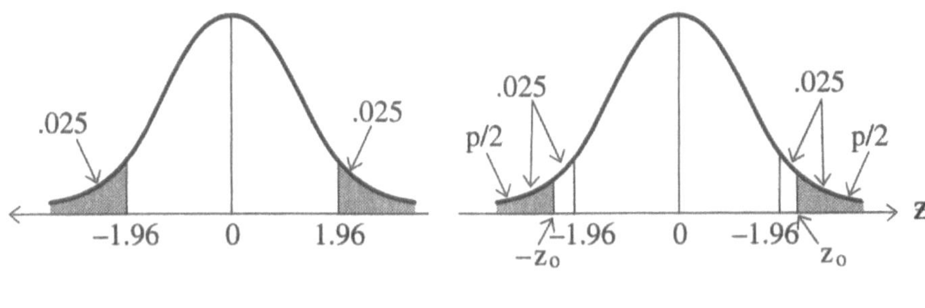

Figure 30 Figure 31

Calculating *p*-values: Summary

> The *p*-value associated with a hypothesis test depends on three factors:
>
> (1) The nature of the alternative hypothesis H_a; whether H_a is one-tailed or two-tailed;
>
> (2) The test statistic employed (*z*, *t*, *F*, etc.);
>
> (3) The magnitude of the computed test statistic that results from carrying out the test.

If H_a is one-sided, with one tail, and the test statistic has the value *K*, let us say, then the *p*-value is the area of the tail determined by *K*. If H_a is two-sided, with two tails, and the test statistic has the value *K*, let us say, then the *p*-value is twice the value obtained by finding the area of the tail determined by *K*.

p-values are easily determined when the standard normal curve is involved because area values for *z* are readily available from the standard normal curve area table. This is not the case for other test statistics and the calculation of *p*-values in these cases is more involved.

Reporting and Interpreting p-values

Many technical journals and software packages report p-values as part of their findings. If the p-value exceeds the level of significance α of the test, this means that the calculated value of the test statistic falls within the accept/reserve judgment region. If the p-value is less than α, then the calculated value of the test statistic falls within the reject region: how deeply it falls within the reject region is indicated by the magnitude of p; if p is "small" compared with α, the test statistic is deep in the reject region, whereas if p is "close" to α, the test statistic is over the edge in the reject region. Some have argued that p-values should be reported as standard procedure so that the reader can make an independent decision on whether or not to reject the null hypothesis.

While p-values may be a useful part of the information picture for decision makers at various levels, there is a slippery slope between enhancing the information picture and adjusting the level of significance after seeing the p-value to justify a "desired" conclusion. Our "friend" from the 0.01 Significance Level Society, encountered in Exercise 20(f) of Section 11.2 (page 327), and his similar-minded colleagues would welcome this slippery slope with glee.

EXERCISES

1. Determine the p-value for the following test situations:

   (a) $H_o: \mu = 5000, \quad H_a: \mu > 5000, \quad z = 1.41$;

   (b) $H_o: \mu = 5000, \quad H_a: \mu \neq 5000, \quad z = 1.41$;

   (c) $H_o: \mu = 70, \quad H_a: \mu < 70, \quad z = -1.89$;

   (d) $H_o: \mu = 300, \quad H_a: \mu < 300, \quad z = -1.50$;

   (e) $H_o: \mu = 300, \quad H_a: \mu \neq 300, \quad z = 1.50$;

   (f) (i) $H_o: \mu = 10,\ H_a: \mu > 10,\ n = 40,\ \overline{x} = 13.41,\ s = 8.43$;

   (ii) In terms of the p-value obtained, what action would be taken on H_o for 0.05? Explain.

   (g) (i) $H_o: \mu = 35,\ H_a: \mu \neq 35,\ n = 50,\ \overline{x} = 33.6,\ s = 4$;

(ii) In terms of the p-value obtained, what action would be taken on H_o for $\alpha = 0.02$? Explain.

2. In Exercise 22 of Section 11.2 (page 328) it was noted that a group of labor economists put out a report stating that the mean income of factory workers in the midwest does not exceed \$25,000. The Midwest Economic Development Association undertook to test this claim at the 0.01 significance level. A randomly drawn sample of size 35 yielded $\bar{x} = 25,300$ with $s = \$800$. Find the p-value for this test.

3. In Exercise 1 of Section 11.5 (page 339) it was noted that a medication which has become standard for the treatment of rheumatoid arthritis is effective in 75% of the cases treated. It is claimed that a newly developed, less expensive medication for this condition is at least as effective as the standard. To test this claim 100 patients were chosen at random and administered the new medication; it proved effective with 72 of the 100 patients.

(a) Find the p-value;

(b) In terms of the p-value obtained, what action would be taken on H_o for $\alpha = 0.02$? Explain.

11.13 ■ Hypothesis Testing and Confidence Intervals

Ramunė's Gourmet Coffee stands behind its claim that the mean weight of the 8 ounce jars of its Deluxe Blend is 8 ounces, we noted in Example 1 of Section 11.2 (page 314). This, we also noted, led to the following hypothesis test framework:

$$H_o : \mu = 8$$
$$H_a : \mu \neq 8$$
$$\alpha = 0.05$$

An envisioned large sample analysis leads to consideration of the z-statistic:

$$z = \frac{x - 8}{\sigma / \sqrt{n}}$$

The decision bounds ± 1.96 emerge in terms of the two-sided nature of the test and $\alpha = 0.05$ (see Figure 32).

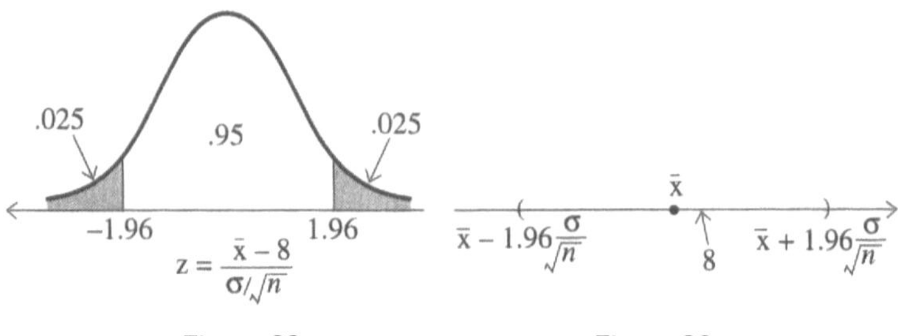

Figure 32 Figure 33

If *z* falls within ± 1.96, which holds with probability 0.95, that is,

$$\pm 1.96 < \frac{\bar{x} - 8}{\sigma / \sqrt{n}} < 1.96, \tag{1}$$

with probability 0.95, we accept H_o or reserve judgment. We reject H_o and accept H_a if *z* falls outside of this interval, under one of the tails defined by ± 1.96.

To solve (1) for 8, the hypothesized value of μ, we follow the same sequence of steps employed in Section 10.2 to isolate μ (page 258). This yields,

$$\bar{x} - 1.96\frac{\sigma}{\sqrt{n}} < 8 < \bar{x} + 1.96\frac{\sigma}{\sqrt{n}} \text{ with probability 0.95,}$$

the end-values of which,

$$\bar{x} - 1.96 < \frac{\sigma}{\sqrt{n}}, \quad \bar{x} + 1.96\frac{\sigma}{\sqrt{n}},$$

define the 95% confidence interval for μ (see Figure 33).

Thus, having the *z*-statistic fall within the accept/reserve judgment bounds ± 1.96 with probability 0.95 is equivalent to 8, the hypothesized value of μ under H_o, lying in the 95% confidence interval of μ (see Figures 32-33). Having the *z*-statistic fall within the reject region with probability 0.05, is equivalent to 8 lying outside the 95% confidence interval for μ, with probability 0.05.

More generally, the hypothesis test framework

$$H_o : \mu = \mu_o$$
$$H_a : \mu \neq \mu_o$$
$$\alpha = 0.05$$

is equivalent to the confidence interval framework

$$\bar{x} - 1.96 < \frac{\sigma}{\sqrt{n}} < \mu_o < \bar{x} + 1.96 \frac{\sigma}{\sqrt{n}} \text{ with probability 0.95}$$

in the case of large sample analysis, and the confidence interval framework

$$\bar{x} - t_{.025} \frac{s}{\sqrt{n}} < \mu_o < \bar{x} + t_{.025} \frac{s}{\sqrt{n}} \text{ with probability 0.95}$$

in the case of small sample analysis, assuming that the underlying population is approximately normal.

These hypothesis test and confidence interval frameworks are equivalent in the sense that the z (or t) statistic falling within the accept/reserve judgment bounds ± 1.96 (or $\pm t_{.025}$) is equivalent to μ_o falling within the 95% confidence interval for μ with probability 0.95.

If the significance level $\alpha = 0.05$ is changed, then the confidence level coefficients ± 1.96 and $\pm t_{.025}$ and the degree of confidence change accordingly.

It comes as no surprise that hypothesis testing and confidence interval analysis are related branches from the same statistical roots, namely, the sampling distribution of the underlying sample statistic. Whether one elects to follow the hypothesis testing or confidence interval branch in an applied situation depends on the circumstances surrounding the applied situation and judgment.

EXERCISES

1. Concerning the large sample hypothesis test framework $H_o : \mu = \mu_o$, $H_a : \mu \neq \mu_o$, state the confidence interval equivalent corresponding to the following α:

   (a) $\alpha = 0.10$,

   (b) $\alpha = 0.05$,

   (c) In what sense are the confidence interval and hypothesis test frameworks corresponding to $\alpha = 0.10$ equivalent?

2. In Sections 10.5 (page 287) and 11.5 (page 337) we noted the following central limit theorem for the sampling distribution of the sample proportion $p = x/n$. If a

sufficiently large sample of size n is chosen at random from a large population, the sampling distribution of $p = x/n$ is approximately normal with mean μ_p and standard deviation σ_p given by,

$$\mu_p = \pi, \quad \sigma_p = \sqrt{\frac{\pi(1-\pi)}{n}},$$

where π is the population proportion, n is the sample size, and x is the number of members of the sample who favor the option or situation in question.

(a) Concerning the hypothesis test framework $H_o : \pi = \pi_o$, $H_a : \pi \neq \pi_o$, $\alpha = 0.05$, state the confidence interval equivalent to this hypothesis test framework.

(b) In what sense are confidence interval and hypothesis test frameworks equivalent?

3. In Sections 10.8 (page 303) and 11.6 (page 342) we noted that the sampling distribution of

$$\chi^2 = \frac{(n-1)s^2}{\sigma^2}$$

is the χ^2 distribution with $n-1$ degrees of freedom, assuming that n is the sample size of a sample chosen at random from an approximately normal population with variance σ^2; s^2 is the sample variance.

(a) Concerning the hypothesis test framework $H_o : \sigma^2 = \sigma^2_o$, $H_a : \sigma^2 \neq \sigma^2_o$, $\alpha = 0.05$, state the confidence interval equivalent to this hypothesis test framework.

(b) In what sense are these confidence interval and hypothesis test frameworks equivalent?

11.14 ■ Robustness

As we have seen, the hypothesis tests considered require that **assumptions** of one sort or another be satisfied. Some of these **assumptions**, such as those requiring the normality of the population(s) involved and equality of population variances, are rather severe.

> An important question concerning applications is, to what extent can we relax the required assumptions and still apply the test? A test that remains reliable under strong modifications of the assumption on which it is based is said to be **robust**.

The test employing a chi-square variable to test the hypothesis $H_o: \sigma^2 = \sigma_o^2$ is an example of a test that lacks robustness with respect to the assumed normality of the underlying populations. Hoel, Port, and Stone [1; pp. 182-84] take up the case where the underlying population is distributed according to $y = xe^{-x}$, $x > 0$ (see Figure 34), where e is approximately 2.71828. They show that the accept/reserve judgment

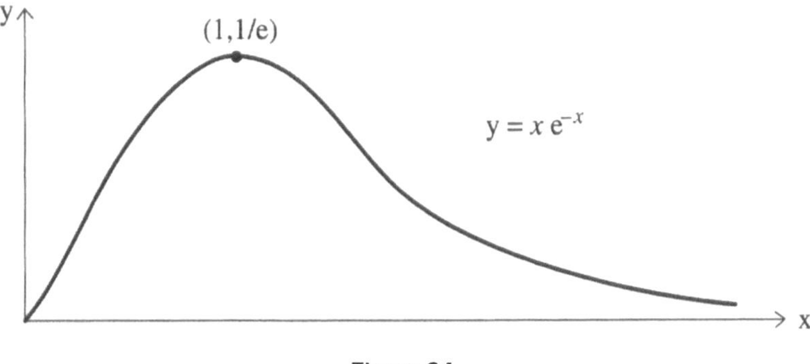

Figure 34

interval is approximately 1.58 times as long as the one obtained under the incorrect assumption of normality, so that the null hypothesis $H_o: \sigma^2 = \sigma_o^2$, versus the two-sided alternative, would be rejected much more often than need be.

What do we do when we are confronted by an applied situation where the conditions required for the use of the appropriate test are not satisfied or it not clear whether they are "reasonably well" satisfied? There are a class of hypothesis tests, called nonparametric tests, which do not require conditions on the underlying population(s) or its parameters and it would make good sense to look for one that is appropriate for the situation at hand. A number of nonparametric tests are discussed in Chapter 16.

REFERENCE

1. P. Hoel, S. Port, C. Stone, *Introduction to Statistical Theory* (Boston: Houghton Mifflin, 1971).

EXERCISES

1. (a) What do we mean when we say that a hypothesis test is robust?

   (b) Why is the robustness of a test an important issue?

11.15 ▪ Hypothesis Testing: A Decision Making Procedure?

One view of examination grades, be they course examination, SAT, GMAT, GRE, New York State Regents, bar, or some other type of examination grade(s), is that they in their own right are decision makers. If you obtain this or that examination grade, then your fate is determined; you pass the course or not, you are admitted to the college of your choice or not, you are admitted to the bar or not.

Another view of examination grades is that they are part of the information picture which assists the decision maker, whoever or whatever that may be, reach a decision on the situation at hand. Your course examination grades are part of the information picture which help your instructor make a decision about your course grade, but your instructor is the decision maker who will use the examination grades and other factors such as level of class participation, your grades compared to those of others in the class, homework, and "extra effort" to arrive at a grade decision. Both approaches are employed.

The same might be said about hypothesis testing in particular and statistical inference in general in connection with decision making. Does the outcome of the test determine the decision when it comes to the implementation follow-up in the next stage, or does the decision maker use the result of the test as part of the information picture to help him arrive at a decision? In many cases, it will most certainly be the latter. If, for example, the test concerning Ramunė's Gourmet Coffee's claim that the mean weight of the 8 ounce jars of its Deluxe Blend is 8 ounces yielded a test statistic which fell into the reject region (Example 1, Section 11.2), it is most unlikely that management would act on this by immediately ordering production stopped and the plant machinery adjusted or overhauled. Management might look at the *p*-value as part of the information picture, weigh the cost of shutting down production against the cost of not doing so, and possibly carry out another hypothesis test to see if consistent results are obtained. Shutting down production is an option that might be chosen, but it is unlikely to be the immediate reaction.

Before carrying out a *t*-test on the equality of population means a researcher conducts an *F*-test on the equality of population variances which leads to the conclusion that they are not equal, let us say. Now what? There is a strong red light which says stop, indications are that a most important prerequisite for application of the *t*-test is not satisfied. Drivers have been known to ignore red lights, but the cost of doing so here is that the conclusion of the *t*-test would be seriously open to question. The circumstances of this situation strongly favor not carrying out the *t*-test, which requires equality of variances, and seeking another test which does not require this condition.

11.16 ■ Strings Attached vs. Robustness

The strings attached to theorems are a fundamental concern not only in the application of statistics but, more generally, all of applied mathematics. Robustness has to do with the degree to which we can stray from the strings attached to a theorem in an application setting and still be able to invoke its valid conclusion. In brief, how "close" to satisfying the strings attached is "good enough".

Two of the robustness issues that must be faced in statistical applications concern the robustness of random sampling (see Achieving Random Sampling in Practice, Sec. 2.3; Deviation in Practice from the Requirements of Theory, Sec. 2.4; Random Selections in Statistical Analysis; Sec. 6.6) and the normality of random variables and populations.

SELF-TESTS FOR PART THREE

Allow 90 or so minutes for each self-test. Go over each one before going on to the next.

Self-Test I

1. In connection with hypothesis testing,

   (a) When is a Type I error committed? Explain.

   (b) When is a Type II error committed? Explain.

2. In a random sample of 300 eligible voters, 180 were for Randell W. Jones for dog catcher. If we estimate the true proportion of voters who are for Jones for dog catcher as being 0.60, what could we say with probability 0.95 about the size of the error of the estimate?

   Null and alternative hypotheses are stated for Exercises 3-10. Also given are the sample size(s) and level of significance. State the test statistic to be examined, the accept/reserve judgment and reject regions, the probability of committing a type I error, when type I and II errors would occur, and condition(s) that must be imposed on the population(s) for the test to be viable.

3. $H_o: \mu = 800,\ H_a: \mu \neq 800,\ n = 36,\ \alpha = 0.05$.

4. $H_o: \mu = 60,\ H_a: \mu > 60,\ n = 10,\ \alpha = 0.05$.

5. $H_o: \pi = 0.40,\ H_a: \pi \neq 0.40,\ n = 300,\ \alpha = 0.01$.

6. $H_o: \sigma = 3,\ H_a: \sigma < 3,\ n = 10, \alpha = 0.01$.

7. $H_o: \mu_1 - \mu_2 = 0,\ H_a: \mu_1 - \mu_2 \neq 0,\ n_1 = 32,\ n_2 = 35,\ \alpha = 0.02$.

8. $H_o: \mu_1 - \mu_2 = 0,\ H_a: \mu_1 - \mu_2 < 0,\ n_1 = 8,\ n_2 = 11,\ \alpha = 0.05$.

9. $H_o: \sigma_1 - \sigma_2 = 0,\ H_a: \sigma_1 - \sigma_2 < 0,\ n_1 = 12,\ n_2 = 10,\ \alpha = 0.01$.

10. $H_o: \sigma_1 - \sigma_2 = 0,\ H_a: \sigma_1 - \sigma_2 \neq 0,\ n_1 = 13,\ n_2 = 9,\ \alpha = 0.01$.

Self-Test 2

1. (a) State the 90% confidence interval for the population mean μ, where it is understood that the sample drawn is large.

   (b) How is this confidence interval interpreted in relative frequency terms?

   (c) The 90% confidence interval for a specific estimation problem is found to be $25 < \mu < 30$, with probability 0.90.

   Does the population mean lie in this interval or not? Explain. How is the probability value of 0.90 to be interpreted in the context of this problem? Explain.

   (d) A sample size of 10 is to be chosen from the population being studied. State the 90% confidence interval for the mean of this population.

2. Arthur Levey, the quality control supervisor for Ajax Electronics, wants to estimate the mean running time of its T54 camcorders to the point at which a breakdown occurs. The running time is assumed to be approximately normally distributed. A random sample of 5 camcorders yielded running times of 310, 330, 300, 290 and 280 hours.

   (a) In constructing a 98% confidence interval for the mean running time of the camcorders should a normal or t distribution be employed? Explain.

   (b) Find 98% confidence limits for the mean running time of the camcorders.

   (c) How is the result obtained in answer to (b) to be interpreted? Explain.

3. An independent testing agency has been contracted to determine whether there is any difference in gasoline mileage output of two different gasolines on the same

model automobile. Super premium A was tested on 200 cars and produced a sample average of 18.5 miles per gallon with a sample standard deviation of 4.6 miles per gallon. Premium B was tested on a sample of 100 cars and produced a sample average of 19.34 miles per gallon with a sample standard deviation of 5.2 miles per gallon.

At the 0.05 level of significance, is there evidence of a difference in performance of the two gasolines? Explain.

4. Marla Jennings, a statistical investigator, plans to test the null hypothesis $H_o : \sigma = 10$ against $H_a : \sigma \neq 10$ at the 0.05 level. A sample size 15 is to be used.

   (a) Are there underlying conditions that Ms. Jennings should give thought to before undertaking the test?

   (b) What decision criterion emerges for accepting/reserve judgment or rejecting the null hypothesis?

5. To estimate the average length of service of its faculty Ecap University took a random sample of 50 faculty and found that they have an average of 10 years of service with a standard deviation of 5 years. If 10 years is taken as an estimate of the number of years of service of the faculty, what is the error in this estimate with a probability of 0.98?

6. A random sample of 12 specimens of a certain material is tested for tensile strength. The variance computed from these data is 3.

   (a) Construct the 95% confidence interval for the population variance.

   (b) Construct the 90% confidence interval.

   (c) What assumption underlies the construction of these confidence intervals?

7. The Bax Company claims that the mean lifetime of its batteries is 500 hours. As a staff member of a consumer protection organization you have been assigned the task of testing this claim at the 0.05 level.

   (a) What alternative hypothesis would be appropriate? Explain.

(b) Your next step is to choose a sample of batteries at random from the manufacturer's output and determine their lifetimes. Should a large or small sample be chosen? On what basis would you make this decision? Explain.

(c) Suppose that a random sample of 81 batteries was found to have a mean lifetime of 450 hours with a standard deviation of 70 hours. Would this tend to support or contradict the manufacturer's claim? Explain.

(d) Suppose that a random sample of 16 batteries yielded the afore results. Would this tend to support or contradict the manufacturer's claim? Explain.

Self-Test 3

1. Much of the business done by the Valenti Appliance Discount Company is generated by repeat customers who were satisfied with the service they received and the quality of the appliances they purchased. To obtain a sense of what proportion of the Company's customers had appliances which needed repairs during the first year after purchase questionnaires were sent to 480 customers selected at random from the Company's files. Of the 480, 408 responded, and of those 150 replied that no repair problems had arisen in the first year of their appliance's life.

   (a) Determine a 95% confidence interval for the proportion of the Valenti Company's customers who expressed satisfaction with the service and appliances they obtained.

   (b) How is the result obtained in answer to (a) to be interpreted?

   (c) Not all of the customers sent questionnaires responded. Does this fact call into question the reliability of the confidence interval obtained in answer to (a)? Explain.

2. The Ardley Company claims that the mean lifetime of its Super-100 bulbs is 2000 hours. A consumer testing service plans to test this claim at the 0.05 level of significance.

(a) State the null hypothesis.

(b) What alternative hypothesis would be appropriate? Explain.

(c) What is the probability of a type I error in this situation? Explain.

How could a type I error occur? Explain.

(d) A random sample of 60 bulbs is found to have a mean lifetime of 1850 hours with a standard deviation of 200 hours. Does this evidence tend to support or refute the manufacturer's claim? Explain.

(e) Suppose that a random sample of 10 bulbs yielded the aforenoted results. Does this evidence tend to support or refute the manufacturer's claim? Explain.

(f) Are there any conditions that must be met for us to be able to draw a random sample of 10 bulbs in the first place and use the results obtained? Explain.

3. Huxley College's Student Council wishes to estimate the average amount paid for books by students last year. The standard deviation of the amount paid is believed to be $30, based on previous studies of a similar kind. The Council wishes to be "95% certain" that the sample mean will differ from the population mean by no more than $10. What is the smallest sample size that can be used for this study?

4. From a preliminary study of the length of telephone calls the James County Telephone Company was led to the following null and alternative hypotheses:

$H_o : \mu = 4$ (minutes)
$H_a : \mu \neq 4$
$\alpha = 0.02$

A random sample of 100 calls yielded a mean of 3.4 minutes with a standard deviation of 2.8 minutes. Determine the probability of a type II error if

(a) The population mean μ is 4.5 minutes;

(b) The population mean μ is 3 minutes.

5. The Atwood Institute, a management training company, runs two types of seminar programs for middle management level managers. At the conclusion of the training seminars both groups are given the same examination. Robert Nielsen, the program director, wants to test whether or not the mean test scores are the same for the two groups. Two groups, one consisting of 16 managers and the other of 13, were randomly and independently chosen and put through the programs. The first group had a mean score of 85 with a variance of 70.3; the second had a mean score of 91 with a variance of 90.1. It has been established that the underlying populations of test scores are normally distributed.

   (a) What other condition must be satisfied in order for us to conduct a test for equality of means?

   (b) Test for this condition at the 0.10 level.

   (c) Do we have a green light to undertake a test for equality of means? Explain.

   (d) Assuming that the answer to (c) is yes, test for equality of means at the 0.05 level.

   (e) What conclusion do you reach?

6. Determine the p-value for the following test situations:

   (a) $H_o : \mu = 350,\ H_a : \mu > 350,\ z = 1.72$.

   (b) $H_a : \mu = 200,\ H_a : \mu \neq 200,\ z = 2.00$.

Self-Test 4

1. To test whether a newly developed wheat hybrid, WH-100, gives larger yields than an older variety, WH-50, the Paulius Agricultural Institute selected 9 test plots at random which they planted with WH-50 and, independently, another 9 test plots at random which they planted with WH-100. The following results (in bushels per acre) were obtained:

   WH-50: 75, 54, 62, 87, 82, 71, 88, 66, 59

WH-100: 77, 57, 64, 89, 84, 73, 91, 67, 60

(a) Formulate null and alternative hypotheses.

(b) What is the basis for your alternative hypothesis?

(c) It has been established that the underlying populations of wheat yield values are normally distributed. What other condition must be satisfied in order for us to conduct this hypothesis test?

(d) Test for this condition at the 0.10 level.

(e) Do we have a green light to proceed? Explain.

(f) Assuming that we have a green light to proceed, carry out the hypothesis test for equality of means at the 0.05 level.

(g) What conclusion are you led to?

(h) Instead of proceeding as described, would it be preferable to generate paired yields obtained by selecting 9 plots at random, planting half of each plot with WH-50 and the other half with WH-100, and consider the results obtained from the pairings?

(i) Suppose the study were carried out along the lines indicated in (h) and that 9 plots were selected at random and halved, with WH-50 being planted on one of the halves and WH-100 being planted on the other. Suppose further that the same results (in bushels per acre) were obtained from the paired halves, which are summarized in Table 1.

Table 1

Plot	1	2	3	4	5	6	7	8	9
WH-50	75	54	62	87	82	71	88	66	59
WH-100	77	57	64	89	84	73	91	67	60

Let μ_1 and μ_2 denote the population means of wheat yields (bushels per acre) obtained from WH-50 and WH-100, respectively. Test $H_o : \mu_d = \mu_1 - \mu_2 = 0$ against $H_a : \mu_d < 0$ at the 0.05 level.

(j) How does the result obtained in (i) compare with the one noted in (g)?

(k) How might this be explained?

2. (a) What do we mean when we say that an hypothesis test is robust?

(b) Why is the robustness of a test an important issue?

PART FOUR

LINEAR REGRESSION AND CORRELATION

- Linear Regression
- Linear Correlation

12

Linear Regression

12.1 ■ Problems of Prediction

One of the most important problems facing business people, social, life and physical scientists, economists and, to one extent or another, us all, is that of predicting the behavior of a variable of interest. The marketing manager of a firm, for example, might seek to predict the sales volume of the new line of women's clothing, an economist might seek to predict the level of foreign trade in the coming year or a physicist might find it necessary to determine the pressure of a gas. In all situations of this sort we are dealing with a variable of interest, called a **dependent variable** (sales volume, foreign trade level, pressure), whose values continually fluctuate. Sales level records for the last 60 months might well show 60 different values and foreign trade levels have been jumping all over the place over the last decade, to take two illustrations.

If the predictions we are seeking to make about a dependent variable are to be reliable, then we must obtain an understanding of what accounts for their fluctuation. The first step in achieving this understanding is in deciding which variables play a "significant" role in determining the value of the dependent variable. Such variables are called **independent variables.** In the sales volume problem the advertising expenditure for the product, advertising expenditures of competing firms for similar products, the state of the economy, and consumer taste emerge as independent variables to be considered. In studying the pressure of the gas in question the physicist would be led to consider the independent variables gas temperature and gas volume.

In general a number of independent variables emerge for consideration, but sometimes, under appropriate conditions, it is possible to realistically focus on one of them. We then seek to express the dependent variable in terms of that one independent variable.

Function Relationships

Sometimes it is possible to obtain a relationship which states the dependent variable in terms of the independent variable in such a way that for any specified value of the independent variable an unique value of the dependent variable is determined. Such a relationship is called a **function.** Many situations in the sciences lend themselves to description in terms of function relationships.

Concerning the pressure p exerted by a gas in terms of the volume v of the gas (the independent variable), Boyle's law says that

$$p = \frac{k}{v},$$

where k is a constant whose value depends on the gas in question, provided that the temperature t of the gas (a second independent variable) is not changing. If, for example, the pressure (in suitable units) of a mixture of gasoline and air in a motorcycle engine is given by

$$p = \frac{245.2}{v},$$

then the pressure p when the volume v of the cylinder is 22.1 milliliters at the end of the compression stroke is given by

$$p = \frac{245.2}{22.1} = 11.1,$$

provided that the temperature of the gas in the cylinder has remained constant or has varied very little.

Statistical Relationships

Function relationships allow us to determine a specific value of the dependent variable for a specified value of the independent variable. Often there is a relationship between two variables y and x, but it is not as unequivocal as a precise function relationship. The relationship between the advertising expenditure x for a product and its sales volume y is not precise in the function sense. The same advertising expenditure will not always bring the same sales volume, but larger advertising expenditures do yield larger average sales volume figures.

Our objective in this kind of situation is to develop what is termed a **regression equation** which allows us to estimate the mean value of y for a specified value of x. The simplest possible regression equation is the linear equation of the form

$$\hat{y} = a + bx$$

where a and b are constants and $\hat{y}$ (read y hat) is the estimated mean value of the y's for a given value of x.

When a strong **linear relationship** between x and y is indicated the appropriate fit to the (x, y) data points is a suitably defined linear equation, or line $\hat{y} = a + bx$. Our specific focus is on this situation. Concerning the sales volume problem, of particular interest is the question, under what conditions can we obtain a **regression line** $\hat{y} = a + bx$ to predict the mean sales volume in terms of the level of advertising expenditure?

12.2 ■ Scatter Diagrams

To obtain a sense of whether a relationship between a dependent and chosen independent variable is indicated, we obtain data for these variables, plot the points, and examine the graph. Such a graph is called a **scatter diagram**.

Example 1 The Daniel Company

The marketing manager of the Daniel Company, a new entry in the fine baked goods market, wants to predict sales volume on the basis of advertising expenditure. To establish a quantitative relationship between these variables data were obtained from reports for eight monthly periods chosen at random from the company's files as shown in Table 1. Advertising expenditure, denoted by x, is given in thousands of dollars and associated sales volume, denoted by y, is given in hundreds of thousands of dollars. Plot and comment on the scatter diagram for these data points.

Table 1

Advertising (x)	6	4	9	14	12	16	10	21
Sales volume (y)	15	17	14	27	21	28	22	24

The scatter diagram is shown in Figure 1.

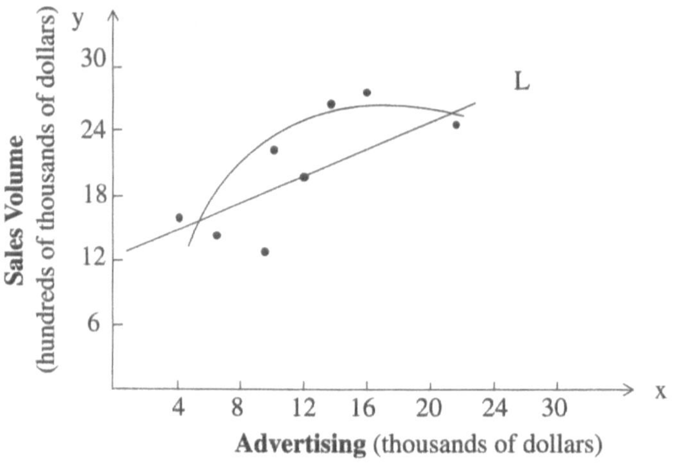

Figure 1

Many curves offer a satisfactory fit to the scatter plot shown in Figure 1, including line L. A line is a desirable fit because of its simplicity, and if it yields a "reasonably" good fit, it's to our advantage to go with a line.

The line L has positive slope, in line with the property that as advertising expenditure increases, generally, so does sales volume. Such a relationship between two variables is called a **direct** relationship and the line describing it is called **direct linear**.

Example 2 Student Absentee Rates vs. Mean Temperature

A school principal is interested in finding out how, if at all, the student absentee rate relates to the mean temperature of the day. The mean Fahrenheit temperature, x, and number of students absent, y, were obtained for a randomly chosen sample of 9 school days, as shown in Table 2. Draw a scatter diagram for these data and comment on the apparent relationship between the variables.

Table 2

x	12	17	23	27	39	49	61	68	74
y	10	7	9	8	5	6	7	8	12

The scatter diagram is shown in Figure 2.

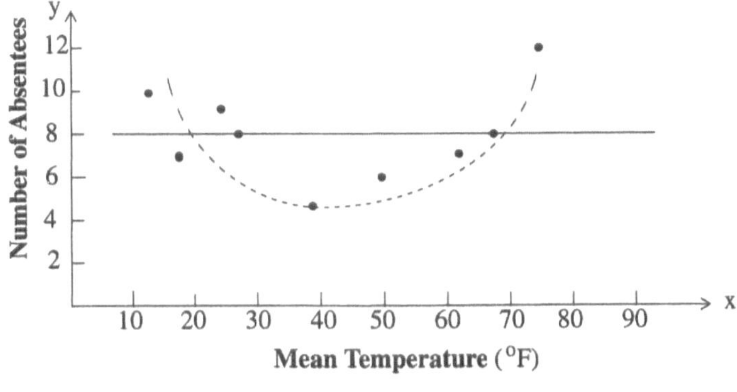

Figure 2

While there does appear to be a marked relationship between *x* and *y*, the underlying nature of this relationship is clearly non-linear. Its nature is suggested by the curve shown in Figure 2.

Example 3 Demand vs. Price

The management of the Tons of Toys Company wants to estimate the weekly demand for its cuddly teddy bear for different price levels. They collected data on quantity sold, *y*, in hundreds per week, and price, *x*, in dollars, for a randomly chosen sample of 7 price levels at stores in the northeast, shown in Table 3. Construct the scatter diagram for these data and comment on the resulting relationship between the variables.

Table 3

x	18	19	17.5	16.5	19.5	17	18.5
y	13	9	12.5	15	10	12	11.5

The scatter diagram is shown in Figure 3.

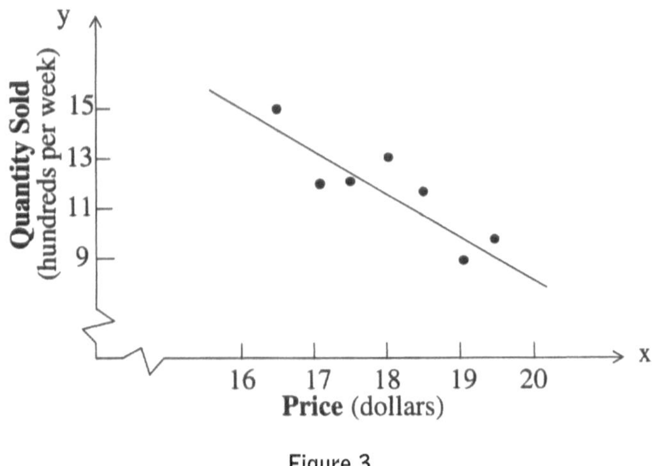

Figure 3

Generally speaking the quantity sold decreases as price increases. Such a relationship between two variables is called an **inverse relationship.** A line of the kind shown in Figure 3, with negative slope, seems a good fit to the data. It is called **inverse linear.**

In many applied situations more than one independent variable is needed to realistically explain the variability in the dependent variable. Here, other factors such as personal taste and the availability of substitutes play a role in influencing consumer demand.

We conclude by examining the main types of scatter diagrams that one may encounter, as shown in Figures 4 (a) through 4 (d).

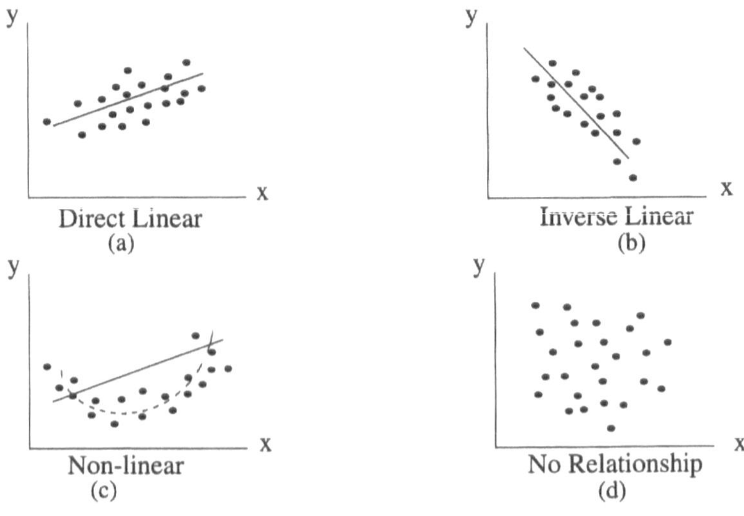

Figure 4(a)-(d)

In Figure 4 (a) a line with positive slope provides a good fit to the scatter points, suggesting a strong direct linear relationship between x and y, as is the case in Figure 1 of Example 1. In Figure 4 (b) a line with negative slope provides a good fit to the scatter points, suggesting a strong inverse linear relationship between x and y, as is the case in Figure 3 of Example 3. Figure 4 (c) suggests that the variables have a strong non-linear relationship. Finally, there does not appear to be a discernible pattern to the scatter points in Figure 4 (d); this graph suggests that there is no relationship between the variables.

EXERCISES

In the following situations samples chosen are understood to be chosen at random.

1. Increases in temperature well beneath the Earth's surface are of great interest to geologists and mining engineers. Table 4 shows such temperature increases for various depths at a well near Pittsville. The first point for which a temperature increase (in degrees Celsius) is measured is 28 meters below the Earth's surface. (This depth is taken as a starting point because the temperature at higher levels varies with the seasons and is too unstable to yield a consistent reading.) Temperatures are recorded as increases over the temperature at this basepoint.

Table 4

Meters below surface (x)	28	67	179	246	293	314
Degrees above basepoint temperature (y)	0	1.5	4.3	9.2	10.4	10.7

Draw the scatter diagram and comment on the relationship between the variables suggested by the scatter diagram.

2. Gerry Bond, Chairwoman of the Huxley College Statistics Department, chose a sample of 10 statistics students to see if there was a relationship suggested between the number of hours of study x in the week immediately preceding a statistics final exam and the grade y received on the exam. She obtained the data shown in Table 5.

Table 5

x	25	12	18	26	19	20	23	15	22	8
y	93	57	55	90	82	95	95	80	85	61

Plot and comment on the scatter diagram for these data.

3. The Superior Accounting firm wishes to investigate whether there is a relationship between years of service and "inefficiency" rate (that is, the average number of accounting mistakes per week) of its assistant accountants. Noted in Table 6 are the results obtained for nine selected accountants over the last 26 weeks.

Table 6

Years of service (x)	3	5	6	6.5	9	10	14.5	17	23
Inefficiency rate (y)	14.5	11	14	9	13.5	7	8	6.5	7.5

Draw the scatter diagram and comment on the relationship suggested between x and y.

4. A group of physicians wants to better understand the relationship between age (in years) and systolic blood pressure (SBP). They examined a selected sample of 10 people and obtained the data shown in Table 7.

Table 7

Age (x)	16	18	22	29	38	46	51	59	64	65
SBP (y)	128	127	121	125	139	138	141	140	147	158

Draw the scatter diagram and comment on the relationship suggested between x and y.

5. Researchers with the Tasty Manufacturing Company conducted an experiment to study the relationship between storage temperature x and the shelf life y of their new Heavenly-Fat-Free Yogurt. Ten containers of Heavenly-Fat-Free Yogurt were stored at ten temperature levels (in degrees Fahrenheit) and the shelf life (in days) was recorded. The results are shown in Table 8.

Table 8

x	35	38	35	42	44	47	51	53	55	56
y	43	45	41	27	23	12	6	4	2	1

Draw the scatter diagram and comment on the relationship suggested between x and y.

6. Many physicians have been recommending exercise for their patients, particularly for those who are overweight. Often one of the benefits of exercise is a reduction in one's blood cholesterol, which may lead to a lower risk of heart disease. To analyze the relationship between exercise and cholesterol level Dr. Edward Jones took a sample of 8 patients who did little exercise. He measured their cholesterol levels and started them on regular exercise programs. After four months he recorded the number of minutes per week each patient exercised on average and measured each one's cholesterol level. The results are shown in Table 9.

Table 9

Weekly exercise level (x)	43	65	100	43	16	120	30	10
Reduction in cholesterol level (y)	38	13	32	20	-5	27	23	-9

A negative "reduction" in cholesterol level means that the patient's cholesterol level increased. Draw the scatter diagram and comment on the relationship suggested between x and y.

7. The data in Table 10 gives the increase in weight y (in decigrams) of 7 male friendly furry creatures over a three month period during which they were given selected doses, x, of Vitamin B (in milligrams).

Table 10

x	0.2	0.3	0.5	0.7	1.0	1.2	1.5
y	25	27	32	39	38	34	26

Draw the scatter diagram and comment on the relationship suggested between x and y.

8. In a survey it conducted, Big Bargains Department Store asked 9 selected credit card holders to state their yearly store charges y (in tens of dollars) and yearly gross income x (in thousands of dollars). Table 11 shows the survey results obtained.

Table 11

x	27	38	46	30	71	89	19	124	146
y	21	23	26	28	35	37	34	45	68

Draw the scatter diagram and comment on the relationship suggested between x and y.

12.3 ■ Fitting the "Best" Regression Line: The Method of Least Squares

If it appears from a scatter diagram that a line provides a good fit to the data points, the question then arises as to which line provides the "best fit." The most widely employed criterion for defining the best fitting line is provided by the **principle of least-squares**, first introduced by A.M. Legendre and C.F. Gauss.

To describe the idea that underlies the least-squares principle in this setting, consider the three points (x_1, y_1), (x_2, y_2) and (x_3, y_3) and the line $\hat{y} = a + bx$ shown in Figure 5.

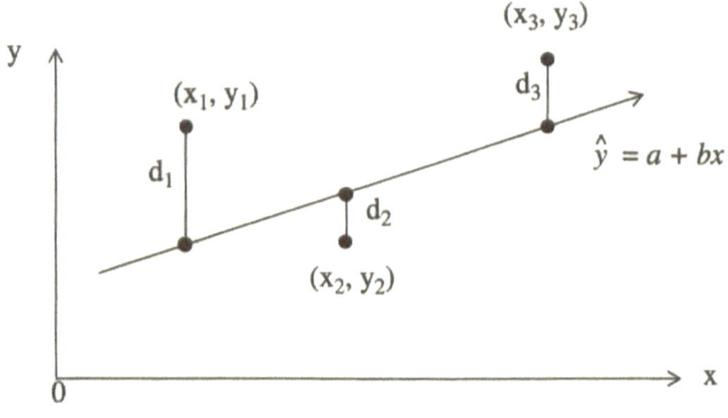

Figure 5

The criterion defining line of best fit should, in some sense, minimize the deviation of the points from the line. The least-squares principle does this by defining the **least-squares line of best fit** as that line for which the sum of the squares of the vertical distances from the data points to the line is smallest. That is, that line which yields the smallest value of

$$d_1^2 + d_2^2 + d_3^2 .$$

It can be shown that for the **least-squares line of best fit**,

$$\hat{y} = a + bx ,$$

a and *b* satisfy the following system of equations, called the normal equations:

$$3a+(\sum x_i)b=\sum y_i$$
$$(\sum x_i)a+(\sum x_i^2)b=\sum x_i y_i$$

More generally, if there are *n* data points, the same system of equations, with 3 replaced by *n*, is obtained. That is, we have:

$$na+(\sum x_i)b=\sum y_i$$
$$(\sum x_i)a+(\sum x_i^2)b=\sum x_i y_i$$

The general solution to this system is given by the following formulas for *a* and *b*:

$$b=\frac{n(\sum x_i y_i)-(\sum x_i)(\sum y_i)}{n(\sum x_i^2)-(\sum x_i)^2}$$
$$a=\frac{\sum y_i - b\cdot\sum x_i}{n}$$
$$=\frac{\sum y_i}{n}-b\frac{\sum x_i}{n}$$

Thus,

$$a=\overline{y}-b\overline{x},$$

where

$$\overline{y}=\frac{\sum y_i}{n},\quad \overline{x}=\frac{\sum x_i}{n}$$

are the means of the *y* values and *x* values, respectively.

We can recognize the denominator of *b* as the numerator of the sample variance of the *x*'s. Therefore, this quantity must be positive. However, the numerator of *b* is not similarly restricted in value and may be positive, negative or zero.

The line $\hat{y}=a+bx$ which has been fitted to the *n* points $(x_1,y_1),(x_2,y_2),\ldots,(x_n,y_n)$ by the method of least squares is called the **regression line of *y* on *x***, or more briefly the **regression line**. The value of $\hat{y}$ provides us with an estimate for the mean of the possible *y* values for any given value of *x*.

Example I The Daniel Company

We return to the Daniel Company's advertising-expenditure sales-volume problem of Example 1 (page 324) of the previous section. Determine the regression line for these data, which are reproduced in Table 12. Use the regression line to predict the mean sales volume for monthly advertising budgets of (a) $9,000, (b) $17,000.

Table 12

Advertising (x)	6	4	9	14	12	16	10	21
Sales volume (y)	15	17	14	27	21	28	22	24

The simplest way to calculate *a* and *b* for the regression line is to employ a column arrangement as shown in Table 13. List the *x* and corresponding *y* values in the first two columns and calculate the mixed products *xy* and squared *x* values in columns 3 and 4. The sums of these columns are the values needed to determine *b* and then *a*.

Table 13

x	y	xy	x^2	
6	15	90	36	
4	17	68	16	
9	14	126	81	
14	27	378	196	
12	21	252	144	
16	28	448	256	
10	22	220	100	
21	24	504	441	
$\sum x = 92$	$\sum y = 168$	$\sum xy = 2086$	$\sum x^2 = 1270$	$n = 8$

We have:

$$b = \frac{8(2086) - 92(168)}{8(1270) - (92)^2} = \frac{1232}{1696} = 0.726$$

$$a = \frac{168 - 0.726(92)}{8} = 12.65$$

Thus, the Daniel Company's regression line is:

$$\hat{y} = 12.65 + 0.726x$$

This regression line has positive slope, consistent with the corresponding scatter diagram of Figure 1 in Section 12.2 (page 412), which suggested a direct linear relationship between the variables.

(a) To predict sales for an advertising budget of \$9,000 per month, take $x = 9$. This yields:

$$\begin{aligned}\hat{y} &= 12.65 + 0.726(9)\\ &= 19.2\end{aligned}$$

For an advertising expenditure 9 thousand dollars per month, we predict a mean monthly sales volume of 19.2 hundred-thousand dollars, or \$1,920,000.

The value of $\hat{y}$ expresses a predicted mean. We are not suggesting that \$1,920,000 in sales will consistently be obtained in each month that \$9,000 is spent on advertising. If \$9,000 is invested in advertising for many months, an **approximate** mean of \$1,920,000 in sales will result, this regressive equation predicts, although the sales volume for any one month may differ slightly or considerably from this predicted mean.

(b) For the monthly advertising expenditure of \$17,000, take $x = 17$. This yields:

$$\begin{aligned}\hat{y} &= 12.65 + 0.726(17)\\ &= 25\end{aligned}$$

As a result, we predicted a mean monthly sales volume of 25 hundred-thousand dollars or \$2,500,000 for a monthly advertising expenditure of \$17 thousand.

Example 2

We return to the Tons of Toys price-demand problem considered in Example 3 (page 413) of the previous section. Determine the regression line for their data, which are reproduced in Table 14. Use this regression line to predict the mean weekly demand (in hundreds) for teddy bears priced at \$18.25.

Table 14

x	18	19	17.5	16.5	19.5	17	18.5
y	13	9	12.5	15	10	12	11.5

The calculation of the sums needed for *a* and *b* of $\hat{y} = a + bx$ is shown in Table 15.

Table 15

X	y	xy	x^2
18	13	234	324
19	9	171	361
17.5	12.5	218.75	306.25
16.5	15	247.5	272.25
19.5	10	195	380.25
17	12	204	289
18.5	11.5	212.75	342.25
$\sum x = 126$	$\sum y = 83$	$\sum xy = 1483$	$\sum x^2 = 2275$

We obtain:

$$b = \frac{7(1483) - 126(83)}{7(2275) - (126)^2} = \frac{77}{49} = -1.571$$

$$a = \frac{83 - (-1.571)(126)}{7} = 40.14$$

Therefore, the regression line is:

$$\hat{y} = 40.14 - 1.571x$$

That the slope of this regression line is negative is consistent with Figure 3 of Section 12.2 (page 414), which suggests an inverse linear relationship between price and demand. For $x = 18.25$ we obtain the predicted mean weekly demand:

$$\hat{y} = 40.14 - 1.571(18.25)$$
$$= 11.5$$

For a unit selling price of $18.25 the predicted average weekly demand is 11.5 hundred or 1150 toys. The demand for different weekly periods may, however, differ considerably from one another.

In using $\hat{y}=a+bx$ as a tool for predicting the mean value of y for a given value of x we have been careful to restrict x to lie between the smallest and largest x values of the data that $\hat{y}=a+bx$ is based on. The smallest and largest x values that arise in the Daniel Company's advertising-expenditure sales-volume problem (Example 1), for example, are 4 and 21, representing advertising budgets of \$4,000 and \$21,000, respectively. We have been careful to employ the regression line $\hat{y}=12.65+0.726x$ to predict mean sales volume activity for advertising budgets which fall between \$4,000 and \$21,000. Can we o beyond this range and use the regression line to predict mean sales volume for advertising budgets of \$2,000 or \$25,000? This and other related issues which require a caution sign are discussed in the next section.

EXERCISES

1. Refer to Exercise 1 of Section 12.2 (page 415) which gives data comparing the increase in temperature y (in degrees Celsius) against the distance beneath the Earth's surface x (in meters). These data are reproduced in Table 16.

Table 16

x	28	67	179	246	293	314
y	0	1.5	4.3	9.2	10.4	10.7

(a) Determine the regression line for these data.

(b) Determine the predicted value for a depth of 179 meters and interpret the result.

(c) Compare your predicted result to the observed temperature increase at the 179 meter depth. How would you explain the difference in these two values?

2. Refer to Exercise 2 of Section 12.2 (page 415) which gives data comparing the grade for a statistics final exam y against the study time x (in hours) for the exam. These data are reproduced in Table 17.

Table 17

x	25	12	18	26	19	20	23	15	22	8
y	93	57	55	90	82	95	95	80	85	61

(a) Determine the regression line for these data.

(b) Determine the predicted grade for students who study 19 hours during the week preceding the final exam and interpret the result.

(c) Compare your predicted result to the observed grade of a statistics student who studied 19 hours for the exam. How would you explain the difference in these two values?

3. Refer to Exercise 3 of Section 12.2 (page 416) which gives data comparing the "inefficiency" rate (average number of accounting mistakes per week) y of its assistant accountants against the number of years of service x at the Superior Accounting Firm. Data gathered from this study are reproduced in Table 18.

Table 18

x	3	5	6	6.5	9	10	14.5	17	23
y	14.5	11	14	9	13.5	7	8	6.5	7.5

(a) Determine the regression line for these data.

(b) Predict the mean inefficiency rate of accountants having 9 years experience and interpret the result obtained.

4. Refer to Exercise 4 of Section 12.2 (page 416) which gives data comparing systolic blood pressure (SBP) against age (in years). These data are reproduced in Table 19.

Table 19

Age (x)	16	18	22	29	38	46	51	59	64	65
SBP (y)	128	127	121	125	139	138	141	140	147	158

(a) Determine the regression line for these data.

(b) Use your line to predict the SBP of a 51 year old person and interpret the result obtained.

(c) Compare your predicted SBP to the observed SBP for the seventh person in the list of data. How would you explain the difference in these two figures?

5. Refer to Exercise 5 of Section 12.2 (page 416) which gives data comparing the shelf life y (in days) of Heavenly-Fat-Free Yogurt against its storage temperature x (in degrees Fahrenheit). The data are reproduced in Table 20.

Table 20

x	35	38	35	42	44	47	51	53	55	56
y	43	45	41	27	23	12	6	4	2	1

(a) Determine the regression line for these data.

(b) Predict the average shelf life for yogurt stored at 35°F and interpret the result obtained.

(c) Compare your predicted result from (b) with the shelf lives of the first and third containers of yogurt that were also stored at 35°F. How do you explain the discrepancies in these figures?

6. Refer to Exercise 8 of Section 12.2 (page 417) which gives data comparing a customer's annual credit card charges y (in tens of dollars) at Big Bargains Department Store to the customer's gross annual income x (in thousands of dollars). The data are reproduced in Table 21.

Table 21

x	27	38	46	30	71	89	19	124	146
y	21	23	26	28	35	37	34	45	68

(a) Determine the regression line for these data.

(b) Predict the mean annual charge for customers whose gross annual income is \$71 thousand and interpret the result.

(c) Compare your predicted result from (b) with the annual charges of the fifth customer, who also has a gross annual income of \$71 thousand.

How do you explain the difference between the predicted and observed annual charges?

12.4 ■ Cautions

Extrapolation Beyond the Data

> **GREAT CAUTION** must be exercised in choosing values of the independent variable x to be substituted into the regression equation $\hat{y} = a + bx$. The further removed x is from the x-values of the data points, the less reliable can we expect the predicted average value of y to be.

Wheat yields, for example, improve as rainfall increases and one could find a regression line for given data. But it is well known that wheat yields improve up to a certain point and that too much rain has a destructive rather than beneficial effect. If the regression line were used to predict average wheat yields for very large rainfall values, the predicted average yield would differ considerably from the average yield actually obtained. The hazard of extrapolating beyond the range of the data is illustrated geometrically in Figure 6. The regression line $\hat{y} = a + bx$

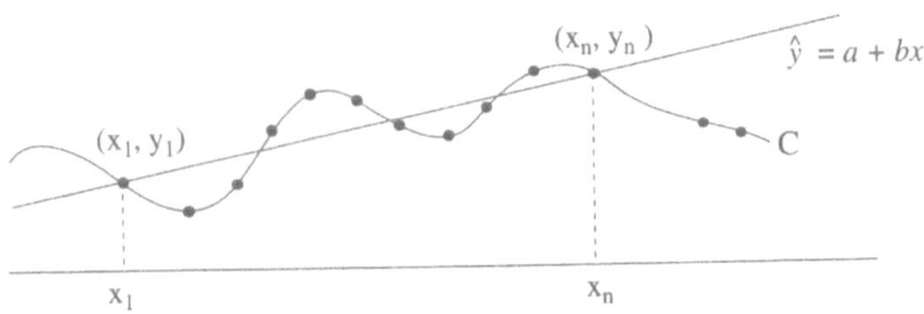

Figure 6

provides a good fit to the data points $(x_1, y_1), \ldots, (x_n, y_n)$, but outside of the range determined by x_1 and x_n it differs considerably from curve C which more accurately expresses the relationship between y and x.

Cause and Effect

> Data $(x_1, y_1), \ldots, (x_n, y_n)$ arise from a study, indications are that there is a strong linear relationship between the variables, the regression line $\hat{y} = a + bx$ is

determined, and it is found to give on-target predictions of the mean value of y for given values of x.

Does this imply that the factor described by x is a **cause** of the factor described by y? **The answer is an unequivocal NO.** It may or may not be that the factor described by x is a cause of the factor described by y, but regression analysis takes no position on this, one way or the other; proof of causality would have to come from other sources.

To take an example, the institutional research department of Ecap University seeks to predict the mean income of graduates five years after graduation on the basis of the graduate's grade point average (GPA). For this purpose the department obtained a sample of 10 alumni who had graduated five years earlier and found a strong linear relationship between GPA and income. Clearly, GPA is not a cause of income. In this case personality and other behind-the-scene characteristics influence an individual's GPA and earning performance in such a way that one factor may be used as a prediction tool for the other.

Similar observations may be made about SAT, GMAT and other examination scores as predictors for academic success. The examination score itself does not cause academic success. At best there are characteristics behind the scene which influence the examination score and academic success in such a way that one may be used as a successful prediction tool for the other.

EXERCISES

1. Refer to the subterranean temperature problem in Exercise 1 of Section 12.2 (page 415) and your analysis of it from Exercise 1 of Section 12.3 (page 423). Would you use your sample regression line to predict the average increase in temperature at a depth of (a) 90 meters, (b) 475 meters below the surface? Explain.

2. Refer to the study-time exam-grade problem in Exercise 2 of Section 12.2 (page 415) and your analysis of it from Exercise 2 of Section 12.3 (page 423).

   (a) Would you use your sample regression line to predict the average grade of students who studied (i) 3 hours, (ii) 17 hours for the exam? Explain.

   (b) Could we conclude that increased study time causes higher exam grades? Explain.

3. Refer to the job-experience inefficiency-rate problem in Exercise 3 of Section 12.2 (page 416) and your analysis of it from Exercise 3 of Section 12.3 (page 424).

   (a) Would you use your sample regression line to predict the average number of accounting mistakes per month for assistant accountants having (i) 12 years, (ii) 28 years of experience? Explain

   (b) Could we conclude that increased experience on the job causes the inefficiency rate to decline? Explain.

12.5 ■ The Population Regression Line

The sample based regression line $\hat{y} = a + bx$ is an approximation for an envisioned **population regression line,**

$$\mu_{y/x} = \alpha + \beta x,$$

where $\mu_{y/x}$ is the mean of y for the given value of x. The parameters α and β (which used in this context have nothing to do with type I and type II errors in hypothesis testing) are called the **regression coefficients** and the sample values a and b serve as **estimated regression coefficients.**

Beware the Assumptions

The following **strings attached** underlie this point of view:

1. For each x the dependent variable y, whose mean is being estimated by $\hat{y} = a + bx$, has mean $\alpha + \beta x$. Thus, the mean of y for $x = x_1$, is $\alpha + \beta x_1$, the mean of y for $x = x_2$, is $\alpha + \beta x_2$, etc.

2. For each x, y is a normally distributed random variable with mean $\alpha + \beta x$ (see Figure 7).

3. These normal distributions have the same standard deviation σ (see Figure 7). Thus, while their means differ, they all have the same measure of spread about the mean.

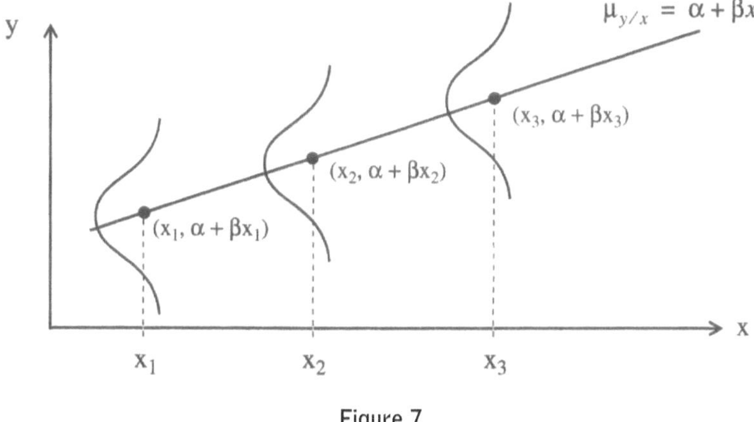

Figure 7

Confidence Intervals for the Regression Coefficients

How well does $\hat{y} = a + bx$ approximate $\mu_{y/x} = \alpha + \beta x$? To address this question we obtain confidence interval estimates for α and β in terms of *a* and *b*. The results, in principal, are not difficult to state, but messy computations that emerge may obscure this feature.

Subject to the assumptions noted, it can be shown that the 95% confidence interval for α is given by:

$$a - t_{.025} s_a < \alpha < a + t_{.025} s_a$$

Here:

$$s_a = s_e \sqrt{\frac{1}{n} + \frac{\overline{x}^2}{(n-1)s_x^2}}$$

$$s_e = \sqrt{\frac{\sum y^2 - a\sum y - b\sum xy}{n-2}}$$

$$s_x^2 = \frac{n(\sum x^2) - (\sum x)^2}{n(n-1)}$$

n is the size of the randomly drawn sample, s_x^2 is the variance of the *x*-values of the data, $\overline{x}$ is the mean of the *x*'s, and $t_{.025}$ is the *t*-value for the 95% confidence interval for $n-2$ degrees of freedom.

Geometrically speaking, the 95% confidence interval for α is centered at a with radius $t_{.025}s_a$ (see Figure 8).

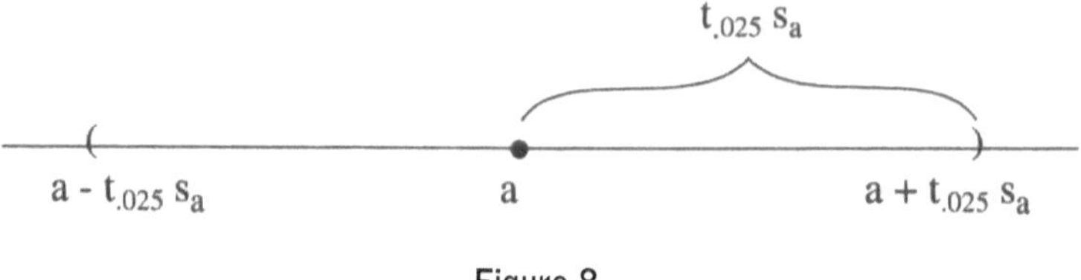

Figure 8

Example 1 Return to the Daniel Company

We return to the Daniel Company's advertising-expenditure sales-volume problem considered in Example 1 of Section 12.2 (page 411) to determine the 95% confidence interval for α.

From that analysis we have the following:

$$n = 8 \qquad \sum y = 168 \qquad a = 12.65$$

$$\sum x = 92 \qquad \sum xy = 2086 \qquad b = 0.726$$

$$\sum x^2 = 1270$$

To compute s_e we need $\sum y^2$. Squaring the y values and adding gives us:

$$\begin{aligned}\sum y^2 &= 15^2 + 17^2 + 14^2 + 27^2 + 21^2 + 28^2 + 22^2 + 24^2 \\ &= 3724\end{aligned}$$

Thus:

$$\begin{aligned}s_e^2 &= \frac{3724 - 12.65(168) - 0.726(2086)}{8 - 2} \\ &= 14.03\end{aligned}$$

We have:

$$s_e = 3.75$$

We also have:

$$s_x^2 = \frac{8(1270)-(92)^2}{8(7)}$$
$$= 30.29$$

In summary, we have:

$n = 8$ $\quad \overline{x} = 11.5$ $\quad s_x^2 = 30.29$

$\overline{x}^2 = 132.3$ $\quad s_e = 3.75$

Therefore:

$$s_a = 3.75\sqrt{\frac{1}{8}+\frac{132.3}{7(30.29)}}$$
$$= 3.75(0.865) = 3.24$$

Since $t_{.025} = 2.447$ for $\text{d.f.} = n-2 = 6$ and $a = 12.65$, we obtain the following 95% confidence interval for α:

$$12.65 - 2.447(3.24) < \alpha < 12.65 + 2.447(3.24)$$
$$4.72 < \alpha < 20.58 \text{ with probability } 0.95$$

Geometrically speaking, α lies in the interval centered at $a = 12.65$ with radius 7.93, and endpoint values 4.72 and 20.58, with probability 0.95 (see Figure 9).

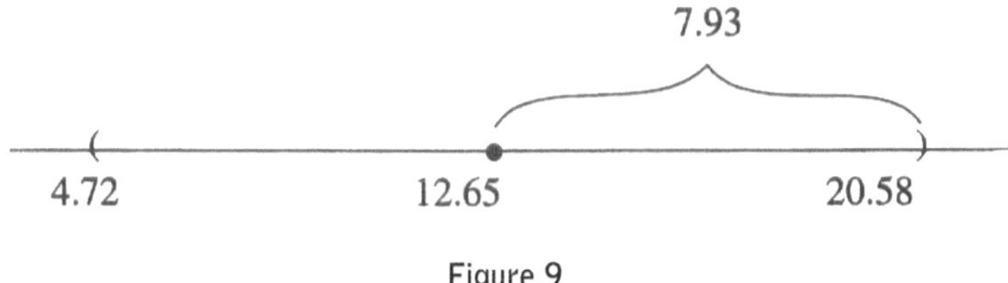

Figure 9

In terms of a point estimate, if we take $a = 12.65$ as an estimate for α, then the error in this estimate does not exceed 7.93 with probability of 0.95.

Beware the Assumptions

Subject to the **assumptions** noted, it can be shown that the 95% confidence interval for β is given by

$$b - t_{.025}s_b < \beta < b + t_{.025}s_b$$

where:

$$s_b = \frac{s_e}{s_x\sqrt{n-1}}$$

s_x is the standard deviation of the x-values of the data, and $t_{.025}$ is the t-value for the 95% confidence interval for $n-2$ degrees of freedom.

Geometrically speaking, the 95% confidence interval for β is centered at b with radius $t_{.025}s_b$ (see Figure 10).

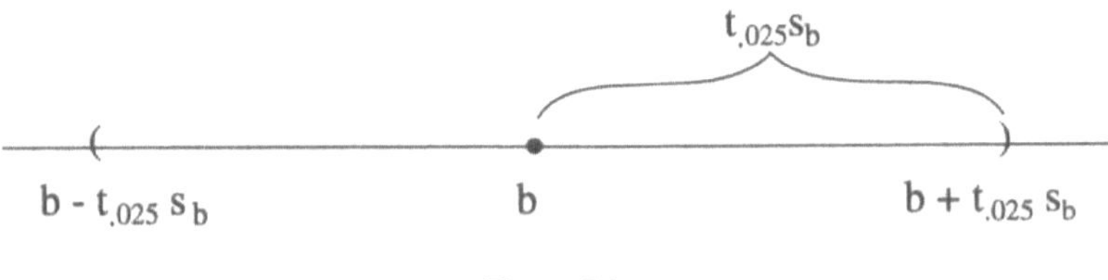

Figure 10

Example 2 Second Return to the Daniel Company

We again return to the Daniel Company's advertising-expenditure sales-volume problem considered in Example 1 of Section 12.2 (page 411) to determine the 95% confidence interval for β.

We have the following:

$$n = 8 \qquad s_x = \sqrt{30.29} = 5.50$$

$$|F(x) - S(x)| \qquad s_e = 3.75$$

$$b = 0.726$$

Thus,

$$s_b = \frac{3.75}{5.50\sqrt{7}} = 0.26$$

and we obtain:

$$0.726 - 2.447(0.26) < \beta < 0.726 + 2.447(0.26)$$
$$0.10 < \beta < 1.36 \text{ with probability 0.95}$$

Geometrically speaking, β lies in the interval centered at $b = 0.726$ with radius 0.63, and endpoints 0.10 and 1.36, with probability 0.95 (see Figure 11).

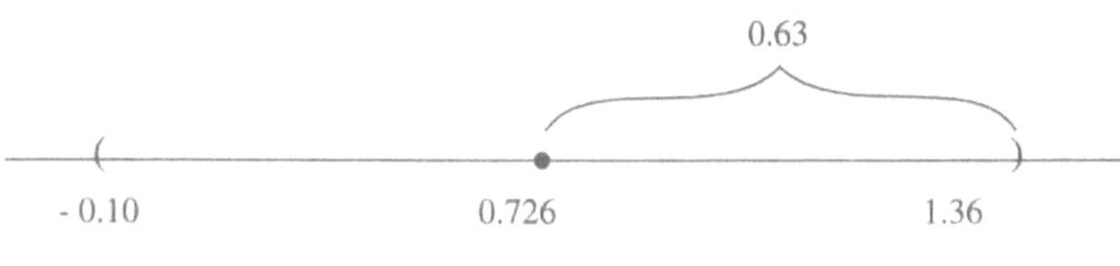

Figure 11

In terms of a point estimate, if we take $b = 0.726$ as an estimate for β, then the error in this estimate does not exceed 0.63 with probability 0.95.

In summary then, the population regression coefficients α and β of the regression line $\mu_{y/x} = \alpha + \beta x$ for the Daniel Company's advertising-expenditure sales-volume situation lie in the confidence intervals

$$4.72 < \alpha < 20.58$$
$$0.10 < \beta < 1.36$$

with probability 0.95. These confidence intervals for α and β give us a sense of the reliability of the estimated regression line $\hat{y} = 12.65 + 0.726x$ as an estimate of the population regression line.

If 98% or 99% confidence intervals, for example, are to be constructed for α and β, then $t_{.025}$ should be replaced by $t_{.01}$ and $t_{.005}$, respectively, for $n-2$ degrees of freedom.

Keep in Mind

The afore results, impressive and convincing as they may seem to be, **are founded on demanding strings attached.** If the situation differs "significantly" from them, these results may simply be nonsense.

EXERCISES

1. In connection with Examples 1 and 2, do α and β lie in the 95% confidence intervals constructed for them or not? Explain.,

2. (a) In connection with the Daniel Company's advertising-expenditure sales-volume problem, determine 98% confidence intervals for α and β of $\mu_{y/x} = \alpha + \beta x$.

   (b) Can we be certain that α and β lie in the confidence intervals obtained in answer to (a)? Explain.

3. Refer to the subterranean temperature problem in Exercise 1 of Section 12.2 (page 415) and your analysis of it from Exercise 1 of Section 12.3 (page 423).

   (a) Find 95% confidence intervals for α and β.

   (b) How is the 0.95 probability value to be interpreted in this situation?

4. Refer to the study-time exam-grade problem in Exercise 2 of Section 12.2 (page 415) and your analysis of it from Exercise 2 of Section 12.3 (page 423).

   (a) Find 98% confidence intervals for α and β.

   (b) Can we be certain that α and β lie in these confidence intervals? Explain.

5. Refer to the job-experience inefficiency-rate problem in Exercise 3 of Section 12.2 (page 412) and your analysis of it from Exercise 3 of Section 12.3 (page 424).

   (a) Find 90% confidence intervals for α and β.

   (b) How is the 0.90 probability value to be interpreted in this situation?

6. Refer to the age systolic-blood-pressure problem in Exercise 4 of Section 12.2 (page 416) and your analysis of it from Exercise 4 of Section 12.3 (page 424).

   (a) If a and b of $\hat{y} = a + bx$ are used as estimates for α and β of $\mu_{y/x} = \alpha + \beta x$, what errors might arise with probability 0.99?

   (b) Can we be certain that the errors that arise do not exceed the bounds determined in answer to (a)? Explain.

13

Linear Correlation

13.1 ■ The Coefficient of Determination

In our consideration of the Daniel Company's advertising-expenditure sales-volume problem in Chapter 12 we observed that advertising expenditure was one of several independent variables that had an impact on sales volume. A natural question that arises is: what proportion of the variability in sales can be explained by advertising expenditure in terms of the estimated regression line?

> More generally, if $y_1, y_2, \ldots, y_n$ are values of the dependent variable y, what proportion of the variability in the y's can be explained by the independent variable x in terms of the estimated regression line $\hat{y} = a + bx$?

We now describe a sample statistic whose determination will help answer this query. The variation in the observations $y_1, y_2, \ldots, y_n$ of the dependent variable y is measured by deviations from its mean: $y_1 - \bar{y},\ y_2 - \bar{y}, \ldots, y_n - \bar{y}$. Squaring each deviation and adding yields

$$\sum_{1}^{n} (y_i - \bar{y})^2 , \tag{1}$$

which we recognize as the numerator of the sample variance of the observed values of the random variable y. More important, however, (1) gives the total of the squared deviations of n observations from their mean $\bar{y}$; in essence, it expresses the total variation of the observed values about their mean. We call this quantity the **total sum of squares** and denote it by:

$$\mathrm{SST} = \sum_{1}^{n} (y_i - \bar{y})^2$$

Example 1 SST for the Daniel Company's Problem

For the Daniel Company's advertising sales-volume problem of Example 1 of Section 12.2 (page 411), $\bar{y} = 21$ and the total sum of squares is:

$$\begin{aligned}\mathrm{SST} &= (-6)^2 + (-4)^2 + (-7)^2 + (6)^2 + (0)^2 + (7)^2 + (1)^2 + (3)^2 \\ &= 196\end{aligned}$$

Variation in the Observed *y* Values

Some of the variation in the observed *y* values can be explained by the relationship between *x* and *y*, whereas a part of SST cannot. As Figure 1 makes clear, we can subdivide SST into these two parts as follows:

$$(y_i - \bar{y}) \quad = \quad (\hat{y}_i - \bar{y}) \quad + \quad (y_i - \hat{y}_i) \qquad (2)$$

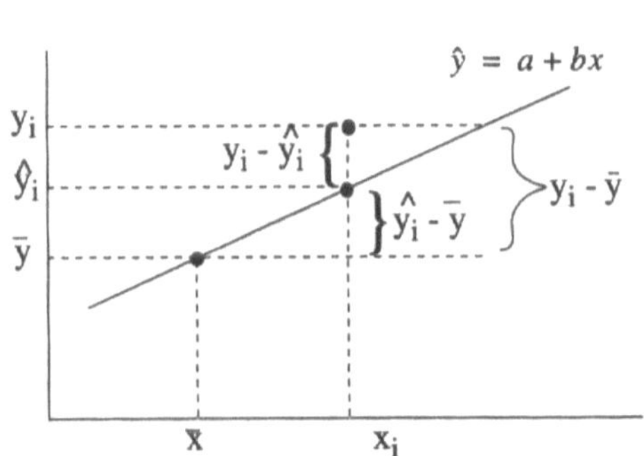

$y_i - \bar{y}$: total variation of y_i, from the mean of y's.

$y_i - \hat{y}_i$: variation of y_i from $\hat{y}_i$, unexplained by regression.

$\hat{y}_i - \bar{y}$: variation of $\hat{y}_i$ from $\bar{y}$, explained by regression.

Figure 1

From (2) and Figure 1 we see that the difference between y_i and $\bar{y}$, namely, $y_i - \bar{y}$, is made up of two components. One component is the difference between $\hat{y}_i$ and $\bar{y}$, namely, $\hat{y}_i - \bar{y}$, and is accounted for by the difference $x_i - \bar{x}$. That is, $\hat{y}_i - \bar{y}$ is accounted for by the regression line $\hat{y} = a + bx$.

The other component, the difference between y_i and $\hat{y}_i$, namely $y_i - \hat{y}_i$, called a **residual**, is not explained by the regression line, which is based on the independent variable *x*. The residual is a reflection of variables other than *x*.

In the advertising-expenditure sales-volume problem these other variables would include advertising expenditures for competing products, prices of competing products, consumer taste, and sampling error.

If we now square each side of (2) and add, the middle terms that would be obtained from squaring the right side most thankfully cancel out and we obtain the analogous relationship, called the:

Fundamental Equation of Regression Analysis:

$$\sum_{1}^{n}(y_i-\bar{y})^2=\sum_{1}^{n}(\hat{y}_i-\bar{y})^2+\sum_{1}^{n}(y_i-\hat{y}_1)^2$$

SST SSR SSE

That is:

$$\text{SST}=\text{SSR}+\text{SSE}$$

SSR is the **sum of squares due to "regression"** and expresses that portion of SST that is explained by the linear relationship between x and y given by the regression line, whereas the **residual or unexplained variation** is given by SSE.

We define the sample **coefficient of determination** as the ratio of the amount of variation explained by the regression line to the total variation in the dependent variable. It is denoted by d^2 and given by:

$$r^2=\frac{\text{SSR}}{\text{SST}}=\frac{\text{SST-SSE}}{\text{SST}} \tag{3}$$

The coefficient of determination gives that proportion of the total variability in the values $y_1, y_2, \ldots, y_n$ of the dependent variable y that is explained by the independent variable x in terms of the regression line.

A useful computation formula for SSR is given by:

$$\text{SSR}=(n-1)b^2s_x^2$$

where

$$s_x^2=\frac{n\sum x^2-(\sum x)^2}{n(n-1)}$$

denotes the sample variance of the x's and b is the slope of the regression line $\hat{y} = a + bx$. Then SSE may be determined by

$$\begin{aligned} \text{SSE} &= \text{SST} - \text{SSR} \\ &= (n-1)[s_y^2 - b^2 s_x^2] \end{aligned}$$

where

$$s_y^2 = \frac{\sum (y_i - \bar{y})^2}{n-1}$$

expresses the sample variance of the y's.

Example 2 SSR, SSE and r^2 for the Daniel Company's Problem

From the advertising-sales data we determine:

$$\text{SSR} = (7)(0.726)^2(30.29) = 111.74$$
$$\text{SSE} = \text{SST} - \text{SSR} = 196 - 111.74 = 84.26$$

Thus,

$$r^2 = \frac{111.74}{196} = 0.57,$$

which is interpreted to mean that advertising expenditures explain approximately 57% of the total variation in the sales data.

Each of SST, SSR, and SSE is non-negative since all arise from a squaring operation. SST must be at least as large as SSR and SSE since SST = SSR + SSE. In particular,

$$0 \le \text{SSR} \le \text{SST}$$

which, from (3), implies that r^2 must take on a value between 0 and 1 inclusive; that is:

$$0 \le r^2 \le 1$$

r^2 could equal zero if SSR = 0, which would require that all of the variation in the data be unexplained by regression or SST = SSE. On the other hand,

r^2 could equal one if SSR = SST, while SSE = 0. In this case, all of the variation in y would be explained by regression.

Geometrically speaking, if $r^2 = 1$, then all of the data points must lie on the fitted regression line. However, if $r^2 = 0$ then b of $\hat{y} = a + bx$ is 0 as well, and the fitted regression line is horizontal, namely $\hat{y} = a = \overline{y}$, the line determined by the mean of the y data values.

> In general, we should anticipate getting a value of r^2 between the two extremes of 0 and 1. As r^2 increases, the data points in the scatter diagram tend to "cluster" about the sample regression line. The closer r^2 is to one, the closer are the data points to their least-squares line, irrespective of whether the relationship between the x's and the y's is a direct or inverse one.

This situation is illustrated by Figures 2 (a)-2 (d)

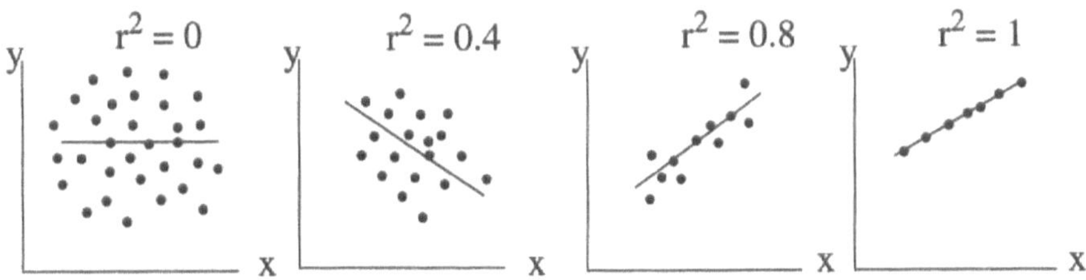

Figure 2

In practice, the value of r^2 is not very high, generally 0.6 or less. This is because problems arising from applications tend to be complex and require more than one independent variable to realistically describe their nature. One independent variable may not be sufficient to explain very much of the variability in the dependent variable.

EXERCISES

1. For the data points (1/2, 0), (1,1), (3/2, 2), (2, 3), (3, 7) and (4, 7).

   (a) Determine the regression line.

   (b) Sketch the scatter diagram, including the regression line. Use your results from (a) and (b) to find:

   (c) The total sum of squares SST;

(d) the regression sum of squares SSR;

(e) the residual sum of squares SSE;

(f) the coefficient of determination r^2.

(g) To what extent does your scatter diagram suggest the value of r^2?

2. For the data points (0, 5), (-1, 4), (1, 3), (2, 1), (-2, 0), (5/2, -1) and (-5/2, -2):

(a) Determine the regression line.

(b) Plot the scatter diagram, including the regression line. How well do you think the line fits the data?

Determine:

(c) SST,

(d) SSR,

(e) SSE,

(f) r^2

3. For the data points (1/2, 5), (1,4), (3/2, 4), (2, 3), and (3, 1):

(a) Determine the regression line.

(b) Plot the scatter diagram and the regression line. Do you think that a linear fit is appropriate? Evaluate:

(c) the total sum of squares;

(d) the sum of squares explained by regression;

(e) the unexplained sum of squares;

(f) the coefficient of determination.

(g) What does the coefficient of determination measure?

4. The data in Table 1 shows the "hardness" y (in "Rockwell" units) of an alloy when 7 randomly chosen specimens were treated at 500°C for randomly chosen time periods, x (in hours).

Table 1

x	4	5	6	8	10	11	13
y	41	48	50	53	54	53	52

(a) Plot the scatter diagram. Do you think that a linear fit is appropriate? Explain.

(b) Determine the regression line that expresses hardness in terms of time.

(c) Compute the coefficient of determination.

(d) How is the result obtained in answer to (c) to be interpreted?

5. The Hitest Oil Company has collected data over a number of years regarding the average daily temperature x (in F°) and the average daily fuel oil consumption y in single family houses (in gallons). A random sample from this data is given in Table 2.

Table 2

x	10	15	20	25	30	35	40	45	50	55
y	8	8.4	7.5	7.2	7.0	6.1	5.0	4.6	3.5	2.7

(a) Plot the scatter diagram. Does a linear fit appear to be appropriate? Comment.

(b) Determine the regression line.

(c) Compute the proportion of the total variability in oil consumption that is not explained by temperature, in terms of the regression line.

6. Refer to Exercise 1 of Sections 12.2 (page 415) which gives data comparing the increase in temperature y (in °C) against the distance beneath the Earth's surface x (in meters).

   (a) Determine the proportion of the total variability in the dependent variable that is accounted for by the independent variable in terms of the regression line.

   (b) How is the result to be interpreted?

7. Refer to Exercise 2 of Section 12.3 (page 423) which gives data comparing the grade for a statistics final exam y against the study time x (in hours) for that exam. Determine the proportion of the total variability in the final exam grade that is explained by studying time.

8. Refer to Exercise 3 of Section 12.3 (page 424) which gives data comparing the average number of accounting mistakes per week y of its assistant accountants against their years of service x at the Superior Accounting Firm.

   (a) Find the coefficient of determination.

   (b) How is the value obtained in answer to (a) to be interpreted?

9. Refer to Exercise 4 of Section 12.3 (page 424) which gives data comparing systolic blood pressure (SBP), y, against age x (in years).

   (a) Find the coefficient of determination.

   (b) How is the answer to the preceding to be interpreted?

10. Refer to Exercise 6 of Section 12.2 (page 417) which gives data comparing the reduction y in blood cholesterol level against average weekly exercise time x (in minutes).

    (a) Find r^2.

(b) How is the answer to the preceding to be interpreted?

11. The value of r^2 for given data is 0.6 and the unexplained component of variation is 50. Find the other components of variation as well as the percentage of the total variability in the dependent variable that is explained by the independent variable in terms of the regression line.

12. The scatter diagram of randomly selected x and y values shows a pronounced negative trend. If the coefficient of determination is 0.75 and $\sum(y_i - \bar{y})^2 = 472$, determine:

(a) SST,

(b) SSR,

(c) SSE,

(d) The proportion of the total variability in y that is not explained by x in terms of the regression line.

13.2 ■ The Sample Correlation Coefficient

The square root r of the coefficient of determination, called the **sample correlation coefficient**, measures the "strength" of the linear relationship between two variables. It can be shown that r is also given by:

$$r = \frac{n(\sum x_i y_i) - (\sum x_i)(\sum y_i)}{\sqrt{n(\sum x_i^2) - (\sum x_i)^2}\sqrt{n(\sum y_i^2) - (\sum y_i)^2}}$$

r may be positive, negative or zero.

If the relationship between x and y is direct, resulting in a regression line with positive slope, then r will be positive; if the relationship between x and y is an inverse one, resulting in a regression line with negative slope, r will be negative. If the numerator of the preceding expression for r is 0, r will be 0.

The sample correlation coefficient is not as difficult to compute as it may first appear to be. Its formula bears a striking resemblance to that for the slope b of the regression line,

$$b = \frac{n(\sum x_i y_i) - (\sum x_i)(\sum y_i)}{n(\sum x_i^2) - (\sum x_i)^2},$$

so that in the course of computing b we are well on our way to computing r. In particular, the numerators of r and b are equal and the denominator of b, once evaluated, may be used to help determine the denominator of r. Since their respective denominators are both positive, we may conclude that while r and b are generally unequal, they will always have the same sign; they are either both positive, both negative or both zero.

Since r^2, it follows that:

$$-1 \le r \le 1$$

The statistic r serves as a **measure of the strength of a linear relationship between the variables.** A value of r close to 1 or -1 indicates a strong linear relationship between x and y and that the regression line is a good fit to the data. A value of r close to 0 indicates a weak linear relationship between x and y and that the regression line is not a good fit to the data.

We illustrate the computation and interpretation of r with the following examples.

Example 1

Consider the following data relating x and y, shown in Table 3:

Table 3

x	1/2	1	3/2	2	3
y	4	3	2	1	-1

(a) Determine the regression line that expresses $\hat{y}$ in terms of x.

(b) Compute r.

(c) Interpret the value of r obtained in answer to (b).

(d) Sketch the scatter diagram and the regression line. Comment on the extent of the proximity of the data points to the regression line and what this pictorially suggests about r.

To compute b and r we require the values of a variety of sums that are part of the formula for r. Since the formulas for b and r are algebraically similar, we compute the necessary sums by means of a column arrangement similar to the one used in the examples of Section 12.3 that dealt with the determination of the regression line. The only difference is that we now require another sum, namely Σy^2, so that we add a y^2 column to the previous framework, as shown in Table 4.

Table 4

x	y	xy	x^2	y^2
1/2	4	2	1/4	16
1	3	3	1	9
3/2	2	3	9/4	4
2	1	2	4	1
3	-1	-3	9	1
$\Sigma x = 8$	$\Sigma y = 9$	$\Sigma xy = 7$	$\Sigma x^2 = 16.5$	$\Sigma y^2 = 31$

These sums give us:

(a)

$$b = \frac{5(7) - 8(9)}{5(16.5) - (8)^2} = -2$$

$$a = \frac{9 - (-2)(8)}{5} = 5$$

Therefore:

(b)

$$\hat{y} = -2x + 5$$

$$r = \frac{5(7) - 8(9)}{\sqrt{5(16.5) - (8)^2}\sqrt{5(31) - (9)^2}}$$

$$= \frac{-37}{\sqrt{18.5}\sqrt{74}} = -1$$

(c) $r = -1$ indicates there is a very strong, inverse, linear relationship between x and y.

(d) The scatter diagram and regression line are shown in Figure 3.

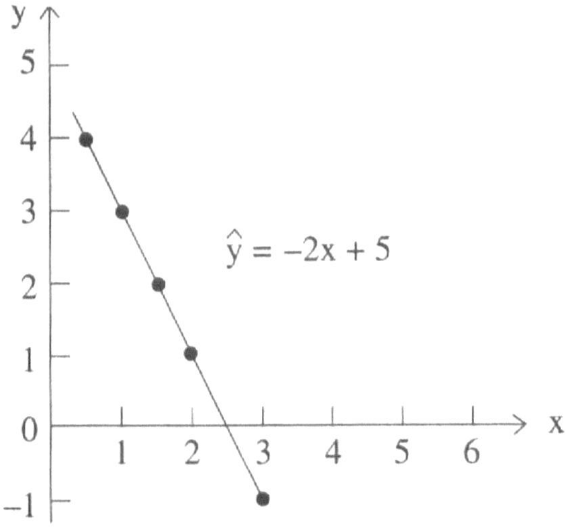

Figure 3

All of the data points lie on the sample regression line. SSE = 0 and all variation in the y values is explained by the independent variable.

The strongest linear association between variables occurs when all of the data points lie on their regression line, as shown in Figure 3. When such a "perfect linear fit" of data to their regression line is achieved we say that the variables are perfectly (**linearly**) correlated.

Example 2

Answer questions (a), (b), (c), and (d) of Example 1 for the data given in Table 5.

Table 5

X	-2	-1	0	1	2
Y	5	2	1	2	5

The required sums are shown in Table 6.

Table 6

X	y	xy	x^2	y^2
-2	5	-10	4	25
-1	2	-2	1	4
0	12	0	0	1
1	2	2	1	4
2	5	10	4	25
$\sum x = 0$	$\sum y = 15$	$\sum xy = 0$	$\sum x^2 = 10$	$\sum y^2 = 59$

(a) We obtain:

$$b = \frac{5(0) - 0(15)}{5(10) - 0^2} = 0$$

$$a = \frac{15 - 0(0)}{5} = 3$$

Thus, our regression line is $\hat{y} = 3$.

(b) $r = \dfrac{0}{\sqrt{50}\sqrt{70}} = 0$

(c) The value $r = 0$ suggests that there is a very weak or non-existent linear relationship between *x* and *y*.

(d) The scatter diagram for the data and regression line are shown in Figure 4.

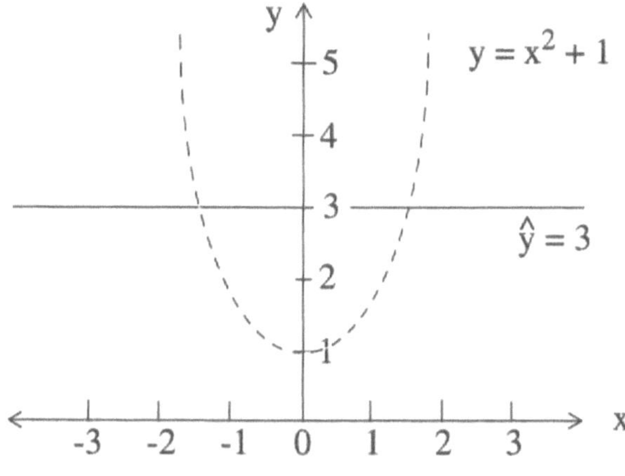

Figure 4

> **The value $r = 0$ does not necessarily mean that there is no relationship between the variables.** While this is a possibility, it is also possible that there is a non-linear relationship between the variables. In this case there is a non-linear relationship between *x* and *y* described by $y = x^2 + 1$.

Example 3 Return to The Daniel Company

As we have seen, the marketing manager of the Daniel Company, a new entry in the fine baked goods market, wants to predict sales volume on the basis of advertising expenditure. To establish a quantitative relationship between these variables data were obtained from reports for eight monthly periods chosen at random from the company's files as shown in Table 7. Advertising expenditure, denoted by *x*, is given in thousands of dollars and sales volume, denoted by *y*, is given in hundreds of thousands of dollars. (This situation is first considered in Example 1 of Sections 12.2 (page 411) and 12.3 (page 420)).

Table 7

Advertising: x	6	4	9	14	12	16	10	21
Sales volume: y	15	17	14	27	21	28	22	24

Answer questions (b), (c) and (d) of Example 1 for the Daniel Company in terms of the data in Table 7.

From Example 1 of Section 12.3 (page 420), we have: $\hat{y} = 12.65 + 0.726x$ is the regression line; $\Sigma x = 92$, $\Sigma y = 168$, $\Sigma x^2 = 1270$, $\Sigma xy = 2086$, and $n = 8$.

(b) To compute *r*, we require:

$$\begin{aligned}\Sigma y^2 &= (15)^2 + (17)^2 + (14)^2 + (27)^2 + (21)^2 + (28)^2 + (22)^2 + (24)^2 \\ &= 3724\end{aligned}$$

Therefore:

$$\begin{aligned}r &= \frac{8(2086) - 92(168)}{\sqrt{8(1270) - (92)^2}\sqrt{8(3724) - (168)^2}} \\ &= \frac{1232}{\sqrt{1696}\sqrt{1568}} \\ &= 0.76\end{aligned}$$

(c) The value $r = 0.76$ suggests that there is a moderately strong, direct, linear relationship between x and y.

(d) The scatter diagram and regression line are shown in Figure 5.

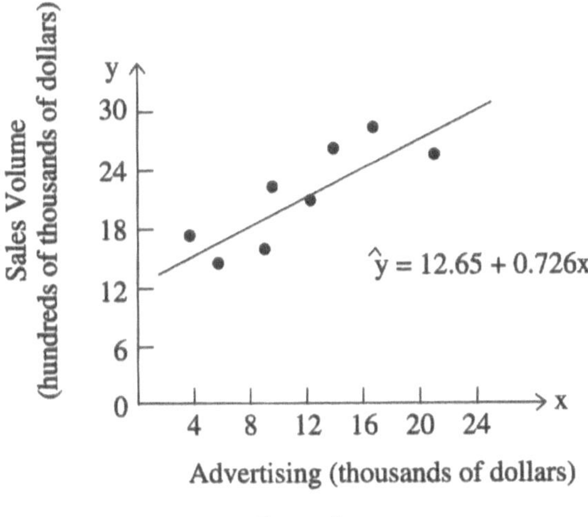

Figure 5

The data points generally lie near but not on the regression line. SSE is positive and a major part of the variation in the y values is not explained by the independent variable, as we noted in Example 2 of Section 13.1 (page 438).

Cause-and-Effect

We commented on this in Section 12.4, but the issue warrants some further observations.

> A correlation coefficient r close to 1 or -1 indicates that there is a strong **linear relationship between the variables**, but this by itself does **not** establish that one is a cause of the behavior of the other. It is possible that some third factor influenced either or both variables. It is also possible that the correlation occurred by chance and is nonsensical or spurious.

An interesting example of a correlation caused by a third factor arose from a study in the Scandinavian countries in the early 20th century. We have all heard the popular myth that storks fly in the newborns, but it was shown that a high positive correlation actually existed between the yearly human birthrate and the stork population. One cause-and-effect "explanation" was that there were more storks to fly in the babies! On further analysis, a third variable was shown to be affecting each of the first two, namely the severity of the winter. In very cold winters people didn't leave their homes as often and fireplaces were continually in use. Since storks tend to make their nests near chimneys, this kept them warm so that fewer newborn storks perished. Since people were indoors more often in warmth, but without well-known modern distractions, the human population increased as well.

Numerous studies have also exhibited spurious correlations between variables that are obviously unrelated, such as the high positive correlation between the production of cars in America in the 1950's and the population of Latin America. Both variables increased during that decade, but it would be extraordinary to conclude that the increased car production in the United States caused an increase in the Latin American population, or vice-versa, or even, for that matter, that these variables relate in any way at all.

Serious examples involving highly correlated variables abound today. There is a high correlation between heavy smoking of long duration and the incidence of lung cancer. Does this imply that heavy smoking over a long period of time "causes" lung cancer? Not by itself. At the same time, we should note that studies which have formed the basis for the claim that smoking is a cause of lung cancer go beyond the correlation analysis discussed here. A recent study [1] or men under 55 established that there is a correlation between baldness on top, called vertex baldness, and heart attacks. No one is jumping to the conclusion that baldness causes heart attacks, but the need for further studies is indicated. They may fail to support such an association, or find some common factor for both conditions, such as male hormones.

REFERENCE

[1] L. Altman, "A Bald Spot on Top May Predict Heart Risk," *The New York Times*, Feb 24, 1993.

EXERCISES

1. (a) Compute the sample correlation coefficient for the data (1/2, 0), (1, 1), (3/2, 2), (2, 3), (3, 7), (4, 7) as per Exercise 1 of Section 13.1 (page 439).

   (b) How is the value of r to be interpreted?

2. (a) Compute the sample correlation coefficient for the data (0, 5), (-1, 4), (1, 3), (2, 1), (-2, 0), (5/2, -1), (-5/2, -2), as per Exercise 2 of Section 13.1 (page 440).

(b) The scatter diagram that you previously plotted for these data show a marked relationship between the variables. In light of this observation, can you explain why the sample correlation coefficient is so low?

3. (a) Compute the sample correlation coefficient for the data (1/2, 5), (1, 4), (3/2, 4), (2, 3), (3, 1), as per Exercises 3 of Section 13.1 (page 440).

(b) How is the value of r to be interpreted?

4. Refer to Exercise 4 of Section 13.1 (page 441), which shows the "hardness" y (in "Rockwell" units) of an alloy when 7 specimens were treated at 500°C for different time periods, x (in hours). These data are reproduced in Table 8.

Table 8

x	4	5	6	8	10	11	13
y	41	48	50	53	54	53	52

(a) Compute the sample correlation coefficient.

(b) May we conclude that the hardness of the alloy is caused by the length of the treatment? Explain.

5. Refer to Exercise 5 of Section 13.1 (page 441), which gives data comparing the average daily fuel consumption y in single family houses (in gallons) against the average daily temperature x (in °F). These data are reproduced in Table 9.

Table 9

x	10	15	20	25	30	35	40	45	50	55
y	8	8.4	7.5	7.2	7.0	6.1	5.0	4.6	3.5	2.7

(a) Compute the sample correlation coefficient.

(b) May we conclude that the changes in temperature cause fuel oil consumption to vary? Explain.

6. Refer to Exercise 2 of Section 12.3 (page 423), which gives data comparing the grade for a statistics final exam y against the study time x (in hours) for that exam.

(a) Compute the sample correlation coefficient.

(b) May we conclude that the variation in the statistics final exam grade is caused by studying time? Explain.

7. Refer to Exercise 3 of Section 12.3 (page 424), which gives data comparing the average number of accounting mistakes per month y of a random sample of assistant accountants against their years of service x at the Superior Accounting Firm.

(a) Compute the sample correlation coefficient.

(b) May we conclude that more experience on the job causes the average number of mistakes per month to decline? Explain.

8. Refer to Exercise 4 of Section 12.3 (page 424), which gives data comparing systolic blood pressure (SBP), y, against age x (in years).

(a) Compute the sample correlation coefficient.

(b) Is it reasonable to claim that the aging process causes the SBP to rise? Explain.

9. Refer to Exercise 6 of Section 12.2 (page 417), which gives data comparing the reduction y in blood cholesterol level against average weekly exercise time x (in minutes).

(a) Determine the sample correlation coefficient.

(b) May we conclude that changes in blood cholesterol are caused by variation in average weekly exercise time? Explain.

10. (a) Determine the sample correlation coefficient for (-2, -3), (-1, 0), (0, 1), (1, 0), (2, -3).

(b) May we conclude that there is no relationship between x and y? Explain.

(c) If your answer to (b) is no, do you see a relationship between x and y? Explain.

13.3 ■ The Population Correlation Coefficient

The value of the sample correlation coefficient r computed from a randomly chosen sample of n pairs of observations depends on the sample chosen. The statistic r serves as an estimate of the underlying **population correlation coefficient**, denoted by the Greek letter ρ (rho).

As we previously noted in Section 12.4, linear regression analysis **presupposes** that y is a random variable. The values of x are not assumed to be randomly chosen. In linear correlation analysis both x and y are random variables. Moreover, both random variables are assumed to be normally distributed with mean and standard deviation μ_x, σ_x and μ_y, σ_y, respectively.

13.4 ■ A Hypothesis Test for ρ

If we obtain an r value which is "very close" to 0 or very close to 1 or -1, then it would seem reasonable to conclude that there is a very weak or non-existent linear relationship between the variables in the former case and a very strong linear relationship between the variables in the latter case. The question is, how do such values of r reflect on ρ? An r value very close to 0 with one sample size may not be very close to 0 in terms of another sample size and judgments about the meaning of "very close" differ from person to person. Moreover, there are intermediate values of r, such as -0.4 or 0.65; what sense is to be made of them? How do they reflect on ρ?

To help answer some of these questions we look at one of the most important hypothesis tests for ρ.

Hypothesis Test for $\rho = 0$

$H_o: \rho = 0$: There is no linear relationship between x and y.

$H_a: \rho \neq 0$: There is some linear relationship between x and y. This does not take a position on how strong that linear relationship is.

The test statistic to carry out the test of H_o is

$$t = r\sqrt{\frac{n-2}{1-r^2}},$$

which, assuming H_o is true, satisfies the t distribution with $n-2$ degrees of freedom.

This analysis **requires** that the values of x and y be taken from **normally distributed populations.**

Example 1

Consider the following data relating x and y, as shown in Table 10.

Table 10

x	1	1	2	3	4	4	5
y	9	1	8	1	10	1	2

Test the null hypothesis $H_o: \rho = 0$ at the 0.10 level of significance.

We have:

$$H_o: \rho = 0$$
$$H_a: \rho \neq 0$$
$$\alpha = 0.10$$

The test **assumes** that the x and y values are chosen from normally distributed populations. The question of whether a population is normal or not is addressed in Sections 8.4, 16.4, and 17.3. This test is two-tailed with decision bounds $\pm t_{.05} = \pm 2.015$ for 5 degrees of freedom (see Figure 6).

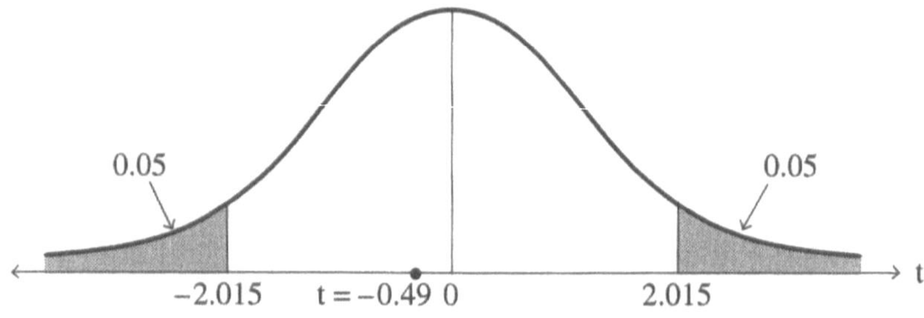

Figure 6

We need r for the test statistic. We have the sums $\sum x = 20$, $\sum y = 32$, $\sum xy = 83$, $\sum x^2 = 72$ and $\sum y^2 = 252$. This yields:

$$\begin{aligned} r &= \frac{7(83)-20(32)}{\sqrt{7(72)-(20)^2}\sqrt{7(252)-(32)^2}} \\ &= \frac{-59}{\sqrt{104}\sqrt{740}} \\ &= -0.213 \end{aligned}$$

Since

$$t = -0.213\sqrt{\frac{7-2}{1-(-0.213)^2}} = -0.49,$$

we accept the null hypothesis that there is no linear relationship between x and y, or reserve judgment.

Example 2 Another Return to the Daniel Company

Refer to Example 3 of Section 13.2 (page 448) where we determine $r = 0.76$ for the Daniel Company's advertising-expenditure sales-volume data. Test the null hypothesis of no linear relationship between advertising expenditure and sales volume at the 5% significance level.

We have:

$$\begin{aligned} &H_o : \rho = 0 \\ &H_a : \rho \neq 0 \\ &\alpha = 0.05 \end{aligned}$$

This test presupposes that the values of x and y are taken from normally distributed populations. It is two-tailed with decision bounds $\pm t_{.025} = \pm 2.447$ for 6 degrees of freedom (see Figure 7).

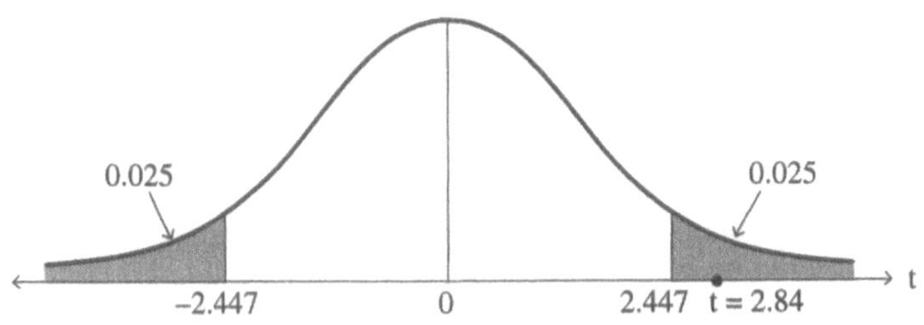

Figure 7

Our test statistic t is:

$$t = 0.76\sqrt{\frac{8-2}{1-0.57}} = 2.84$$

Since $2.84 > 2.447$, we reject the null hypothesis of no linear relationship between the variables and conclude that one exists between advertising expenditure and sales volume.

The result of this hypothesis test comes as no surprise since $\rho = 0$ does seem extreme in terms of the sample statistic $r = 0.76$ and the coefficient of determination $r^2 = 0.57$, which is interpreted as saying that about 57% of the variation in sales volume is explained by advertising expenditures.

This suggests that the sample regression line $\hat{y} = 12.65 + 0.726x$ determined in Example 1 of Section 12.3 may be used, with caution, to predict an average monthly level of sales for a given monthly advertising expenditure. In addition, the cautions noted in Section 12.3 regarding the accuracy of these predictions still apply.

EXERCISES

1. Increases in temperature beneath the Earth's surface are of great interest to geologists and mining engineers. Table 11 shows such temperature increases y for various randomly chosen depths x at a well near Pittsville. Temperatures are recorded as increases over the temperature at a basepoint 28 meters below the Earth's surface. (This situation is first considered in Exercise 1 of Section 12.2, page 415.)

Table 11

x	28	67	179	246	293	314
y	0	1.5	4.3	9.2	10.4	10.7

(a Test the null hypothesis $H_o : \rho = 0$ at the 1% level of significance.

(b) What **assumption** do we make in carrying out this test?

(c) Would you be inclined to use your regression equation to make predictions about the mean of the dependent variable? Explain.

2. Gerry Bond, Chairwoman of the Huxley College Statistics Department, chose at random a sample of 10 statistics students to see if there was a relationship suggested between the number of hours of study x in the week immediately preceding a statistics final exam and the grade y received on that exam. She obtained the data shown in Table 12. (This situation is first considered in Exercise 2 of Section 12.2 page 415.)

Table 12

x	25	12	18	26	19	20	23	15	22	8
y	93	57	55	90	82	95	95	80	85	61

(a) Test the null hypothesis of no linear relationship between studying time and final exam grade at the 0.05 level of significance.

(b) (i) On what **statistical assumptions** is this test based?

(ii) How might we explore or determine whether these assumptions are satisfied? Explain.

(c) Would you be inclined to use the regression line to predict mean statistics final exam grades? Explain.

3. The Superior Accounting firm wishes to investigate whether there is a relationship between years of service x and "inefficiency" rate y (that is, the average number of accounting mistakes per week) of its assistant accountants. Noted in Table 13 are the results obtained for nine randomly selected accountants over the last 26 weeks. (This situation is first considered in Exercise 3 of Section 12.2, page 416.)

Table 13

x	3	5	6	6.5	9	10	14.5	17	23
y	14.5	11	14	9	13.5	7	8	6.5	7.5

(a) Test the null hypothesis of no linear relationship between experience level and accounting mistakes committed at the 0.02 level of significance.

(b) Suppose it was found that the underlying populations on which x and y are defined are U shaped. What effect, if any would this have on the hypothesis test for $\rho = 0$? Explain.

(c) In light of your result from (a), can we conclude that there is no linear relationship between the variables? Explain.

(4) A group of physicians wants to better understand the relationship between age (in years) and systolic blood pressure (SBP). They examined a randomly selected sample of 10 people and obtained the data shown in Table 14. (This situation is first considered in Exercise 4 of Section 12.2, page 416.)

Table 14

Age (x)	16	18	22	29	38	46	51	59	64	65
SBP (y)	128	127	121	125	139	138	141	140	147	158

(a) Test the null hypothesis $H_o : \rho = 0$ at the 1% level of significance.

(b) In light of your result from (a), can we conclude that there is a strong linear relationship between the variables? Explain.

5. Researchers with the Tasty Manufacturing Company conducted an experiment to study the relationship between storage temperature x and the shelf life y of their new Heavenly-Fat-Free Yogurt. Ten randomly selected containers of Heavenly-Fat-Free Yogurt were stored at 10 randomly selected temperature levels (in degrees Fahrenheit) and the shelf life (in days) was recorded. The results are shown in Table 15. (This situation is first addressed in Exercise 5 of Section 12.2, page 416.)

Table 15

x	35	38	35	42	44	47	51	53	55	56
y	43	45	41	27	23	12	6	4	2	1

(a) Test the null hypothesis of no linear relationship between x and y at the 0.05 significance level.

(b) (i) On what **assumptions** is this test based?

(ii) How might we explore or determine whether these assumptions are satisfied? Explain.

(c) Would you be inclined to use the regression line obtained in Exercise 5 of Section 12.3 to predict the average shelf life at a storage temperature of 30°F? Explain.

6. Many physicians have been recommending exercise for their patients, particularly for those who are overweight. To analyze the relationship between exercise and cholesterol level, Dr. Edward Jones took a random sample of 8 patients who did little exercise. He measured their cholesterol levels and started them on regular exercise programs. After four months he recorded the number of minutes per week each patient exercised on average and measured each one's cholesterol level. The results are shown in Table 16. A negative "reduction" in cholesterol level means that the patient's cholesterol level increased. (This situation is first considered in Exercise 6 of Section 12.2, page 417.)

Table 16

Weekly exercise level (x)	43	65	100	43	16	120	30	10
Reduction in cholesterol level (y)	38	13	32	20	-5	27	23	-9

(a) Test the null hypothesis $H_o : \rho = 0$ at the 0.05 level of significance.

(b) In light of your result from (a), may we conclude that there is no linear relationship between the variables? Explain.

7. The data in Table 17 gives the increase in weight y (in decigrams) of 7 randomly selected male friendly furry creatures over a three month period during which they were given randomly selected doses, x, of vitamin B (in milligrams). (This situation is first considered in Exercise 7 of Section 12.2, page 417.)

Table 17

x	0.2	0.3	0.5	0.7	1.0	1.2	1.5
y	25	27	32	39	38	34	26

(a) Test the null hypothesis of no linear relationship between Vitamin B dosage and weight gain in friendly furry creatures at the 10% significance level.

(b) On what statistical premise is this test based?

(c) In light of your result from (a), may we conclude that there is a non-linear relationship between the variables? Explain.

SELF-TESTS FOR PARTS ONE THROUGH FOUR

Allow 100 or so minutes for each self-test. Go over each one before going on to the next.

Self-Test I

1. A government agency investigating the claims of advertisers decides to compare two brands of deodorants. Each brand is advertised as being the most effective in eliminating body odor. A random sample of 50 people is chosen and asked to apply deodorant A, and another random sample of 50 is given deodorant B. Each person is then asked to stand in front of a sensitive aroma-detecting machine used in many TV commercials, and a reading is made. The average reading for deodorant A is found to be 5.2 with a standard deviation of 2, and the average for deodorant B is 4.7 with a standard deviation of 2.3.

   Do these results indicate any significant difference in body odor-reducing ability between the two deodorants at a 0.05 level of significance? Explain.

2. For a sample of randomly selected mothers who are smokers Table 1 shows the number of packs of cigarettes smoked per day during pregnancy, x, and weights of newborn babies, y.

Table 1

x	0	0	0.5	1.1	1.5	2.3	2.5	5.0
y	8.2	8.0	7.9	7.6	7.4	7.1	7.3	6.9

(a) Calculate the coefficient of correlation for these data. What is your interpretation of the result obtained?

(b) Find the least squares line of best fit to these data.

(c) Find the predicted value of y for $x = 3$. How would you interpret this value?

(d) What is the coefficient of determination? How is this value interpreted?

(e) May we conclude that the weight of new born babies is determined by the number of packs of cigarettes smoked per day during pregnancy? Explain.

3. In a random sample of 200 houses in Keith City 20 were found to have faulty wiring. If we estimated the true proportion of houses in the city with faulty wiring as being 0.10, what could we say with a probability of 0.98 about the size of the error of the estimate?

4. The Kaskin Company claims that the mean lifetime of its Glaston steel-belted tires is 40,000 miles. As a staff member of a consumer testing service, you have been asked to test this claim at the 0.01 level of significance.

(a) What alternative hypothesis would be appropriate? Explain.

(b) Before doing the testing a decision must be made as to whether to use a large or small sample. On what basis would you make this decision? Explain.

(c) A random sample of 100 tires was found to have a mean lifetime of 39,850 miles with a standard deviation of 500 miles. Would you accept/reserve judgment or reject the manufacturer's claim? Explain.

(d) Assuming that it is feasible to work with a small sample, suppose that a random sample of 15 tires yielded the results cited in (c), would you accept/reserve judgment or reject the manufacturer's claim? Explain.

5. It has been argued that a presentation supported by figures is more credible than one that is not because figures add substance to a presentation by providing the reader with tangible evidence that can be used to evaluate the presentation in concrete terms.

Discuss this point of view, citing specific examples in support of your discussion.

6. "80 percent of the A grades in my math class were earned by Plutonians," commented Bob Levy to his colleague Amy Flores. "This means that they have an innate ability for math." Are there other interpretations of this figure? Are there other figures that might be relevant? Discuss.

Self-Test 2

1. In connection with hypothesis testing,

   (a) When is a type I error committed?

   (b) When is a type II error committed?

2. For a randomly chosen sample of 25 automobile insurance claims paid out during a one-year period it is found that the average is $400 per claim with a standard deviation of $100. Find the 90% confidence interval for the mean claim payment.

3. For the following distribution of the amount of overtime pay received in a certain week by 50 workers in the accounting department of the Andrius Company determine:

Amt. Earned ($)	No. of Workers
10.00-19.99	16
20.00-29.99	9
30.00-39.99	10
40.00-49.99	7
50.00-59.99	8

   (a) The third quartile,

   (b) the mean,

   (c) the standard deviation.

   (d) On what **assumption** is your standard deviation calculation based? Explain.

4. The parachutist Ogden Stiles has a probability of 0.40 of landing on a preset target. Consider a sequence of 50 jumps.

   (a) Define a probability model for this situation.

   (b) On what **assumption** is this model based? Is this assumption realistic? Discuss.

(c) State, but do not evaluate, the probability that Ogden will land on target 22 times.

(d) Can a normal curve approximation be used to estimate the probability that Ogden lands on target 22 times? Explain.

(e) Assuming that the answer to (d) is yes, find a normal curve approximation for the probability that Ogden lands on target 22 times.

5. Ann Curtis, quality control manager at Lixfield Electrical Products, wants to test whether or not the mean life time of the company's premium light bulbs is the same for the two plants where they are manufactured. It has been established that the underlying populations of bulb life times at the two plants are normally distributed.

(a) What other **condition** must be satisfied in order for Ann to conduct a small sample test for equality of means?

(b) A randomly drawn sample of 10 bulbs drawn from plant A had a mean lifetime of 427 hours with a standard deviation of 14 hours while a randomly drawn sample of 10 bulbs drawn from plant B had a mean lifetime of 400 hours with a standard deviation of 9 hours. Test for the required condition at the 0.10 level.

(c) Do we have a green light to undertake a test for equality of means? Explain.

(d) Assuming that the answer to (c) is yes, test for equality of means at the 0.05 level.

(e) What conclusion do you reach?

5. State and discuss in appropriate detail the problems that arise from newspaper polls and their modern equivalents.

6. The Harold Institute for Urban Studies is seeking to obtain a sense of the opinion of Baxter City's residents about the problems of the poor. To obtain this they interviewed a random sample of Baxter City's residents whose names appeared in a list of those who voted in last year's municipal elections. Will doing this give the Harold Institute what it seeks? Explain.

Self-Test 3

1. Bell City's Medical Emergency Response Team (**MERT**) claims that at least 35% of its calls are life threatening emergency calls. The problem is to test this claim at the 0.05 level.

   (a) Formulate appropriate null and alternative hypotheses.

   (b) What is the basis for your alternative hypothesis? Explain.

   (c) Of a random sample of 200 calls taken from MERT's extensive files, 64 were for life threatening emergencies. What conclusion do you reach from this? Explain.

2. Population $Q = \{3, 7, 9, 11, 13\}$. Consider the process of drawing a sample of size 2 at random from Q.

   (a) Define a probability model for this process.

   (b) On what assumption is this model based?

   (c) Define the sampling distribution of the mean $\overline{x}$.

   (d) Determine the mean of $\overline{x}$.

   (e) Determine the standard deviation of $\overline{x}$.

   (f) Find $P(6 \le \overline{x} \le 8)$.

3. Table 2 specifies a random sample of waiting times (in minutes) of patients seeking treatment in the emergency room of Lister Hospital during the last six months.

Table 2

20	13	24	15
24	13	20	11
30	25	16	13
16	27	24	13

(a) Find: (i) the mean, (ii) median, (iii) mode, (iv) range, (v) variance, (vi) standard deviation.

(b) Construct the frequency distribution for the data using classes 10-14, 15-19, 20-24, 25-29, 30-34.

(c) For the class 15-19, specify its (i) class limits, (ii) class mark, (iii) class width, (iv) class boundaries.

(d) Construct the histogram for the frequency distribution.

(e) Construct the cumulative less than distribution for the data.

(f) Sketch the ogive for the cumulative less than distribution.

4. Dr. Joan Zuber, Director of Lister Hospital, wants to obtain confidence intervals for the mean waiting time and variance of the population of emergency room waiting times. Based on the sample given in Table 2,

(a) Construct a 95% confidence interval for the population variance;

(b) Construct a 95% confidence interval for the population mean.

(c) What condition underlies these constructions?

5. (a) What is the hypothesis of elementary errors?

(b) What is its significance?

6. It has been found that computers elicit more honest response about sex behavior than human beings. See: A. Harmon, "Underreporting Found on Male Teen-Ager Sex," *The New York Times*, May 8, 1998; A20. C. Turner, L. Ku, S. Rogers, L. Lindberg, J. Pleck, F. Sonenstein, "Adolescent Sexual Behavior, Drug Use, and Violence: Increased Reporting with Computer Survey Technology," *Science*, vol. 280, May 8, 1998; 867-871.

Does this raise questions about previous methodologies used to obtain data on "sensitive" matters? Comment.

7. A National Household Survey on Drug Abuse stated that 9% of those 12 to 17 "reported" using an illicit drug in the survey taken in 1999 compared with 11.4% in the one taken in 1997. For the 18 to 25 group 18.8% "reported" illicit drug use in the 1999 survey compared with 14.7% in the one in 1997. It was noted that the 1999 study did not cover active-duty military personnel, people in prison or drug-treatment centers or homeless people not in shelters when the survey was conducted. D. Stout, "Use of Illegal Drugs Is Down Among Young, Survey Finds," *The New York Times*, Sept. 1, 2000; A18.

   (a) Dr. Donna E. Shalala, Secretary of Health and Human Services, interpreted the results as meaning the 'most of our young people are obviously getting the message that drugs are not the stuff of dreams, but the stuff of nightmares.' Would you agree with this interpretation? Explain.

   (b) Do you have reservations about the accuracy of the afore figures? Explain.

8. Observing that it was 80 degrees in Miami and 20 degrees in New York, Kristin commented to her friend Laura, "I think I'll fly down to Miami; it's four times as warm there as it is here." Any problem with this statement (not the sentiment)?

Self-Test 4

Null and alternative hypotheses are stated for Exercises 1-5. Also given are the sample size(s) and level of significance. State the test statistic to be examined, the accept/reserve judgment and reject regions, the probability of committing a type I error, when type I and II errors would occur, and condition(s) that must be imposed on the population(s) for the test to be viable.

1. $H_o: \mu = 250,\ H_a: \mu \neq 250,\ n = 8,\ \alpha = 0.05$.

2. $H_o: \pi = 0.30,\ H_a: \pi > 0.30,\ n = 500,\ \alpha = 0.01$.

3. $H_o: \sigma^2 = 25,\ H_a: \sigma^2 < 25,\ n = 12,\ \alpha = 0.01$.

4. $H_o : \mu_1 - \mu_2 = 0,\ H_a : \mu_1 - \mu_2 < 0,\ n_1 = 32,\ n_2 = 36,\ \alpha = 0.05$.

5. $H_o : \sigma_1^2 - \sigma_2^2 = 0,\ H_a : \sigma_1^2 - \sigma_2^2 < 0,\ n_1 = 11,\ n_2 = 15,\ \alpha = 0.05$.

6. A random sample of 30 of 850 graduating seniors of State University received an average starting salary of $25,000 a year, with a standard deviation of $3200. Find 95% confidence limits for the mean starting salary for all 850 graduating seniors.

7. The Last National Bank wishes to estimate the average amount of time that a client must wait before being served by its special services division. A random sample of 6 clients yielded waiting times of 40, 50, 45, 65, 60 and 75 minutes.

   (a) Determine 90% confidence limits for the mean waiting time of clients dealing with special services.

   (b) How is the result obtained in answer to (a) to be interpreted?

   (c) If the sample mean is used as an estimate for the population mean, what is the error inherent in the estimate at the 95% confidence level?

   (d) What assumption underlies the results obtained in anwer to (a), and (c)?

   (e) Suppose the **assumption** given in answer to (d) was later found to be dead wrong; what affect would this have on the results obtained in answer to (a), (b), and (c)? Explain.

8. Table 3 gives the number of accidents involving deaths, serious passenger injuries or substantial damage to plane for nine airlines in the most recent five year period.

Table 3

Airline	A	B	C	D	E	F	G	H	I
Accidents	14	12	12	10	9	5	3	2	2

Based on this data Janet Reed, a spokeswoman for airline H, claimed that no airline had a better safety record than H; "fly H for safety" became the airline's motto. Is this data a suitable measure of airline safety?

9. "Statisticians Occupy Front Lines In Battle over Passive Smoking" by Jerry Bishop, which appeared in *The Wall Street Journal* a few years ago, reported that: "In the controversy over passive smoking, the difference between 90% and 95% has become a matter of life and death."

The U.S. Environmental Protection Agency says that there is a 90% probability that the risk of lung cancer for passive smokers is somewhere between 4% and 35% higher for those who aren't exposed to environmental smoke.
This statement has as its basis a 90% confidence interval. A tobacco company consultant countered with the statement: 'Ninety-nine percent of all epidmiological studies use a 95% confidence interval.'

Assuming large sample size, compare the 90% and 95% confidence intervals in terms of reliability, and precision.

Self-Test 5

1. According to some estimates the population of basic statistics books currently in print numbers 350 titles, all awaiting eager minds to grab them. A student organization, interested in determining the mean price of these titles, arranged for a random sample of 10 titles to be chosen from available lists. This sample yielded a mean price of $75, with a standard deviation of $15.

   (a) Determine 95% confidence limits for the population mean.

   (b) How is the result obtained in answer to (a) to be interpreted?

   (c) Is it necessary to introduce any special assumptions about the population being sampled from? Explain.

2. The operations manager of Torez Taxis wants to estimate the mean age of the company's 800 car fleet. A random sample of 36 car records yielded a mean of 5.5 years with a standard deviation of 2.5 years.

   (a) If the sample mean is used as an estimate for the mean age of a the company's fleet, what is the error inherent in the estimate at the 0.98 confidence level?

   (b) How is the result obtained in answer to (a) to be interpreted?

   (c) What size sample would have to be employed if the operations manager wanted to estimate the mean age of the company's fleet so as to be able to assert with probability 0.98 that the error in the estimate does not exceed 1/2 years?

3. On reviewing the two preceding confidence interval questions we see a small versus large sample analysis being employed, and the 95% versus the 98% confidence level being employed.

   Discuss the pros and cons of these characteristics in doing a confidence interval analysis in terms of (a) reliability, (b) precision, and (c) assumptions that must be made.

4. Seven students who had Professor Arthur A. for an introductory science course gave him ratings shown in Table 4, where 50 corresponds to Outstanding, 20 to Good, 10 to Satisfactory, 0 to Poor, and -50 to Terrible.

Table 4

Rating	No. of Responses
50	1
20	2
10	2
0	1
-50	1

What numerical value would you use to express the class' overall view of Professor A.'s teaching performance? Explain. How is this numerical value interpreted?

5. Table 5 shows a sample of advertising expenditures, x, in thousands of dollars, and the associated sales volumes, y, in thousands of dollars, for the Rasa Company during ten randomly selected months.

Table 5

x	1.2	0.8	1.0	1.3	0.7	0.8	1.0	0.6	0.9	1.1
y	101	92	110	120	90	82	93	75	91	105

(a) Find the least squares line of best fit to these data.

(b) If the Rasa Company has budgeted $1000 for advertising in a month, what is the predicted sales volume? How would you interpret the predicted value?

(c) Calculate the coefficient of correlation for these data. What is your interpretation of the result obtained?

(d) What is the coefficient of determination? How is this value interpreted?

(e) May we conclude that sales volume is caused by advertising expenditures? Explain.

(f) Would you use the regression line to predict the average sales volume for an advertising expenditure of $3000 per month? Explain.

6. In a survey it conducted, Big Bargains Department Store asked 9 randomly selected credit card holders to state their yearly store charges y (in tens of dollars) and yearly gross earned income x (in thousands of dollars). Table 6 shows the survey results obtained.

Table 6

x	27	38	46	30	71	89	19	124	146
y	21	23	26	28	35	37	34	45	68

(a) Test the null hypothesis of no linear relationship between annual income and annual credit card charges at the 0.05 level of significance.

(b) Suppose it were found that the populations which underlie *x* and *y* are J shaped. What effect, if any, would this have on the analysis you carried out? Explain.

(c) In light of your work in (a), would you be inclined to use your regression line to predict the mean values of the dependent variable? Explain.

7. The Association for the Advancement of Business Research (AABR) gives its accreditation to colleges and universities that have a strong business research dimension. Ecap University applied for AABR accreditation, citing as evidence of its business research dimension the fact that 600 papers were published by its business faculty over the last three years for an average of two papers per faculty member per year. Is this the kind of data that an evaluation team should focus on as an indication of a strong business research dimension? Explain.

8. Determine the p-value for the following test situations:

(a) $H_o : \mu = 120,\ H_a : \mu \neq 120,\ z = 1.74$

(b) $H_o : \mu = 150,\ H_a : \mu < 150,\ z = -2.23$

(c) In terms of the p-value obtained in answer to (b), what action would be taken on H_o for $\alpha = 0.02$? Explain.

9. (a) What does it mean to say that an hypothesis test is robust?

(b) Why is the robustness of a test an important issue?

10. "Don't cancel the course in Modern Lithuanian Literature," said the Chairman of the East European Languages Department to the Dean of Academic Affairs of Ecap University. "The course registration increased by 25% over last week and with such rapid growth I'm optimistic that we'll have a strong registration," he further pointed out. Is the percentage growth in the course's enrollment the figure that should be focused on in making a decision on whether or not to run the course?

Self-Test 6

1. Janet Wood, reporter for *The Arley College Informer*, has been asked by her editor to interview a sample of 12 of Arley's 520 faculty about the challenges and rewards of college life. Arley College consists of six schools, Arts and Sciences, Business, Computer Science and Information Systems, Education, Medicine, and Law. The schools are broken into departments in the usual way. Describe the sampling method open to Janet, their advantages and disadvantages, and how she might implement them.

2. Describe the characteristics of nominal, ordinal, interval, and ratio data and give examples of each kind.

3. The following is a probability model for one of Dapper Dan's dice: $S = \{1, 2, 3, 4, 5, 6\}$,

$$P(1) = \frac{1}{10}, \quad P(2) = \frac{2}{10}, \quad P(3) = \frac{1}{10}, \quad P(4) = \frac{3}{10}, \quad P(5) = \frac{1}{10}, \quad P(6) = \frac{2}{10}$$

Let A denote the event than an even number shows.

(a) Find $P(A)$.

(b) State the relative frequency interpretation of the conclusion obtained in (a).

(c) In tossing Dan's die 2000 times an even number showed 800 times. Does this evidence establish that the conclusion reached in (a) is not valid? Explain.

(d) Is the conclusion reached in (a), interpreted in relative frequency terms, true? Explain.

(e) Are the events "a number greater than 2 shows" and "an even number shows" independent? Explain.

4. Arnold Williamson believes that the probability that he will receive an A in the statistics course he is taking this semester is 0.95. How is this probability value to be interpreted? Explain.

5. (a) The 98% confidence interval, large sample case, for the population mean μ is:

$$\overline{x} - 2.33\frac{\sigma}{\sqrt{n}} < \mu < \overline{x} + 2.33\frac{\sigma}{\sqrt{n}}$$

How is this confidence interval interpreted in relatively frequency terms?

(b) The 90% confidence interval obtained for a population mean μ is $30 < \mu < 40$.

(i) Is μ in this interval or not? Explain.

(ii) How is the 0.90 probability value to be interpreted in this situation?

6. In the article "Powell in the Middle", by Michael Hirsh and Roy Gutman (*Newsweek*, Oct. 1, 2001, 26-29), the following assertion is made in boldface without further information being given:

"71% favor striking terrorist bases even if civilians die, but 59% say we should take time to plan a response that will work."

(a) Would you accept the figures as accurate without additional information being given? Explain the basis for your view.

(b) If you answered No, what additional information would you want provided. Explain.

7. The votes are in. Adolf Hitler is the most evil person of the millennium and Thomas Edison is the most influential, according to a poll of *The Post's* online users."

(A. Soltis, "Post Readers: Hitler Was Most Evil," *New York Post, Nov. 17, 1999; 2).*

Could the results of this online poll be considered representative of the population in general? Explain.

PART FIVE

SELECTED TOPICS

- Index Numbers and Index Number Modeling
- Time Series Analysis and Forecasting
- Nonparametric Statistics
- Additional Tests of Hypotheses

14

Index Numbers and Index Number Modeling

14.1 ■ The Nature of Index Numbers

Index numbers are statistics used to measure the change in a quality that has occurred over time, or how a quantity compares with another. Index numbers are widely used to study fluctuation in business and economic activity. The initial time frame against which changes are measured is called the **base period** and the end of the time period is called the **given** or **current period.** Typically, the base and current periods are specific years and we will illustrate the various index number constructions using this format. However, index numbers are often calculated month by month, as well as a year by year.

The value of the index at the base year is set at 100.0 so that the corresponding value of the index at any future time is either above, below or 100.0, depending on whether the given year measurement has increased, decreased or remained the same.

For instance, from January 1, 1996 to January 1, 1998 the Dow Jones Industrial Average rose from 5117 to 7908 or 2791/5117 = 54.5%. If base year (January 1 1996) aggregate stock prices are defined as 100.0, then aggregate January 1, 1998 prices equal 154.5. Between 1987 and 1989 the average price of fuel oil in New York declined by 8.9%. Thus, on average, the price of fuel oil in New York in 1989 is 91.1 relative to an initial value of 100 two years previous.

In comparison to the cost of living in Denver, Colorado in 2005, the cost of living in Milwaukee, Wisconsin in the same year was estimated as 6.6% higher, whereas the cost of living in the Tampa-St. Petersburg, Florida area was estimated as 11.1% lower. Consequently, we would assign a 2005 cost of living index of 106.6 for Milwaukee and

89.9 for Tampa-St. Petersburg, relative to the Denver, Colorado (base) index of 100, indicated by writing 2005 = 100.

Example 1 Index Number for Stock

The average price per share of James Company stock was $14.23 in 2000, $17.37 in 2002, and $15.04 in 2004.

(a) Write an index comparing the 2004 and 2002 average price per share of James Company stock to that of 2000.

(b) Using 2002 = 100, determine an index that states the 2004 James Company's average price per share.

(a) For 2004, $\frac{15.04}{14.23} \cdot 100 = 105.7 \cdot$

For 2002, $\frac{17.37}{14.23} \cdot 100 = 122.1$

Therefore, we can say that relative to the 2000 average price, the 2004 and 2002 average share prices of James Company stock were 5.7% and 22.1% higher, respectively.

(b) For 2004, $\frac{15.04}{17.37} \cdot 100 = 86.6 \cdot$

Therefore, the average James Company stock price was 13.4% lower in 1997 than it was in 2002.

Index Number Types

There are three important types of economic index numbers:

(a) **Price indexes** measure the changes in prices that are paid or received by producers and consumers.

(b) **Quantity indexes** measure changes in production or consumptions of goods and services.

(c) **Value indexes** measure changes in the monetary worth of various commodities and services. That is, they show a change in the overall dollar value of the items being measured by combining price and quantity changes. When feasible, one may attempt to factor in changes in the quality of the items measured in order to obtain a more realistic figure, but such assignments are judgmental and prone to dispute.

Index numbers may be further categorized based on the mathematical approach used in their determination. We may construct "weighted" or "unweighted" indexes, depending upon whether or not we wish to attach a measure of importance to the price, quantity or value of each commodity used in constructing the index. For each case we may obtain either an "aggregative" or "average-of-relatives" type of index.

Finally, we adopt the following notation: Let P represent the price of an item, Q the quantity consumed or produced, and PQ the monetary value of the Q units of the given commodity. We use the subscript o as a reference to the base year and the subscript n for the given year.

EXERCISES

1. What is an index and what is the general purpose behind its construction?

2. Describe the three kinds of economic indexes.

3. Suppose that in Tulsa a typical basket of food items cost \$54 in 2000. At the same time the same basket of goods cost \$61.50 in New Orleans, but only \$50.25 in Omaha. Using the Tulsa 2000 rate as a base, construct an index that shows the relative overall food prices in New Orleans and Omaha at that time.

4. The price of a quart of milk in Smithville was \$0.90 in 2001. It rose to \$0.98 in 2003 and \$1.02 in 2005. Relative to its base year (2001) price, write an index describing the 2003 and 2005 prices of a quart of milk in Smithville.

5. Relative to the cost of living in Heightstown in 2002, the cost of living was 13.8% higher in Billstown, but 9.6% lower in New City. Construct an index for

Billstown and New City that states the 2002 cost of living in these two towns in comparison to Heightstown's 2002 cost of living.

14.2 ■ Unweighted Index Numbers

Simple Aggregative Index

The simplest index number to construct is the **simple** or **unweighted aggregative index**, defined by

$$I = \frac{\sum P_n}{\sum P_o}, \tag{1}$$

where $\sum P_n$ is the sum of the given-year prices and $\sum P_o$ is the sum of the base year prices. Their ratio is then multiplied by 100 to express the index as a percent.

Example 1 Index Number for Clothing Items

Consider a collection of basic clothing items used by a "typical" Ecap University student, listed in Table 1. Construct a simple aggregative index comparing the 2000 prices of these items with their prices three years earlier.

Table 1

Item	Unit Price	
	1997	2000
Shirt/blouse	$15	$25
Socks (pair)	1	2
Jeans (pair)	23	38
Shoes (pair)	50	65
Sweater	16	25
Total:	105	155

Given two years of data for comparison, we generally consider the earlier year to be the base year and the later year to be the given or current year. Therefore, from the data the simple aggregative index for 2000 is given by:

$$\begin{aligned} I &= \frac{\sum P_{2000}}{\sum P_{1997}} \cdot 100 \\ &= \frac{155}{105} \cdot 100 = 147.6 \end{aligned} \tag{2}$$

We can interpret this result in two ways. We can say that it cost 47.6% more in 2000 than in 1997 to purchase the group of items listed in Table 1. Equivalently, the 2000 aggregative cost of these goods is 147.6% of their previous value.

This procedure is simple to use, but has two major drawbacks. First, the index may be unduly affected by items having large price fluctuations. Suppose, for example, the Table 1 had listed a sixth article of clothing, a jacket whose price had dropped from $120 to $90 over the three year time span. Then the index would have been

$$\frac{245}{225} \cdot 100 = 108.9,$$

which suggests that there was only a small increase in overall prices, even though five of the six items in the group increased in price. The sharp change in the aggregative index was due to this sixth item having a much higher price in 2000 and 1997 than the other articles of clothing in the survey. Suppose, instead, that the sixth clothing item, underwear, diminished in price from $5 to $2, which represents a larger percentage drop in price (60% drop) than the previous change from $120 to $90 per jacket (25.0% drop), but which involves smaller absolute dollar values. For this case the aggregative index would be:

$$\frac{157}{110} \cdot 100 = 142.7$$

This value is much closer to the original aggregative index, which is due to the use of smaller dollar values for the underwear and despite the steeper percentage drop in the price of the underwater compared to that for the jacket.

Another problem is that the aggregative index may be changed by altering the units of the items being compared; the more dramatic the unit change and the greater the number of commodities whose units are changed, the greater the potential change in the aggregative index. If in Table 1, for example, we had expressed the unit price of jeans in terms of every ten pairs, then the 1997 and 2000 prices per unit would have been listed as \$230 and \$380, respectively. Our aggregative index would then be:

$$I = \frac{25+2+380+65+25}{15+1+230+50+16} \cdot 100 ,$$
$$= \frac{497}{312} \cdot 100 = 159.3$$

suggesting an overall price increase of 59.3% over the three year period in question, as opposed to the more modest estimate of a 47.6% increase obtained earlier for the same articles of clothing.

Thus, we see that the simple aggregative index may be easily, artificially and perhaps drastically altered to reflect different levels of price changes depending on the insights or whims of those conducting the analysis. For these reasons the simple aggregative index is not widely used today.

Arithmetic Mean of Price Relatives

We are able to eliminate the units problem, although not the high price fluctuation problem, by employing **price relatives**. For each commodity we compute the ratio of its price for the given year to that of the base year. The value of this ratio is fixed, irrespective of the unit used for the commodity. Then we use any measure of central tendency to arrive at a "typical" ratio. The arithmetic mean is generally used for this purpose. On carrying out these steps we obtain the **arithmetic mean of price relatives**, defined by

$$I = \frac{\sum \frac{P_n}{P_o} \cdot 100}{k} \qquad (3)$$

where k is the number of items whose price relatives are being averaged to form the index.

Example 2 Arithmetic Mean of Price Relatives

Based on the data of Table 1, construct an arithmetic mean of price relatives measuring the overall change in the prices of the given articles of clothing from 1997-2000. Use 1997 = 100.

After dividing the 2000 price of each item by its 1997 price and multiplying by 100, we obtain Table 2.

Table 2

Clothing	Price Relative
Shirt/blouse	$\frac{25}{15}\cdot 100 = 166.7$
Socks (pair)	$\frac{2}{1}\cdot 100 = 200.0$
Jeans (pair)	$\frac{38}{23}\cdot 100 = 165.2$
Shoes (pair)	$\frac{65}{50}\cdot 100 = 130.0$
Sweater	$\frac{25}{16}\cdot 100 = 156.3$
Total	818.2

Accordingly, we find the arithmetic mean of price relatives index to be:

$$I = \frac{818.2}{5} = 163.6 \tag{4}$$

We interpret this result as meaning that prices have increased 63.6% on average for the relevant goods over the course of the three year period. Not surprisingly, this result differs from that obtained by using the simple aggregative index.

The value of an index depends on the method employed, the years chosen for comparison, the commodities chosen for the index, and the regions chosen as a source of data.

As before, if one or few commodities undergo significant price changes, this may significantly alter the value of the mean of the price relatives index. For instance, suppose that the price of a pair of shoelaces tripled from $0.50 to $1.50 between 1997 and 2000. If we include this accessory item in our price relatives computation, the index becomes

$$I = \frac{818.2 + 300}{6} = 186.4,$$

which is significantly larger than 163.6, the arithmetic mean of price relatives index obtained from (3) without the shoelaces.

On the other hand, if we include the shoelaces, the revised simple aggregative index becomes

$$I = \frac{156.5}{105.5} \cdot 100 = 148.3$$

which differs little from 147.6, the simple aggregative index obtained from (1) prior to inclusion of the shoelaces.

Many would consider the change in the latter index to be the more realistic in this case. Even tripling the cost of a pair of shoelaces should have a minimal effect on the overall change in the price of clothing because its cost is still low relative to that of other articles.

The addition of this low cost item into the price relatives calculation had a disproportionate effect on the index. If we took into account the quantities used of these articles, we might get a more realistic picture of overall price changes. We shall have more to say about this later on.

The defining formulas in this section express price indexes. We may replace P's by Q's to obtain quantity indexes. Such numbers compare quantities produces or used in one time period with those produced or used in another time period.

Example 3 Simple Aggregative and Mean of Quantity Relatives Indexes

Table 3 lists the average annual purchases, per Ecap University student, of the five basic articles of clothing listed in Table 1. Using 1997 = 100, construct the simple aggregative and mean of quantity relatives indexes comparing 1997 with 2000 average purchases.

Table 3

Item	Average Annual Purchases	
	1997	2000
Shirt/blouse	3	5
Socks	4	6
Jeans	3	4
Shoes	2	3
Sweater	2	1
Total	14	19

The simple aggregative index is given by

$$I = \frac{19}{14} \cdot 100 = 135.7,$$

while the arithmetic mean of relatives index is:

$$I = \frac{\frac{5}{3} + \frac{6}{4} + \frac{4}{3} + \frac{3}{2} + \frac{1}{2}}{5} \cdot 100 = 130.0 \cdot$$

Thus, according to our two indexes, usage by Ecap University students of the types of clothing listed in Table 3 has increased by 35.7% and 30.0%, respectively, between 1997 and 2000.

Unweighted index numbers are straightforward to compute. However, aside from the drawbacks we have noted, a major disadvantage of unweighted price index is that they make no distinction between high and low turnover items, so that a large price change in even one or two low turnover commodities could significantly distort the index. Suppose, for example, that we add a raincoat to our previous list of clothing items and its price dropped from \$105 to \$50 between 1997 and 2000. From (1) and Table 1, and (3) and Table 2, this would yield the following revised values of the aggregative and price relatives indexes, respectively:

$$I = \frac{155 + 50}{105 + 105} \cdot 100 = 97.6$$

$$I = \frac{818.2 + \frac{50}{105} \cdot 100}{6} = 144.3$$

The revised price relative value of 144.3 suggests an average price increase of 44.3%, rather than the previous one of 63.6% (see (4)). The revised aggregative index of 97.6 suggests that, on the whole, prices have actually declined by 2.4%, rather than the previous one which suggests a 47.6% (see (2)) overall price increase!

Over the course of a year a "typical" student would purchase the other articles more often than a raincoat, and a more realistically designed price index would have hardly changed despite the 52% drop in the price of the raincoat. Furthermore, even a relatively small price change in a high usage item may have a significant impact which is not reflected by unweighted indexes.

Because of the ways in which unweighted indexes may give misleading results, they are not commonly used today in important price analyses. Prior to 1914 the Wholesale Price Index of the Bureau of Labor Statistics was computed as an arithmetic mean of price relatives of about 250 commodities. Around this time it was changed to a weighted index. We consider weighted indexes in the next section.

EXERCISES

1. Table 4 gives the average retail prices in Helenville of selected foods during the years 1992, 1997 and 2002.

Table 4

	Average Prices		
Product	1992	1997	2002
Chicken (lb.)	$0.38	$0.45	$0.58
Cheese (lb.)	0.63	0.81	1.15
Bread (lb.)	0.72	0.91	0.84
Coffee (lb.)	1.93	2.26	2.84

(a) Using 1992 as the base year, calculate a simple aggregative index comparing 1997 and 2002 prices.

(b) Construct a 2002 simple aggregative price index with 1997 = 100.

(c) Find the arithmetic mean of price relatives comparing:

(i) 1992 with 2002 prices. Use 1992 = 100.

(ii) 1997 with 2002 prices. Use 1997 as base year.

In each part, interpret your results.

2. Table 5 lists 2001 and 2004 prices in hundreds of dollars per year of some basic items needed to attend Ecap University.

Table 5

Item	Average Price	
	2001	2004
Tuition	75	96
Books	5	7
Transportation	3	4
Health Insurance	20	30

Construct the following and interpret your results:

(a) An arithmetic mean of the price relatives, using 2001 = 100.

(b) A simple aggregative index using 2001 as base year.

3. A furniture manufacturer uses five raw materials in the production of chairs. Table 6 shows the average price per unit of these raw materials in 1990, 1992 and 1995.

Table 6

Raw Material	Average Prices		
	1990	1992	1995
Wood	$9.90	$12.70	$13.81
Screw	0.42	0.50	0.54
Upholstery	18.53	20.97	22.10
Fabric	12.18	13.49	13.87
Springs	6.71	8.27	9.05

(a) Find the 1992 and 1995 values of a simple aggregative index using 1990 = 100.

(b) Using 1992 as the base year, construct the 1995 simple aggregative index and 1995 arithmetic mean of price relatives.

(c) Compute the arithmetic mean of price relatives comparing 1990 to 1995 prices, using 1990 = 100.

Interpret your results in each case.

4. Table 7 indicates sales of domestic and foreign automobiles (in hundreds) in Fastown in 2000, 2002, 2004 and 2006.

Table 7

Car Type	Car Sales			
	2000	2002	2004	2006
Domestic	68	87	84	91
Foreign	83	91	85	79

(a) With 2000 = 100, compute the arithmetic mean of quantity relatives for 2002 and 2004.

(b) Compute the aggregative quantity index comparing 2002 and 2006 quantities, using 2002 as the base year.

(c) Compute the arithmetic mean of relatives comparing 2004 and 2006 quantities, with 2004 = 100.

Interpret your results in each case.

5. The Precise Publishing Company wants to compare sales (in hundreds) of college textbooks in the last few years. An analysis of company records is summarized in Table 8.

Table 8

Subject	Textbooks Sold		
	2001	2003	2005
Science	78	97	105
Math	56	58	63
English	24	22	25
History	31	33	27
Philosophy	9	7	6

(a) Using 2001 as the base year, calculate the unweighted aggregative quantity index for 2003 and 2005.

(b) Calculate the arithmetic mean of quantity relatives comparing 2003 and 2005 quantities.

In each case, interpret your results.

6. The Sweet Passion Bakeshop lists yearly expenditures (in hundreds of dollars) for basic ingredients as shown in Table 9.

Table 9

Ingredient	Cost		
	2002	2004	2006
Flour	32	35	35
Sugar	23	22	24
Eggs	35	36	37
Frosting	38	44	51

(a) With 2002 = 100, construct the arithmetic mean of price relatives, comparing

(i) 2002 and 2004 costs.

(ii) 2002 and 2006 costs.

(b) Calculate an aggregative index comparing 2004 and 2006 prices, using 2004 as the base year.

In each case, interpret your results.

14.3 ■ Weighted Index Numbers

Weighted Aggregative Index: Base Year Weights

We can modify our aggregative index by first weighting (that is, multiplying) the current and base year price of each commodity in the index by the quantity of each that was consumed or produced in a specific year. We obtain a weighted aggregative index by dividing the total of the weighted current year prices by the sum of the weighted base period prices. Different indexes are obtained, depending on which quantities are used as weights.

If we use base-year weights, the resulting index is called a **Laspeyres Index**, after the statistician and economist Etienne Laspeyres, who first proposed its use. It is defined by:

$$I_L = \frac{\sum P_n Q_o}{\sum P_o Q_o} \cdot 100 \qquad (5)$$

Weighted Aggregative Index: Current Year Weights

We may instead construct a price index by weighting all prices with current year quantities. By doing so we obtain a **Paasche Index**, after the economist Herman Paasche, who played a key role in its development. This weighted aggregative index is defined by:

$$I_P = \frac{\sum P_n Q_n}{\sum P_o Q_n} \cdot 100$$

We will compare the relative merits and shortcomings of these weighted indexes, but we first look at their determination.

Example 1 Laspeyres and Paasche Indexes for Food

Consider a collection of five foods consumed by a "typical family" in Hightstown, listed in Table 10. Compute the values of the Laspeyres and Paasche indexes, comparing 2004 and 2007 prices.

Table 10

Food	Unit Price		Annual Consumption	
	2004	2007	2004	2007
Meat	$2.10	$3.60	80	58
Fish	1.80	1.90	40	56
Milk	1.70	2.40	70	74
Eggs	0.90	1.35	39	23
Bread	0.60	0.92	62	67

As we have noted, given any two years for comparison we assume, unless otherwise specified, that the base year is the earlier year and the given year is the later year.

Therefore, we compute our indexes using 2004 as the base year and 2007 as the current year. Consequently, the Laspeyres index is given by,

$$I_L = \frac{\sum P_{2007} Q_{2004}}{\sum P_{2004} Q_{2004}} \cdot 100$$

$$= \frac{3.60(80) + 1.90(40) + 2.40(70) + 1.35(39) + 0.92(62)}{2.10(80) + 1.80(40) + 1.70(70) + 0.90(39) + 0.60(62)} \cdot 100$$
$$= 148.8,$$

indicating a 48.8% overall increase in the price of the given foods over the three year period.

The Paasche index is,

$$I_P = \frac{\sum P_{2007} Q_{2007}}{\sum P_{2004} Q_{2007}} \cdot 100$$

$$= \frac{3.60(58) + 1.90(56) + 2.40(74) + 1.35(23) + 0.92(67)}{2.10(58) + 1.80(56) + 1.70(77) + 0.90(23) + 0.60(67)} \cdot 100$$
$$= 143.0$$

indicating a 43.0% overall increase in the price of the given foods over the three year period.

Note that while the prices of meat, fish, and eggs increased, the price of fish increased less sharply. The consumption of meat and eggs declined in 2007 relative to 2004, whereas consumption of fish increased. Consumers' exchange of less expensive fish for meat played some role in leading to the smaller value of the Paasche index compared to the Laspeyres index. The Paasche index used 2007 quantities as weights, reflecting the above-noted change in the consumption pattern. On the other hand, the Laspeyres index employs the quantities for the earlier time period, which gives greater weight to the higher priced meat and less weight to the relatively cheaper fish.

In general, the Paasche index gives greater weight to the relatively less expensive items while the Laspeyres index gives greater weight to the higher priced items. As a result, the Laspeyres index tends to show a higher overall price increase than does the Paasche. Basically, their imperfections in describing price changes reflect their limitations in accounting for the behavior of supply and demand.

> The Laspeyres index uses fixed weights which could be a disadvantage if the index is computed over a long period of time and the quantities consumed of the surveyed items are no longer indicative of current consumption patterns.

In this situation a "moving" Laspeyres index could be computed by periodically changing the base year, thereby keeping the set of quantity weights current. Changing the base year, however, is not a simple matter in practice because of the significant, perhaps enormous, work involved in recalculating the new weights. The complexity of such revisions depends on their frequency and on the number of items in the survey.

On the other hand, a point favoring the use of the Laspeyres index is that base-year weights will not change from year to year. Thus, we can make meaningful comparisons of the change in prices and purchasing power over time since we are considering only the change in the price per unit of each commodity and not the number of units. If the Laspeyres index is used to compare only "essential" items whose quantities do not vary much from year to year, the effect of using the same weights in each new calculation should be relatively small.

The consumption pattern is always up to date in the Paasche index since it employs current year quantities. The Paasche index will quickly reflect the diminished purchasing of the higher priced items and the increased consumption of the lower priced items and thus may be viewed as somewhat more realistic indication of the overall change in prices to consumers.

On the other hand, a factor potentially affecting the realism of the Paasche index in describing change is the increase in the commodities purchased in the current year due to population growth, which may be small over the short term but not necessarily small in the long run.

In addition, it is awkward to interpret values of the Paasche index from one year to the next, since not only prices but also their weights change for each new computation. A major difficulty in using the Paasche index consistently in practice is the expense of data collection since new weights must be prepared for each new period. For this reason the Paasche index is not widely used in situations where the number of commodities to be compared is large. In such situations, it is generally preferable to use the Laspeyres index with one set of base-year weights, thereby trading off some (hopefully small) degree of accuracy for relative manageability.

Typical-Period Price Index

The typical-period price index is an aggregative price index in which quantities for a specified year, usually between the base and given years, are used as weights.

The Bureau of Labor Statistics uses this technique to compute the Producer Price Index for which they periodically revise the base year and assignment of weights. This type of index is especially appropriate for those situations where the quantities used during the base or given year are viewed as unrepresentative of the market situation. Sometimes a "typical" year's weights are computed by averaging the data of several recent years.

The **typical-period price index** is defined by,

$$I_t = \frac{\sum P_n Q_t}{\sum P_o Q_t} \cdot 100,$$

where period t typically represents either a year between the base year and given year or an average of the weights for two or more of these years.

Example 2 Laspeyre, Paasche, and Typical-period Indexes

Use Table 11 to determine the following indexes for the town of Splitsburg.

(a) A Laspeyres index, comparing 2000 and 2003 prices.

(b) A Paasche index, comparing 2003 and 2000 prices.

(c) A typical-period index, comparing 2000 and 2006 prices, using 2003 quantities as weights.

(d) A typical-period index, comparing 2003 and 2006 prices, using the average of the 2003 and 2006 quantities as weights.

(e) Interpret the results obtained.

Table 11

	Unit Price (dollars)			Quantity Purchased		
Commodity	2000	2003	2006	2000	2003	2006
Cotton (lb.)	2	3	4.80	18	32	20
Wool (lb.)	3	4	6.20	15	22	16
Silk (lb.)	1.50	2.25	1.75	32	34	56

(a) $$I_L = \frac{\sum P_{2003}Q_{2000}}{\sum P_{2000}Q_{2006}} \cdot 100 = \frac{3(18)+4(15)+2.25(32)}{2(18)+3(15)+1.50(32)} \cdot 100$$
$$= 144.2$$

(b) $$I_P = \frac{\sum P_{2006}Q_{2006}}{\sum P_{2003}Q_{2006}} \cdot 100 = \frac{4.80(20)+6.20(16)+1.75(56)}{3(20)+4(16)+2.25(56)} \cdot 100$$
$$= 117.3$$

(c) $$I_t = \frac{\sum P_{2006}Q_{2003}}{\sum P_{2000}Q_{2003}} \cdot 100 = \frac{4.80(32)+6.20(22)+1.75(34)}{2(32)+3(22)+1.50(34)} \cdot 100$$
$$= 193.1$$

(d) Let $\bar{Q}$ denote the average of the 2003 and 2006 quantities. Then the values of $\bar{Q}$ for cotton, wool and silk are given by 26, 19, and 45, respectively. Therefore:

$$I_t = \frac{\sum P_{2006}\bar{Q}}{\sum P_{2003}\bar{Q}} \cdot 100 = \frac{4.80(26)+6.20(19)+1.75(45)}{3(26)+4(19)+2.25(45)} \cdot 100$$
$$= 125.9$$

Note that the results for parts (b) and (d) sharply differ, despite the fact that prices for the same two years (2003 and 2006) are compared in each calculation. This is an illustration of the potential impact on the overall index that a change in the weights may bring, as we have discussed.

(e) As to be expected, we interpret out results to mean that in the years surveyed prices increased 44.2%, 17.3%, 93.1% and 25.9% in parts (a) through (d), respectively.

Weighted Arithmetic Mean of Price Relatives Index

In Section 14.2 we defined the simple mean of price relatives index. We can combine this approach with the notion of weights, thereby obtaining a **weighted arithmetic mean of price relatives index.** It is defined by,

$$I_R = \frac{\sum \left(\frac{P_n}{P_o} \right) \cdot w}{\sum w} \cdot 100 \tag{6}$$

where the w's are "appropriate weights" assigned to the individual price relatives of the given commodities.

It is often argued that the relative change in the price of a commodity is most accurately reflected by the total amount of money spent on it. Therefore, it is usual to use value weights for the w's of (6). The total value associated with a given commodity equals the product of its unit price P and the number of units consumed Q, namely PQ.

If base-year values P_oQ_o are used as weights, the weighted mean of price relatives formula is given by,

$$I_R = \frac{\sum \left(\frac{P_n}{\cancel{P_o}} \right) \cancel{P_o} Q_o}{\sum P_o Q_o} \cdot 100 = \frac{\sum P_n Q_o}{\sum P_o Q_o} \cdot 100 = I_L$$

namely the Laspeyres index (5).

Example 3 Weighted Arithmetic Mean Index

Using the information provided in Table 12, compute a weighted arithmetic mean of price relatives index, comparing:

(a) 2002 and 2007 prices, using the average of the 2002 and 2005 values as weights.

(b) 2005 and 2007 prices, using 2005 values as weights.

(c) Interpret the results obtained.

Table 12

Commodity	Unit Price (dollars)			Quantity Purchased		
	2002	2005	2007	2002	2005	2007
Cotton (lb.)	2	3	4.80	18	32	20
Wool (lb.)	3	4	6.20	15	22	16
Silk (lb.)	1.50	2.25	1.75	32	34	56

(a) The 2002 values of the three commodities are \$2(18) = \$36, \$3(15) = \$45 and \$1.50(32) = \$48, respectively. The corresponding 2005 values are given by \$3(32) = \$96, \$4(22) = \$88 and \$2.25(34) = \$76.50, respectively. Therefore, the means of the 2002 and 2005 values of these commodities are,

$$\frac{36+96}{2} = \$66, \quad \frac{45+88}{2} = \$66.50 \text{ and } \frac{48+76.50}{2} = \$62.25,$$

respectively. As a result, the weighted mean of price relatives index is given by:

$$I_R = \frac{\frac{4.80}{2}(66) + \frac{6.20}{3}(66.50) + \frac{1.75}{1.50}(62.25)}{66 + 66.50 + 62.25} \cdot 100$$
$$= 189.2$$

(b) We found the 2005 values to be \$96, \$88 and \$76.50, respectively, for the three commodities. This yields the weighted mean of price relatives index:

$$I_R = \frac{\frac{4.80}{3}(96) + \frac{6.20}{4}(88) + \frac{1.75}{2.25}(76.50)}{96 + 88 + 76.50} \cdot 100$$
$$= 134.2$$

(c) We interpret our results to mean that in the years surveyed prices increased 89.2% and 34.2% in parts (a) and (b), respectively.

EXERCISES

1. The shellfish catch (in thousands of pounds) in Cape Crab during 2001 and 2006 is listed in Table 13.

Table 13

Type of Shellfish	Price (\$/lb.)		Quantity	
	2001	2006	2001	2006
Clams	0.23	0.56	26	43
Crabs	0.12	0.26	108	124
Lobsters	0.67	1.28	19	17
Oysters	0.51	0.62	29	27

Construct and interpret the following:

(a) (i) The 2006 simple aggregative price index.

(ii) The 2006 unweighted mean of quantity relatives index.

(b) A weighted aggregative index comparing 2001 and 2006 prices, using 2006 quantities as weights.

(c) A weighted arithmetic mean of price relatives, using current values as weights.

(d) With 2001 = 100, a weighted mean of price relatives, using base year values as weights.

(e) A Laspeyres aggregative price index.

2. Price and quantities consumed of several basic commodities in Anglesburg during 2002 and 2007 are given in Table 14.

Table 14

Commodity	Unit Price		Quantity	
	2002	2007	2002	2007
Butter (lb.)	$1.18	$1.78	450	510
Sugar (lb.)	0.62	0.87	380	318
Flour (lb.)	1.28	1.71	702	786
Bread (lb.)	0.53	0.72	630	578

Calculate and interpret the values of the following indexes:

(a) A Laspeyres index comparing 2002 and 2007 prices.

(b) A weighted mean of price relatives, using 2007 quantities as weights.

(c) A weighted mean of price relatives, using base year values as weights.

(d) An aggregative index comparing 2002 and 2007 prices, using the average of the 2002 and 2007 quantities as weights.

3. Table 15 gives the prices (in dollars) and sales figures for the L and C Hardware Company for 1990 and 1996.

Table 15

Product	Price		Sales	
	1990	1996	1990	1996
Screwdriver (each)	2.40	3.67	2340	2860
Cleaning Compound (gallon)	1.98	2.17	20500	17600
Nails (pound)	0.36	0.43	11200	13800

Determine and interpret the following indexes:

(a) A Paasche index comparing 1990 and 1996 prices.

(b) A weighted mean of price relatives, using the average of the 1990 and 1996 values as weights.

(c) A Laspeyres aggregative index.

(d) A simple arithmetic mean of price relatives and an unweighted aggregative quantity index.

4. The average price per share and number of shares for both the Volatile and Lovar Corporations are listed in Table 16 for 2001, 2003, and 2005.

Table 16

Corporation	Price Per Share			Shares Outstanding (Thousands)		
	2001	2003	2005	2001	2003	2005
Volatile	\$12.37	\$18.71	\$15.87	423	694	548
Lovar	10.21	11.96	12.52	271	297	303

(a) With 2001 = 100, construct a weighted mean of price relatives comparing 2003 and 2005 prices. Use base year values as weights.

(b) Construct a weighted aggregative index comparing 2001 and 2005 prices, using the average of the 2001 and 2005 quantities as weights.

(c) Determine a weighted aggregative index comparing 2001 and 2003 prices, with current year quantities as weights.

(d) Interpret these indexes.

5. Table 17 states the cost of upkeep, in dollars, of an average sized car in Fast City in 1993 and 1997.

Table 17

Category	Price		Quantity Consumed	
	1993	1997	1993	1997
Gasoline (gallon)	1.27	1.38	900	1100
Oil (quart)	1.12	1.96	12	13
Maintenance (hour)	15.00	21.00	70	60

Calculate and interpret the following indexes:

(a) A weighted arithmetic mean of price relatives, using 1997 values as weights.

(b) The Laspeyres index, comparing 1993 and 1997 prices.

(c) The weighted aggregative index comparing 1993 and 1997 prices. Use the average of the 1993 and 1997 quantities as weights.

14.4 ■ Index Number Models

An index number, as the afore examples illustrate, does not stand alone. It has a family in that it is a valid consequence of a number of assumptions about the items chosen, the years chosen for comparison, regions chosen as a source of data, and the mathematical method used to make the comparison; unweighted aggregative method vs. unweighted arithmetic mean of price relatives method vs. weighted aggregative method vs . . . .

In his treatise *The Making of Index Numbers* (1922) Irving Fisher describes 134 approaches to constructing weighted index numbers. Each such structure together with the background and underlying assumptions made, and the index number obtained as a valid conclusion is called an **index number model**.

14.5 ■ Problems Faced By Index Number Model Builders

In many applied situations index number models arise from a comparison of a large "basket" of goods and services over a long period of time. As a result, problems in their construction arise and we must be wary of potential pitfalls.

It is sometimes difficult to find data that is suitable for the kind of comparison that one would like to make. This may occur if the only available data are in a measurement unit that is either difficult or impossible to convert to the desired one. It would be impossible, for example, to compute an index to analyze seasonal variation of data if the data values are reported annually. Comparison of sales data per department of a company may not be feasible if sales are reported, say, per floor of the building, and several floors contain more than one department, or several departments are located on parts of more than one floor. Analysis of insurance payments per claim cannot be made if they are listed per policy written on per week, per branch, etc.

Data comparisons are not possible if data for a desired time period are nonexistent or have been lost. Consider the dilemma in trying to compare prices of commodity items in 2005 with those in 1920. A number of commonly used items today did not exist in 1920, such as contact lenses, frozen foods, air conditioners, computers, televisions, VCR's, radios, and hand calculators.

Changes in the quality of the items to be compared or the use of definitions may make comparisons difficult. It prices of automobiles in 1960 and 2005 are compared, it would be found that prices have increased substantially. But this direct comparison does not take into account technological advances that were achieved over that time period. Various government agencies may publish different indexes that appear to be the same. Reports that the money supply in August 2001 increased 2% from the previous year's figure, but also decreased 1% from August 2000 would seem contradictory. Not necessarily so; the government measures different types of money supply. One type, M1, measures currency in circulation plus money deposited in checking accounts. Another, M2, involves a slightly broader definition which includes savings accounts and some other types of bank time deposits. The government may also announce different indexes for the same quantity at different points in time due to an error in the first figure or a seasonal adjustment.

A decision on the scope of the data required for a meaningful comparison is a matter of judgment. If an index is intended to measure the change in price of a commodity at a specific locale at two points in time, there is little question as to the figures to be included. The situation, however, is more complex if the index is intended to be more general and reflect changes over a wide geographic area. A basic problem arises in the construction of a "general-purpose" index designed to measure changes in consumer prices. It is physically impossible to include in such an index all commodities and all services in any given city, let alone include the various prices of each commodity and service throughout the entire country. Those who construct such an index must take a sample of the nation's cities. Those charged with constructing indexes take samples with the intention that the cities sampled from and items included "reasonably" reflect the overall phenomenon being described. Of course, such judgments are always subject to debate.

Many figures in index number construction cannot be meaningfully compared without paying attention to the importance of each item. Suppose, for example, price changes of two items are examined for 1981-91. If their indexes are 120% and 180%, respectively, one may prematurely conclude that the "overall" average price change for these commodities is 150%. But if the first time was used, say, 100 times as often as the second item in 1981, and 500 times as often in 1991, then the two price changes would have to be weighted to obtain a realistic measure of the overall price change. We have discussed various weighting schemes and their advantages and disadvantages, so that a decision to use a particular technique is always subject to some disagreement.

Finally, care must be given to the choice of a base year. To best compare data the chosen year should represent a period of relatively recent economic stability. The base period should be recent so that price and production data for relatively new products and services are available. An economically stable time period should be used to avoid making misleading comparisons of current prices with those occurring at either a peak or a trough in the economic cycle. At these times, such as during a war, depression or time of great prosperity, consumer prices may be "abnormally" high or low and a disproportionate size of the population may be rich or poor, resulting in buying habits that are skewed, relative to other recent years. If, for example, the base year chosen was one during which a depression or recession was well underway, such as 1981, current prices would appear to have increased inordinately. On the other hand, if the base year was one in which growth and prosperity were at their peak, current prices may appear to be low. Therefore, a middle-of-the-road year, economically speaking, is the most desirable choice for a base year. Alternatively, prices and consumption patterns for several recent years may be averaged, in order to have a "composite" base year.

14.6 ■ Some Important Index Number Models and Their Indexes

The government is continually releasing new data about the economic health of the country. The **Consumer Price Index (CPI)**, for example, is one of the most generally accepted ways to measure the behavior of inflation. It charts the average change in prices over time of a fixed market basket of goods and services. The Bureau of Labor Statistics releases this information for two different population groups: (1) A CPI for All Urban Consumers (CPI-U), which covers about 80 percent of the total population, and (2) A CPI for Urban Wage Earners and Clerical Workers (CPI-W), which covers about 32 percent of the total population. Of course, in light of these percentages, there is some overlap in the groups covered by these indexes.

The CPI notes changes in the price of such items as food, clothing, housing, energy, transportation, medical and dental services, medical drugs, and other goods and services that people require for day-to-day living. Different urban areas, housing units and retail establishments around the country are taken into account in compiling this index. Then price changes for the various categories in each region are weighted to take into account the relative amount spent by that particular locale. Finally, local data is combined to obtain an overall average.

Movement of the CPI affects the lives of millions of Americans. Many workers have union collective-bargaining contracts which provide for automatic wage increases, usually called **cost-of-living adjustments**, when the appropriate CPI index rises by a specified amount. Similarly, food stamp recipients, retired military and federal civil service personnel, and postal workers have their incomes or pensions tied to the consumer price index. The CPI may also be used to adjust such factors as alimony and child support payments, attorney's fees, worker compensation payments, apartment, home and office rentals, and welfare payments. In addition to the preceding, movement of the CPI influences Congress on setting the personal tax exemption level and thereby touches most of us in some way.

There are numerous other national measures of economic activity, such as the **Industrial Production Index** and the **Producer Price Index (PPI)**. The former measures changes in the physical volume or quantity of output of the nation's mines, factories, and electric and gas utilities. This monthly overall index shows changes in overall United States industrial production by combining 252 individual monthly series into a number of groups and then combining the group measures into one complete index.

The PPI measures the average change in prices received by domestic producers or commodities in all stages of processing. The information is obtained by sampling nearly every industry in the manufacturing and mining sectors of the economy, as well as obtaining information from other sectors, such as agriculture, fishing, forestry, gas, and electricity. Currently, the sample contains about 3,200 commodities and 75,000 quotations per month.

Also watched very closely by economists is a group of economic statistics collectively called the **Index of Leading Economic Indicators.** Each quantity in the collection is thought to go down before the peak of the business cycle and go up before the bottom or trough of the business cycle. Some important leading indicators are number of business failures, new orders for durable goods, average work week, building contracts, stock prices, and new incorporations.

There are good economic reasons why these values tend to turn down before a peak or up before a trough. Sometimes they indicate changes in spending in strategic areas of the economy, whereas in other instances they indicate changes in business persons' and investors' expectations. For purposes of economic forecasting, to both aid the government in determining appropriate economic policies and to aid firms in planning it is important to estimate turning points in the business cycle—peaks and troughs—in advance. If a large number of leading indicators turn down, this is believed to signal an impending downturn. If a large number of such indicators go up, this is viewed as a sign of a coming trough, followed by better economic times.

Unfortunately, the leading indicators are not always reliable. It is true that each time the economy has turned down in recent years, the leading indicators have warned us some months ahead. However, these indicators have gone down on several occasions—1952 and 1962, for example—when the economy did not subsequently follow suit. Therefore, the indicators occasionally give us unreliable signals. In periods of expansion they sometimes go down too far in advance of the real peak, and in periods of recession they may go up only a very short time before the trough. Nevertheless, these indicators are watched closely and are used to supplement other sophisticated forecasting techniques.

14.7 ■ Determining "Real" Dollar Amounts

As we have all experienced, what a dollar bought yesterday is not what it buys today. What it buys today is not what it can be expected to buy tomorrow. If comparisons of a dollar's worth over time are to be meaningful, we must have a mechanism for adjusting dollar values to reflect "real buying power" compared to some suitable based used as a point of reference. Price indexes provide such a mechanism. The process of using price indexes to make these dollar value adjustments is called **deflation** and the price index used as a divisor is called a **deflator**. The dollar figure that results from this deflation procedure represents "constant dollars" or "real wages"—that is, the purchasing power of the obtained (deflated) dollar value in the base year. Formally expressed, we have:

$$\textbf{Real Dollar Value} = \frac{\text{Current Dollar Value}}{\text{Price Index}} \cdot 100$$

Example 1 Jane's Real Buying Power

Jane Felder, a management science major, entered Huxley College in January 2002 under a fellowship which pays her $7000 for each of her four years. She, in turn, is required to perform such services to the department as tutoring and homework grading. The local price index for the years 2002 through 2005 is listed in Table 18.

Table 18

Year	Price Index
2002	100.0
2003	104.8
2004	107.6
2005	109.3

Determine Jane's real buying power for the years 2003, 2004, and 2005.

To do this we must deflate her fellowship income for each of these years. Relative to 2002, Jane's real fellowship value in 2003 is:

$$\frac{7000}{104.8} \cdot 100 = \$6679.39$$

Similarly, her 2004 and 2005 real fellowship values are, respectively:

$$\frac{7000}{107.6} \cdot 100 = \$6505.58, \quad \frac{7000}{109.3} \cdot 100 = \$6404.39$$

Note that as the price index increases each year, Jane's real fellowship values, based on a fixed annual income, continue to diminish.

Example 2 Real Average Weekly Wage

In 2001 the average weekly paycheck for workers in a factory in Hightstown was \$352. In 2005 the average salary for the same group of wage earners was \$465. In 2005 the price index for Hightstown was 127, relative to a base year index of 100 in 2001. Determine the real average weekly wage in 2005.

The real wage is given by

$$\frac{465}{127}\cdot 100 = \$366,$$

so that current wages buy more than wages in 2001, but \$99 less in real purchasing value than the \$465 figure would suggest.

Another way to analyze these figures is to observe that in the given time span wages increased by

$$\frac{465-352}{352}\cdot 100 = \text{ 0.32, which translates to 32\%,}$$

whereas consumer prices increased by only 27%. Therefore, wages rose at a higher rate than prices so that these wages allow the wage earner to purchase more at 2005 prices than did the 2001 wages. Having one's real wage equal to or exceed the wage earned in a previous year is commonly called "keeping up with inflation."

Example 3 Real Gross Domestic Product

Table 19 shows unadjusted values of the Gross Domestic Product (GDP), in billions of dollars, for the United States in 1992 and 1995 as well as the GDP price deflators (aggregative price index numbers) for these years, in terms of 1990. Calculate their percentage growth in actual (unadjusted) GDP. Determine the real GDP in 1990 prices and the percentage growth in real GDP.

Table 19

	Year	
	1992	1995
GDP	6244.4	7253.8
Price Deflator (1990 = 100)	106.8	114.9

Percentage growth in actual GDP between 1992 and 1995 is:

$$\frac{7253.8-6244.4}{6244.4}\cdot 100 = 0.162, \text{ which translates to } 16.2\%$$

Real GDP in 1990 prices for the years 1992 and 1995, respectively, are:

$$\frac{6244.4}{106.8}\cdot 100 = \$5846.82, \quad \frac{7253.8}{114.9}\cdot 100 = \$6313.14$$

Therefore, percentage growth in real GDP during 1992-1995 is:

$$\frac{6313.14-5846.85}{5846.82}\cdot 100 = 0.08, \text{ which translates to } 8.0\%$$

We may conclude from this analysis that although actual GDP grew 16.2% in this time span, only 49.4% of this increase (namely 8.0/16.2 times 100) may be attributable to real growth in GDP, whereas the remaining 50.6% of the actual growth is attributable to inflation.

EXERCISES

1. Anna Lopez, accounting major at Hilevel University, pays her school costs by means of a part-time school job. Her wages were $80 per week in 2000 and $85 per week in 2001. However, college costs have increased 6% in 2001.

   (a) What is the price index for 2001 relative to 2000 = 100?

   (b) Calculate Anna's real weekly wage relative to 2000.

(c) Does her 2001 salary give her more buying power, as compared to her 2000 salary? Explain.

2. The consumer price index (CPI) and the average national hourly earnings in selected industries are listed in Table 20 for a sample of recent years.

Table 20

Year	CPI (2000 = 100)	Average Hourly Earnings		
		Service	Retail Trade	Manufacturing
2000	100.0	7.47	5.69	8.54
2002	107.3	8.16	6.03	8.92
2004	113.4	8.47	6.12	9.78
2006	120.0	8.81	6.27	10.77

(a) Relative to 2000, determine the real hourly earnings of a typical worker in each of the three selected industries in 2004. What happened to the purchasing ability of each type of worker in 2004, relative to 2000? Explain.

(b) Determine the equivalent 2002 hourly wage of a typical worker in each of the industries in 2006. In each case, does the 2006 wage earner have more or less purchasing power than he did in 2002? Explain.

(c) Relative to 2004, determine the real hourly wage of a typical worker in each of the three fields in 2006. In each case, did average wages rise to a faster rate than did average prices during the 2004-2006 time span? Explain.

3. The Gross Domestic Product (GDP) is often used as a measure of the economic activity of a country. The annual GDP of the United States (in $ billions) and the corresponding CPI for selected years is shown in Table 21.

Table 21

Year	GDP	CPI (1989 = 100)
1989	5438.7	100.0
1991	5916.7	109.8
1993	6553.0	116.5
1995	7253.8	122.9

(a) Use the CPI to deflate the GDP for each of the years 1991, 1993, and 1995 to equivalent 1989 dollars.

(b) Determine the percentage growth in real GDP during the time interval stated: (i) 1989 to 1993; (ii) 1991 to 1995; (iii) 1989 to 1995.

(c) Compare the responses to part (b) with the corresponding growth rates of actual GDP during each of these time spans. What part may be attributable to inflation and what part expresses real growth in GDP?

4. The average dividend in dollars per share of the Hirise stock fund for five consecutive years is shown in Table 22 along with an index of average dividend per share for competing stock funds of a similar kind, relative to 2003.

Table 22

	2003	2004	2005	2006	2007
Average Dividend Per Share	1.8	2.1	2.3	2.0	2.1
Index of Average Dividend Per Share (2003 = 100)	100	107	113	112	118

(a) Compute the equivalent 2003 dividend per share for the Hirise stock fund in 2004. Was the increase in Hirise's 2004 dividend equal to the overall industry dividend increase, relative to 2003? Explain.

(b) Determine the equivalent 2005 dividend per share for Hirise in 2006. Relative to 2005, was the decrease in Hirise's 2006 dividend in line with the overall industry average decrease? Explain.

(c) Relative to 2003, did Hirise increase their dividend in 2006 to a larger extent than the overall industry average? Explain.

14.8 ▪ The Consumer Price Index Model: Realism vs. Politics

In 1995 center stage of government business was taken up by the problem of eliminating the federal budget deficit. As part of their plan to accomplish this, House

Republicans recommended that the annual-cost-of-living adjustment for Social Security and other benefits tied to increases in the Consumer Price Index be reduced starting in 1999.[25] Washington wisdom circulating at the time held that the CPI overstated inflation by as much as 1.5 percentage points and that a reduction in the CPI's value was not only justified, but defensible.

The problem was to give legitimacy to the Washington wisdom. Since the Bureau of Labor Statistics (BLS) moved cautiously and, from the point of view of the Washington establishment, unreliably in this matter, in June 1995 the Senate Finance Committee appointed a five-member panel of economists, chaired by former President Bush economic advisor Michael Boskin, to study the CPI and make recommendations on revisions. All members of what came to be called the Boskin Commission had respectable credentials and some might be described as eminent, but all had previously given Congressional testimony that the CPI exaggerated inflation. Economists who took a different view, such as former Commissioner of BLS Janet Norwood, were not invited to join the panel.

The Boskin Commission released its report in early December 1996, claiming that consumer inflation was being overstated by the CPI by about 1.1% a year, arguing that the index did not adequately reflect the improving quality of goods, did not take into account new products quickly enough, did not properly reflect consumers' tendency to purchase cheaper alternatives when the price of goods rose, and did not properly take into account consumer shift toward discount stores.[8]

A number of questions arise: What does it mean to say that the CPI overstates the reality of inflation by about 1.1% a year? Many interpret this to mean that there is an ideal standard for measuring the reality of inflation which is known by the Boskin Commission and that in comparing the BLS's CPI against this ideal standard, the BLS's CPI overstated inflation by about 1.1% a year. **This is utter nonsense; there is no ideal standard.** The BLS's CPI is a valid conclusion of a math model based on data, accepted procedures, and assumptions made by the agency's economists. The Boskin Commission's proposed 1.1% per annum adjustment is based on the same data, same procedures, but with somewhat different assumptions. In effect, what the Boskin Commission was saying was that if you employ our assumptions rather than your's, then you have to make a 1.1% per annum downward adjustment in your CPI value. If your CPI increased by 3.3% over 1995, then 2.2% would be a more accurate description of the reality of inflation over that year, based on our assumptions.

Did the Boskin Commission consider the possibility that the BLS's CPI understates inflation? No; for discussion of this situation see [3, 15, 17, 24].

Is there good reason to prefer the Boskin Commission's assumptions over those made by the BLS? If politics were put aside, it becomes "experts" versus "experts." At a panel session at the annual meeting of the American Economic Association held in New Orleans in January 1997 Boskin and his four commission colleagues presented their views followed by BLS Commissioner Katherine Abraham who stated that 'she agreed with some of the Boskin Commission's recommendations, including that the CPI should be as close to a measure of the cost of living as possible.' She added, however, that her agency would not and should not produced a CPI based partly on subjective judgments.'[7] Abraham later further elaborated to the Senate Finance Committee: 'If we get into the business of making judgments about things that are not measurable—guessing, even if it's . . . a best guess—we really, I think, would be undermining the credibility of all of the data we produce.'

The planting of an idea which developed into a BOLÉRO drum roll to reduce the CPI was probably done by the Congressional Budget Office when it asserted in late 1994 that the CPI exaggerated inflation by an amount between 0.2 and 0.8 percentage points a year. Federal Reserve Board head Alan Greenspan expressed the view that the CPI exaggerated inflation by an amount between 0.5 and 1.5 percentage points a year at a joint meeting of the House and Senate Budget Committees in January 1995. Greenspan also noted that correcting these estimates could save the government $150 billion over five years and suggested the possibility that Congress pass a law that would lower the CPI by a percentage point or half a percentage point for determining benefits tied to the CPI. It's a short hop from this plateau to the establishment of the Boskin Commission and what became the "official" view that the CPI overstated inflation by about 1.1 percentage points a year.

The Boskin Commission's report set BOLÉRO into motion in the form of a rash of calls to "fix" the way inflation is measured. Testifying before the Senate Finance Committee on January 30, 1997, Alan Greenspan recommended that an independent commission be established to set cost-of-living adjustments for federal receipts and outlays each year. Economist Martin Feldstein, who had been President Reagan's top economic advisor, suggested that Greenspan's proposed committee should recommend an "appropriate" inflation adjustment factor through informed judgment, apart from any adjustment made to the CPI by the Bureau of Labor Statistics through its normal work. Senators William Roth and Daniel Patrick Moynihan introduced a sense-of-the-Senate resolution that urged an accurate cost-of-living index. With momentum at a peak to push through a CPI fix, it all came apart. President Clinton, faced by strong opposition within his own party and constituencies like Labor and the elderly, decided not to pursue a CPI fix outside of the highly professional, non-political machinery of the Bureau of Labor Statistics. Republican enthusiasm for a "fix" waned with the discovery of a two-year old White House memorandum on how Democrats could use the issue against Republicans.

Within two years talk of engineering a CPI fix to help eliminate the budget deficit had turned to arguments over what to do with the projected budget surplus.

These were the salvos in a drive to "strongly" modify the math model which is the foundation stone on which the CPI rests. A number of questions emerge:

(1) Were there concerns that the economic assumptions that are part of the CPI are unrealistic? Yes.

(2) Was there a political objective tied to modifying the CPI downward so as to reduce Social Security payments and other cost-of-living payments tied to the CPI? No doubt about it.

(3) When it is argued that the CPI exaggerated inflation by 1.5 percentage points (or by any figure for that matter), does this mean that there is an

"ideal" standard against which the CPI is being compared and found wanting to the extent of 1.5 percentage points? **No**; for more on this point see Adams [1] or [2].

(4) It is possible that the CPI underestimates inflation? Yes.

The following readings will be helpful to anyone who wishes to obtain insight into this CPI brouhaha, and mathematical modeling.

SUGGESTED READINGS ON THE CPI BROUHAHA

1. W.J. Adams, "The CPI and You," *The Christian Science Monitor*, Dec. 10, 1996. This letter was in response to [11].

2. W.J. Adams, "Numbers Crunch," *The Economist*, Jan. 11, 1997. This letter is similar to [1].

3. D. Baker, "The Inflated Case Against the CPI," *The American Prospect*, Winter 1996.

4. D. Baker, "Hyping Hypo-inflation," *In These Times*, Jan. 6, 1997.

5. D. Baker, *Getting Prices Right: A Methodologically Consistent Consumer Price Index*, Economic Policy Institute, April 12, 1996.

6. J. Berry, "The CPI's Wide Reach," *The Washington Post National Weekly Edition*, December 23, 1996—Jan. 5, 1997.

7. J. Berry, "A Numbers Game Played for High Stakes: The BLS chief is standing fast," *The Washington Post National Weekly Edition*, March 17, 1997.

8. The Boskin Commission, *Toward a More Accurate Measure of the Cost of Living*, Final Report to the Senate Finance Committee from the Advisory Commission to Study the Consumer Price Index, Dec. 4. 1996.

9. M. Boskin, "More Accurate C.P.I," *The New York Times*, March 27, 1997, This letter is in response to [27].

10. J. Chait, "Revolution of '96", *The New Republic*, Dec. 30, 1996.

11. "An Enticing Number," *The Christian Science Monitor*, Dec. 10, 1996.

12. "Fix the CPI Yardstick Now," *The Christian Science Monitor*, March 7, 1997.

13. C. Duff, "Fix in Consumer Price Index Falls Short of Surgery Some Believe is Needed," *The Wall Street Journal*, April 1, 1996.

14. D. Francis, "Senators Eager to Fix Cost-of-Living Index," *The Christian Science Monitor*, Feb. 14, 1997.

15. D. Francis, "Fixing the Inflation Index-But Is It Really Broken?" *The Christian Science Monitor*, March 6, 1997.

16. D. Francis, "The Numbers Say Americans Made More . . ." *The Christian Science Monitor*, March 14, 1997.

17. D. Francis, "Poking Holes in the CPI Balloon," *The Christian Science Monitor*, March 14, 1997.

18. G. Grier, "Is US Economy Too Complex For Inflation Sleuths to Track?" *The Christian Science Monitor*, Dec. 5, 1996.

19. R. Hershey, Jr., "Labor Statistics Chief Takes Issue with Critics of Inflation Index," *The New York Times*, Dec. 20, 1996.

20. R. Kuttner, "Boskin's Magic Wand," *The Washington Post National Weekly Edition*, Dec. 16-22, 1996.

21. R. Kuttner, "The Fix is In on the C.P.I.," *The Washington Post National Weekly Edition*, March 17, 1997.

22. J. Madrick, "The Cost of Living: A New Myth," *The New York Review of Books*, March 6, 1997.

23. B. Moulton, "Bias in the Consumer Price Index: What is the Evidence?" *Journal of Economic Perspectives*, Fall 1996.

24. P. Passell, "Some Experts Say Inflation Is Understated," *The New York Times*, Nov. 6, 1997.

25. R. Pear, "G.O.P. Suggests Smaller Benefit Adjustments," *The New York Times*, May 11, 1995.

26. S. Pearlstein, "Fine-Tuning the Consumer Price Index," *The Washington Post National Weekly Edition*, Dec. 23, 1996—Jan 5, 1997.

27. J. Popkin, "Why Play a Numbers Game," *The New York Times*, Feb. 27, 1997.

28. R. Stevenson, "Moves Continue on Price Index Changes," *The New York Times*, March 5, 1997.

29. L. Uchitelle, "Balancing Quantity, Quality and Inflation," *The New York Times*, Dec. 18, 1996.

30. L Uchitelle, "Measuring Inflation: Can't Do It, Can't Stop Trying," *The New York Times*, March 16, 1997.

31. L. Uchitelle, "The Negotiators Forgo a Cut in Inflation Index," *The New York Times*, May 3, 1997.

14.9 ■ Keep in Mind: Limitations of Index Number Models

Beware the Assumptions

It is important to always keep in mind that **index numbers are valid conclusions of assumptions of mathematical models** of our making to help us describe and predict the movement of highly complex phenomena.

Sometimes these valid conclusions do the job well and sometimes they miss reality's mark by a wide margin. The more complex the phenomenon, the more difficult it is to realistically capture its essence in terms of a math model and predict its behavior in terms of index numbers.

GENERAL REFERENCES

1. R. Hershey, Jr., "How Good is Key U.S. Index," *The New York Times*, Sept. 28, 1984.

2. R. Samuelson, "How well does the Consumer Price Index measure the inflated prices we pay for dog food and doctors, parking lots and paperbacks?" *The New York Times Magazine*, Dec. 8, 1974.

3. P. J. McCarthy, "The Consumer Price Index", *Statistics: A Guide to the Unknown*, ed. J. M. Tanur *et al* (San Francisco: Holden Day, 1972, 266-275).

15

Time Series Analysis and Forecasting

15.1 ■ Introduction

A time series is a sequence of quantitative measurements of some phenomenon taken over a period of time at regular intervals, such as each day, week, month, quarter of a year, etc. Examples of time series data include the daily closing price of a stock on the New York Stock Exchange, the weekly sales volume of a department store, the monthly measurement of the Consumer Price Index (CPI), the quarterly statement of Gross Domestic Product (GDP), and the annual number of births in a given state. The data will typically change from one period to the next and business persons, economists, and social scientists, among others, study these changes in an attempt to understand the "forces" that underlie these changes. The ultimate goal of time-series analysis is to isolate the various patterns and characteristics of historical data to predict or forecast future data values.

"Good forecasts" are important for both short-term and long-term business planning. Having good forecasting techniques can be most useful, for example, in properly planning production levels, capital investment, and the hiring of new personnel.

15.2 ■ Components and Models of a Time Series

Fundamental Assumption

> The usual approach to time series analysis rests on the **assumption** that time series data are a composite of four underlying quantities acting independently of one another, called **trend** (*T*), **seasonal variation** (*S*), **cyclical variation** (*C*) and **error** (or **random or irregular**) **variation** (*E*).

> The long-term trend of a time series is the "smooth" component of a time series that indicates the general behavior of a given quantitative variable over a long period of time. Thus, we can characterize the data as showing either long-term growth ("upward" trend), long-term contraction ("downward" trend) or possibly neither, in the case of a stationary time series.

As an illustration, consider Figure 1 which shows the net asset value (NAV), in dollars per share, of the High Yield Junk Bond Fund over the last several years. As the figure suggests, although the NAV has fluctuated upward and downward over time, it has generally tended to decrease. Therefore, we would conclude that the NAV of this fund exhibits long-term contraction.

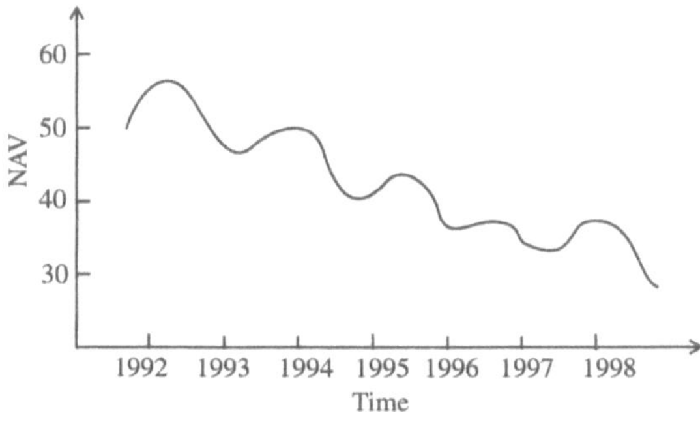

Figure 1

If observations are recorded several times a year we may note seasonal variation in the data, which is short-term periodic variation that "completes itself" within the year and then repeats in approximately the same periodic pattern in subsequent years. For instance, in the northeastern United States sales of bathing suits and air conditioners tend to peak from May through July, whereas sales of home heaters are relatively low. However, from December through February these seasonal effects tend to be reversed as people concentrate on keeping warm rather than cool. Recreation activities also vary with the seasons in a generally predictable manner. Consider Example 1 defined by Table 1 and Figures 2(a) and 2(b).

Example 1 The Elemental Industrial Chemical Company

Table 1 lists the sales revenue (in millions of dollars) of the Elemental Industrial Chemical Company for each season (quarter of a year) between 2002 and 2007 inclusive. Describe the seasonal fluctuation of this time series, as shown in Figures 2(a) and 2(b). Figure 2(a) best exhibits the long-term trend of the time series and Figure 2(b) focuses on its seasonal variation.

Table 1

Season	W	Sp	Su	F	W	Sp	Su	F	W	Sp	Su	F
Year	02	02	02	02	03	03	03	03	04	04	04	04
Revenue	24	27	24	23	26	28	26	25	27	31	27	28

Season	W	Sp	Su	F	W	Sp	Su	F	W	Sp	Su	F
Year	05	05	05	05	06	06	06	06	07	07	07	07
Revenue	30	34	31	28	30	33	30	29	33	38	36	34

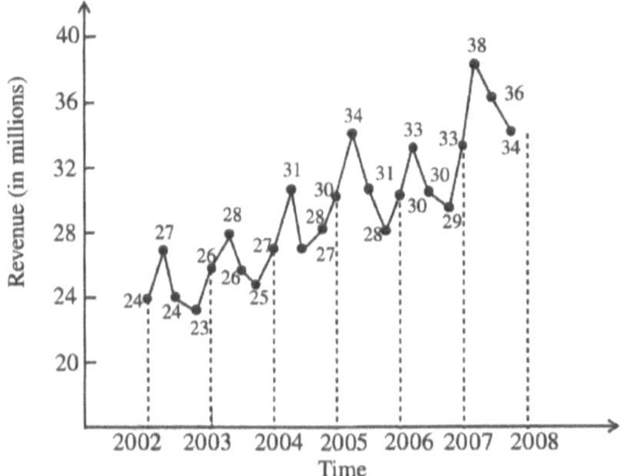

Figure 2(a)

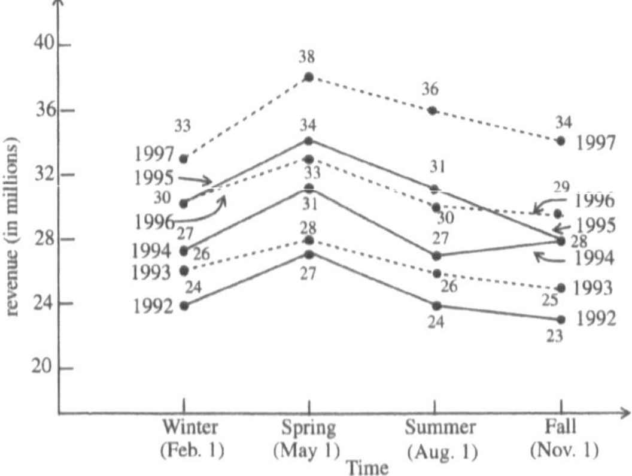

Figure 2(b)

It is clear from both graphs that there is a long-term upward trend in the data. From Figure 2(b) we also see a definite seasonal pattern within each year. There is a tendency for the sales revenue of this company to peak during spring, decline from May to November, stabilize from November to February, and begin its upward movement again as spring approaches.

The cyclical component of a time series shows the upward and downward movements of the data over extended periods of time. These fluctuations, called **business cycles** in a business of economic setting, may last from two to ten years or longer when measured from peak to peak of from trough to trough (see Figure 3). Cyclical variation is different from seasonal variation in that the former covers longer periods of time, has different causes, and is generally less predictable.

Error variation expresses changes in the data that are not accounted for by either the trend, seasonal or cyclical factors. It is not systematic as are the other factor and is sometimes called the **random or irregular or residual variation.**

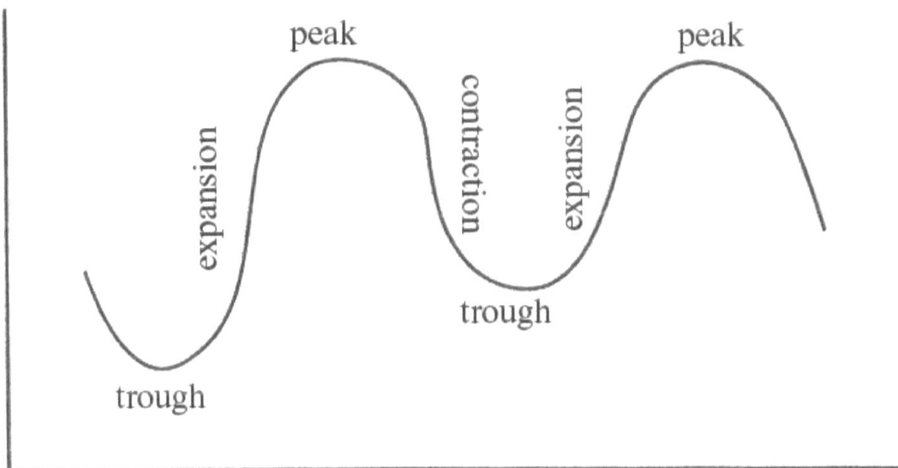

The four phases of a business cycle: trough, expansion, peak, contraction.

Figure 3

Such variation can occur from a variety of unpredictable factors, such as hurricanes, nuclear accidents, wars, uncertain political "climate," and the like. These factors may be identified with a cause or may be random occurrences which are not clearly explainable. It is assumed that the error variation will make up a relatively unimportant proportion of the data and will thus have a negligible influence on future predictions.

Multiplicative Model

From these four components of a time series we may construct models that seek to describe how they interact to yield the resulting time series data. The most widely used model of a time series is the multiplicative model in which the time series is described as the product of the four components. This general model may be expressed by the equation

$$y = T \cdot S \cdot C \cdot E, \tag{1}$$

where y is the observed value of the variable of interest and T, S, C and E represent the trend, seasonal, cyclical and error components, respectively. In this model only the trend is expressed in its original units, while the other components are expressed as percentages or "relatives". Consequently, this model is most suited to situations in which percentage changes best describe the movement in the series, such as in a wide variety of economic data.

Additive Model

It the four components interact in what seem to be an additive fashion to produce the given data, then an additive model of the form

$$y = T + S + C + E \tag{2}$$

is more appropriate.

Beware the Model and the Assumption on which It Rests

Our focus is on the multiplicative model in analyzing time series data, but we should keep in mind that not all time series are best described by this model; if an examination of the data suggests that the components do not interact in a multiplicative fashion, then use of a different model should be investigated.

Since the components are assumed to act independently of each other, in the next few sections we illustrate how the values of each component may be separately determined. We then combine them, via multiplication, to yield a predicted result. This technique is called the "classical decomposition" method of forecasting time series data. First, we examine three techniques of trend determination.

15.3 ■ Moving Averages

Trend determination allows us to more easily determine the general behavior of the dependent variable over time by "smoothing" the seasonal, cyclical and erratic effects.

One smoothing method is the method of moving averages. To compute an N **period moving average** of a time series, or moving average of length N, we must first decide on a value for N, which describes the number of consecutive data values to be averaged. The first moving average is obtained by averaging the first N data values. The second smoothed average is the average of N consecutive data values beginning with the second data value, and so on. In general, the Nth moving average is the average of N consecutive data values beginning with the Nth data value. Once we have obtained the N period moving average we may use the last moving average figure as a predictor of the trend value one period in the future.

Example 1 3 Month and 5 Month Moving Averages for VCR Repairs

Table 2 shows the number of requests y for television and VCR repairs at the Fixem Electronics Company over a period of fourteen recent consecutive months.

(a) Calculate the three-month and five-month moving averages for these data.
(b) Predict the trend value of the number of television and VCR repairs in the fifteenth month using the three-month and five-month moving averages.
(c) Plot the data and the three-month and five-month moving averages.
(d) Comment on these plots

Table 2

Month	1	2	3	4	5	6	7	8	9	10	11	12	13	14
y	13	12	14	17	19	18	17	20	21	23	27	24	22	25

(a) The first three-month moving average is:

$$\frac{13+12+14}{3} = 13$$

The other three-month moving averages are similarly computed, culminating with:

$$\frac{24+22+25}{3}=24$$

The three-month moving averages are summarized in Table 3.

Table 3

Month	Number of requests	Three-month moving total	Three-month moving average
1	13		
2	12	39	13
3	14	43	14
4	17	50	17
5	19	54	18
6	18	54	18
7	17	55	18
8	20	58	19
9	21	64	21
10	23	71	24
11	27	74	25
12	24	73	24
13	22	71	24
14	25		

The first five-month moving average is given by

$$\frac{13+12+14+17+19}{5}=15,$$

and the last one is:

$$\frac{23+27+24+22+25}{5}=24$$

The five-month moving averages are summarized in Table 4.

(b) We use the last moving average to forecast the trend value of the number of repairs in the fifteenth month. The last three-month and five-month moving averages are 24 and 24.

(c) Figure 4(a) shows the plot of the data and the three-month moving averages, while Figure 4(b) compares the plots of the data, the three-month, and the five-month moving averages.

Table 4

Month	Number of requests	Five-month moving total	Five-month moving average
1	13		
2	12		
3	14	75	15
4	17	80	16
5	19	85	17
6	18	91	18
7	17	95	19
8	20	99	20
9	21	108	22
10	23	115	23
11	27	117	23
12	24	121	24
13	22		
14	25		

(d) In Figure 4(a) we see that the plot of the three-month moving average is "smoother" than that of the data. That is, the peaks of the three-month average are not as high as that of their non-smoothed counterpart and the troughs not as low. In Figure 4(b) it is apparent that the plot of the five-month moving averages is smoother than that of the three-month averages. In general, the greater the value of N, the smoother is the corresponding plot.

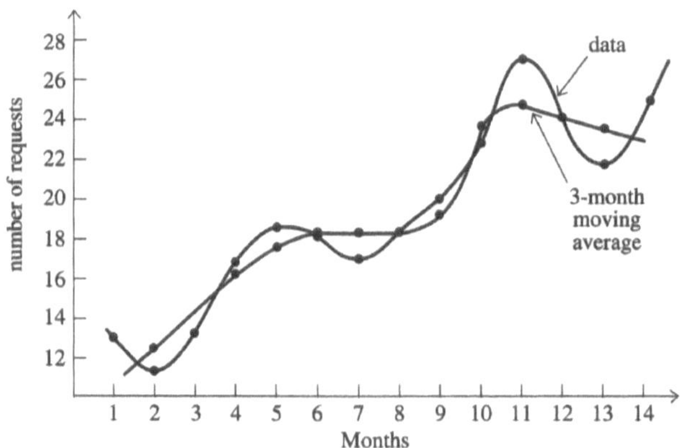

Figure 4(a)

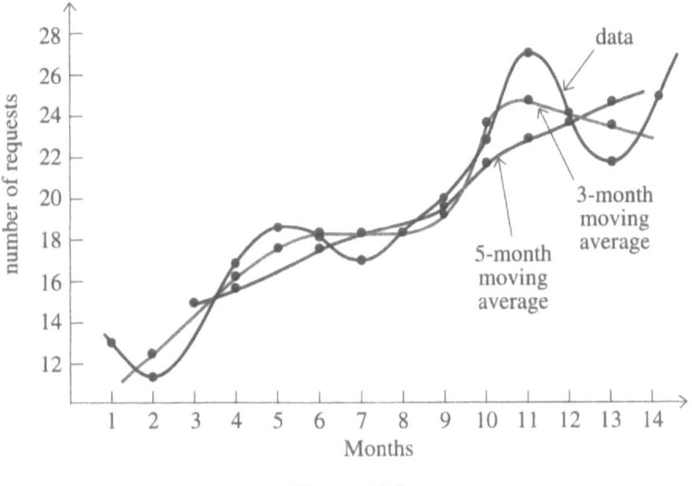

Figure 4(b)

To decide on an appropriate value of N we must have some sense of the extent of smoothing required to reasonably eliminate the seasonal, cyclical and random effects from the data. Also, the choice of N may depend on the method of calculation of the seasonal effect, which is discussed further in Section 15.6.

A moving average is arithmetically simple to calculate, but it also has **shortcomings.** We "lose" moving average values at the beginning and end of the series. For three-period averages, two values are lost, one at each end of the series. (In Example 2, we had fourteen data values and twelve three-month averages.) For five-period averages, four values are lost, two at each end of the series. (We had only ten five-month averages.)

Another drawback is that there is no particular equation to use as a moving average model, so that forecasting of future trend values is awkward if not impossible. As previously noted, one might use the last moving average as a forecast of the next trend value. The moving average smoothing method employs only a small part of the time series in any one particular calculation, and weighs these values equally.

EXERCISES

1. Average interest rates (in percent) for the Prudent National Bond Fund for each month of 2005 are as follows: 6.3, 6.4, 6.6, 6.7, 6.6, 6.4, 6.3, 6.5, 6.7, 6.8, 6.9, 6.8.

   (a) Develop a three-month moving average for this data.

   (b) Develop a five-month moving average for this data.

   (c) Plot the data and each moving average on the same graph. Which plot appears to be the smoothest in the sense that the peaks are not as high and the troughs not as low? Why do you think this occurred?

   (d) Let y_t denote the original data and let $\hat{y}_t$ denote its moving average counterpart. The **aggregate prediction error** is defined as $\Sigma(y_t - \hat{y}_t)^2$ for all values of t for which a moving average $\hat{y}_t$ exists. Compute the aggregate prediction error for the (i) three-month average, (ii) five-month moving average.

   (e) Which method yields better forecasts (i.e., a lower aggregate error)?

   (f) Predict the trend value of the January 2006 interest rate using the three-month and five-month moving averages.

2. Table 5 shows the annual sales (in millions of dollars) of the Associated Department Store during the ten year period 1996-2005:

Table 5

Year	1996	1997	1998	1999	2000	2001	2002	2003	2004	2005
Sales	23	20	18	21	22	24	26	28	24	27

(a) Develop a three-year moving average for this data.

(b) Develop a five-year moving average for this data.

(c) Plot the data and each moving average on the same graph. Which plot appears to be smoothest? Can you explain why this occurred?

(d) Compute the aggregate forecast error for each type of moving average. (See part (d) of Exercise 1.) Which method yields better forecasts?

(e) Predict the trend value of sales for this company in 2006 using the (i) three-year moving average, (ii) five-year moving average.

3. Average attendance figures (in thousands) at home football games for Ecap University over the last nine years are shown in Table 6.

Table 6

Year	1	2	3	4	5	6	7	8	9
Attendance	47	52	55	51	54	59	62	60	57

(a) Compute a three-year moving average for this data.

(b) Compute a five-year moving average for this data.

(c) Plot the data and each moving average on the same graph. Which plot appears to be smoothest? Can you explain why this occured?

(d) Using each moving average, predict the trend value of the average attendance at an Ecap University football game during the tenth year.

4. Automobile sales revenue (in hundreds of thousands of dollars) at Safe Motors show the time series in Table 7 over the last eleven years:

Table 7

Year	1	2	3	4	5	6	7	8	9	10	11
Sales	342	363	371	367	352	360	370	389	365	385	397

(a) Compute a three-year moving average for this data.

(b) Develop a five-year moving average for this data.

(c) Draw a graph of the data and each moving average. Which plot appears to be smoothest? Was the answer predictable?

(d) Predict the trend value of sales in year 12 using each type of moving average.

(e) Compute the aggregate prediction error using the (i) three-year moving average, (ii) five-year moving average.

5. Table 8 lists the number y of postcards and letters (in hundreds) received by radio station WXYZ in response to listener contests conducted between 2000 and 2007 inclusive.

Table 8

Year	2000	2001	2002	2003	2004	2005	2006	2007
y	65	82	70	73	68	76	79	83

(a) Construct a three-year moving average for this data.

(b) Construct a five-year moving average for this data.

(c) Predict the trend value of the volume of mail the radio station will receive in 2008, using each type of moving average.

6. The number y of passengers flown by National Airlines (in thousands) in the time period 1998 to 2007 is shown in Table 9.

Table 9

Year	1998	1999	2000	2001	2002	2003	2004	2005	2006	2007
y	65	67	72	70	75	79	73	68	65	71

(a) Develop a three-year moving average for the time series.

(b) Develop a five-year moving average for the time series.

(c) predict the trend value of the volume of passengers to be flown by this airline in 2008, using each type of moving average.

15.4 ■ Exponential Smoothing

> **Exponential smoothing** is another smoothing technique available to us which is computationally simple to use and **does not have the disadvantages of the moving average method.** This smoothing method is an exponentially weighted moving average of data values of the time series, weighted in such a way that recent observations are given greater importance in the sense of being more heavily weighted than older time series data.
>
> The weights assigned to the time series diminish over time so that when a calculation is made the most recently observed value receives the highest weight, the previously observed value receives the second highest weight, and so on, culminating with the initial value, which has the least weight.

The idea behind this method is the same as that of the moving average technique, namely to reduce fluctuations in the data that are attributable to seasonal, cyclical, or random variability and thereby isolate the trend component of the time series.

To smooth the data by exponential smoothing, we must first choose a smoothing constant, denoted by α, where $0 < \alpha < 1$. (This use of α in no way relates to the level of significance of a hypothesis test.) Let:

y_t = data value for time period t

= the smoothed value for time period t

We compute S_t as a weighted average of y_t and the previously smoothed value S_{t-1} as follows:

$$S_t = \alpha y_t + (1 - \alpha) S_{t-1} \tag{3}$$

In words this says:

$$\begin{pmatrix}\text{The smoothed}\\ \text{value for the}\\ \text{current time period}\end{pmatrix} = \alpha\begin{pmatrix}\text{The most recently}\\ \text{observed time}\\ \text{period}\end{pmatrix} + (1-\alpha)\begin{pmatrix}\text{The smoothed}\\ \text{value for the}\\ \text{previous time}\\ \text{period}\end{pmatrix}$$

Replacing S_{t-1} in (3) by

$$S_{t-1} = \alpha y_{t-1} + (1-\alpha)S_{t-1}$$

yields:

$$\begin{aligned} S_t &= \alpha y_t + (1-\alpha)[\alpha y_{t-1} + (1-\alpha)S_{t-2}] \\ &= \alpha y_t + \alpha(1-\alpha)y_{t-1} + (1-\alpha)^2 S_{t-2} \end{aligned}$$

This gives us S_t in terms of the data value y_t, the previous data value y_{t-1} and the smoothed value S_{t-2} for time $t-2$, the time prior to the one which gave us data value $y_{t-1} \cdot S_{t-2}$ may, in turn, be expressed in terms of data values y_{t-2} and S_{t-3}, and so son, so that this chain reaction followed through to the initial time series value y_1, gives us the weighted sum:

$$S_t = \alpha y_t + \alpha(1-\alpha)y_{t-1} + \alpha(1-\alpha)^2 y_{t-2} + \cdots + (1-\alpha)^{t-1} y_1$$

The weight assigned to each time series value y_t, $y_{t-1}, \ldots, y_1$ decreases exponentially. The largest weight α is assigned to the most recent time series value y_t, the next largest, $\alpha(1-\alpha)$, is assigned to the next most recent time series value y_{t-1}, down the line, with the smallest weight $(1-\alpha)^{t-1}$ being assigned to the initial time series value y_1

Once a sequence of smoothed values are determined one may forecast the trend for the next time period by taking a weighted average of the most recently available information, y_t and S_t.

This predicted value $\hat{y}_{t+1}$, is given by:

$$\hat{y}_{t+1} = \alpha y_t + (1-\alpha)S_t \qquad (4)$$

Example 1 Exponential Smoothing to Estimate and Forecast Trend

The revenue (in $ millions) from the sale of clothing at the Perfect Fit Company for each of the first nine months of 2004 are as follows: 49, 51, 48, 45, 50, 53, 58, 60 and 57. Use exponential smoothing, with $\alpha = 0.4$, to estimate the trend, and forecast the trend value for October 2004.

As we undertake the smoothing calculations, we see that there is no previously smoothed value; let $S_{\text{Jan}} = S_1 = y_1 = 49$. Now we determine the succeeding smoothed values as follows:

$$\begin{aligned}
S_2 &= 0.4(51) + 0.6(49) = 49.8, \\
S_3 &= 0.4(48) + 0.6(49.8) = 49.1, \\
S_4 &= 0.4(45) + 0.6(49.1) = 47.5, \\
S_5 &= 0.4(50) + 0.6(47.5) = 48.5, \\
S_6 &= 0.4(53) + 0.6(48.5) = 50.3, \\
S_7 &= 0.4(58) + 0.6(50.3) = 53.4, \\
S_8 &= 0.4(60) + 0.6(53.4) = 56.0, \\
S_9 &= 0.4(57) + 0.6(56.0) = 56.4.
\end{aligned}$$

The most recent information available to us are the values $y_9 = 57$, and $S_9 = 56.4$. Thus, the predicted trend value $\hat{y}_{10}$ for the tenth period (October 2004) is given by:

$$\begin{aligned}
\hat{y}_{10} &= \alpha y_9 + (1-\alpha)S_9 \\
&= 0.4(57) + 0.6(56.4) \\
&= 57
\end{aligned}$$

Our predicted trend value $\hat{y}_{10}$ for October 2004 is 57 million dollars.

Figure 5 compares the plots of the original data and the exponentially smoothed series with $\alpha = 0.4$. Note that the plot of the smoothed series does not fluctuate as much as that of the original time series.

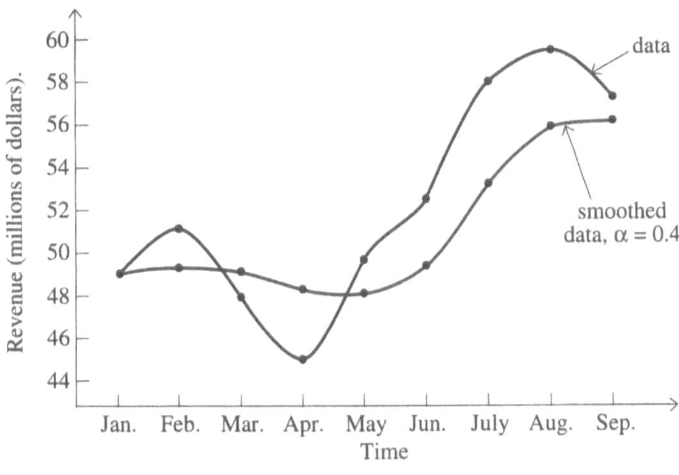

Figure 5

Choosing the Smoothing Constant

Exponentially smoothed time series data is a weighted sum of the actual data, with the weights decreasing exponentially over time. The rate at which the reduction in weights occurs depends on the value of α. The most recent data value is assigned the weight α and each succeeding weight equals $(1-\alpha)$ times the preceding one, which ultimately yields an assigned weight of $(1-\alpha)^{t-1}$ to the initial data value y_1. Suppose that α is relatively high, say 0.8, so that $(1-\alpha)$ is relatively low, 0.2 in this case. Then the major part of the weight is given to the most recent value y_t, leaving little for the remaining data. But if α is small, say 0.1, then the multiplier $(1-\alpha)$ is high, so that the next weight is not very different from the previous one. Therefore, we may find that the final weight $(1-\alpha)^{t-1}$ assigned to y_1, is not insignificant. As α is taken smaller and smaller, the current data value y_t is given diminished importance and the weights α, $\alpha(1-\alpha),\ldots(1-\alpha)^{t-1}$ assigned to y_t, $y_{t-1},\ldots,y_1$, respectively, become closer and closer to each other, and the resulting smoothed series becomes more and more stable.

The disparity in the weight assignment for the most recent five time series values $y_t,\ldots,y_{t-4}$ for three choices of smoothing constant are shown in Table 10.

Table 10

Time period	$\alpha = 0.2$	$\alpha = 0.4$	$\alpha = 0.8$
t	0.2	0.4	0.8
$t-1$	0.16	0.24	0.16
$t-2$	0.128	0.144	0.032
$t-3$	0.1024	0.0864	0.0064
$t-4$	0.08192	0.05184	0.00128

Weights for the most recent five time series for specified choices of α

Example 2 The Effect of α on Smoothing

Show the effect α has on the degree of smoothing of a time series by:

(a) Smoothing the data of Example 1 using both $\alpha = 0.2$ and $\alpha = 0.8$;

(b) Plotting the original time series and the smoothed series for the two choices of α.

(c) Comment on the results.

(a) The smoothed time series for each of $\alpha = 0.2$ and $\alpha = 0.8$ is listed in Table 11.

Table 11

Month	Sales y_i	S_i $\alpha = 0.2$	S_i $\alpha = 0.8$
January	49	49	49
February	51	49.4	50.6
March	48	49.1	48.5
April	45	48.3	45.7
May	50	48.6	49.1
June	53	49.5	52.2
July	58	51.2	56.9
August	60	53.0	59.4
September	57	54.0	57.5

(b) The plots of the original data and each smoothed series is shown in Figure 6.

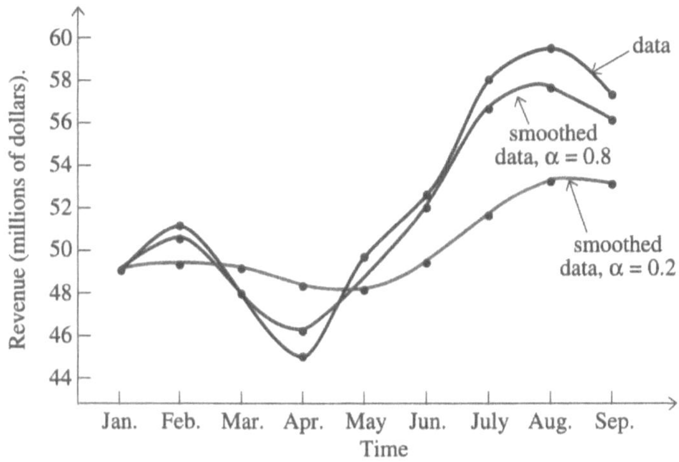

Figure 6

(c) We see that the graph with $\alpha = 0.8$ is somewhat more subdued than the graph of the original data, that is, the peaks are not as high and the trough is not as low. Moreover, the graph with $\alpha = 0.2$ is "smoother" than both of its companions.

How does one determine the best choice of α for a particular time series? First of all, suppose we observe a "big" change in the time series data from one time period to the next. It would be helpful to know whether this change was caused by random fluctuation or whether it suggests a change in the underlying forces that affect the variable of interest, such as a change in consumer demand when the variable of interest is sales volume as in Example 2. If our sense is that the random component is large, then the trend component is relatively small. To obtain a more stable graph we would choose a small value of α so that large random fluctuations in the data are given minimal weight. However, we would choose a large value of α if the random component is judged to be small, so that the smoothed results are not too different from the original data.

If we are unsure about the extent of randomness in the time series, we could proceed by computing smoothed values for several values if α. Would could then use a predetermined criterion to gauge the accuracy of the smoothed results, such as the aggregate prediction error $\Sigma(y_t - \hat{y}_t)^2$ defined in Exercise 1(d) of Section 15.3. We would choose that value of α which yields the smallest aggregate prediction error.

EXERCISES

1. The time series shown in Table 12 represents the yearly number of defective items returned to the Federated Department Store.

Table 12

Year	1996	1997	1998	1999	2000	2001	2002	2003	2004	2005	2006
Returns	118	129	116	137	131	128	125	130	120	115	112

(a) Compute a five-year moving average for this data.

(b) Using a smoothing constant of 0.3, exponentially smooth this data.

(c) Using a smoothing constant of 0.7, exponentially smooth this data.

(d) Draw a graph of the original data, the five-year moving average and the exponentially smoothed data with $\alpha = 0.3$ on the same coordinate system.

(e) In general, does the time series get more or less smooth as the value of the smoothing constant α is increased?

(f) Predict the trend value of the number of defectives returned in 2007 using (i) five-year moving average, (ii) exponential smoothing with $\alpha = 0.3$, (iii) exponentially smoothing with $\alpha = 0.7$.

2. The annual dividend (in dollars per share) of stock in the Big Enterprise Company over the last ten years is as follows: 2.06, 1.97, 1.81, 2.01, 2.04, 2.12, 1.93, 2.00, 2.07, and 2.15.

(a) Determine exponentially smoothed values using $\alpha = 0.5$.

(b) Determine exponentially smoothed values using $\alpha = 0.8$.

(c) Draw a graph of the original data and both sets of exponentially smoothed values on the same coordinate system. Which plot appears to be smoothed? Which is the least smooth? How do you explain this?

(d) Predict the trend value of next year's dividend per share using (i) $\alpha = 0.5$, (ii) $\alpha = 0.8$.

3. Table 13 states the average daily percentage of its assets that a conservative stock fund invested in government securities for the years 1993 through 2005.

Table 13

Year	1993	1994	1995	1996	1997	1998	1999	2000	2001	2002	2003	2004	2005
Assets	12.4	13.5	13.6	11.7	11.9	11.3	10.8	12.2	11.9	11.7	12.0	12.4	11.8

(a) Construct a three-year moving average for this time series.

(b) Exponentially smooth the data using a smoothing constant of 0.4.

(c) Exponentially smooth the data using a smoothing constant of 0.6,

(d) Draw a graph of the original data and both sets of exponentially smoothed values on the same coordinate system.

(e) Predict the trend value of the average daily percent of assets that the stock fund will invest in government securities in 2006 using (i) three-year moving average, (ii) exponential smoothing with $\alpha = 0.4$, (iii) exponential smoothing, with $\alpha = 0.6$.

4. Refer to Exercise 2 of Section 15.3 (page 525), which shows the annual sales (in millions of dollars) of the Associated Department Store during the ten year period 1996-2005. These data are reproduced as Table 14.

Table 14

Year	1996	1997	1998	1999	2000	2001	2002	2003	2004	2005
Sales	23	20	18	21	22	24	26	28	24	27

(a) Exponentially smooth this time series using $\alpha = 0.2$.

(b) Plot the data, the three-year moving average, and the exponentially smoothed data on the same coordinate system.

(c) Predict the trend value of sales in 2006 using exponentially smoothing with $\alpha = 0.2$.

15.5 ■ Trend Determination by the Method of Least Squares

We have examined two smoothing methods for determining linear trend. Neither method, however, allows predictions to be made for more than one period in the future. But if our data analysis suggests an approximately linear trend, we can use the method of (linear) least squares, discussed in Section 12.3, to estimate the trend component. Once we obtain the best available line (in the least squares sense), projections of future trend are easily calculated.

Example 1 Return to the Elemental Industrial Chemical Company

We return to the Elemental Industrial Chemical Company situation described by Table 1 (page 518). We assign integer values, beginning with $t = 1$, to the seasons beginning with Winter 2002 and running in consecutive order, so that Winter 2007 is assigned the value 24. Running time t is our independent variable and industrial chemical sales (in $ millions) y is the time series data value for t. This yields Table 15.

Table 15

Season	W 02	Sp 02	Su 02	F 02	W 03	Sp 03	Su 03	F 03	W 04	Sp 04	Su 04	F 04
t	1	2	3	4	5	6	7	8	9	10	11	12
y	24	27	24	23	26	28	26	25	27	31	27	28

Season	W 05	Sp 05	Su 05	F 05	W 06	Sp 06	Su 06	F 06	W 07	Sp 07	Su 07	F 07
t	13	14	15	16	17	18	19	20	21	22	23	24
y	30	34	31	28	30	33	30	29	33	38	36	34

Determine the best linear trend equation for the quarterly data of Table 15. Project the future trend for each season (quarter) of 2010.

Let us note that $\sum t = 300$, $\sum t^2 = 4900$, $\sum y = 702$, $\sum ty = 9314$ and $n = 24$. For b and a of the trend line $T_t = a = bt$ we have:

$$b=\frac{24(9314)-300(702)}{24(4900)-(300)^2}$$
$$=\frac{12936}{27600}$$
$$=0.469$$
$$a=\frac{1}{24}[702-(0.469)(300)]$$
$$=23.39$$

Thus, the trend line is given by,

$$T_t = 23.39 + 0.469t\ ,$$

where T_t is the least-squares estimate of the trend at time period t. To forecast the trend for 2010 let us observe that the values of t for 2010 are 33 through 36. Substitution of each of these values for t in our trend equation yields the following estimates of trend for 2010:

$$T_{33} = 23.39 + 0.469(33) = 38.9$$
$$T_{34} = 23.39 + 0.469(34) = 39.3$$
$$T_{35} = 23.39 + 0.469(35) = 39.8$$
$$T_{36} = 23.39 + 0.469(36) = 40.3$$

The slope b measures the change in the dependent variable per unit change in the independent variable. Since each change in t is one unit in the trend projections for 2005 above, we could calculate the last three trend values T_{34} through T_{36} by simply adding b to each previous trend projection. That is,

$$T_{34} = b + T_{33}$$
$$= 0.469 + 38.9$$
$$= 39.3,$$

$$T_{35} = b + T_{34}$$
$$= 0.469 + 39.3$$
$$= 39.8,$$

$$\begin{aligned} T_{36} &= b + T_{35} \\ &= 0.469 + 39.8 \\ &= 40.3 \end{aligned}$$

The graph of the trend line, superimposed over the times series values for 2002-2007, is shown in Figure 7. The computed trend line seeks to express the long-term growth exhibited by the time series.

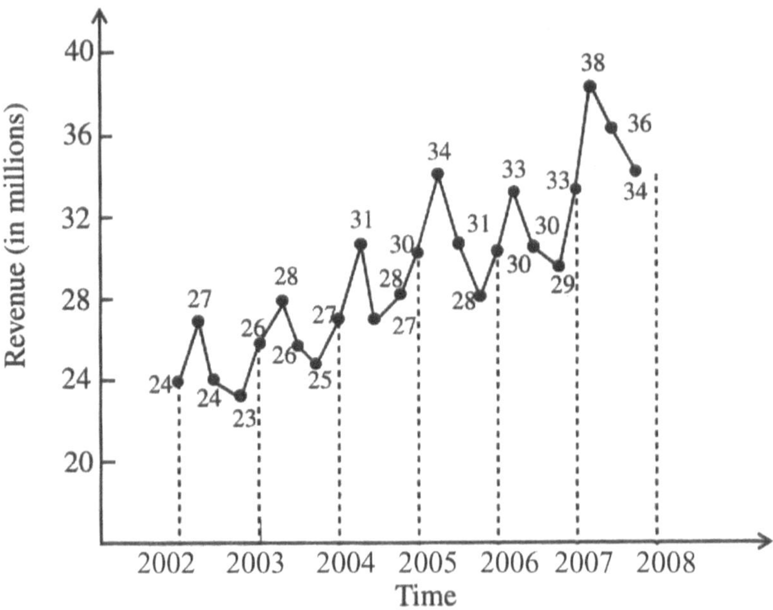

Figure 7

Beware the Assumption

We have been working under the **assumption** that the trend can be realistically represented by a line, which led us to define the least-squares line as the line of best fit. Not all trends can realistically be described by lines. We do not take up here the more general question of what curves should be fitted to various types of nonlinear trends.

As we saw in Section 15.2, seasonal variation within a year plays a role in determining the value of the time series. This is also clear from observation of Figure 2(b) (page 398). In the next section we take up the problem of isolating and determining the seasonal component of a time series by means of the ration-to-moving average method.

EXERCISES

1. Table 16 lists the number of books checked out of the Pittstown public library for the last fourteen weeks, where week 14 denotes the most recent week.

Table 16

Week	1	2	3	4	5	6	7	8	9	10	11	12	13	14
Books	373	398	409	418	392	437	452	411	436	472	503	486	507	523

(a) Develop the least-squares equation which serves as the linear trend component for this time series.

(b) Generally speaking, with the passage of time, is there an increase or decrease in the number of books checked out of the Pittstown library?

(c) Predict the trend component of this time series for week (i) 16, (ii) 19.

2. The annual enrollment data (in thousands) for State University during the period 1997-2004 is shown in Table 17.

Table 17

Year	1997	1998	1999	2000	2001	2002	2003	2004
Enrollment	24.6	25.3	24.5	23.8	23.1	22.9	22.4	22.7

(a) Find the least-squares line which shows the linear trend component of this time series.

(b) What is happening, in general, to enrollment at this institution? Is your conclusion consistent with your trend line? Explain.

(c) Predict the trend component of the data for the year (i) 2007, (ii) 2010.

3. Table 18 lists the annual advertising expenditures (in thousands of dollars) by the Perfect Fit Company for the period 1996-2005.

Table 18

Year	1996	1997	1998	1999	2000	2001	2002	2003	2004	2005
Advertising	746	761	803	822	831	791	849	886	914	940

(a) Determine the least-squares line that shows the trend component of this data.

(b) Generally speaking, are advertising expenditures increasing or decreasing with the passage of time? Is your conclusion consistent with your trend line? Explain.

(c) Predict the trend component of advertising expenditures for this company for the year (i) 2009, (ii) 2011

4. The quarterly sales data (number of copies sold) of an advanced economics textbook over the past four years are given in Table 19.

Table 19

		Quarter			
		Winter	Spring	Summer	Fall
	1	803	700	761	820
Year	2	891	720	740	850
	3	916	843	821	957
	4	964	897	867	981

(a) Determine the linear trend component of sales by the least-squares method.

(b) Predict the trend component of sales for each quarter of the (i) sixth year, (ii) ninth year.

5. The number of customers served at Bob's Bistro from Monday through Thursday of the last six weeks is listed in Table 20.

Table 20

		Day			
		Mon.	Tues.	Wed.	Thurs.
Week	1	67	75	82	81
	2	72	77	79	73
	3	70	81	83	82
	4	77	85	89	90
	5	75	81	87	93
	6	81	89	86	92

(a) Find the least-squares line which shows the trend component of the time series.

(b) Predict the trend component of this data for Monday through Thursday of week (i) 9, (ii) 11, (iii) 12.

(c) Is the general trend of customers served upward or downward? Is your conclusion consistent with your trend line? Explain.

6. *For Your Health* magazine is published bimonthly and sales (in thousands of copies) for the years 2005 through 2007 are cited in Table 21.

Table 21

		Issue					
		A (Jan./ Feb.)	B (Mar./ Apr.)	C (May/ Jun.)	D (Jul./ Aug.)	E (Sep./ Oct.)	F (Nov./ Dec.)
Year	2005	39	45	41	37	44	49
	2006	42	44	47	42	48	53
	2007	43	42	49	43	50	57

(a) Determine the least-squares line which shows the trend component of sales.

(b) Is the general trend of sales upward or downward? Is this conclusion consistent with your trend line? Explain.

(c) Predict the trend component of sales for the year (i) 2011, (ii) 2012.

7. Table 22 lists the number of new housing starts in Heightston for each quarter of the years 2001-2005.

Table 22

		Quarter			
		Winter	Spring	Summer	Fall
Year	2001	24	34	36	29
	2002	21	35	38	27
	2003	19	28	32	26
	2004	20	31	34	23
	2005	16	24	25	21

(a) Find the linear trend component of housing starts by the least-squares method.

(b) Is the general trend of housing starts upward or downward? Is this conclusion consistent with your trend line? Explain.

(c) Predict the trend component of housing starts for 2008.

15.6 ■ Measuring Seasonal Variation

Seasonal variation involves those fluctuations in the time series data that are associated with the change in climate and/or seasonally related activities at different points in time during the given year. We should keep in mind that a "season" need not represent a quarter of a year. For a business a season is the amount of elapsed time before the organization can reasonably expect a change in the time series values caused by a change in the time period. For instance, over a paid of time, sales at the Chateau Rouge restaurant generally varies in a fairly consistent manner, depending on the day of the week. Sales during the weekend tend to be higher than average, sales on Mondays tend to be lower than average and sales during the middle of the week are generally close to average. Thus, the proprietor of this dining establishment would consider each day (Sunday through Saturday) to be a different "season," which implies 7 different seasons before the next week begins. A publisher of a monthly magazine would be more concerned with monthly fluctuation in sales throughout the year. In this case, a month would represent a season, so that 12 seasons pass before the next year begins.

In more general terms, we may define a seasonal cycle as a set (i.e., a fixed number) of consecutive time periods for which the value of the data at any point in time is related to its position in the set. In the aforementioned illustrations, a seasonal cycle would be a week for the French restaurant and a year for the monthly magazine. Let us denote by n the number of periods in a seasonal cycle. In the preceding illustrations the values of n are 7 and 12, respectively.

To efficiently estimate seasonal effects it is desirable to have data for numerous seasonal cycles. To make the following presentation manageable, however, we limit our discussion and analysis to six years of quarterly data; that is, a season will be one quarter of a year and six seasonal cycles will be presented. We now outline the ratio-to-moving-average method for determining seasonal variation, using the data of the Elemental Industrial Chemical Company of Example 1 of Section 15.5 (page 534). For convenience we reproduce this in Table 23 along with Figure 2(b) of Section 15.1, which we reproduce as Figure 8.

Table 23
Further Consideration of the Elemental Industrial Chemical Company

Season	W	Sp	Su	F	W	Sp	Su	F	W	Sp	Su	F
Year	02	02	02	02	03	03	03	03	04	04	04	04
Revenue	24	27	24	23	26	28	26	25	27	31	27	28

Season	W	Sp	Su	F	W	Sp	Su	F	W	Sp	Su	F
Year	05	05	05	05	06	06	06	06	07	07	07	07
Revenue	30	34	31	28	30	33	30	29	33	38	36	34

The ratio-to-moving-average method has as its basis the simple idea that if a quarterly time series is affected by seasonal fluctuations due to the four seasons of the year, then a moving average of length 4 will remove these fluctuations.

The following sequence of steps generates the column organization shown in Table 24.

Step 1:

List the years for which the data was collected (column 1 of Table 24) and the four seasons of the year (column 2). List the running time values assigned to the reasons in the year (column 3) and the time series values corresponding to these running times (column 4).

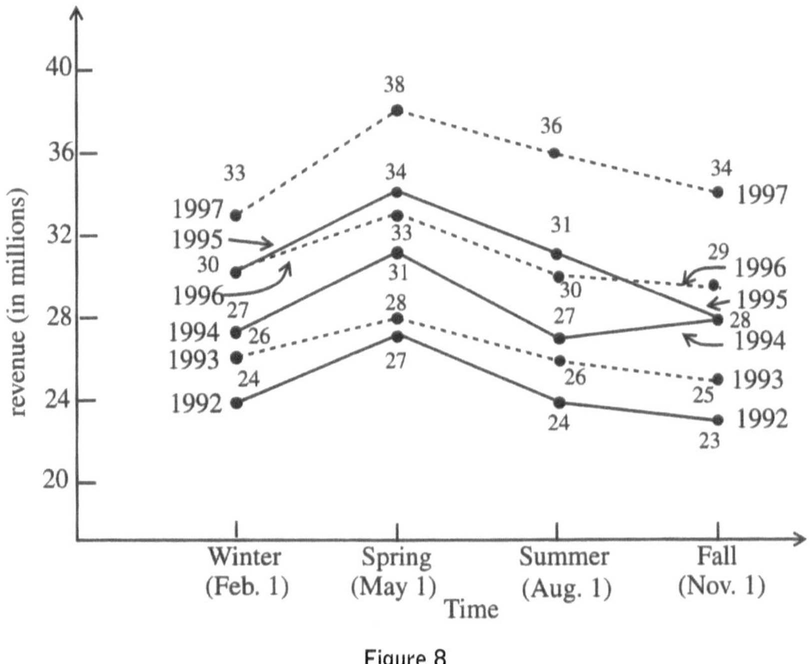

Figure 8

Step 2:

Determine the 4-period moving totals. These values are listed in column 5. (In general, we determine an n-period moving total for a seasonal cycle of length n). The first listed value, 98, is the total for the four seasons of 2002 ($98 = 24 + 27 + 24 + 23$). The second value, 100, is the total for the values from Spring 2002 to Winter 2003 ($100 = 27 + 24 + 23 + 26$), and so on.

There is a question of where 98 and the other twenty season totals should be placed. We want to place them in the middle of the year in question. For 98 this means midway between $t = 2$ and $t = 3$, which yields a line value of 2.5. This line number is assigned to 98, and successive line values are listed in column 6.

Step 3:

Determine the 4-period moving averages by dividing the moving totals in column 5 by 4. This yields column 7. The first value 24.5 = 98/4.

Table 24
Ratio-to-Moving-Average Method

(1) Yr	(2)	(3) t	(4) y	(5) Σ	(6) Line #	(7) $MA = \Sigma/4$	(8) $T \cdot C$	(9) Line #	(10) $S \cdot E$
'02	W	1	24						
	Sp	2	27						
				98	2.5	24.5			
							24.75	3	0.970
				100	3.5	25.0			
	F	4	23				25.125	4	0.915
				101	4.5	25.25			
'03	W	5	26				25.5	5	1.020
				103	5.5	25.75			
	Sp	6	28				26.0	6	1.077
				105	6.5	26.25			
	Su	7	26				26.375	7	0.986
				106	7.5	26.5			
	F	8	25				36.875	8	0.930
				109	8.5	27.25			
'04	W	9	27				27.375	9	0.986
				110	9.5	27.5			
	Sp	10	31				27.875	10	1.112
				113	10.5	28.25			
	Su	11	27				28.625	11	0.943
				116	11.5	29.0			
	F	12	28				29.375	12	0.953
				119	12.5	29.75			
'05	W	13	30				30.25	13	0.992
				123	13.5	30.75			
	Sp	14	34				30.75	14	1.106
				123	14.5	30.75			
	Su	15	31				30.75	15	1.008
				123	15.5	30.75			
	F	16	28				30.625	16	0.914
				122	16.5	30.5			
'06	W	17	30				30.375	17	0.988
				121	17.5	30.25			
	Sp	18	33				30.375	18	1.086
				122	18.5	30.5			
	Su	19	30				30.875	19	0.972
				125	19.5	31.25			
	F	20	29				31.875	20	0.910
				130	20.5	32.5			
'07	W	21	33				33.25	21	0.992
				136	21.5	34.0			
	Sp	22	38				34.625	22	1.097
				141	22.5	35.25			
	Su	23	36						
	F	24	24						

The sums in column 5 and their related averages in column 7 contain no seasonal effect. We conclude, then, that these averages contain only the trend and cyclical components, namely $T \cdot C$ in terms of our **multiplicative model**.

Analogous to the way one (wisely) begins a card game, we have "shuffled the deck" to blur the seasonal effects and we will shuffle our deck one more time for good measure in our next step, which generates column 8. Once we are confident in having a best estimate for $T \cdot C$ we can return to our data y in column 4. Cancellation gives us

$$\frac{y}{T \cdot C} = \frac{\text{column 4}}{\text{column 8}},$$
$$= \frac{T \cdot S \cdot C \cdot E}{T \cdot C}$$

so that:

$$\frac{y}{T \cdot C} = S \cdot E \qquad (5)$$

We next do some arithmetic to minimize the error and determine our best estimates for the values of S.

Step 4:

We "shuffle the deck" for the second and final time to generate column 8 by averaging each consecutive pair of values in column 7.

Note that we average the line numbers as well, which allows us to place each response on an integer numbered line. For this reason the column 8 quotients are called centered moving averages. For example, the first such average, (24.5 + 25)/2 = 24.75 and is placed on line 3. The last entry in column 8 is (34 + 35.25)/2 = 34.625 and is placed on line 22. These entries are out $T \cdot C$ values so that column 8 is headed by $T \cdot C$. Column 9 shows the line numbers of the $T \cdot C$ values.

Step 5:

We now have the capability to determine the values of $S \cdot E$, which are called **seasonal relatives.** Recall from (5) that:

$$S \cdot E = \frac{\text{column 4}}{\text{column 8}}$$

This yields column 10. Thus, for example, the first $S \cdot E$ quotient is 24/24.75 = 0.970, which is placed on line 3. Ideally, had no error component been present in the seasonal relative computation, four consecutive, repeating values of S would have been generated. However, all analyses of this sort contain some error component. To minimize this component we first organize the column 10 entries by placing the results for each season in a separate column. This is summarized in Table 25.

Table 25
Determination of Seasonal Indexes from Seasonal Relatives

	Winter	Spring	Summer	Fall
			0.970	0.915
	1.020	1.077	0.986	0.930
	0.986	1.112	0.943	0.953
	0.992	1.106	1.008	0.914
	0.988	1.086	0.972	0.910
	0.992	1.097		
Median	0.992	1.097	0.972	0.915
Seasonal Index = (1.006)(median)	$0.998 = S_W$	$1.104 = S_{Sp}$	$0.978 = S_{Su}$	$0.920 = S_F$

We next take the median of the results of each column to minimize the influence of extreme values. These medians are noted in the last row of Table 25.

There is one final step in this procedure which involves converting the seasonal relatives into our final product, called the **seasonal indexes.** However before we discuss and carry out this final step, it is important to understand how to interpret these indexes with respect to our original data y.

Since this analysis presupposes the multiplicative model $y = T \cdot S \cdot C \cdot E$, if our seasonal index is less than 1 our data are expected to show a decline because of the effect of that particular season on the time series data. That is, arithmetic tells us that whenever we multiply a data value by a quantity less than one, we get a product that is less than the given data value. On the other hand, a seasonal index that is greater than 1 implies an increase in the data value during that particular season. Clearly, a seasonal index that equals one implies that there is no seasonal effect whatsoever during that particular season.

Step 6:

As we discussed prior to step 4, over the course of a year the seasonal effects should "even out." Indeed, if time series data are observed only once each year, it is impossible to determine any seasonal variation for that data. Thus, over the course of the year the average seasonal effect, per season, should equal one. Thus, for quarterly data the total of the four seasonal indexes should equal 4; the sum of seasonal indexes for monthly data should be 12. Invariably, we do not exactly get the required sum; in the case of our time series analysis the sum is 3.976.

We must therefore **adjust each seasonal relative** by multiplying each by 4.000/3.976 = 1.006, which finally yields the four seasonal indexes for our quarterly data. These indexes are noted on the last row of Table 25.

Note that this last adjustment has also helped to minimize our error. Steps 5 and 6 have (essentially) "eliminated" the error component, leaving us with the four values of S for the winter, spring, summer and fall of each year, respectively. They are, in order, 0.998, 1.104, 0.978 and 0.920. These results are consistent with the general pattern of the data, which is to rise in the spring of each year $(S_{\text{Sp}} > 1)$ and decline in the summer and fall $(S_{\text{Su}} < 1 \text{ and } S_{\text{F}} < 1)$ and stabilize in the winter (S_{W} is about 1).

As we shall see, these figures will be needed in Sections 15.7 and 15.9 to do further analysis and forecasting.

EXERCISES

1. Refer to Exercise 4 of Section 15.5 (page 540), which shows the quarterly sales (number of copies sold) of an advanced economics textbook over the past four years. These data are reproduced as Table 26.

Table 26

		Quarter			
		Winter	Spring	Summer	Fall
Year	1	803	700	761	820
	2	891	720	740	850
	3	916	843	821	957
	4	964	897	867	981

(a) Construct the four seasonal indexes for this time series.

(b) As you constructed the centered moving average, which component of the time series data did you eliminate?

2. Refer to Exercise 5 of Section 15.5 (page 540), which lists the number of customers served at Bob's Bistro from Monday through Thursday during each of the last six weeks. These data are reproduced as Table 27.

Table 27

		Day			
		Mon.	Tues.	Wed.	Thurs.
Week	1	67	75	82	81
	2	72	77	79	73
	3	70	81	83	82
	4	77	85	89	90
	5	75	81	87	93
	6	81	89	86	92

(a) What is a "season" in the context of this situation? How long is any one seasonal cycle?

(b) Compute the five seasonal indexes for these data.

3. Refer to Exercise 6 of Section 15.5 (page 541), which gives the number of copies sold (in thousands) of *For Your Health* magazine for the period 2005-2007. These data are reproduced as Table 28.

Table 28

		Issue					
		A (Jan./ Feb.)	B (Mar./ Apr.)	C (May/ Jun.)	D (Jul./ Aug.)	E (Sep./ Oct.)	F (Nov./ Dec.)
Year	2005	39	45	41	37	44	49
	2006	42	44	47	42	48	53
	2007	43	42	49	43	50	57

(a) What is a "season" in the context of this situation?

(b) Compute the six seasonal indexes for these data.

(c) As you constructed the centered moving average, which component of the time series data did you eliminate?

4. Refer to Exercise 7 of Section 15.5 (page 542), which shows the number of new housing starts in Heightston for each quarter of the years 2001-2005. These data are reproduced as Table 29.

Table 29

		Quarter			
		Winter	Spring	Summer	Fall
	2001	24	34	36	29
	2002	21	35	38	27
Year	2003	19	28	32	26
	2004	20	31	34	23
	2005	16	24	25	21

(a) Construct the four seasonal indexes for these data.

(b) As you constructed the centered moving average, which component of the time series data did you eliminate?

5. Table 30 lists the number of kitchen appliances sold each quarter for the years 2002-2006 by the Federated Department Store.

Table 30

		Quarter			
		Winter	Spring	Summer	Fall
	2002	348	323	327	351
	2003	340	328	337	359
Year	2004	353	336	334	350
	2005	341	332	338	356
	2006	347	341	355	374

Compute the four seasonal indexes.

6. Table 31 states the number of clients that an unemployment agency found jobs for during each two month period in the years 2002-2005.

Table 31

		(Jan./ Feb.)	(Mar./ Apr.)	(May/ Jun.)	(Jul./ Aug.)	(Sep./ Oct.)	(Nov./ Dec.)
	2002	114	118	110	114	131	119
Year	2003	119	123	113	120	137	124
	2004	117	119	107	115	129	108
	2005	121	130	123	127	145	125

(a) What is a "season" in the context of this situation? How long is any one seasonal cycle?

(b) Compute the six seasonal indexes for this time series.

7. Table 32 lists the number of passengers served (in thousands) by the Heightston transit system during each quarter of the years 2005-2007.

Table 32

	Quarter			
	Winter	Spring	Summer	Fall
2005	43	41	37	40
2006	46	45	38	42
2007	47	40	31	34

Determine the four seasonal indexes for these data.

15.7 ■ Deseasonalizing Data

Having obtained measures of seasonal variation, we can make use of this information in two ways. In this section we shall show how seasonal indexes can be used to analyze past data and in Section 15.9 we combine trend, seasonal and cyclical factors to predict future data values.

We remove seasonal variation from a time series to see how the data would have looked had there not been seasonal fluctuation. The results of such calculations yield deseasonalized data. For example, let us briefly return to the Elemental Industrial Chemical Company as per Table 24 (page 546), where we note that the data for the four seasons of 2003 are 26, 28, 26, and 25, respectively. But let us keep in mind that these data contain, among other factors, seasonal effects.

Had we been able to eliminate these periodic and predictable upturns and downturns in the overall result, what would these values have been? This question is not hard to answer since, at least theoretically, we can eliminate each seasonal effect S by simply dividing each data value y, **assumed to be given by** $y = T \cdot S \cdot C \cdot E$, by S, so that each deseasonalized data value is given by:

$$\hat{y}_i = \frac{y_i}{S} \tag{6}$$

Example 1 Deseasonalized Data for the Elemental Industrial Chemical Company

Using (6), compute the deseasonalized data for 2003.

$$\hat{y}_5 = \frac{y_5}{S_{\text{w}}} = \frac{26}{0.998} = 26.05$$

$$\hat{y}_6 = \frac{y_6}{S_{\text{Sp}}} = \frac{28}{1.104} = 25.36$$

$$\hat{y}_7 = \frac{y_7}{S_{\text{Su}}} = \frac{26}{0.978} = 26.59$$

$$\hat{y}_8 = \frac{y_8}{S_{\text{F}}} = \frac{25}{0.920} = 27.17$$

Deseasonalized results often prove to be most illuminating. For instance, on examining the original data one is inclined to conclude that the Spring 2003 value was the largest and the Fall 2003 value was the smallest. However, a significant rise (of 10.4%, since $S_{\text{Sp}} = 1.104$ in our data is expected during each spring, so that by factoring out this explainable rise we see that the deseasonalized Spring 2003 value is actually the smallest of the four deseasonalized values. Also, after taking seasonal fluctuation into account, the largest value is the one for Fall 2003, even though the actual data suggested the opposite view.

> We have just seen that the deseasonalized data can look very different from their unadjusted counterparts. For this reason, the Federal Government is continually releasing "seasonally adjusted" economic data, so that the country's economists can give the "best possible" reading of the economic state of our country.

EXERCISES

1. Refer to Exercise 1 of Section 15.6 (page 549) and Exercise 4 of Section 15.5 (page 540), which shows the quarterly sales (number of copies sold) of an advanced economics textbook over the past four years.

   (a) Deseasonalize the data for the third year.

   (b) The largest and smallest data values in the third year occur in the fall and spring, respectively. Do the deseasonalized data repeat this pattern? Would you expect them to? Explain.

2. Refer to Exercise 2 of Section 15.6 (page 550) and Exercise 5 of Section 15.5 (page 540), which gives the number of customers served at Bob's Bistro from Monday through Thursday during each of the last six weeks.

   (a) Deseasonalize the data for weeks 4 and 6.

   (b) Do the largest and smallest data values in weeks 4 and 6 occur on the same days as their deseasonalized counterparts? Explain.

3. Refer to Exercise 3 of Section 15.6 (page 550) and Exercise 6 of Section 15.5 (page 541), which shows the number of copies sold (in thousands) of *For Your Health* magazine for the years 2005-2007.

   (a) Deseasonalize the data for 2006.

   (b) Do the largest and smallest data values in 2006 occur at the same two month time period as their deseasonalized counterparts? Explain.

4. Refer to Exercise 4 of Section 15.6 (page 551) and Exercise 7 of Section 15.5 (page 542), which shows the number of new housing starts in Heightston for each quarter for the years 2001-2005.

   (a) Deseasonalize the data for 2002 and 2004.

   (b) Do the largest and smallest data values in 1992 and 1994 occur during the same quarter as their deseasonalized counterparts? Explain.

5. Refer to Exercise 5 of Section 15.6 (page 551), which gives the number of kitchen appliances sold each quarter for the years 2002-2006, at the Federated Department Store. These data are reproduced as Table 33.

Table 33

		Quarter			
		Winter	Spring	Summer	Fall
	2002	348	323	327	351
	2003	340	328	337	359
Year	2004	353	336	334	350
	2005	341	332	338	356
	2006	347	341	355	374

Deseasonalize the data for 2003 and 2006.

6. Refer to Exercise 6 of Section 15.6 (page 552), which shows the number of clients that an unemployment agency found jobs for during the years 2002-2005. These data are reproduced as Table 34.

Table 34

		(Jan./ Feb.)	(Mar./ Apr.)	(May/ Jun.)	(Jul./ Aug.)	(Sep./ Oct.)	(Nov./ Dec.)
	2002	114	118	110	114	131	119
Year	2003	119	123	113	120	137	124
	2004	117	119	107	115	129	108
	2005	121	130	123	127	145	125

Deseasonalize the 2002 and 2005 data.

7. Refer to Exercise 7 of Section 15.6 (page 560), which gives the number of passengers served (in thousands) by the Heightston transit system during each quarter of the years 2005-2007. These data are reproduced as Table 35

Table 35

	Quarter			
	Winter	Spring	Summer	Fall
2005	43	41	37	40
2006	46	45	38	42
2007	47	40	31	34

(a) Deseasonalize the 2006 and 2007 data.

(b) Do the largest and smallest data values in 2005 and 2006 occur during the same quarter as their deseasonalized counterparts? Explain.

15.8 ■ Determination of Cyclical Indexes

Once we have a trend equation, such as $T_t = 23.39 + 0.469t$ for the Elemental Industrial Chemical Company (Example 1, Section 15.5, page 536), and a collection of $T \cdot C$ values, as in Table 24 (page 546), we are ready to compute cyclical relatives. These relatives will be converted into cyclical indexes in much the same way as was done to the seasonal relatives. To best follow the sequence of steps involved, please refer to Tables 36 (page 557) and 37 (page 559).

Example 1 Cyclical Relatives and Indexes for the Elemental Industrial Chemical Company

In Table 36, columns 1 through 5 were copied Table 24 (page 546), while the trend values in column 6 are obtained by evaluating the above-noted trend line using $t = 3$ through 22. The cyclical relatives in column 7 may be computed by using the following approach:

On each line, take the $T \cdot C$ and T values located in columns 5 and 6, respectively. Then:

$$C = \frac{T \cdot C}{T} = \frac{\text{column 5 entry}}{\text{column 6 entry}} \qquad (7)$$

Table 36
Determination of Cyclical Relatives

(1) Yr	(2) Season	(3) t	(4) y	(5) $T \cdot C$	(6) T	(7) C
'02	W	1	24			
	Sp	2	27			
	Su	3	24	24.75	24.795	0.998
	F	4	23	25.125	25.264	0.994
'03	W	5	26	25.5	25.733	0.991
	Sp	6	28	26.0	26.202	0.992
	Su	7	26	26.375	26.671	0.989
	F	8	25	26.875	27.14	0.990
'04	W	9	27	27.375	27.609	0.992
	Sp	10	31	27.875	28.078	0.993
	Su	11	27	28.625	28.547	1.003
	F	12	28	29.375	29.016	1.012
'05	W	13	30	30.25	29.485	1.026
	Sp	14	34	30.75	29.954	1.027
	Su	15	31	30.75	30.423	1.011
	F	16	28	30.625	30.892	0.991
'06	W	17	30	30.375	31.361	0.969
	Sp	18	33	30.375	31.83	0.954
	Su	19	30	30.875	32.299	0.956
	F	20	29	31.875	32.768	0.973
'07	W	21	33	33.25	33.237	1.000
	Sp	22	38	34.625	33.706	1.027
	Su	23	36			
	F	24	24			

How are the cyclical relatives in column 7 of Table 36 to be interpreted? Unlike their seasonal counterparts, cyclical variation is thought of as being very long-term, not just from period to period.

Beware the Assumptions

We make two **simplifying assumptions:**

(1) It is more meaningful to have one cyclical index for an entire year than to have a different index for each period within the year.

> (2) The business cycle for any given time series analysis has a duration equal to the number of years that encompass that data.

Therefore, we **assume that the business cycle for our data is of six years duration.**

In summary, to continue our data analysis we shall **assume** that a cyclical index exists for each of six consecutive years and that these values will recur during every six year time frame.

Thus, if our "first" business cycle contains the years 2002 through 2007 and thereby yield cyclical indexes (let's call them) C_1 through C_6, respectively, the next cycle would be 2008 through 2013, having the same cyclical indexes C_1 through C_6 respectively. The third cycle would be 2014 through 2019, having cyclical indexes C_1 through C_6, respectively, and so on.

Based on these assumptions, the cyclical relatives are noted on the second through (at most) fifth lines of Table 37, where 2002 corresponds to year 1, 2003 to year 2, etc., culminating with 2007 as year 6. Note that years 1 and 6 contain only two entries each, since these years contain only two values of $T \cdot C$ in Table 36. The median of the cyclical relatives for each year is noted on the sixth line of Table 37.

According to our assumptions, we should be at the same point in our business cycle in exactly six years as we are now. That is, the average cyclical index for all six years should be 1 or equivalently, the total of six consecutive cyclical indexes should be 6.

However, the sum of our six medians on the sixth line of Table 37 is only 5.981. Therefore, analogous to the seasonal indexes computation, multiplication by the appropriate constant (6/5.981 = 1.003) yields the cyclical indexes on the seventh line of Table 37.

We shall use the cyclical indexes in the next section to help forecast values for future time periods.

Table 37
Determination of Cyclical Indexes from Cyclical Relatives

Year i in Business cycle	1	2	3	4	5	6	
	0.998	0.991	0.992	1.026	0.969	1.000	
	0.994	0.992	0.993	1.027	0.954	1.027	
		0.989	1.003	1.011	0.956		
		0.990	1.012	0.991	0.973		
Median	0.996	0.991	0.998	1.019	0.963	1.014	Σ medians = 5.981
C_i = cyclical index = (1.003)(median)	0.999	0.994	1.001	1.022	0.966	1.017	

EXERCISES

1. Refer to Exercise 1 of Section 15.6 (page 549) and Exercise 4 of Section 15.6 (page 540), which shows the quarterly sales (number of copies sold) of an advanced economics textbook over the past four years. Compute the cyclical indexes for the time series. Assume a four year business cycle.

2. Refer to Exercise 2 of Section 15.6 (page 550) and Exercise 5 of Section 15.5 (page 540), which gives the number of customers served at Bob's Bistro from Monday through Thursday during each of the last six weeks. Assuming a six week business cycle, compute the cyclical indexes for this time series.

3. Refer to Exercise 3 of Section 15.6 (page 550) and Exercise 6 of Section 15.5 (page 541), which shows the number of copies sold (in thousands) of *For Your*

Health magazine for the years 2005-2007. Assuming a three year business cycle, compute the cyclical indexes for this time series.

4. Refer to Exercise 4 of Section 15.6 (page 549) and Exercise 7 of Section 15.5 (page 542), which gives the number of new housing starts in Heightston for each quarter for the years 2001-2005. Determine the cyclical indexes for this time series, assuming a five year business cycle.

5. Refer to Exercise 5 of Section 15.6 (page 551), which gives the number of kitchen appliances sold during 2002-2006 at the Federated Department Store.

   (a) Find the least-squares line which shows the trend component of this time series.

   (b) Construct the cyclical indexes for the time series, assuming a five year business cycle.

6. Refer to Exercise 6 of Section 15.6 (page 552), which shows the number of clients that an unemployment agency found jobs for during the years 2002-2005.

   (a) Find the least-squares line which shows the trend component of the time series.

   (b) Assuming a four year business cycle, determine the cyclical indexes for this time series.

7. Refer to Exercise 7 of Section 15.6 (page 552), which gives the number of passengers served (in thousands) by the Heightston transit system during each quarter of the years 2005-2007.

   (a) Find the least-squares line which shows the trend component of the time series.

   (b) Assuming a three year business cycle, construct the cyclical indexes for the time series.

15.9 ■ Forecasting from Time Series Data

Before the Assumptions

Under the **assumption of the multiplicative model** that $y = T \cdot S \cdot C \cdot E$, we proceeded to "dissect" the data into its component parts and carry out the computations for *S* and *C* so as to minimize the error.

Under the **assumption** that the random error component *E* is of little significance, we illustrate how to use our linear trend equation $T_t = 23.39 + 0.469t$ and seasonal and cyclical indexes *S* and *C* to forecast data values for any future time period. This forecast is obtained by merging the three measurable quantities *T*, *S*, and *C*.

Under the **multiplicative model's assumption**, our forecast $\hat{y}$ is obtained from:

$$\hat{y} = T \cdot S \cdot C \tag{8}$$

Example 1 Sales Forecasts for the Elemental Industrial Chemical Company

Forecast sales for the Elemental Industrial Chemical Company for each season of the years 2010 and 2013.

Let us recall that we obtained the trend projection for each quarter of 2010 in Example 1 of Section 15.5 (page 536), namely, $T_{33} = 38.9$, $T_{34} = 39.3$, $T_{35} = 39.8$ and $T_{36} = 40.3$. With respect to *S*, each year yields the four seasonal indexes $S_{\text{w}} = 0.998$, $S_{\text{Sp}} = 1.104$, $S_{\text{Su}} = 0.978$ and $S_{\text{F}} = 0.920$. As for the cyclical factor *C*, 2010 corresponds to the third year of a business cycle, so that we assign to 2010 the cyclical index $C_3 = 1.001$. This yields:

$$\hat{y}_{33} = T_{33} S_{\text{w}} C_3 = 38.9(0.998)(1.001) = 38.8$$
$$\hat{y}_{34} = T_{34} S_{\text{Sp}} C_3 = 39.3(1.104)(1.001) = 43.5$$
$$\hat{y}_{35} = T_{35} S_{\text{Su}} C_3 = 39.8(0.978)(1.001) = 39.0$$
$$\hat{y}_{36} = T_{36} S_{\text{F}} C_3 = 40.3(0.920)(1.001) = 37.1$$

Thus, we predict that sales revenue for the Elemental Industrial Chemical Company will be 38.8, 43.5, 39.0 and 37.1 millions of dollars in the winter, spring, summer and fall, respectively, of 2010.

Our data consists of the years 2002 through 2007, so that 2013 represents the twelfth year for the variable of interest. Since 11 years (44 quarters) have elapsed, we use $t = 45$ through 48 for 2013. Therefore, we need $T_{45} = 44.5$. The remaining trend values are obtained by adding 0.469 to the previous result. The year 2013 represents the last year in a business cycle so that we use $C_6 = 1.017$. This gives the following projections for 2013:

$$\begin{aligned}
\hat{y}_{45} &= T_{45}S_{\mathrm{w}}C_6 = 44.5(0.998)(1.017) = 45.2 \\
\hat{y}_{46} &= T_{46}S_{\mathrm{w}}C_6 = 45.0(1.104)(1.017) = 50.5 \\
\hat{y}_{47} &= T_{47}S_{\mathrm{w}}C_6 = 45.4(0.978)(1.017) = 45.2 \\
\hat{y}_{48} &= T_{48}S_{\mathrm{w}}C_6 = 45.9(0.920)(1.017) = 43.0
\end{aligned}$$

Therefore, we predict that sales revenue for the Elemental Industrial Chemical Company will be 45.2, 50.5, 45.2 and 43.0 millions of dollars in the winter, spring, summer and fall, respectively, of 2013.

BEWARE THE ASSUMPTIONS

We should be **cautious** in the use of this forecasting method because of the rather **large number of assumptions on which the methodology is based.**

We have **assumed**, for example, that the multiplicative model is preferable to the additive model. We have defined and computed a "best" trend line by employing the **least squares principle**.

We have analyzed six years worth of data and have **assumed** that our business cycle is of six years duration. Moreover, rather than having a cyclical index for each quarter, we employed a single cyclical index for each year of our business cycle. In calculating the *C*'s and *S*'s, we have (hopefully) "minimized" the unexplained component *E*, but there is no guarantee that we have literally eliminated it.

Moreover, **realism** requires that over the course of time random perturbations of varying degrees may still have an impact on the accuracy of our determinations.

Last, but certainly not least, the **accuracy** of our predictions should be expected to diminish as we predict sales for periods of time further and further into the future. This is because our projections are based on the **assumption that the economic "atmosphere" that prevailed between 2002 and 2007 will be maintained.** For the short term, this is **hopefully realistic.** But the long term is another matter.

EXERCISES

1. Refer to Exercise 1 of Section 15.8 (page 559), Exercise 1 of Section 15.6 (page 549) and Exercise 4 of Section 15.5 (page 540), which shows the quarterly sales (number of copies sold) of an advanced economics textbook over the past four years. Forecast sales of this textbook during each quarter of

   (a) year 5,

   (b) year 10.

   (c) Why should we be careful about the use of these predictions?

2. Refer to Exercise 2 of Sections 15.8 (page 559) and 15.6 (page 550) as well as Exercise 5 of Section 15.5 (page 540), which gives the number of customers served at Bob's Bistro from Monday through Thursday of the last six weeks. Predict the number of customers served from Monday through Thursday of

   (a) week 7,

   (b) week 11.

   (c) Why must we exercise caution about the accuracy of these predictions?

3. Refer to Exercise 3 of Section 15.8 (page 559) and 15.6 (page 550) as well as Exercise 6 of Section 15.5 (page 541), which shows the number of copies sold (in thousands) of *For Your Health* magazine for the years 2005-2007. Forecast sales of this magazine (in thousands) during each two month period of the year

   (a) 2010,

   (b) 2012.

(c) Are there any cautions that should be exercised in the use of these predictions?

4. Refer to Exercise 4 of Section 15.8 (page 560) and 15.6 (page 551) as well as Exercise 7 of Section 15.5 (page 542), which gives the number of new housing starts in Heightston for each quarter of the years 2001-2005. Forecast the number of new housing starts during each quarter of the year

(a) 2008,

(b) 2011.

(c) Why must we be careful in the way we use these predictions?

5. Refer to Exercise 5 of Section 15.8 (page 560) and 15.6 (page 551), which gives the number of kitchen appliances sold during 2002-2006 at the Federated Department Store. Predict the sales of kitchen appliances by the store during each quarter of the year

(a) 2009,

(b) 2013.

(c) Why must we exercise caution in the use of these predictions?

6. Refer to Exercise 6 of Section 15.8 (page 560) and 15.6 (page 552), which shows the number of clients that an unemployment agency found jobs for during the years 2002-2005. Predict the number of clients who the agency will find jobs for during each two month period of the year

(a) 2007,

(b) 2010.

(c) Why must we be careful in the way we use these predictions?

7. Refer to Exercise 7 of Section 15.8 (page 560) and 15.6 (page 552), which gives the number of passengers served (in thousands) by the Heightston transit system during each quarter of the years 2005-2007. Forecast the number of

passengers served (in thousands) by the Heightston transit system during each quarter of the year

(a) 2011,

(b) 2014.

(c) Are there cautions that should be exercised in the use of these predictions?

16

Nonparametric Statistics

16.1 □ Parametric Versus Nonparametric Methods

The methods of statistical inference that we discussed have a common thread; we first identify a population parameter (μ, σ^2, or ρ, for example) about which we wish to draw inferences. We then select an estimator of the parameter in question ($\bar{x}$, s^2 or r) and determine its value once a random sample is taken. Finally, we use the sampling distribution of the estimator to construct hypothesis testing decision rules or confidence interval results. However, to obtain information about the approximate sampling distribution of the sample statistic to be used we often require conditions on the "parent" populations from which we sample. For instance, in many cases we must require that the populations be normally distributed or approximately so, that the variances be equal, or the samples be independently chosen. The Central Limit Theorems give us some leeway since they allow us to work with random samples from non-normal populations. When the appropriate conditions that underlie the statistical analyses are met, including knowledge of the values of parameters as required, the methodology is called **parametric statistics.**

In many application situations some of these conditions are not met or it is not clear whether they are met. Statistical tests that do not require conditions on the population or its parameters are referred to as **nonparametric** or **distribution-free** tests.

Pros

Nonparametric techniques are appealing for a number of reasons. In practice we may be unable to check whether the required parametric assumptions hold.

Most nonparametric methods are computationally easier to explain and carry out than their parametric counterparts. Many nonparametric tests can be applied to situations where classical parametric tests cannot be used because of the nature of the data. This is the case when the data is ordinal, consisting of numbers assigned as ranks, such as those used in judging a contest.

Cons

These techniques are not, however, without disadvantages. In some instances we must ignore the actual numerical value of an observation and take into account only its relative standing (or ranking) in relation to other observations in the sample. As a result, acquired information is not fully utilized. Furthermore, if conditions under which a parametric test can be applied do indeed hold, then a nonparametric test is not as "efficient" as its classical counterpart. This makes sense because a nonparametric test assumes little or nothing about the underlying population and therefore cannot be expected to be as "strong" as a test that makes use of known properties of the population. This lack of efficiency is seen by the user in terms of a higher probability of erroneously accepting H_o (i.e., an increased type II error probability), as compared with the parametric approach when all statistical assumptions are met.

In the sections that follow we examine some commonly applied nonparametric techniques.

16.2 □ A One-Sample Sign Test

This test can be used to test the hypothesis that a set of n randomly obtained values came from a population having a specified median. For symmetric populations the mean and median are equal, and this median test is equivalent to a test on the population mean for such populations. The sign test is the nonparametric counterpart of the t-test described in Section 11.2. However, the t-test requires that the population being sampled from be normally distributed, whereas our median test imposes no such requirement.

The one-sample sign test is applicable to populations whose data scale is ordinal or higher.

The hypotheses for this test are:

$$H_o: \text{The population median equals } M,$$

versus one of the following alternative hypotheses:

H_a : The population median is not equal to *M*, or

H_a : the population median exceeds *M*, or

H_a : the population median is less than *M*,

depending on the context.

If the null hypothesis is true, it should be equally likely that a sample value would be either less than or greater than the median. Consequently, we replace all sample values less than the hypothesized median *M* with a minus sign and all data values greater than *M* with a plus sign. If a sample value equals *M*, we simply discard it. We now have a sequence of plus and minus signs and, assuming the median equals *M*, this sequence could be interpreted as a random sample from a population in which the probabilities of either plus or minus are each 0.5. In this case the observations constitute a random sample for which the probability π of plus is 0.5. Thus, given the sequence of signs, our original null hypothesis is equivalent to

$$H_o : \pi = 0.5,$$

to be tested against the appropriate one-sided or two-sided alternative.

We should recall that the normal approximation to the binomial may be used whenever both $n\pi$ and $n(1-\pi)$ are at least 5. Since π is assumed to equal 0.5 under the null hypothesis of our sign test, we may use the *z*-statistic to carry out the sign test whenever $0.5n \geq 5$, that is whenever $n \geq 10$.

If this condition holds, the technique is continued by calculating:

$$p = \frac{\text{number of plus signs in the sequence}}{\text{total number of signs in the sequence}}$$

Let us recall that the appropriate *z*-statistic for testing hypotheses about proportions is:

$$z = \frac{p - \pi}{\sqrt{\frac{\pi(1-\pi)}{n}}}$$

We compute this test statistic and compare it to the appropriate standard normal curve value or values.

Example 1 Weight Loss Claim

The Take It Off weight loss clinic advertises that their over-sized (more than fifty pounds overweight) clients lose, on average, 40 pounds within the first year. A consumer interest group is concerned about the possibility that this clinic may be exaggerating the extent of its clients' first year weight loss. It decided to use the sign test with significance level $\alpha = 0.10$ to test the null hypothesis that the median weight loss for the indicated population is 40 pounds, against the alternative that the median weight loss is less than 40 pounds.

A random sample of 13 over-sized clients showed the following number of pounds lost during their first year:

24, 40, 42, 29, 57, 31, 12, 18, 37, 46, 31, 38, 43.

The sequence of plus and minus signs is:

$$- + - + - - - - + - - +$$

Note that the sequence consists of only 12 signs since $M = 40$ is the second of our data values. Since $n\pi = n(1-\pi) = 12(0.5) = 6 \geq 5$, it is appropriate to use the z-statistic. Our null and alternative hypotheses are:

$H_o : M = 40$, or equivalently: $\pi = 0.5$
$H_a : M < 40$, or equivalently: $\pi < 0.5$

There are 4 plus out of 12, so that $p = 4/12 = 0.333$.

Since $\alpha = 0.10$, our left tail begins at -1.28 (see Figure 1).

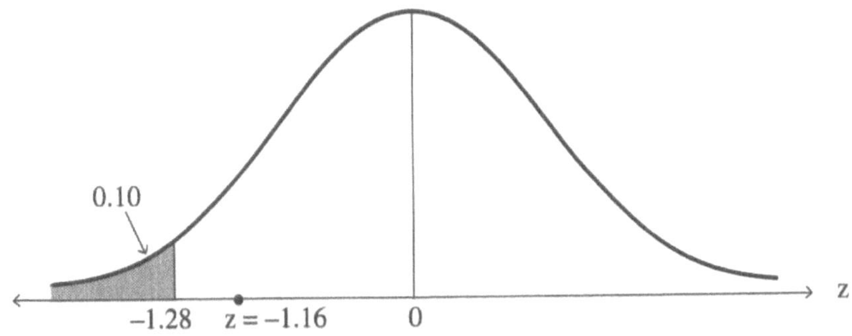

Figure 1

However:

$$z = \frac{0.333 - 0.5}{\sqrt{\frac{(0.5)(0.5)}{12}}} = -1.16 > -1.28$$

Thus, at the 0.10 level, the consumer interest group cannot reject H_o and must either conclude that the true median weight loss for the population of oversized clients is 40 pounds or reserve judgment. In either case, they have no basis for challenging the clinic's claim.

Example 2 Median Age Claim

Brides magazine recently reported that the median age of New York state women at the time of their first marriage is 24.5 years. At the 5% level of significance, use the sign test to decide whether or not it is reasonable to state that the median age for first time New York brides is 24.5 years. Base your decision on a recent randomly drawn sample of 40 first marriages in New York which revealed 27 cases where the bride was more than 24.5 years old and 13 cases where the bride was less than 24.5 years old.

$H_o : M = 24.5$, or equivalently: $\pi = 0.5$
$H_a : M \neq 4.5$, or equivalently: $\pi \neq 0.5$.

We have a sequence of 40 signs containing 27 plus signs, so that $p = 27/40 = 0.675$. The z-statistic may be used since $n\pi = n(1-\pi) = 40(0.5) = 20 \geq 5$. We have a two-sided test, with $\alpha = 0.05$, so that our tails begin at ± 1.96 (see Figure 2).

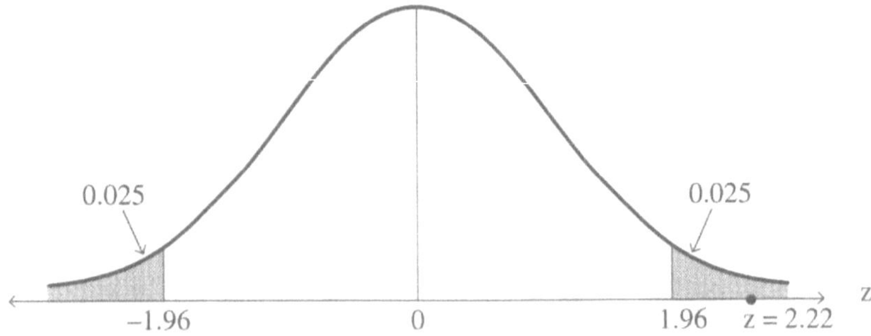

Figure 2

We have:

$$z = \frac{0.675 - 0.5}{\sqrt{\frac{(0.5)(0.5)}{40}}} = 2.22 > 1.96$$

Therefore, at the 0.05 level we reject H_o and accept H_a, that the median age for first time New York brides is not 24.5 years.

Example 3 Sulfur Dioxide Emissions Claim

A large industrial plant of the Hitest Gas Company emits sulfur oxides daily. Federal regulations require that the mean level of sulfur dioxide emissions for such a plant not exceed 20 tons per day. A Federal regulatory agency wishes to verify that Hitest is not exceeding the legal limit at this plant. The agency measured the plant's daily sulphur dioxide emissions (in tons) on 32 randomly chosen days, which yielded the results listed in Table 1.

Table 1

19	17	21	27	18	28	24	28	16	8	25
13	24	22	29	30	27	31	15	20	25	26
18	22	13	31	29	17	20	23	15	27	

Test, at the 0.025 significance level, the null hypothesis that the plant's mean daily sulphur dioxide emission is 20 tons against the alternative hypothesis that the gas company is violating Federal regulations by having a mean daily emissions level higher than this value. We assume that the population of sulfur dioxide emission values is symmetric (so that mean = median).

$H_o : M = 20$, or equivalently: $\pi = 0.5$
$H_a : M > 20$, or equivalently: $\pi > 0.5$.

There are two data values equal to 20, so that we have a sequence containing 19 plus signs out of 30. Therefore, $p = 19/30 = 0.633$. Use of the z-statistic is appropriate since $n\pi = n(1-\pi) = 30(0.5) = 15 \geq 5$. Since $\alpha = 0.025$, our right tail begins at 1.96 (see Figure 3).

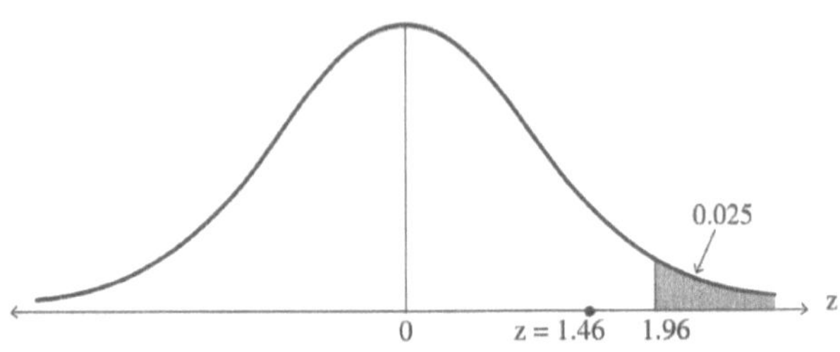

Figure 3

We calculate:

$$z = \frac{0.633 - 0.5}{\sqrt{\frac{(0.5)(0.5)}{30}}} = 1.46 < 1.96$$

Thus, at the 0.025 level we cannot reject H_o, and we either conclude that the true daily mean level of sulfur dioxide emission is 20 tons or reserve judgment. Either way, the Federal regulatory agency does not have sufficient cause to close the plant or impose a warning or fine.

We conclude by summarizing the nonparametric one-sample sign test technique.

(1) Test $H_o : \pi = 0.5$,
versus one of:

$$H_a : \pi < 0.5$$
$$H_a : \pi > 0.5$$
$$H_a : \pi \neq 0.5$$

(2) Determine the sequence of signs and the number of plus signs.

(3) Check the conditions $n\pi \geq 5, n(1-\pi) \geq 5$. If satisfied, proceed.

(4) Compute the test statistic

$$z = \frac{p - \pi}{\sqrt{\frac{\pi(1-\pi)}{n}}}$$

and compare it with the appropriate z-bound or bounds.

EXERCISES

1. The goal of the Carterville Ambulance Service is to reach the scene of a medical emergency in 7 minutes or less. Test, at the 0.05 level of significance, the hypothesis that the median waiting time is 7 minutes, against the alternative that a patient, more often than not, must wait longer for the ambulance to arrive. During a recent year, the number of minutes it took an ambulance service to reach the scene of thirteen randomly selected Carterville emergencies were:

   9, 5, 13, 8, 7, 9, 10, 11, 4, 13, 6, 12, and 11.

2. The Dean of Graduate Studies of Ecap University claims that the median Graduate Management Admissions Test (GMAT) score for its admitted students is 630.

   (a) Use the sign test and a 10% significance level to decide whether the dean's claim is warranted, using the randomly selected sample of 18 scores shown in Table 2 that were obtained from the files of admitted students for a recent semester.

   Table 2

672	570	596	523	607	486
595	604	653	720	758	629
586	619	737	585	630	564

   (b) What **assumption** would have to be made about the underlying population in order to test the null hypothesis that the mean GMAT score of admitted students is 630, using the same median test?

3. From past experience Overseas Airlines has concluded that the median weight of a suitcase on its transoceanic flights is 40 pounds. A random sample of 22 suitcases carried by this airline last year consisted of 15 suitcases weighing less than 40 pounds and 7 suitcases weighing in excess of 40 pounds. Test the null hypothesis that the median weight of a suitcase is 40 pounds against the alternative that most suitcases are not as heavy. Use an 0.05 level of significance.

4. Some years ago homes in a Queens neighborhood were appraised as having a median price of $245,000. More recently, however, a recession has caused the real

estate market to be "soft" and homeowners in this neighborhood wonder whether the recession has had a softening effect on their property. A random sample of 14 homes in this neighborhood were re-appraised last month, revealing property values (in thousands of dollars) as follows: 217, 223, 248, 231, 218, 243, 251, 227, 239, 262, 249, 233, 230 and 240. At the 0.025 level of significance, does the evidence suggest that the median price of homes in the locale is still \$245,000, or do we have good reason to concluded that the purchase of real estate in this locale is currently less expensive, on the whole?

5. The nationwide median hourly wage for construction workers is \$17.50 per hour. Of a sample of 108 randomly selected New Jersey construction workers fifty eight had an hourly wage rate greater than \$18.50, forty two had an hourly wage rate less than \$17.50, and eight had a wage rate of \$17.50. At the 5% level of significance, test the null hypothesis that the median hourly wage for New Jersey construction workers is the same as the nationwide hourly wage against the alternate hypothesis that the New Jersey median wage exceeds the nationwide median wage.

6. The law firm of Kent, Kent Rosen, and Levy wishes to hire a secretary, a position for which it requires a mean typing speed of 60 words per minute (wpm). Susan Smith's typing abilities were tested and her rate on ten randomly selected samples was: 67, 58, 70, 60, 59, 65, 51, 56, 63, 72, 55, and 57 wpm.

   (a) At the 10% significance level, test the null hypothesis that Susan's mean typing speed is 60 wpm against the alternative that her typing is not up to that requirement.

   (b) What **assumption** are we making in carrying out this test?

16.3 ■ A Paired-Sample Sign Test

By slightly extending the one-sample sign test we obtain the paired-sample sign test. This sign test can be used to analyze data that consist of matched pairs. The null hypothesis for matched pairs is that the population of within-pairs differences has a median equal to zero, versus some appropriate one or two sided alternative hypothesis.

The nonparametric paired sample approach is virtually identical to that of the one-sample method, except that we replace each pair of data values (rather than each data value) with a plus sign if the first component exceeds the first. As before, no ties are allowed, so that those pairs for which the two components are equal are discarded from the sample. We may then proceed as before with the z-statistic.

> The paired-sample sign test is applicable to populations whose data scale is ordinal or higher.

Example 1 Return to the Mathematics Review Tape Situation

We return to Example 2 of Section 11.9 (page 368) which describes a paired sample study comparing final exam test scores of pairs of ten randomly selected students. One member of each pair participated in a mathematics videotape review program and the other did not. The underlying statistical analysis requires that the populations of final exam grades be approximately normal.

If we are **uncomfortable with this assumption of normality**, the parametric methods of Section 11.8 should not be used. Instead we may employ the nonparametric Paired-Sample Sign Test which **does not require assumptions** about the distributions of the populations involved. For the comparison of tests to be meaningful we must assume that the population of paired differences is symmetric so that its mean and median are equal.

Does the mathematics review program help increase students' median exam performance? To start, we reproduce Table 7 of Section 11.8 as Table 3, which lists the final exam scores of the ten randomly selected students who participated in the study.

Table 3

Pair	1	2	3	4	5	6	7	8	9	10
Took video program	85	72	66	59	67	91	73	72	73	88
No video program	76	64	67	61	65	83	72	74	77	80

If the video review program helps increase the students' median exam performance, we would expect the median exam grade of the students who participated in the program to "significantly" exceed that of their counterparts who did not participate in the program. That is, most of the test score differences (video program minus no video program) should be positive. We state our null and alternative hypotheses as follows:

$$H_o : \pi = 0.5$$
$$H_a : \pi > 0.5$$

The final exam scores of students who did not participate in the video review program exceeded that of their counterparts in pairs 3, 4, 8 and 9, while in the other six pairs, the student participating in the program achieved a higher final exam score. We have the sequence of ten signs

$$+ + - - + + + - - +$$

six of which are plus, so that $p = 6/10 = 0.6$. The z-statistic may be used since $n\pi = n(1-\pi) = 5 \geq 5$. We use the same 5% level of significance that was previously employed (Section 11.9, Example 2, page 368) so that our right tail begins at 1.65 (see Figure 4).

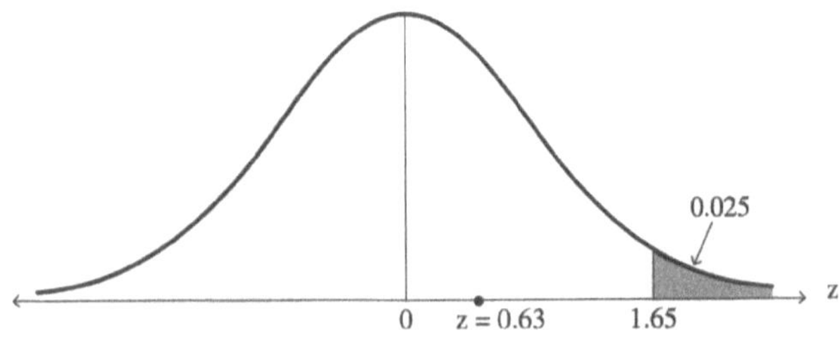

Figure 4

Our z-statistic is:

$$z = \frac{0.6 - 0.5}{\sqrt{\frac{(0.5)(0.5)}{10}}} = 0.63 < 1.65$$

Thus, at the 0.05 significance level we cannot reject H_o. Either we conclude that the videotape review program is not of benefit in preparing freshmen students for their initial study of mathematics, or reserve judgment.

The fact that we were led to reject the null hypothesis when applying parametric methods to the mathematics videotape review situation considered in Example 2 of Section 11.9 (page 368) and cannot reject it here illustrates a general trade-off between parametric and nonparametric methods. In applying a parametric test we make fuller use of the available information in that we took into account the magnitude of the differences in paired test scores, whereas in the nonparametric test we only made use of the direction of the differences (plus or minus) and not their magnitude. The nonparametric method is weaker than its parametric counterpart in that it does not make full use of available information. On the other hand it **requires less in the way of assumptions** to be applicable; it does not require that the underlying population satisfy a specific distribution.

In general, in requiring less you get less in return. Alas, free statistical lunches are few and far between.

It is worthy of note that we cannot perform this hypothesis test on the data of Example 1 of Section 11.8 where we compared corn yields obtained from two different types of fertilizer. This is because we have five pairs of data while our nonparametric procedure requires a minimum sample size of ten in order to justify the use of the z-statistic.

Example 2 The Effect of Temperature on Skiing Time

To study the effect of temperature on skiing time fourteen randomly selected skiers were clocked on a course (in minutes) on two days, once when the temperature was 20°F and once when it was just above the freezing mark. Wind and cloud conditions were similar on the two days. The skiing times are shown in Table 4.

Table 4

Skiers	Esther	Joshua	May	Bruce	Carol	Bob	Darice	Barry	Helen	Lou	Sharon	Heather	Rick	Donna
20°F	8.1	7.9	10	8.6	7.8	9.2	8.6	8.4	8.5	9.1	8.5	8.9	7.9	8.8
35°F	8.4	8.2	9.7	8.8	7.9	9.7	8.2	8.3	8.6	9.1	8.4	9.2	8.1	9.2

The feeling was that the increase in temperature might result in an increase in the median skiing time. Test, at the 0.05 level of significance, the hypothesis that temperature has no effect on the median skiing time against the alternative hypothesis that the warmer temperature adversely affects is.

If the warmer weather does "adversely" affect skiing performance time, then we would expect the time in the warmer temperature to be the higher of the two times for most skiers, so that it is reasonable to expect most of the differences (colder temp.—warmer temp.) to be negative. Then there should be relatively few positive signs, which leads us to formulate our null and alternative hypotheses as follows:

$$H_o : \pi = 0.5$$
$$H_a : \pi < 0.5$$

May, Darice, Barry and Sharon took longer in the lower temperature than in the higher temperature. Lou's times in the two temperatures were the same, while all other skiers took less time in the lower temperature. As a result, the sequence of signs is:

$$- - + - - - + + - + - - -$$

Our sequence of 13 signs contains 4 plus signs, so that $p = 4/13 = 0.308$. The *z*-statistic may be used since $n\pi = n(1-\pi) = 13(0.5) = 6.5 \geq 5$. Since $\alpha = 0.05$, our left tail begins at -1.65 (see Figure 5).

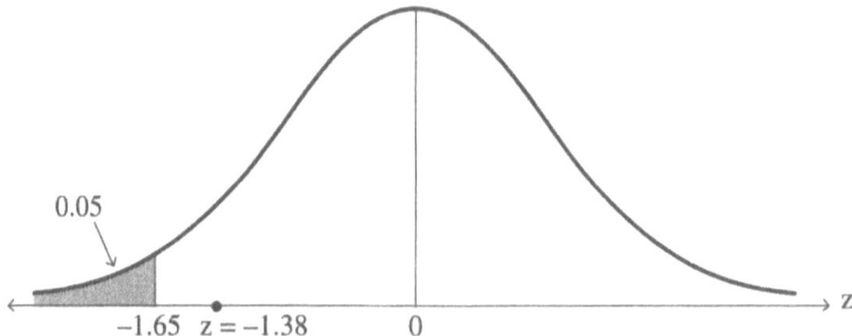

Figure 5

The *z*-statistic is:

$$z = \frac{0.308 - 0.5}{\sqrt{\frac{(0.5)(0.5)}{13}}} = -1.38 > -1.65$$

Consequently, at the 0.05 significance level we cannot reject H_o. Instead, we accept H_o that the warmer temperature does not have an adverse affect on skiing time or reserve judgment.

EXERCISES

1. The Arnold Clinic wants to test the effectiveness of a proposed diet plan. Does the plan help to reduce the weight of those who undertake it? Weight reduction data for a random sample of 15 dieters who completed the plan are given in Table 5.

Table 5

Weight Before	137	156	163	174	160	207	184	245
Weight After	142	147	158	180	160	196	173	231

Weight Before	185	172	133	253	129	175	197
Weight After	178	174	138	237	126	164	191

Use the paired-sample sign test to test the null hypothesis that the diet is not effective in reducing a person's median weight against the alternative hypothesis that it is effective in reducing the median weight. Use the 0.025 significance level.

2. At Huxley College an honors student can take a course in statistics in the conventional lecture-problem solving framework or as a personalized system of instruction course (PSI), where the student would set his own pace. Sixteen randomly selected professors who have taught both the conventional and PSI courses were asked to give their evaluation of the effectiveness of the two methods by assigning scores from 0 to 5 to each method. Table 6 gives the instructors' ratings.

Table 6

PSI	4	3	3	0	2	1	3	5	2	1	2	3	0	3	1	3
Lecture	2	4	2	4	3	5	1	1	5	3	4	5	3	2	4	5

At the 0.10 level of significance, test the null hypothesis that there is no difference in terms of effectiveness between one method of instruction and the other.

3. The productivity of eleven randomly chosen workers was measured for one week when they were under strict supervision and one week when they were not. The general

belief is that workers generally work harder and producer more when they are under supervision. Table 7 provides measures of their productivity for both approaches.

Table 7

Worker	John	Harry	Alice	May	Ralph	Mark	Lou	Bill	Andy	Jean	Paula
With Supervision	46	83	59	76	51	58	66	87	64	54	53
Without Supervision	47	75	54	82	37	47	68	79	60	57	49

(a) At the 5% level of significance, does the data indicate that strict supervision results in a higher median level of productivity?

(b) How does the outcome of this test compare with the outcome of the matched pairs' mean difference parametric test employed in Exercise 4 of Section 11.9 (page 373)?

(c) Under what conditions are the two tests comparable?

(d) What are the advantages of the paired-sample sign test over its parametric counterpart?

(e) What are the advantages of the matched pairs' mean difference parametric test over its nonparametric counterpart?

(f) Assuming that the tests are comparable, what might explain the different outcomes obtained in this case from the two tests?

4. An experiment was conducted to determine whether yoga mediation plays a role in blood pressure reduction. Table 8 lists the systolic blood pressure of sixteen randomly selected subjects before and after yoga meditation.

Table 8

	Systolic Blood Pressure							
Before	153	164	148	145	152	136	166	151
After	147	151	140	149	148	132	157	153

	Systolic Blood Pressure							
Before	167	133	159	129	183	136	152	134
After	153	138	159	123	174	141	137	129

At the 0.10 level of significance, has the median systolic blood pressure lowered?

5. Under a peer tutoring program students at Silver University who wish to obtain extra help are provided tutorial assistance with their studies. The Dean of Academic Affairs wants to know if this program has had a positive effect in helping students raise their semester quality point averages (QPA's). The academic achievement of a randomly chosen sample of 68 students enrolled in this program was examined. At the end of the program 46 showed a marked improvement, 4 did not show a change, and 18 deteriorated.

   At the 0.01 level of significance, test the null hypothesis that the program is not beneficial to the students in general, against the alternative hypothesis that the program helps to improve the median grade of its student participants.

6. In considering the Mathematics Review Tape situation we saw in Example 1 that application of the paired-sample sign test leads us to accept H_o or reserve judgment. Application of the matched pairs' mean difference test of Section 11.9 leads us to reject H_o (Example 2, page 368).

   (a) It is possible that H_o is true and that the weaker nonparametric test led us to uphold a true statement where its stronger parametric counterpart can lead us to reject a true hypothesis? If so, how could this happen?

   (b) It is possible that there is no contradiction in the test results because the tests are testing different things? Explain.

7. The Effect of Temperature on Skiing Time problem is addressed in Example 2 of this section and in Exercise 5 (page 374) of Section 11.9. In both cases the tests lead to accepting or reserving judgment on the null hypothesis formulated.

   (a) Suppose it were found that the population of paired ski time differences is J shaped. (i) What effect would this have on the hypothesis test conclusion reached in Example 2? Explain. (ii) What effect would this have on the hypothesis test conclusion reached in Exercise 5 of Section 11.9? Explain.

(b) It is possible for the conclusion reached from one of the tests to be true and the other false? Explain.

16.4 ■ Tests for Normality

Our experience with a number of estimation and hypothesis testing situations makes clear the importance of the normality condition. In this section we consider two nonparametric techniques that can be used to test population distributions. The first, called the Kolmogorov-Smirnov test (or more briefly, the K-S test), can be used to test for any given distribution. The second, called the Lilliefors test, is specifically designed to test for normality.

To begin, we define a cumulative distribution function for a random variable X by:

$$F(x) = P(X \le x)$$

That is, $F(x)$ is the probability that the random variable X takes on a value that is no larger than x. If the random variable were normal, with mean μ and standard deviation σ, then $F(x)$ would geometrically express the area of that region under that normal curve which lies to the left of the number x, as shown by the shaded region in Figure 6.

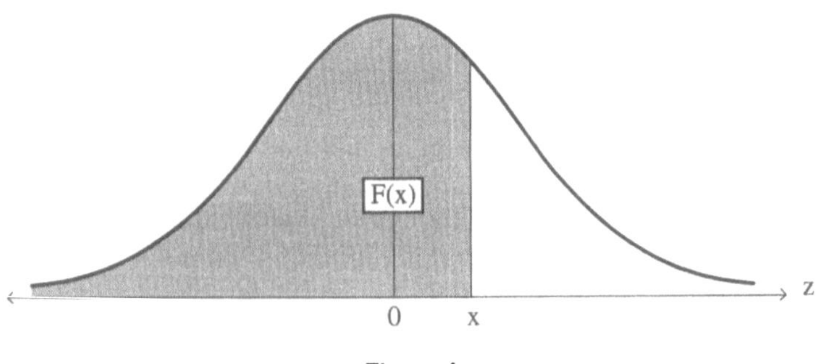

Figure 6

Kolmogorov-Smirnov Test

The heart of the K-S test is a comparison of the hypothesized and sample cumulative distribution functions. For this test the parameters of the hypothesized distribution must be specified. We illustrate the K-S test with the following example.

Example 1 The "Heavy" Basic Statistics Text

A recent edition of *Basic Statistics* was the focus of a wrist strength test at Ecap University. The time lengths, in minutes, that the population of 156 students at Ecap U. who used the text were able to hold it in one sitting were found to have mean $\mu = 4.6$ and standard deviation $\sigma = 1.6$ (see Sections 5.2 and 5.3, Examples 1 and 1, page 127 and page 138). It is believed that the population of wrist strength times is approximately normal with mean $\mu = 4.6$ and $\sigma = 1.6$

Use the K-S test at the 5% level to test the hypothesis that the population of wrist strength times is normally distributed with $\mu = 4.6$ and $\sigma = 1.6$.

The null and alternative hypotheses are:

H_o: The population of wrist strength times is normally distributed with $\mu = 4.6$ and $\sigma = 1.6$.

H_a: The population of wrist strength times is not normally distributed with $\mu = 4.6$ and $\sigma = 1.6$.

A randomly drawn sample yielded the values 3.2, 7.2, 5.0, 4.7, 2.5 (see Section 2.3, Example 1, page 51). We begin by arranging the data in increasing order.

2.5 3.2 4.7 5.0 7.2

We next construct a sample cumulative distribution function $S(x)$ on the sample values as follows. Since 2.5 is the smallest of the sample values, one-fifth of the five sample values is less than or equal to 2.5. We therefore define $S(2.5) = 1/5$. Two-fifths of the sample values are less than or equal to 3.2, which leads us to define $S(3.2) = 2/5$. The definition of $S(x)$ on all sample values is summarized in Table 9.

Table 9

x	2.5	3.2	4.7	5.0	7.2
$S(x)$	$\frac{1}{5}$	$\frac{2}{5}$	$\frac{3}{5}$	$\frac{4}{5}$	1

Next, we evaluate the hypothesized cumulative distribution $F(x)$ for each sample value. Since, with respect to H_o, X is normal with $\mu = 4.6$ and $\sigma = 1.6$ we have:

$$\begin{aligned} F(2.5) &= P(X \le 2.5) \\ &= P\left(z \le \frac{2.5-4.6}{1.6}\right) \\ &= P(z \le -1.31) \\ &= 0.5000 - 0.4049 \\ &= 0.0951 \end{aligned}$$

Similarly, we obtain $F(x)$ for the other sample values as summarized in Table 10.

Table 10

x	2.5	3.2	4.7	5.0	7.2
F(x)	0.0951	0.1894	0.5239	0.5987	0.9484

If the null hypothesis is true, then $S(x)$ and $F(x)$ should be close for all x. Large differences between $F(x)$ and $S(x)$ for at least one value of x would be evidence against the null hypothesis.

The K-S test makes this precise by defining the test statistic D as the largest of the differences between $F(x)$ and $S(x)$ in absolute value (that is, non-negative difference between $F(x)$ and $S(x)$).

$$D = \text{Largest of } | F(x) - S(x) |$$

The calculations leading to D in this case are summarized in Table 11. If D is "significantly large," we must reject H_o and accept H_a; if not we accept H_o or reserve judgment. The decision bounds for the K-S test in terms of sample size and significance level are given in Table E (page 657), part of which is reproduced as Table 12.

Table 11

x	$F(x)$	$S(x)$	$\|F(x)-S(x)\|$	
2.5	0.0951	0.2000	0.1049	
3.2	0.1894	0.4000	0.2106	□ D
4.7	0.5239	0.6000	0.0761	
5.0	0.5987	0.8000	0.2013	
7.2	0.9484	1	0.0516	

Table 12
Decision Bounds for the K-S Test

Sample size n	Significance Level α			
	0.10	0.05	0.02	0.01
4	.565	.624	.689	.734
5	.509	.563	.627	.669
6	.468	.519	.577	.617

For this situation $n=5$ and $\alpha=0.05$, so that from Table 12 we have that the decision bound is 0.563. If $D>0.563$, we reject H_o and accept H_a; otherwise we accept H_o or reserve judgment.

Since $D=0.2106<0.563$, we have the option of accepting the null hypothesis that the population of wrist strength times is normally distributed with $\mu=4.6$ and $\sigma=1.6$, or reserving judgment.

Andrei N. Kolmogorov (1903-1987), co-author of the K-S test, is one of the towering figures of 20th century mathematics. He left a strong mark on many areas of mathematics, but his single greatest achievement was in providing a rigorous foundation for the theory of probability. Kolmogorov's axioms, formulated in the 1930's, made possible the modern theory of probability. What Euclid was to geometry, Kolmogorov was to probability theory.

Lilliefors Test

The K-S test requires knowledge of the population mean and standard deviation, which are not known in many situations. The Lilliefors test is a refinement of the K-S test which allows us to circumvent this difficulty. The major difference between the two tests is that in the Lilliefors test the sample mean $\overline{x}$ and sample standard deviation s are used to calculate $F(x)$ instead of μ and σ. The sample cumulative distribution function $S(x)$ and the test statistic D are both computed as in the K-S test. With the Lilliefors test we compare the computed value of D with the appropriate Lilliefor's test decision bound.

Example 2 Normality of Fertilizer Yields?

We return to Section 11.8, Example 4 (page 355) which is concerned with comparing corn yields obtained from currently used and a new brand of fertilizer. The analysis of this problem requires that the underlying populations of corn yields obtained from each type of fertilizer be approximately normal. We employ the Liliefors test to test this assumption of normality for the population of corn yields obtained from the current fertilizer, at the 0.05 level. The sample values, taken from Example 4, Section 11.7, (page 355) are 70.5, 80.2, 90.3, 60.2 and 85.4.

The null and alternative hypotheses are:

H_o: The underlying population of corn yields for the currently used fertilizer is normal.

H_a: The underlying population of corn yields for the currently used fertilizer is not normal.

As with the K-S test, we put the sample data in increasing order and calculate $S(x)$. The resulting distribution is shown in Table 13.

Table 13

x	60.2	70.5	80.2	85.4	90.3
$S(x)$	$\frac{1}{5}$	$\frac{2}{5}$	$\frac{3}{5}$	$\frac{4}{5}$	1

From the sample we compute $\bar{x} = 77.3$ and $s = 12.1$. We next calculate the hypothetical cumulative distribution $F(x)$ under H_o. For example:

$$\begin{aligned} F(60.2) &= P(X \leq 60.2) \\ &= P\left(z \leq \frac{60.2 - 77.3}{12.1}\right) \\ &= P(z \leq -1.41) \\ &= 0.0793 \end{aligned}$$

The cumulative distribution function $F(x)$ values are shown in Table 14.

Table 14

x	60.2	70.5	80.2	85.4	90.3
F(x)	0.0793	0.2877	0.5948	0.7486	0.8577

We next calculate $|F(x)-S(x)|$ for the sample values and determine the largest such value *D*, as shown in Table 15.

Table 15

x	F(x)	S(x)	$\|F(x)-S(x)\|$	
60.2	0.0793	0.2000	0.1207	
70.5	0.2877	0.4000	0.1123	
80.2	0.5948	0.6000	0.0052	
85.4	0.7486	0.8000	0.0514	
90.3	0.8577	1	0.1423	← D

The decision bounds for the Lilliefors test in terms of sample size and significance level are given in Table F, (page 658), part of which is reproduced as Table 16.

Table 16
Decision Bounds for the Lilliefors Test

Sample size n	Significance Level α		
	0.10	0.05	0.01
4	.352	.381	.417
5	.315	.337	.405
6	.294	.319	.364
7	.276	.300	.348
8	.261	.285	.331

For $n=5$ and $\alpha=0.05$ the decision bound is 0.337. $D=0.1423<0.337$ and thus we accept the null hypothesis that the population of corn yields for the currently used fertilizer is normal, or reserve judgment.

The test of normality is one part of a preliminary analysis on the populations of corn yields obtained by use of the two fertilizers. Our next task is to test the assumption that the population of corn yields for the new brand of fertilizer is normal (see Exercise 1).

EXERCISES

1. Refer to Example 1 of Section 11.9 (page 364), in which we compare corn yields from currently used and a new brand of fertilizer. The analysis required

the assumption that the underlying populations of corn yields obtained from each type of fertilizer be approximately normal. In Example 2 of this section we employed the Lilliefors test to test this assumption of normality for the population of corn yields obtained from the current fertilizer at the 0.05 level.

(a) Test the assumption of normality for the population of corn yields obtained from the new brand of fertilizer at the 0.05 level. The sample yields (in bushels per acre) taken from 5 random plots so treated are 73.2, 82.1, 93.1, 63.6 and 87.1.

(b) In light of the results obtained from Example 2 and the preceding test, is it feasible to carry out the hypothesis test comparing the mean corn yields obtained from using the two fertilizers? Explain.

2. In Example 1 $\mu = 4.6$ and $\sigma = 1.6$ are available. Suppose that these parameters were not available. Carry out a Lilliefors test by means of the same sample stated in Example 1 to test for the normality of the wrist strength time population. Use the 5% significance level.

3. Refer to Exercise 8 of Section 10.3 (page 280), in which we are asked to find a 90% confidence interval for the actual mean chlorine level of Cameron City's reservoir system on a given day.

(a) At the 5% level test the null hypothesis of normality of the population. A random sample of 8 containers of water, collected on a specified day, yielded chlorine readings of 4.9, 5.2, 5.2, 5.3, 5.8, 5.6, 5.1 and 5.5.

(b) What are the consequences of the preceding test for the construction of 90% confidence limits for the mean chlorine level of the City's reservoir system on the day in question? Explain.

4. Refer to Exercise 4 of Section 11.6 (page 346), in which we are asked to test a null hypothesis regarding the standard deviation of the time to complete a mathematics placement test. It is assumed that the population of exam completion times is normally distributed.

   (a) Test the null hypothesis of normality of the population of completion times at the 0.05 level, given that six randomly chosen students who sat for the exam were observed to complete it is 45, 53, 62, 65, 50 and 54 minutes.

   (b) Do the results obtained in answer to (a) give us a green light to test the null hypothesis concerning the population standard deviation? Explain.

5. Refer to Exercise 8 of Section 11.7 (page 362), in which we compare the average number of runs scored per game in the National and American Leagues. The analysis requires that each population be at least approximately normally distributed with equal variances. Test the null hypothesis that the population of runs scored in the (a) American League, (b) National League, is normally distributed, using a significance level of 0.10 for each test. Random samples of 6 and 7 games chosen independently from the records of the leagues yielded the following number of runs scored:

   National League: 8, 12, 2, 5, 15, 18

   American League: 14, 8, 1, 9, 6, 17, 6

   (c) Do the results obtained in answer to (a) and (b) give us a green light to test the hypothesis that the end-season mean number of runs scored per game in both leagues is the same? Explain.

6. The final exam grades of 165 students who took statistics at the Vilnius Technical Institute in a recent semester are shown in Table 17.

Table 17

76	79	60	58	62	51	34	47	73	33	66
49	59	81	92	52	56	76	76	44	81	70
89	71	88	70	66	41	85	72	77	65	84
54	76	34	53	72	60	72	72	76	69	83
50	90	62	64	92	90	53	53	60	84	71
90	40	58	90	41	59	70	67	56	88	71
75	64	86	77	90	81	30	77	50	79	67
51	65	78	60	54	89	69	84	56	53	57
43	51	52	72	74	85	61	57	68	40	64
56	42	79	78	80	82	87	73	60	82	68
60	72	63	52	90	67	77	60	64	93	91
45	49	79	85	55	61	51	81	51	53	80
35	60	78	78	50	96	41	82	61	91	49
52	66	60	55	64	84	82	47	83	81	61
56	67	77	54	48	51	65	72	89	71	85

(a) By using the procedures discussed in Section 2.3 on using a table of random numbers determine a random sample of 7 grades.

(b) Based on the results obtained in answer to (a), test the grade population for normality with 67.3 and $\sigma = 15.1$. Use the 10% level of significance.

Of interest in connection with this problem are Exercise 4 of Section 3.1 (page 93), Exercise 4 of Section 3.2 (page 99), Exercise 5 of Section 5.2 (page 128), and Exercise 5 of Section 5.3 (page 141).

16.5 ■ Rank Correlation

Let us recall from Section 13.4 that to carry out a test of the population linear correlation coefficient ρ, the two variables involved must be normally distributed. When this requirement is not met we can fall back on the nonparametric rank correlation coefficient test developed by C.E. Spearman in 1906.

This nonparametric test seeks to detect an association, which **may or may not be linear**, between two numerical variables x and y by ranking the x and y values in order of their magnitude. This approach can also be used to analyze the relationship between two variables that cannot be exactly measured but for which ranked data can be obtained. The ranking procedure might be based on some qualitative factor such as taste or appearance.

The Spearman rank correlation test is applicable to populations whose data scale is ordinal or higher.

To illustrate, suppose two people judge a randomly selected set of ten contestants entered in a contest. Each judge ranks the sampled contestants from 1 to 10. These rankings are shown in Table 18.

Table 18

Judge 1: x	2	1	3	8	6	4	5	7	9	10
Judge 2: y	5	3	8	7	6	1	2	10	4	9

When there are no ties in the ranking of either the x or y observations, the **Spearman rank correlation coefficient** r_s is defined by

$$r_s = 1 - \frac{6\sum_{i=1}^{n} d_i^2}{n(n^2 - 1)},$$

where

$d_i = x_i - y_i$, the difference between ranked pairs,

n = number of paired items in the sample.

This result may also be used if the number of ties is small in comparison to the number of data pairs.

The computations for r_s for the data in Table 18 are given in Table 19.

Table 19

x	2	1	3	8	6	4	5	7	9	10
y	5	3	8	7	6	1	2	10	4	9

$d = x - y$	-3	-2	-5	1	0	3	3	-3	5	1
d^2	9	4	24	1	0	9	9	9	25	1

For $n = 10$ and $\sum d_i^2 = 92$ we obtain:

$$r_s = 1 - \frac{6(92)}{10(10^2 - 1} = 0.44 \cdot$$

We should note that, with respect to the possible rankings, there are two possible extremes. If the two rankings are identical, then the rank of *x* equals the rank of *y* for all pairs. Consequently, each $d_i = 0$ and $\sum d_i^2$ achieves its minimum value of 0, yielding $r_s = 1$. If $\sum d_i^2$ is minimized, then 1 is the highest attainable value of r_s. Suppose, on the other hand, that the rankings are diametrically opposite. In this case the smallest rank of *x* is paired with the largest rank of *y*, and so on, until finally the largest rank of *x* will be linked with the smallest rank of *y*. It can be shown that the maximum value $\sum d_i^2$ occurs in this, and only this, situation and that the maximum value of $\sum d_i^2$ is $\frac{n(n^2-1)}{3}$. In this case r_s achieves its minimum value, which is:

$$r_s = 1 - \frac{6\left[\frac{n(n^2-1)}{3}\right]}{n(n^2 - 1} = 1 - \frac{6}{3} = -1$$

In summary, for all possible pairs of rankings of the two data sets, we have

$$-1 \le r_s \le 1.$$

Hypothesis Testing

If the sample size is not too small (even as small as 10), it can be shown that the sampling distribution of r_s is approximately normal with mean equal to the Spearman population rank correlation coefficient, denoted by ρ_s, and standard deviation $\sigma_s = 1/\sqrt{n-1}$. In other words, the statistic

$$z = \frac{r_s - \rho_s}{\sigma_s}$$

is approximately standard normal. Substituting $1/\sqrt{n-1}$ for σ_s yields

$$z = (\sqrt{n-1})(r_s - \rho_s).$$

We can use the Spearman rank correlation coefficient to test the null hypothesis of zero Spearman rank correlation, that is, of no association between rank pairs,

$$H_o : \rho_s = 0$$

against any one of three possible alternative hypotheses:
H_a: There is an association between rank pairs (two-tailed test). That is, $\rho_s \neq 0$.

H_a: The rank correlation between rank pairs is positive (one-tailed test). That is, $\rho_s > 0$.

H_a: The rank correlation between rank pairs is negative (one-tailed test). That is, $\rho_s < 0$.

The first alternative hypothesis implies that large values of *x* tend to be associated with either large values of *y* or small values of *y*. (That is, *x* and *y* are correlated.) The second alternative hypothesis implies that large values of *x* tend to be paired with large values of *y*. (That is, *x* and *y* are positively correlated.) The third alternative hypothesis implies that large values of *x* tend to be paired with small values of *y*, and vice-versa. (That is, *x* and *y* are negatively correlated.)

We carry out this hypothesis test by following these steps:

(1) Be sure that *n* is at least 10, so as to satisfy the condition for the use of the *z*-statistic.

(2) Compute the value of r_s.

(3) Compute the value of the *z*-statistic

$$z = r_s\sqrt{n-1}$$

and compare it to the appropriate bound or bounds (note ρ_s is assumed to equal zero under the null hypothesis).

Example 1 Zero Correlation Between Rankings?

Using the data in Table 18 test, at the 10% significance level, the null hypothesis of zero correlation between the rankings of the judges, against the alternative hypothesis that positive rank correlation exists between the rank pairs.

We have:

$$H_o : \rho_s = 0$$
$$H_a : \rho_s > 0$$

Since $n = 10$, we may use the z-statistic. Our right tail begins at 1.28 (see Figure 7).

We had previously computed $r_s = 0.44$. This yields.

$$z = 0.44\sqrt{9} = 1.32 > 1.28.$$

Therefore, at the 0.10 level we reject H_o and conclude that positive rank correlation exists between the ranked pairs. That is, the two judges tend to give similar ratings to the contestants.

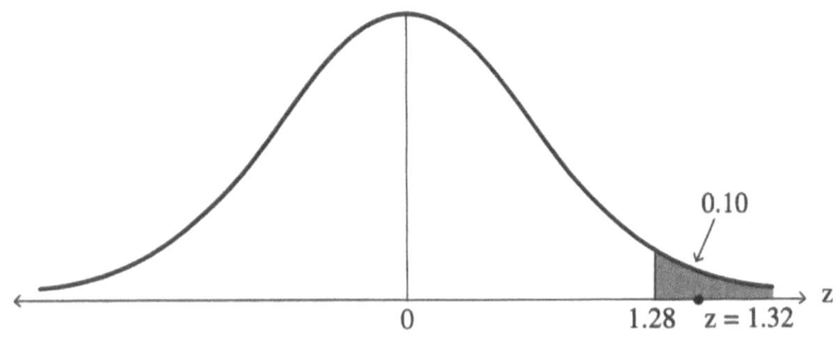

Figure 7

Example 2 Zero Correlation Between Rankings?

Each of two panels of ten "experts" is asked to rank a randomly chosen sample of 12 ideas for television programs on the basis of their relative appeal to a general audience. At the 10% level of significance can we conclude from this data that there is a correlation between the rankings of the two panels? Their rankings are given in Table 20.

Table 20

Program Idea	1	2	3	4	5	6	7	8	9	10	11	12
Panel 1: x	7	4	9	8	3	1	11	5	6	12	2	10
Panel 2: y	8	9	6	12	10	4	5	11	1	3	7	2

We have:

$$H_o: \rho_s = 0$$
$$H_a: \rho_s \neq 0$$

To help calculate r_s we employ Table 21, in which the values of d_i and d_i^2 are added to the information given in Table 20.

Table 21

Program Idea	1	2	3	4	5	6	7	8	9	10	11	12
Panel 1: x	7	4	9	8	3	1	11	5	6	12	2	10
Panel 2: y	8	9	6	12	10	4	5	11	1	3	7	2
$d_i = x_i - y_i$	-1	-5	3	-4	-7	-3	6	-6	5	9	-5	8
d_i^2	1	25	6	16	49	9	36	36	25	81	25	64
										$\sum d_i^2 = 376$		

The Spearman rank correlation coefficient is:

$$r_s = 1 - \frac{6(376)}{12(144-1)} = -0.32$$

Use of the z-statistic is justified since $n = 12$. Our tails begin at ± 1.65 (see Figure 8).

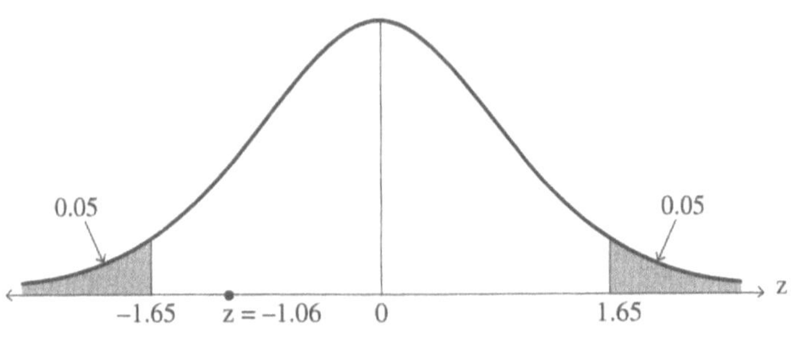

Figure 8

The *z*-statistic is:

$$z = -0.32\sqrt{11} = -1.06 > -1.65.$$

Consequently, at the 0.10 level we cannot reject H_o and either conclude that there is no correlation between the ratings of the two panels, or reserve judgment.

Example 3 Zero Correlation Between Rankings?

In the past two sports commentators, Ralph Albert and Denise Knox, have had similar opinions about the relative abilities of college football teams. They rank a randomly chosen sample of fourteen of these teams, as shown in Table 22. Test the hypothesis that their rankings are not correlated against the alternative hypothesis that these commentators continue to rank football teams in a similar fashion. Use $\alpha = 0.025$.

Table 22

Team	Wildcats	Buffalos	Pioneers	Jokers	Whales	Eskimos	Muskies	Americans	Surfers	Satellites	Scouts	Pilots	Elks	Generals
Albert: x	3	5	9	1	2	8	14	12	7	4	11	6	10	13
Knox: y	6	1	11	5	8	2	12	13	3	7	14	4	9	10

Table 23 adds to the information provided in Table 22 by listing the values of d_i and d_i^2.

Table 23

Team	Wildcats	Buffalos	Pioneers	Jokers	Whales	Eskimos	Muskies	Americans	Surfers	Satellites	Scouts	Pilots	Elks	Generals
Albert: x	3	5	9	1	2	8	14	12	7	4	11	6	10	13
Knox: y	6	1	11	5	8	2	12	13	3	7	14	4	9	10
$d_i = x_i - y_i$	-3	4	-2	-4	-6	6	2	-1	4	-3	-3	2	1	3
d_i^2	9	16	4	16	36	36	4	1	16	9	9	4	1	9
													$\sum d_i^2 = 170$	

We have:

$$H_o : \rho_s = 0$$
$$H_a : \rho_s > 0$$

Since $\sum d_i^2 = 170$, we calculate:

$$r_s = 1 - \frac{6(170)}{14(196-1)} = 0.63$$

Note that $n = 14$, allowing us to use the z-statistic. We must determine where the z-statistic lies, relative to 1.96 (see Figure 9).

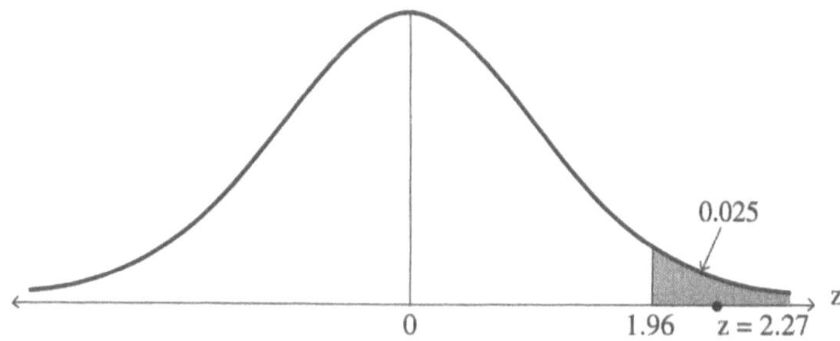

Figure 9

We find that

$$z = -0.63\sqrt{13} = 2.27 > 1.96,$$

so we reject H_o and conclude that the two sports commentators tend to rank the football teams in a similar manner.

The rank correlation test may also be employed when the data consists of actual numerical values. In this case the numerical values for each data set *x* and *y* must be separately ranked before r_s can be computed.

For each data set, assuming all *n* values in that set are different, assign the rank 1 to the largest value, 2 to the second largest value and so on, culminating with the assignment of *n* as the rank of the smallest value.

If two or more numerical values are equal in either data set, each value is assigned the average rank. If, for example, a tie for ranks 7 and 8 occurs, each rank would be 7.5. If a three-way tie occurs, say for ranks 4, 5 and 6, each data value would be given a rank of 5.

Example 4 Zero Correlation Between Rankings?

Peggy Garrett, academic counselor at Huxley College, would like to investigate whether there is a positive correlation between study time and examination grades in statistics. A random sample of fifteen statistics students was chosen for this study. The data in Table 24 show the number of hours these students studied for a statistics exam and the grades they received. At the 1% significance level, does the data suggest that study time and exam score have a positive association?

Table 24

Student	Al	Bill	Karen	Hilary	Mark	Fred	Rosa	Jerry	Ann	Marty	Lyndsey	Judy	Steve	Shara	Rona
Study time: x	8	12	5	10	14	11	6	12	9	11	17	7	8	10	14
Exam score: y	56	75	62	75	87	74	59	68	80	71	96	48	62	90	75

We first rank the *x* data and the *y* data, as described earlier. Ties are assigned average ranks. For example, there is a tie for study time involving Bill and Jerry (both 12 hours). Since three other students study longer than 12 hours, the study time of Bill and Jerry is each ranked 4.5. The values of d_i, d_i^2 and $\sum d_i^2$ are shown in Table 25.

Table 25

Student	Al	Bill	Karen	Hilary	Mark	Fred	Rosa	Jerry	Ann	Marty	Lyndsey	Judy	Steve	Shara	Rona
Study time: x	11.5	4.5	15	8	2.5	6	14	4.5	10	8	1	13	11.5	8	2.5
Exam score: y	14	6	11.5	6	3	8	13	10	4	9	1	15	11.5	2	6
$d_i = x_i - y_i$	-2.5	-1.5	3.5	2	-0.5	-2	1	-5.5	6	-1	0	-2	0	6	-3.5
d_i^2	6.25	2.25	12.25	4	.25	4	1	30.25	36	1	0	4	0	36	12.25
													$\sum d_i^2 = 149.5$		

We have:

$$H_o : \rho_s = 0$$
$$H_a : \rho_s > 0$$

We compute:

$$r_s = 1 - \frac{6(149.5)}{15(225-1)} = 0.73$$

Since $n = 15$, use of the *z*-statistic is justified. Our right tail begins at 2.33 (see Figure 10).

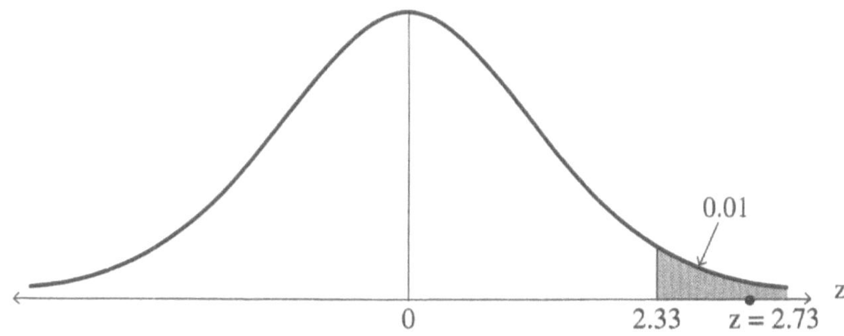

Figure 10

Since

$$z = r_s\sqrt{n-1} = 0.73\sqrt{14} = 2.73 > 2.33$$

we reject H_o and conclude that there is a positive relationship between study time and exam grade. This, of course, comes as no surprise.

EXERCISES

1. Consider the following two illustrations of pairs of rankings for 6 items shown in Table 26 and 27.

Table 26

Case I	A	B	C	D	E	F
First ranking: x	1	4	3	6	2	5
Second ranking: y	1	4	3	6	2	5

Table 27

Case II	A	B	C	D	E	F
First ranking: x	1	2	3	4	5	6
Second ranking: y	6	5	4	3	2	1

Note that in case I the rankings are identical, whereas in case II the rankings are exactly opposite.

(a) What value would you expect for the Spearman rank correlation coefficient in each case above? Explain briefly.

(b) Justify your conclusions by calculating the Spearman rank correlation coefficient for each case.

2. The Bullish Corporation, which sells vacuum cleaners, ranked eleven randomly selected salespersons who were hired on sales potential based on resumes, interviews, academic records, letters or recommendation and experience. Two years later the organization ranked their sales performance to determine whether the predictions made at the time of hiring were reasonably accurate. These data are summarized in Table 28.

Table 28

Initial ranking: x	3	5	1	4	7	10	2	6	11	9	8
Subsequent ranking: y	1	9	4	3	11	5	7.5	10	6	7.5	2

(a) At the 0.025 level of significance, test the null hypothesis that the original ranks have no relationship to actual job performance, against the alternative hypothesis of positive rank correlation between the ranks.

(b) May the management of the Bullish Corporation conclude from the test that it predicted sales potential reasonably well? Explain.

3. In judging the baseball potential of a young player there is a general perception among baseball "experts" that the faster players are below average in physical size and have less power-hitting potential, whereas those who have demonstrated power-hitting ability are above average in size and are relatively slow. Perhaps with these beliefs in mind, a scout ranks twelve randomly selected players in the baseball draft according to speed and power-hitting potential as shown in Table 29.

Table 29

Speed rating: x	6	5	11	12	4	2	3	9	7	10	1	8
Power rating: y	11	9	6	2	8	10	12	3	1	7	5	4

(a) At the 5% significance level, test the hypothesis of no association between the pairs of ranks against the alternative hypothesis that the scout had the good-speed-means-no-power and good-power-means-no-speed view in mind when he assessed the baseball potential of these twelve players.

(b) May we conclude from this test that there is a strong linear relationship between the variables?

4. Sonya Kovalevsky, president of the Mathematics Company, ranked a randomly chosen sample of ten of her vice presidents according to their affability and level of competence as shown in Table 30. Test the null hypothesis that rankings of affability and competence are unrelated. Use $\alpha = 0.02$

Table 30

Vice-president	Steven	Jodi	Carl	Gary	Leon	Jerry	Sandra	Allysa	Rachel	Philip
Affability: x	3	5	4	7	1	9	2	8	10	6
Competence: y	4	1	7	3	6	10	9	2	6	8

5. Movie critics Tom and Jerry viewed a random sample of fifteen current movies and ranked them from 1 to 15. They are known to have different philosophies with respect to the characteristics of high quality cinema and have had disagreements in the past about the quality of various movies. Their rankings are shown in Table 31.

Table 31

Tom: x	9	13	3	14	8	4	7	6	12	15	1	10	5	11	2
Jerry: y	1	11	12	15	4	6	2	3	8	13	7	5	9	14	10

At the 0.025 level of significance test the claim that there is no correlation between the rankings against the alternative hypothesis that the rankings are negatively correlated.

6. Students who finish exams more quickly than others are often thought to be the better students. A math instructor ranked 15 randomly chosen students according to the order in which they finished a statistics exam and the grade received. These rankings are shown in Table 32. At the 0.05 level of significance, does the data support the belief that the students with the higher grades have a tendency to complete an exam relatively quickly?

Table 32

Student	Rhoda	Harry	Zena	James	Bonnie	Carol	Shara	Mike	Sue	Judd	Helen	Charles	Melissa	Sylvia	Marvin
Order of completion: x	12	5	9	8	13	4	2	15	14	1	11	3	7	6	10
Grade ranking: y	8	6	12	5	10	11	3	14	13	9	7	1	2	4	15

7. Table 33 shows the results of a poll of predicted percentages of the vote for eleven randomly chosen candidates for state senator in different states, to their corresponding actual percentages of the vote received. Test, at the 0.10 significance level, the hypothesis that the predicted and actual percentages are unrelated.

Table 33

Predicted percentage: x	23	36	52	21	54	36	43	27	19	49	12
Actual percentage: y	29	42	46	38	43	29	41	29	17	43	7

8. Table 34 shows the number of minutes it took 10 randomly chosen mechanics to assemble a piece of machinery in the morning and afternoon. At the 0.01 significance level test the null hypothesis of no difference in efficiency during the different times of day against the alternative hypothesis that some workers work more efficiently in the morning whereas others do better in the afternoon.

Table 34

Time A.M.: x	13	19	18	14	9	19	7	13	21	8
Time P.M.: y	14	9	15	6	21	11	15	20	12	16

9. Students who show an aptitude for science often do well in mathematics. Table 35 shows the exam grades of a random sample of thirteen college students in science and mathematics, respectively. At the 0.10 level of significance, test the null hypothesis of no association between science and mathematics grades of these students against the alternative hypothesis that there is a positive association between grades in these areas.

Table 35

Student	Al	Bob	Nat	Debra	Jack	Lucille	Ralph	Art	Sam	Charles	Marla	Maria	David
Science grade: x	78	83	87	64	56	91	87	61	98	75	71	80	74
Math grade: y	85	96	88	58	67	80	90	63	92	76	88	85	76

10. Leslie Ernst, Dean of the School of Business of Brite University, believes there is a positive correlation between the age and starting salary for Ph.D. candidates

graduating this year from his school. Table 36 shows the age (in years) and annual starting salary (in thousands of dollars) of a random sample of twelve graduates this year. At the 0.05 significance level, is there sufficient evidence to justify the dean's belief?

Table 36

	Salary											
Age: x	33	28	41	46	35	32	27	58	38	40	28	36
Salary: y	47	41	54	48	46	40	43	53	57	49	35	39

16.6 ▯ Runs: A Test for Randomness

Our discussions of statistical inference have relied heavily on the randomness of the sample drawn. In certain situations questions may arise regarding the non-randomness in a distribution. In this section we analyze specific kinds of non-randomness which can be observed from an examination of the internal structure of a sample. After such an analysis we will be able to decide, statistically, whether patterns that look suspiciously non-random may be attributed to chance. This technique is based on the theory of runs.

> The data to which the runs test for randomness is applied is on a **nominal scale.**

For a sequence of symbols of two kinds a run is defined as a succession of identical symbols. To illustrate, consider the following sequence of 24 homes in a community containing either of two ethnic groups D and E:

DDDDDD/EEEE/DDD/EE/DD/EEEEE/DD

Using slashes to separate each subsequence of consecutive identical letters, we find that there is a first a run of six D's, then a run of four E's, then three D's, then two E's, then two D's, then five E's, and finally a run of two D's. Overall, there are 7 runs, with 4 runs of D's and 3 runs of E's.

The total number of runs appearing in a sequence is often a good indication of a possible lack of randomness, as illustrated by the following sequences (b), (c) and (d), which are extreme cases.

(b) AAAA . . . . AA

(c) AAA . . . . AABBB . . . . BB

(d) ABABAB . . . . ABAB

In each case the randomness of the sample would be seriously questioned, but for different reasons. Sequences (b) and (c) contain very few runs (1 and 2, respectively), where 1 is the minimum number of runs. In general, too few runs suggest a grouping or clustering of the data or perhaps some sort of trend. On the other hand for sequence (d) containing n data entries, there are n runs, which is the largest possible number. In general, too many runs leads us to suspect that some kind of repeated alternating or cyclical pattern may be unfolding.

The null hypothesis H_o to be tested is that the number of runs in a given sequence of observations is consistent with the number to be expected from the random process; that is, the sequence is a random sequence. The alternative hypothesis H_a is that the sequence is not a random sequence. The test statistic in the runs test is the number of runs observed in the given sequence.

This randomness test is based on the distribution of runs. Suppose there are R number of runs observed in the sample and the sequence contains m symbols of one kind and n symbols of another kind. For the sake of discussion, take $m > n$.

It can be shown that the mean μ_R and standard deviation σ_R of R are:

$$\mu_R = \frac{2mn}{m+n} + 1$$

$$\sigma_R = \sqrt{\frac{2mn(2mn - m - n)}{(m+n)^2(m+n-1)}}$$

If m and n are both at least 10, then the sampling distribution of R can be approximated by a normal distribution with mean μ_R and standard deviation σ_R. The z-statistic for the runs test is:

$$z = \frac{R - \mu_R}{\sigma_R}$$

Since we have non-randomness if there are either too few or too many runs, the runs test is two-tailed in all instances.

Example 1 Are Homes of the Two Groups Randomly Mixed?

Let us return to the previous illustration of 24 homes in a bi-ethnic community. Will B. Quick, spokesperson for Quickclose Real Estate, Inc., states that housing units in the community contain a "good mix" of the two ethnic groups. However, a community representative says that the two ethnic groups tend to cluster together rather than mix. Using the aforenoted data and a 5% significance level, can we conclude that the homes of the two ethnic groups are randomly mixed?

We wish to test,

H_o: The homes are, ethnically speaking, randomly mixed,

versus

H_a: The homes are not randomly mixed.

From the data, we see that $R = 7$, $m = 13$, $n = 11$. Since both m and n are at least 10, we may use the *z*-statistic. Our task is to compare the *z*-value of R with the bounds of ± 1.96 (see Figure 11).

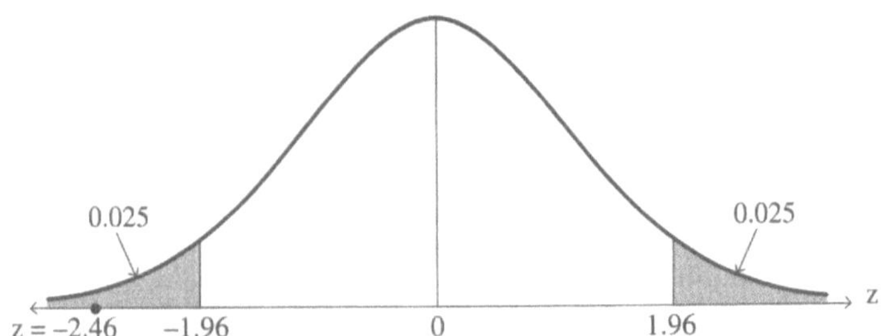

Figure 11

We first compute:

$$\mu_R = \frac{2(13)(11)}{13+11} + 1 = 12.9$$

and

$$\sigma_R = \sqrt{\frac{2(13)(11)[2(13)(11) - 13 - 11]}{(13+11)^2(13+11-1)}} = \sqrt{5.656} = 2.4$$

These lead to:

$$z = \frac{R - \mu_R}{\sigma_R} = \frac{7 - 12.9}{2.4} = -2.46 < -1.96$$

Therefore, we reject H_o and conclude that the housing distribution is not random. This does not imply that the distribution is not a "good mix" as Will Quick claimed, or suggest that some sort of discrimination has taken place. It means that the mix cannot be considered random, which is what we would expect to begin with.

Example 2 Win vs. Did not Win: A Random Sequence?

Bill Blitz, football coach of Huxley college, has a theory that success feeds-on-itself in that winning a football championship in any given year increases the team's motivation to win it the next year. He expressed his feelings to Tom Tester, a statistics student, who asked for records over the last 30 years. The coached provided Tom with data which indicated whether the team had won (W) or did not win (L) the championship that year. These data are given below. At the 2% significance level, does the team win and not win the championship appear in a random sequence?

WWWWLLLWWWWWWLLWWWWLLLLLWWWWWWWWL

H_o : The occurrence of wins and non-wins is a random sequence.

H_a : The occurrence of wins and non-wins is not random.

Inspection of the data shows that $R = 8$, $m = 19$, $n = 11$. Both *m* and *n* are at least 10, which justifies the use of the *z*-statistic. Since $\alpha = 0.02$, we will have to compare the *z*-statistic of *R* with the bounds ± 2.33 (see Figure 12).

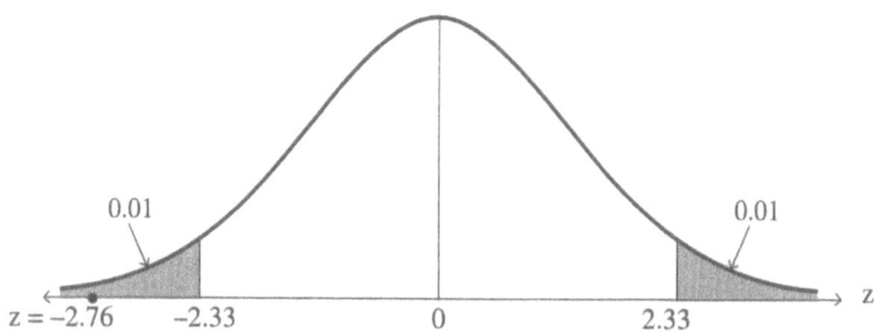

Figure 12

We have:

$$\mu_R = \frac{2(19)(11)}{19+11} + 1 = 14.9$$

This yields:

$$\sigma_R = \sqrt{\frac{2(19)(11)[2(19)(11) - 19 - 11]}{(19+11)^2(19+11-1)}} = \sqrt{6.214} = 2.5$$

Therefore, we reject H_o and conclude H_a, that the sequence of wins and non-wins over the previous 30 years did not occur in a random pattern. However, this does not prove Bill Blitz's theory. While Bill's premise may be that one of the factors leads to the non-random sequence, clearly there are many other factors involved as well. Our analysis simply shows that the string of wins and non-wins cannot be viewed as a random sequence.

EXERCISES

1. Amy Wong owns stock in Safe Motors Corporation and watched the progress of this stock on the stock market for a period of twenty-four days, recording U if the stock price either rose or remained the same and D if the price declined. She obtained the following sequence:

   UUUUDDDDDUUDDDDUUUDUUDDD

   Use the runs test, at the 0.05 significance level, to test the null hypothesis that the sequence of price fluctuations is random.

2. Austin Childs, a commuter, observed the arrival time of his 7:30 A.M. train on 28 consecutive days, writing T if the train was either early or on time and L if the train was late. The sequence turns out to be:

   TTTTLLLTTLLLLTTTLTTLTLLTLLLL

   Austin happens to know that this train is operated by one of two engineers, each working a few days at a time, and he suspects trends regarding the punctuality or tardiness of this train. At the 0.10 significance level, test the null hypothesis of a random sequence of arrival times.

3. The main office of Glaxton Mines, Inc. selects its executives from the management personnel of its two subsidiary companies, the Alfred (A) and Bernice (B) Companies. During the past three years executives have been selected by the parent company from subsidiaries in the following order:

AAABBAABAABAABBBAABBBAABA

   (a) Does this sequence provide sufficient evidence, at the 0.02 level of significance, to suggest non-randomness in the selection of executives by the parent company from its subsidiaries?

   (b) Does the result contradict what you would expect? Explain.

4. Alice Tamm, a golfer, kept records of the last twenty-six games she played. She marked A if she was above par for eighteen holes and B if she was either at or below par. Alice believes that the sequence is non-random in that it may contain trends corresponding to "hot-streaks" and "cold-streaks". The sequence is:

BBAAABBBBAABBABBAABBBAABAA

At the 0.05 significance level, does the result support Alice's belief?

5. The following is the order in which thirty consecutive items came off an assembly line. Defective items are indicated by D and nondefective items by N.

NNNNNDDDDNNDNNDDNNNDDDNNDNNNDN

At the 0.02 significance level, test the null hypothesis that the quality of the item produced fluctuates randomly.

6. Madeline Kane drives along local streets in a large city. She notes upon arriving at a traffic light, whether it is red (R) or green (G), counting a yellow light as green if she gets through the intersection before the light changes and red if she does not. After 27 lights the sequence is:

GRRRGGGRRGGRGGRGGRRGGRRRGRG

Test, at the 0.05 level of significance, the null hypothesis that the sequence of Madeline's arrival's at traffic lights may be viewed as random.

7. The results of a 5 kilometer run sponsored by the Roadrunners Association showed the following order of the first 16 male (M) and female (F) finishers:

MMMFFMMMMFFFFFMMFFFMMMMFFFMMFFMF

(a) At the 0.10 level of significance, did these men and women finish the race in a random order?

(b) Are you surprised by your finding? Explain.

16.7 ■ The Mann-Whitney U Test

In this section we develop a nonparametric alternative to the two independent-sample *t* test for the difference between two means. It is variously called the "*U* Test," the "Wilcoxin test" or the "Mann-Whitney test".

Let us recall that in order to compare means of two populations by a *t* test we **require that the populations be normally distributed with equal variances** (see Section 11.7).

In many situations of interest the two populations may have non-normal distributions or distributions whose nature is unclear, thereby rendering inapplicable this parametric procedure.

We use the *U* test by choosing a random sample from each of two populations of interest, for which each outcome is described by a continuous random variable. Unlike the paired-sample sign test, these data do not consist of matched pairs and the samples should be chosen independently of one another. We also assume that the two populations have the same spread and shape.

The null hypothesis to be tested is that the underlying population medians are equal against the alternative hypothesis that the medians differ. While the Mann-Whitney test can be one-sided, we restrict our attention to a two-sided alternative hypothesis.

The Mann Whitney U test is applicable to populations whose data scale is ordinal or higher.

Example 1 Median Appraisal is the Same?

Two real estate appraisers, Herman and Jean, are responsible for appraising the real estate value of homes in a community. The homeowners in the community have complained that the appraisers tend to appraise homes at different levels. To test this conjecture the town clerk pulled the two appraisers' records and independently took a random sample of appraisals (in thousands of dollars) for each, as shown in Table 37.

At the 10% significance level, he wishes to test the null hypothesis that the median appraisal for each appraiser is the same against the alternative hypothesis that their median appraisals differ.

Table 37

Herman	132	108	103	110	98	86	107	93
Jean	106	93	82	100	113	116	108	97

Herman	114	115	118	112	105	102	104
Jean	91	80	93	87	113		

An outline of the procedures for the *U* test follows:

Assignment of Ranks

(1) Denote the sample sizes of populations 1 and 2 by n_1 and n_2, respectively. In this example, $n_1 = 15$ and $n_2 = 13$

(2) Combine the $n_1 + n_2$ data values into one group and arrange the data values in decreasing order of magnitude.

Table 38

Data Value	132	118	116	115	114	113	113	112	110	108	108	107	106	105
Rank	1	2	3	4	5	6.5	6.5	8	9	10.5	10.5	12	13	14
Source	1	1	2	1	1	2	2	1	1	1	2	1	2	1

Data Value	104	103	102	100	98	97	93	93	93	91	87	86	82	80
Rank	15	16	17	18	19	20	22	22	22	24	25	26	27	28
Source	1	1	1	2	1	2	2	2	1	2	2	1	2	2

(3) Assign a rank number to each data value — the largest is assigned rank 1, the next largest 2 and so on, up to the smallest data value, which is assigned rank $n_1 + n_2$ (28 in this example). If any data values are equal, assign to each one the mean rank of their rank positions.

See Table 38 for the complete list of rankings for the given example. For instance, the tenth and eleventh largest data values are 108. Accordingly, these data values are each given rank 10.5. Also, the twenty-first, twenty-second and twenty-third largest data values are 93. Thus, the ranking (21 + 22 + 23)/3 = 22 is assigned to these values.

Calculation of the *U* Statistic

(1) Gather the individual ranks for the data values from each of the populations. Add up these two sets of ranks. Denote the sums of the ranks of data from populations 1 and 2 by R_1 and R_2, respectively.

In this illustration,

$$R_1 = 1+2+4+5+8+9+10.5+12+14+15+16+17+19+22+26$$
$$= 180.5$$
$$R_2 = 3+6.5+6.5+10.5+13+18+20+22+22+24+25+27+28$$
$$= 225.5$$

(2) The *U* score for each sample is obtained by using the following pair of statistics:

$$U_1 = R_1 - \frac{n_1(n_1+1)}{2}$$
$$U_2 = R_2 - \frac{n_2(n_2+1)}{2}$$

For this example, we have:

$$U_1 = 180.5 - \frac{15(16)}{2} = 60.5$$

$$U_2 = 225.5 - \frac{13(14)}{2} = 134.5$$

(3) The preliminary U statistic is the smaller of U_1 and U_2.

Therefore, $U = 60.5$ in this problem.

Return to the Null Hypothesis

Our null and alternative hypotheses are:

H_o: Herman and Jean's median appraisals are equal, or $M_1 = M_2$.

H_a: Herman and Jean's median appraisals are different, or $M_1 \neq M_2$

If the null hypothesis of equality of the two population medians is true, it can be shown that the distribution of U has mean μ_U and standard deviation σ_U given by:

$$\mu_U = \frac{n_1 n_2}{2}$$

$$\sigma_U = \sqrt{\frac{n_1 n_2 (n_1 + n_2 + 1)}{12}}$$

In our illustration, $n_1 = 15$ and $n_2 = 13$, so that:

$$\mu_U = \frac{15 \cdot 13}{2} = 97.5$$

$$\sigma_U = \sqrt{\frac{15 \cdot 13(15 + 13 + 1)}{12}} = 21.7$$

If both n_1 and n_2 are at least 10, the distribution of U is approximately normal. Consequently, the z-value of our test statistic is:

$$z = \frac{U - \mu_U}{\sigma_U}$$

The normal approximation is appropriate in this case since each of n_1 and n_2 is at least 10. The town clerk desires a 0.10 significance level, so that the tails begin at ± 1.65 (see Figure 13).

Since $U = 60.5$, $\mu_U = 97.5$ and $\sigma_U = 21.7$, we obtain:

$$z = \frac{60.5 - 97.5}{21.7} = -1.71 < -1.65$$

Therefore, we reject H_o and accept H_a, that Herman and Jean's median appraisals are not the same.

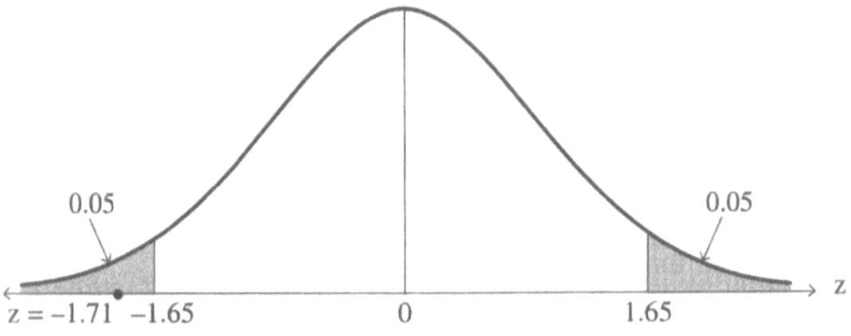

Figure 13

Example 2 Median SAT Scores are the same?

The Board of Regents of Hitech College desires to compare the English SAT scores of students at the Hillsdale and Brookline branches of the institution to see whether their median English SAT scores differ. A random sample of sixteen student files chosen from each branch has produced the data shown in Table 39. At the 0.05 level of significance, what conclusion can be drawn?

Table 39

	SAT Score							
Hillsdale	510	570	540	610	500	450	530	660
Brookline	530	470	600	490	680	620	730	590

	SAT Score							
Hillsdale	470	530	520	490	550	720	640	700
Brookline	520	440	740	580	560	510	560	690

First, we note that $n_1 = n_2 = 16$. We next compose one list of 32 data values and rank them, keeping in mind that equal data values are assigned the average of their rank positions. These rankings are shown in Table 40.

Table 40

Data Value	740	730	720	700	690	680	660	640	620	610	600	590	580
Rank	1	2	3	4	5	6	7	8	9	10	11	12	13
Source	2	2	1	1	2	2	1	1	2	1	2	2	2

Data Value	570	560	560	550	540	530	530	530	520	520	510	510
Rank	14	15.5	15.5	17	18	20	20	20	22.5	22.5	24.5	24.5
Source	1	2	2	1	1	1	1	2	1	2	1	2

Data Value	500	490	490	470	470	450	440
Rank	26	27.5	27.5	29.5	29.5	31	32
Source	1	1	2	1	2	1	2

We compute:

$$R_1 = 3+4+7+8+10+14+17+18+20+20+22.5+24.5+26+27.5+29.5+31 = 282$$
$$R_2 = 1+2+5+6+9+11+12+13+15.5+15.5+20+22.5+24.5+27.5+29.5+32 = 246$$

We next calculate

$$U_1 = 282 - \frac{16(17)}{2} = 146$$

and

$$U_2 = 246 - \frac{16(17)}{2} = 110,$$

so that $U = 110$.

We have:

H_o: The median English SAT scores of students at the two branches are equal.

H_a: The median English SAT scores of students at the two branches differ.

The distribution of U has parameters:

$$\mu_U = \frac{16(16)}{2} = 128$$

$$\sigma_U = \sqrt{\frac{16(16)[16+16+1]}{12}} = 26.5$$

Since $n_1 = n_2 = 16$ are both at least 10, we may use the z-statistic. For $\alpha = 0.05$, our tails begin at ± 1.96 (see Figure 14).

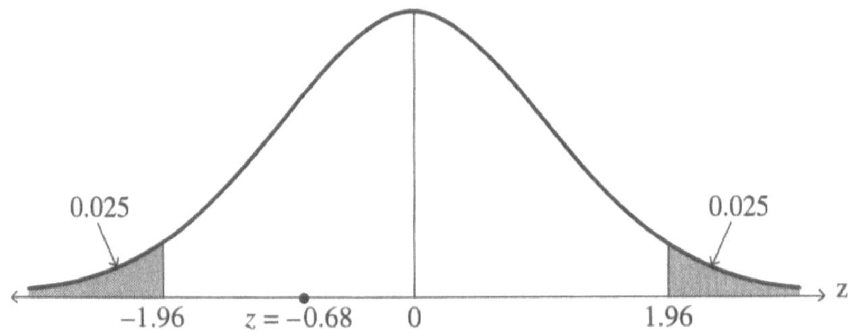

Figure 14

Our z-statistic is:

$$z = \frac{U - \mu_U}{\sigma_U} = \frac{110 - 128}{26.5} = -0.68 > -1.96$$

Consequently, at the 0.05 level of significance we cannot reject H_o. We conclude that the median English SAT scores at Hillsdale and Brookline are equal, or reserve judgment.

EXERCISES

1. Carl Credit, an independent auditor, examined personnel records of randomly chosen senior accountants from two large accounting firms, Balanced, Inc. (B) and Double Entry, Inc. (D). He wished to determine whether a difference exists in their median salaries. Table 41 shows the monthly salaries (in thousands of dollars) of randomly and independently chosen samples of senior accountants from firms B and D. Can Carl infer that a difference exists in senior accountants' median salaries? Use $\alpha = 0.05$.

Table 41

	Salary						
Firm B	4.7	5.8	5.3	6.2	7.1	5.1	6.8
Firm D	4.9	5.1	5.4	4.8	6.0	5.5	6.1

	Salary							
Firm B	7.5	7.2	6.7	6.5	7.0	6.6	6.5	
Firm D	6.3	4.7	6.2	5.9	5.7	5.6	5.0	6.7

2. The Stankus Marketing Company, Inc., wants to compare two methods of communicating information about a new car model. Two groups of people are randomly and independently chosen to take part in the experiment. Those in the first group are informed about the new product by television commercial (method A) and those in the second group are informed about it by newspaper advertisement (method B). At the end of the experiment each person is given the same test to measure knowledge of the new product. The results are shown in Table 42. At the 0.10 level of significance, test the hypothesis of no distinction between the two methods with respect to effective learning.

Table 42

	Results of Examination											
A	78	82	71	67	86	79	91	84	75	81	76	92
B	83	90	81	82	88	93	86	97	78	95	90	94

3. In a spelling bee contest independent random samples of eighteen students were chosen from among the fourth grade children in Irvingtown and Bakersville. In the contest each child was given the opportunity to spell correctly as many words as possible in a five minute time span from a master list. The results are shown in Table 43. At the 0.02 significance level, test the null hypothesis of equal spelling achievement between Irvingtown and Bakersville fourth graders.

Table 43

	Number of Words Spelled Correctly							
Irvingtown	18	16	19	14	25	27	17	16
Bakersville	22	23	18	17	21	24	28	26

	Number of Words Spelled Correctly							
Irvingtown	15	13	11	20	21	18	15	27
Bakersville	23	26	22	29	25	16	12	14

4. Asta Hurkus, a reporter for the *Financial Gazette*, picked ten voters at random from those in favor of a state bond proposal and twelve from those who opposed it to determine whether a difference existed in the median educational level of these two groups. Table 44 lists the number of years of post-elementary school education (starting with high school) of these twenty two people. Carry out this hypothesis test at the 0.05 significance level.

Table 44

	Years of Post-Elementary School Education											
In favor	8	7	9	6	7	6.5	8	11	12	10.5		
Opposed	3	6	7.5	8	2	2.5	4	9	10	7	5.5	4.5

5. Mileage performance tests were conducted for two automobile models, Flash and Sedate, to see whether median miles-per-gallon (m.p.g.) ratings of the two models are equal. Twelve Flashes and 13 Sedates were randomly and independently selected and an m.p.g. rating for each car tested was compiled based on 1000 miles of highway driving over similar terrain. The data obtained are shown in Table 45. At the 10% significance level, test the null hypothesis of equal median m.p.g. ratings for the two models.

Table 45

	Miles Per Gallon												
Flash	20.7	19.6	18.5	19.2	18.4	20.6	18.8	19.8	20.3	21.0	18.6	17.8	
Sedate	20.9	22.3	21.2	19.9	23.7	21.6	20.1	19.5	24.0	22.6	19.4	21.5	22.1

6. The Northchester Police Department wanted to determine, at the 5% level of significance, whether there was a difference in the median number of crime reports in their community during the winter and summer. Carry out this test using Table 46, which lists the number of daily crime reports from random samples of days in the winter and summer months.

Table 46

	Number of Daily Crime Reports										
Winter	18	21	14	15	19	14	13	20	12	17	15
Summer	23	18	24	27	20	23	28	19	24	16	22

7. The Marks Hospital Corporation wants to determine whether its General and Lister hospitals differ in the median length of time that emergency room patients must wait before they can be seen by a doctor. Randomly and independently chosen samples of patients admitted to the two hospitals shows the waiting times in minutes given in Table 47.

Table 47

	Waiting Time											
General	6	8	11	5	12	13	15	16	4	10	9	7
Lister	3	5	9	8	7	10	11	2	6	1		

Can the Marks Hospital Corporation conclude, at the 0.05 level of significance, that there is a difference in the median waiting time for emergency room patients to see a doctor at the two hospitals?

8. Fluorescent and track lighting systems are available for lighting work stations in a factory. Each station is used to light twelve work stations. As part of a study of the effectiveness of the systems, two sets of twelve employees were randomly and

independently selected. Each of the first set of twelve was randomly assigned to a fluorescent lighted station while each of the second set of twelve was randomly assigned to a track lighted station. Each employee in this study was given an efficiency rating by a supervisor at the end of one week. The results are shown in Table 48.

Table 48

	Efficiency Rating											
Fluorescent	33	35	39	41	31	30	35	40	42	30	29	34
Track	42	43	38	44	46	45	36	45	40	36	32	43

Use the 0.10 level of significance to determine whether a difference exists in the median ratings attributed to the two techniques.

16.8 ■ Advantages and Disadvantages of Nonparametric Methods

Nonparametric methods have a number of advantages over parametric procedures:

Pros

(1) They require **few if any assumptions** about the population. We do not, for example, have to establish or assume that the population has a normal distribution or any other specific shape.

(2) In some instances they can be used to analyze data that are not compatible with parametric tests, such as when the data are ordinal. For these reasons, nonparametric procedures are widely applicable.

(3) They are generally easier to understand and carry out than their parametric counterparts. We are often able to replace data values such as 148.06, 119.37, 86.24, 71.63, etc., with either symbols (such as plus or minus signs) or integers (i.e., ranks), such as 1, 2, 3, and so on, for which the resulting analysis involves less work.

On the other hand, nonparametric tests have two main shortcomings:

Cons

(1) They often do not make use of all acquired information. We have seen how plus and minus signs or integers can replace the actual data values in the nonparametric analysis. But if we represent 148.06, say, by 1, we lose information that is contained in the 148.06 value. In our previous ordering of the values 148.06, 119.37, 86.24 and 71.63, the largest data value could just as well have been 1,480.6, 14,806 or larger, and it still would have had rank 1 as the largest value in the list.

Therefore, we usually can analyze the data more thoroughly and learn more about the underlying situation by analyzing the data values than by examining symbols for, or ranks of the data.

(2) For an hypothesis test of any kind the probability of committing a type I error is fixed at some specified level of α. Therefore, the more efficient testing procedure (in the sense of having the best chance of making the proper decision) is the one which has the smaller probability β of committing a type II error; that is, the test which has the highest power (i.e., probability of rejecting a false hypothesis).

If the data actually consists of numerical values and all assumptions necessary to perform the analogous parametric test are satisfied, the parametric test is associated with a smaller value of β and is therefore said to be the "more powerful" test. In other words, in those circumstances where a parametric test is applicable, it should be used. The advantage in doing so is that the parametric test will yield the smaller value of β.

In summary, we have a trade-off in the use of a parametric test over its nonparametric counterpart; we require more conditions and the calculations might be more tedious, but we have a smaller chance of making an error in judgment as a result.

The Mathematics Review situation addressed by a parametric method in Example 2 of Section 11.9 (page 368) and by a nonparametric method in Example 1 of Section 16.3 (page 575) illustrates this trade-off.

17

Additional Tests of Hypotheses

17.1 Difference Between Proportions

In this section we develop a **multi-sample test for the equality of proportions** arising from two or more populations. The null hypothesis is that the population proportions are equal against the alternative hypothesis that they are not all equal. For the case of k populations, let π_i denote the value of the ith population proportion. Our objective is to develop a test for the null hypothesis

$$H_o : \pi_1 = \pi_2 = \cdots = \pi_k$$

versus

$$H_a : \text{not all } \pi_i \text{ are equal}$$

> The multi-sample test for the equality of proportions is applicable to situations where the underlying data are on a **nominal scale.**

Example 1 Equal Proportions of Fatal and Critical injuries

Many insurance companies are beginning to question the policy of offering reduced rates to owners of subcompact cars because they believe their rate of serious and fatal accidents is higher than that for larger sized cars. To investigate this issue John Underwood, an insurance investigator with the Cars-R-Sure Insurance Company, made an analysis of 360 randomly selected accidents to determine the number of cases in which at least one automobile occupant was either critically or fatally injured for automobiles of three sizes—subcompact, compact, and full size. His findings are summarized in Table 1.

Table 1

	Car		
	Subcompact	Compact	Full size
Fatal or critical injury	50	25	15
Not fatal or critical injury	100	90	80
Total	150	115	95

At the 0.025 level of significance, can we conclude that the proportions of fatal or critical injuries are the same for the three car classes, or does the frequency of such accidents differ for at least one of these car sizes?

$$H_o : \pi_1 = \pi_2 = \pi_3$$
$$H_a : \text{not all } \pi_i \text{ are equal}$$
$$\alpha = 0.025$$

The sample proportions of fatal and critical injuries in accidents involving each car class are 50/150 = 0.33, 25/115 = 0.22 and 15/95 = 0.16 for the subcompact, compact, and full size cars, respectively. If the null hypothesis is true, $\pi_1 = \pi_2 = \pi_3 = \pi$, then we can estimate the value of π for these populations by combining the results of the three samples into one ratio, which is called **pooling the data.** This yields an estimate p for π given by:

$$p = \frac{50+25+15}{150+115+95}$$
$$= 0.25$$

In terms of this estimate for π the products 150(0.25) = 37.5, 115(0.25) = 28.75, and 95(0.25) = 23.75, obtained by multiplying the same size for each population (column total) by the estimated proportion p, are called the expected values or expected frequencies. They are measures of the number of fatal and critical injuries to be expected from accidents involving subcompact, compact, and full sized cars, respectively. Therefore, the expected values of less severe accidents involving each car type are 150 - 37.5 = 112.5, 115 - 28.75 = 86.25, and 95 - 23.75 = 71.25, respective. These results are given in Table 2, where the expected values are shown in parentheses below the observed values.

Table 2

	Car		
Accident	Subcompact	Compact	Full size
Fatal or critical injury	50 (37.5)	25 (28.75)	16 (23.75)
Not fatal or critical injury	100 (112.5)	90 (86.25)	80 (71.25)

If the null hypothesis is true, we would expect the differences between the observed frequencies and their corresponding expected values to, in some sense, be small. We sharpen this as follows. Let *o* denote the observed frequency and *e* the expected value for a given cell of Table 2. We define the test statistic χ^2 as:

$$\chi^2 = \sum \frac{(0-e)^2}{e}$$

This χ^2 statistic has an **approximate chi-square distribution with** $k-1$ **degrees of freedom.** The approximation is satisfactory if none of the expected values is less than 5, a problem which does not arise here.

A simple, systematic way of carrying out the computation of the χ^2 test statistic, using a column arrangement similar to that employed in Section 12.3 to compute the slope of a regression line, is shown in Table 3. List the values for *o* and *e* in the

Table 3

o	e	$o-e$	$(o-e)^2/e$
50	37.5	12.5	4.17
25	28.75	-3.75	0.49
15	23.75	-8.75	3.22
100	112.5	-12.5	1.39
90	86.25	3.75	0.16
80	71.25	8.75	1.07
			$\chi^2 = 10.50$

first two columns and the $o-e$ differences in the third column. Finally, determine the fourth column entries by squaring the third column entries and dividing them by the second column entries. The χ^2 test statistic is the sum of the fourth column entries. In this case, $\chi^2 = 10.50$.

The tail region for this one-sided test begins at $\chi^2_{.025} = 7.38$, for $k-1=2$ degrees of freedom (see Figure 1).

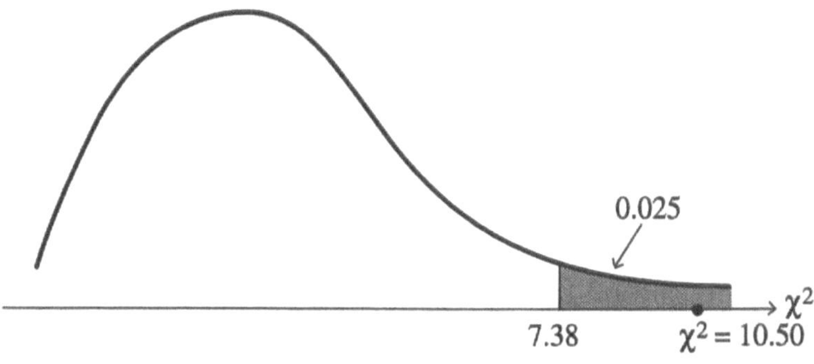

Figure 1

Since $\chi^2 = 10.50 > 7.38$, we reject the null hypothesis of equal population proportions of fatal and critical injuries for the different sized cars and conclude that the actual proportions of such tragedies in car accidents do depend on the size of the car.

Example 2 The Proportion of Fund Managers Favoring Non-Economic Goals in Portfolio Management is the Same?

Some financial experts believe that investment managers for certain fund management groups select securities for their clients' portfolios to further noneconomic social and political goals. To explore this issue Glenn Peterson, a financial researcher, asked a randomly chosen sample of fund managers employed by four major fund management agencies, Rich Return, Stock Up, Upward Mobility, and Capital Advisors, whether they approve or disapprove of selecting securities for a portfolio to further social and political goals. Glenn's findings are summarized in Table 4.

Table 4

	Fund Management Agency			
	Rich Return	Stock Up	Upward Mobility	Capital Advisors
Favor noneconomic goals	6	8	8	14
Do not favor noneconomic goals	19	12	22	31

At the 0.05 level of significance, do these data provide evidence in support of the hypothesis that the proportion of fund managers favoring noneconomic goals in portfolio is the same for these four fund management agencies?

$$H_o : \pi_1 = \pi_2 = \pi_3 = \pi_4$$
$$H_a : \text{not all } \pi_i \text{ are equal}$$
$$\alpha = 0.05$$

Assuming the null hypothesis is true, we may estimate the true proportion π of fund managers who give consideration to noneconomic criteria by:

$$p = \frac{6+8+8+14}{25+20+30+45} = 0.3$$

Under the assumption that H_o is true, the expected values may be determined in a simple, systematic way by referring to the row and column totals for the entries in Table 4. This augmented table with the expected values in parentheses, is listed in Table 5.

To obtain such expected values through a simple routine, note that the expected value e for each cell is given by:

$$e = \frac{\begin{pmatrix}\text{total data for}\\ \text{the cell row}\end{pmatrix}\begin{pmatrix}\text{total data for}\\ \text{the cell column}\end{pmatrix}}{\text{total number of data}}$$

Table 5

	Fund Management Agency				
	Rich Return	Stock Up	Upward Mobility	Capital Advisors	Total
Favor noneconomic goals	6 (7.5)	8 (6)	8 (9)	14 (13.5)	36
Do not favor noneconomic goals	19 (17.5)	12 (14)	22 (21)	31 (31.5)	84
Total	25	20	30	45	120

To illustrate, consider the expected value, 7.5, for the cell defined by the first row and first column. From Table 6 we have:

$$\frac{36.25}{120} = 7.5$$

Table 6

	Fund Management Agency				
	Rich Return	Stock Up	Upward Mobility	Capital Advisors	Total
Favor noneconomic goals	6 →				36
	(7.5) ↓				
Do not favor noneconomic goals					
Total	25				120

For the other cells in the first row we have expected values:

$$\frac{36.20}{120} = 6, \quad \frac{36.30}{120} = 9, \quad \frac{36.45}{120} = 13.5$$

For the cells in the second row we have expected values:

$$\frac{84.25}{120} = 17.5, \quad \frac{84.20}{120} = 14, \quad \frac{84.45}{120} = 31.5, \quad \frac{84.45}{120} = 31.5$$

Since the sum of the expected values in any row equal the total number of data for that row, once the expected values in the first row are computed, those in the second row may be obtained by subtraction. Thus, the expected values for the second row may be computed by 25 - 7.5 = 17.5, 20 - 6 = 14, 30 - 9 = 21, and 45 - 13.5 = 31.5.

The computation of the χ^2 test statistic is summarized in Table 7.

Table 7

o	e	$o-e$	$(o-e)^2/e$
6	7.5	-1.5	0.3
8	6	2	0.67
8	9	-1	0.11
14	13.5	0.5	0.02
19	17.5	1.5	0.13
12	14	-2	0.29
22	21	1	0.05
31	31.5	0.5	0.01
			$\chi^2 = 1.58$

Our χ^2 decision bound, corresponding to 3 degrees of freedom, is given by $\chi^2_{.05} = 7.81$ (see Figure 2).

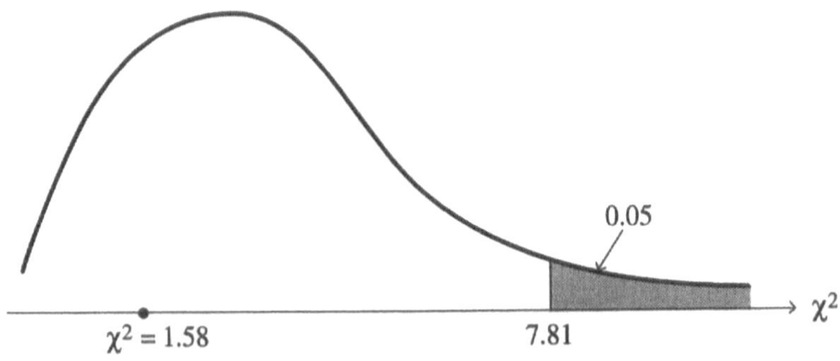

Figure 2

Since $\chi^2 = 1.58 < 7.81$, we accept the null hypothesis that the proportion of fund managers who use noneconomic criteria in determining their portfolios is the same for the four fund management agencies, or reserve judgment.

EXERCISES

1. Janet Wise, professor of psychology at Huxley College, wants to determine whether different groups of female undergraduate college students differ with respect to their career goals. She interviewed 120 randomly selected full-time students who work part-time, 130 full-time students who do not work, and 100 part-time students who work part-time, asking each student whether she plans a career full-time. A summary of the responses is given in Table 8. At the 0.025

Table 8

		Type of student		
		Full-time, work	Full-time, no work	Part-time, work
Career full-time?	Yes	68	90	52
	No	52	40	48

level of significance, can Professor Wise conclude that the proportions of female college undergraduate students in these groups who are planning a career full-time are equal?

2. An experiment was conducted to compare the effectiveness of three new drugs, Extracalm, Superior, and Statrelief, as pain relievers for conditions such as sprains, strains, and dislocations. Each drug was administered to 140 volunteers chosen at random. Table 9 gives the results of the experiment. At the 0.05 level

Table 9

	Drug		
	Extracalm	Superior	Statrelief
Relief	104	112	99
No relief	36	28	41

of significance, test the null hypothesis that the proportions of patients who received pain relief from these drugs are equal.

3. Silver University requires that students in a number of programs take the elementary statistics course. Edward Hickman, Dean of Studies at Silver University, wants to know whether the proportion of students who achieve a grade of at least B in statistics is the same for various majors. He looked up the statistics grades of a random sample of 145 chemistry, 170 computer science, 95 mathematics and 90 biology majors. His findings are summarized in Table 10. At the 0.10 level of significance, what can Dean Hickman conclude?

Table 10

		Type of major			
		Chemistry	Computer Science	Mathematics	Biology
Grade	At least B	75	100	65	45
	Lower than B	70	70	30	45

4. Edith Nettles, president of the Mercurial Market Research firm, wants to compare the proportion of travellers who are satisfied with the cuisine of the three most travelled airlines, Highair, Transatlantic and Truflight. She randomly sampled 250, 300, and 350 recent fliers, respectively, and found the number of satisfied and unsatisfied fliers to be as shown in Table 11.

Table 11

		Airline		
		Highair	Transatlantic	Truflight
Satisfied?	Yes	192	254	274
	No	58	46	76

At the 0.05 level of significance, what can Edith Nettles conclude?

5. Alba Jimenez, president of the Lucrative Corporation, plans to augment her employees' benefits package and wishes to know whether her workers prefer that these extra monies be placed in their pension funds or go towards increased

health benefits. She asked her personal director to conduct a survey of 160 production, 100 administrative, and 80 executive personnel, chosen at random, on this matter. The responses received are given in Table 12.

Table 12

	Employee position		
	Production	Administrative	Executive
Prefer pension	60	58	40
Prefer health	100	42	40

At the 0.01 level of significance, what can Alba Jimenez conclude?

17.2 ■ Contingency Tables

We may extend our analysis of the previous section to a test of whether two attributes of a population are independent. Our null hypothesis is that the variables describing the attributes are independent, against the alternative hypothesis that the variables are dependent, that is, that values of one variable are influenced by values of the other.

Given two attributes, we first choose a random sample of size n and classify the data in a two-way table similar to those employed in the previous section. In the table employed here we are no longer restricted to two rows.

We will use this table, called a **contingency table**, to determine whether the distribution of data for one variable is contingent on the distribution of the other. A table with r rows and c columns, denoted by $r \times c$, is called an **r by c contingency table.**

Once the contingency table is set up, we compute the expected values as we did in Example 2 of the previous section, leading to the same χ^2 test statistic

$$\chi^2 = \sum \frac{(o-e)\ 2}{e},$$

which has an approximate chi-square distribution, assuming that the null hypothesis is true.

Strictly speaking, this is a nonparametric test since it does not assume that the underlying population has a specific distribution. The only restriction is that each value of e be at least 5. In order to render a decision we compare the χ^2 test statistic to the appropriate χ^2 bound determined by $(r-1)(c-1)$ degrees of freedom.

Example 1 Does the Length of Hospital Stay Depend on a Patient's Health Insurance Coverage?

Leonard McNeil, president of the Complete Health Insurance Company, is opposed to national health insurance. He believes that it would be too costly to implement since such a system would, he argues, tend to encourage people to spend more time in hospitals. That is, according to Leonard, one's length of stay in a hospital is dependent on the extent of costs covered by the patient's health insurance plan. He asked Donna Chu, his staff statistician, to investigate the matter. Donna's findings, based on a random sample of 600 hospital stays, are summarized in Table 13.

At the 0.025 level of significance, does Mr. McNeil have sufficient reason to conclude that the length of hospital stay depends on the patient's health insurance coverage?

H_o: Length of hospital stay is independent of health coverage.

H_a: Length of hospital stay depends upon health coverage.

$\alpha = 0.025$

Table 13

		Days in hospital			
		< 5	5-10	> 10	Total
Fraction of costs covered by insurance	< 30%	40	70	40	150
	30-60%	50	80	50	180
	> 60%	60	100	110	270
	Total	150	250	200	600

Under the assumption that H_o is true, we compute the expected values as before:

$$e = \frac{\text{(row total)(column total)}}{n}$$

The expected values in the first row, for example, are:

$$\frac{150.150}{600} = 37.5, \quad \frac{150.250}{600} = 62.5, \quad \text{and} \quad \frac{150.200}{600} = 50$$

The observed and expected values are shown in Table 14, with the expected values in parentheses.

Table 14

		Days in hospital			
		< 5	5-10	> 10	Total
Percentage of costs covered by insurance	< 30%	40 (37.5)	70 (62.5)	40 (50)	150
	30-60%	50 (45)	80 (75)	50 (60)	180
	> 60%	60 (67.5)	100 (112.5)	110 (90)	270
	Total	150	250	200	600

The computation of the χ^2 test statistic is shown in Table 15.

Table 15

o	e	$o-e$	$(o-e)^2/e$
40	37.5	2.5	0.17
70	62.5	7.5	0.9
40	50	-10	2
50	45	5	0.56
80	75	5	0.33
50	60	-10	1.67
60	67.5	-7.5	0.83
100	112.5	-12.5	1.39
110	90	20	4.44
			$\chi^2 = 12.29$

The χ^2 bound is given by $\chi^2_{.025} = 11.14$ for (3 - 1)(3 - 1) = 4 degrees of freedom (see Figure 3).

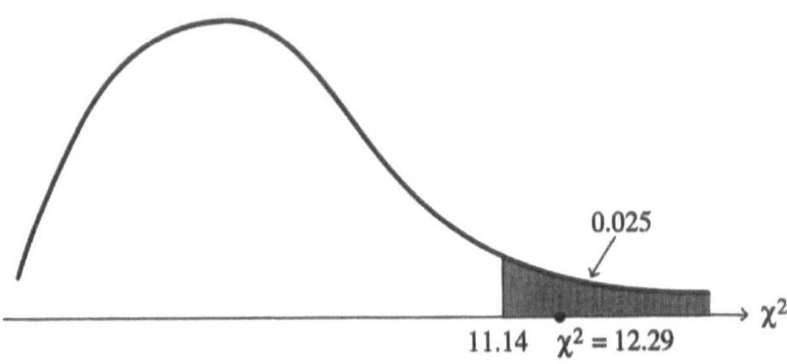

Figure 3

Since $\chi^2 = 12.29 > 11.14$, we reject the null hypothesis of independence and conclude that the length of stay at a hospital depends on the extent of the patient's health insurance coverage.

Example 2 Are Employee Position and Most Valued Job Characteristic Independent?

Hector Gonzalez, Heightston sociologist, wanted to determine if various characteristics of a job (high salary, good fringe benefits, job satisfaction, and opportunity for advancement) are considered equally important in Heightston. He conducted a survey involving a random sample of 190 administrative workers, 170 managers, and 140 salespersons in this locale. Each person was asked to indicate which of the four characteristics he most valued in a job. The results are shown in Table 16.

Table 16

		Position			
		Administrative	Managerial	Salesperson	Total
Most valued job characteristic	High salary	47	53	40	140
	Good fringe benefits	30	28	32	90
	Job satisfaction	63	49	48	160
	Opportunity for advancement	50	40	20	110
	Total	190	170	140	500

At the 0.05 level of significance, test the null hypothesis of independence between employee position and most valued job characteristic.

H_o: Employee position and most valued job characteristic are independent.

H_a: Employee position and most valued job characteristic are dependent.

$\alpha = 0.05$

We use the row and column totals to compute the expected values. For example, the values of *e* for the first row are given by:

$$\frac{140.190}{500} = 53.2, \quad \frac{140.170}{500} = 47.6, \text{and} \frac{140.140}{500} = 39.2$$

The *o* and *e* values are summarized in Table 17.

Table 17

		Position			
		Administrative	Managerial	Salesperson	Total
Most valued job characteristic	High salary	47 (53.2)	53 (47.6)	40 (39.2)	140
	Good fringe benefits	30 (34.2)	28 (30.6)	32 (25.2)	90
	Job satisfaction	63 (60.8)	49 (54.4)	48 (44.8)	160
	Opportunity for advancement	50 (41.8)	40 (37.4)	20 (30.8)	110
	Total	190	170	140	500

The computation of the test statistic follows in Table 18.

We must compare the $\chi^2 = 10.35$ test statistic to the test bound $\chi^2_{.05} = 12.59$, for (4 - 1)(3 - 1) = 6 degrees of freedom (see Figure 4).

Table 18

o	e	$o-e$	$(o-e)^2/e$
47	53.2	-6.2	0.72
53	47.6	5.4	0.61
40	39.2	0.8	0.02
30	34.2	-4.2	0.52
28	30.6	-2.6	0.22
32	25.2	6.8	1.83
63	60.8	2.2	0.08
49	54.4	-5.4	0.54
48	44.8	3.2	0.23
50	41.8	8.2	1.61
40	37.4	2.6	0.18
20	30.8	-10.8	3.79
			$\chi^2 = 10.35$

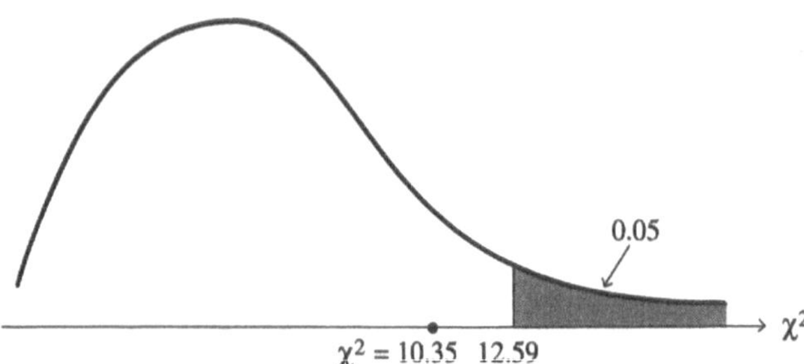

Figure 4

Since $\chi^2 = 10.35 < 12.59$, we accept the null hypothesis that the employee's most valued job characteristic is independent of job position, or reserve judgment.

EXERCISES

1. The number of fatalities resulting from automobile accidents has decreased substantially throughout the country in recent years. This, in part, prompted the Department of Transportation of Fastown to conduct a survey seat belt usage.

The results of this survey for a randomly chosen sample of 175 male adults, 145 female adults, and 130 teenagers are shown in Table 19.

Table 19

	Seat belt use		
	Always	Sometimes	Never
Male adult	75	50	50
Female adult	60	55	30
Teenager	45	45	40

Test, at the 0.5 level of significance, whether passenger classification and seat belt usage in Fastown are independent.

2. Ronald Templeton, Dean of Liberal Arts of Silver University, believes that the grades his students achieve are dependent on the amount of time they spend each week listening to music. To test his theory, he asked his friend Sam Wright, a professor of statistics, to randomly select 200 students to participate in a survey in which they are asked to state their average grade for all liberal arts courses at the end of the semester and an estimate of the average number of hours per week they spend listening to music. The results obtained are shown in Table 20.

Table 20

		Average grade			
		A	B	C	D or F
Hours/week	Less than 5	20	25	10	10
spent	5-10	14	16	17	24
Listening to music	More than 10	11	14	13	26

Do the data support the Dean's hypothesis at the 0.01 significance level?

3. Diane Lockwood, publisher of the *Norwood Daily News*, wonders whether weekly readership is related to the readers' educational background. She surveyed a randomly chosen sample of 300 adults on this matter and obtained the data shown in Table 21.

Table 21

		Level of education			
		Graduate degree	Undergraduate degree	High school graduate	Did not complete high school
Weekly frequency of readership	5-7 days	37	42	21	15
	2-4 days	19	38	31	22
	Less than 2 days	14	20	23	18

At the 0.05 level of significance, is the *Norwood Daily News'* weekly readership independent of the educational level as its readers?

4. Scientific studies have suggested a link between smoking and the state of one's respiratory system. Mercy General Hospital of Healthville decided to conduct their own study. They asked 360 randomly chosen inhabitants about their smoking habits, in return for a free, thorough medical examination. The data obtained are summarized in Table 22.

Table 22

		Condition of lungs		
		Healthy	Below par	Serious illness
Daily frequency of smoking (packs)	1 or more	31	19	10
	Less than 1	40	21	11
	Nonsmoker	169	50	9

At the 0.025 level of significance, can Mercy General Hospital conclude that a relationship exists between frequency of smoking and respiratory health?

5. The Janestown mayoral election is approaching and WWTK, a local television station, conducted a poll comparing the political leaning and yearly gross income (in thousands of dollars) of 540 randomly selected residents. The outcome of this survey is given in Table 23.

Table 23

	Yearly gross income		
	Less than 30	30-60	Over 60
Democrat	100	90	50
Republican	70	67	43
Other	46	32	42

At the 0.10 level of significance, test the null hypothesis that political inclinations are independent of gross annual income in Janestown.

6. The Fantastic Fit Clothing Company wanted to gauge women's attitudes toward cotton and other clothing fibers. They randomly selected 750 women and asked each her age and her preferred fiber, be it cotton, polyester, or silk, for a blouse. The responses are summarized in Table 24.

Table 24

		Fiber		
		Cotton	Polyester	Silk
Age	18-29	122	110	68
	30-39	77	62	41
	40-49	59	57	34
	50 and over	47	41	32

At the 0.01 level of significance, can the Fantastic Fit Clothing Company conclude that a woman's preference for blouse material is not related to her age?

17.3 ■ Goodness of Fit

In Section 16.4 of the previous chapter we considered the Kolmogorov-Smirnov (K-S) and Lilliefors tests to test for specific probability distributions. As we observed, the K-S test may be employed to test for any given distribution, whereas the Lilliefors test is specific for normality.

The χ^2 methods considered in the previous two sections may also be used to test for goodness-of-fit, and we refine them for this purpose in this section.

In general, we wish to test:

> H_o : The random variable of interest has a specified probability distribution.
>
> versus
>
> H_a : The random variable of interest does not have the specified probability distribution.

The computational aspect of the problems considered is similar to those of the previous sections of this chapter. We generate data according to the random variable of interest and compute the probabilities of relevant events, under the assumption that the null hypothesis is true. We use these probabilities to determine expected values and compare them to the observed frequencies as before.

The χ^2 test statistic is again given by

$$\chi^2 = \sum \frac{(o-e)\ 2}{e},$$

which has an approximate chi-square distribution if the sample observations support the assumed distribution and each expected value is at least 5.

We compare our test statistic with the appropriate χ^2 bound for $k-1-m$ degrees of freedom, where k is the number of classes in the frequency distribution and m is the number of parameters that must be estimated to determine the expected values.

Example 1 The Probability Distribution for Dapper Dan 's Die is What He Claims?

Dapper Dan, who we first met in Section 7.1 is known for his collection of unbalanced dice. In his latest adventure Dan presented a die to Cautious Cal, a student of statistics, with the claim that the probabilities with which the various faces show are as given in Table 25.

Table 25

Number of dots: x	1	2	3	4	5	6
Probability: p(x)	$\frac{1}{21}$	$\frac{2}{21}$	$\frac{3}{21}$	$\frac{4}{21}$	$\frac{5}{21}$	$\frac{6}{21}$

To test Dan's claim Cal tossed the die 210 times and observed the outcomes described by Table 26.

Table 26

No. of dots on face: x	1	2	3	4	5	6
Observed frequency: o	16	19	37	48	39	51

Determine, at the 0.10 level, whether the evidence supports Dan's claim.

Cal's problem, then, is to test:

H_o : The probability distribution for Dan's die is the one given in Table 25,

versus

H_a : The probability distribution for Dan's die is not the one given in Table 25.

$\alpha = 0.10$

Since Cal tossed the die 210 times, the expected value for each face of the die showing is obtained by multiplying the probability of that face by 210. Therefore, the expected values for faces 1 through 6 are given, respectively, by:

$$\frac{1}{21}\cdot 210 = 10,\quad \frac{2}{21}\cdot 210 = 20,\quad \frac{3}{21}\cdot 210 = 30,$$
$$\frac{4}{21}\cdot 210 = 40,\quad \frac{5}{21}\cdot 210 = 50,\quad \frac{6}{21}\cdot 210 = 60.$$

The observed frequencies, expected values and computation of the test statistic are summarized in Table 27.

Table 27

o	e	$o-e$	$(o-e)^2/e$
16	10	6	3.6
19	20	-1	0.05
37	30	7	1.63
48	40	8	1.6
39	50	-11	2.42
51	60	-9	1.35
			$\chi^2 = 10.65$

The test bound is $\chi^2_{.10} = 9.24$, for $k-1-m = 6-1-0 = 5$ degrees of freedom. ($m = 0$ Since no parameters had to be estimated to determine the expected values.) See Figure 5.

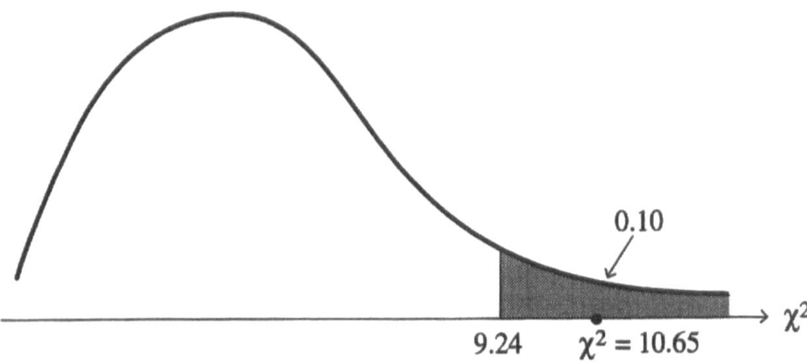

Figure 5

Since $\chi^2 = 10.65 > 9.24$, Cal (and the rest of us) reject the null hypothesis and conclude that the actual probability distribution for Dan's die is not the one given by Table 25.

Example 2 The Probability that a Newborn Baby is a Boy is ½?

Dr. Harold Dunleavy, pediatrician, believes that the probability a newborn baby is a boy is 0.5. To test this hypothesis he interviewed 480 families with four children, chosen at random, and obtained the distribution of boys in a family given in Table 28.

Table 28

Number of sons	0	1	2	3	4
Number of families	23	110	194	114	39

Test, at the 0.05 level of significance, whether the data support Dr. Dunleavy's hypothesis.

To simplify the analysis, we assume that the gender of a given newborn in any family is independent of that of any other child in that family. Then, if Dr. Dunleavy's belief is correct, the underlying gender distribution in families having four children is binomial with parameters $n = 4$ and $\pi = 0.5$. The problem reduces to testing:

H_o: The gender distribution of a family of four children is binomial with $\pi = 0.5$,

versus

H_a: The gender distribution of a family of four children is not binomial with $\pi = 0.5$.

$\alpha = 0.05$

If the null hypothesis is true, we may calculate the binomial probabilities of 0 up to 4 boys as follows:

$$p(x) = C(n, x)(0.5)^x (0.5)^{4-x}$$
$$= C(n, x)/16$$

This yields the probability distribution shown in Table 29.

Table 29

x	0	1	2	3	4
p(x)	$\frac{1}{16}$	$\frac{4}{16}$	$\frac{6}{16}$	$\frac{4}{16}$	$\frac{1}{16}$

We obtain the expected frequencies by multiplying the probabilities by 480, which yields the following expected values for 0, 1, 2, 3, and 4 sons, respectively:

20, 120, 180, 120 and 30.

The computation of the test statistic is shown in Table 30.

The decision bound is $\chi^2_{.05} = 9.49$, for 5 - 1 - 0 = 4 degrees of freedom (see Figure 6).

Table 30

o	e	$o-e$	$(o-e)^2/e$
23	30	-7	1.63
110	120	-10	0.83
194	180	14	1.09
114	120	-6	0.3
39	30	9	2.7
			$\chi^2 = 6.55$

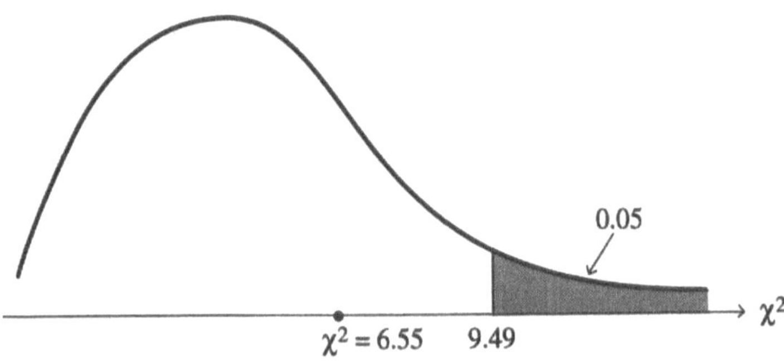

Figure 6

Since $\chi^2 = 6.55 < 9.49$, we accept the null hypothesis that the gender distribution of families of four children is binomial with $\pi = 0.5$, or reserve judgment.

Had Dr. Dunleavy not had a value of π in mind before obtaining his data, we would have had to estimate its value from the data. In this case $m = 1$ and the χ^2 bound is $\chi^2_{.05} = 7.81$ for 5 - 1 - 1 = 3 degrees of freedom.

Example 3 The "Heavy" Basic Statistics Text

As we have noted, statistics texts have not only become more expensive in recent years but bigger and heavier as well. A recent edition of *Basic Statistics* was the focus of a wrist strength test at Ecap University. The time lengths, in minutes, that the population of 156 students at Ecap U. who used the text were able to hold it in one sitting is given in Example 1 of Section 3.1 (page 85). The frequency distribution given in Table 31 for these data was also obtained in that example.

Table 31

Time Interval	Frequency
0.0-0.9	2
1.0-1.9	7
2.0-2.9	13
3.0-3.9	31
4.0-4.9	42
5.0-5.9	32
6.0-6.9	16
7.0-7.9	10
8.0-8.9	3
	156

From this distribution we obtain $\mu = 4.6$ and $\sigma = 1.6$ (see Sections 5.2 and 5.3, Examples 1 and 1, page 127 and page 138).

Is this population normal with $\mu = 4.6$ and $\sigma = 1.6$? This question was addressed in Example 1 of Section 8.4 (page 231) and Example 1 of Section 16.4 (page 583). We return to it here in terms of the χ^2 test.

The problem is to test:

H_o: The population of wrist strength times is normally distributed with mean 4.6 and standard distribution 1.6,

versus

H_a: The population of wrist strength times is not normally distributed with mean 4.6 and standard deviation 1.6

$\alpha = 0.05$

To determine the expected frequencies of the classes listed in Table 31 we must first determine their probabilities with respect to H_o. Geometrically speaking, we want to find the probabilities of the bases of the histogram rectangles which describe the frequency distribution, which in turn are defined by the class boundaries of the classes. For details on obtaining the class boundaries of these classes refer to Example 1 (page 85) of Section 3.1, particularly Table 10. In summary we have the first column of Table 32.

Table 32

Class Boundaries	z-values	Probability	e	o
-0.05-0.95	-2.91, -2.28	0.0094	1.47	2
0.95-1.95	-2.28, -1.66	0.0372	5.80	7
1.95-2.95	-1.66, -1.03	0.1030	16.07	13
2.95-3.95	-1.03, -0.41	0.1894	29.55	31
3.95-4.95	-0.41, 0.22	0.2462	38.40	42
4.95-5.95	0.22, 0.84	0.2124	33.13	32
5.95-6.95	0.84, 1.47	0.1297	20.23	16
6.95-7.95	1.47, 2.09	0.0525	8.19	10
7.95-8.95	2.09, 2.72	0.0150	2.34	3
				156

Our next step is to determine the *z*-values z_1 and z_2 of each pair of class boundaries and find the probability of each class with respect to H_o. For the first class we have:

$$z_1 = \frac{-0.05 - 4.6}{1.6} = -2.91, \quad z_2 = \frac{0.95 - 4.6}{1.6} = -2.28$$

$$\begin{aligned} P(-2.91 \le z \le -2.28) &= 0.4981 - 0.4887 \\ &= 0.0094 \end{aligned}$$

The *z*-values of the class boundaries of the classes and their corresponding probabilities are given in columns 2 and 3 of Table 32. The corresponding expected values are obtained by multiplying each probability in column 3 by 156, the total number of data. The results are shown in column 4. The observed frequencies for each class are given in column 5.

Unfortunately, the expected values for the first and last classes are each less than 5. We combine adjacent classes and their expected values to bring the total expected value of the combined class to the minimum required value of 5. Our new first class has class boundaries -0.05 and 1.95 with expected value 1.47 + 5.80 = 7.27 and observed value 2 + 7 = 9. Our new last class has class boundaries 6.95 and 8.95 with expected value 8.19 + 2.34 = 10.53 and observed value 10 + 3 = 13. Note that only seven classes remain in the modified frequency distribution.

The revised collection of observed frequencies, expected values, and the computation of the test statistic is given in Table 33.

Table 33

o	e	$o-e$	$(o-e)^2/e$
9	7.27	1.73	0.41
13	16.07	-3.07	0.59
31	29.55	1.45	0.07
42	38.40	3.60	0.34
32	33.13	-1.13	0.04
16	20.23	-4.23	0.88
13	10.53	2.47	0.58
			$\chi^2 = 2.91$

The test bound is $\chi^2_{.05} = 12.6$ for 7 - 1 - 0 = 6 degrees of freedom (see Figure 7).

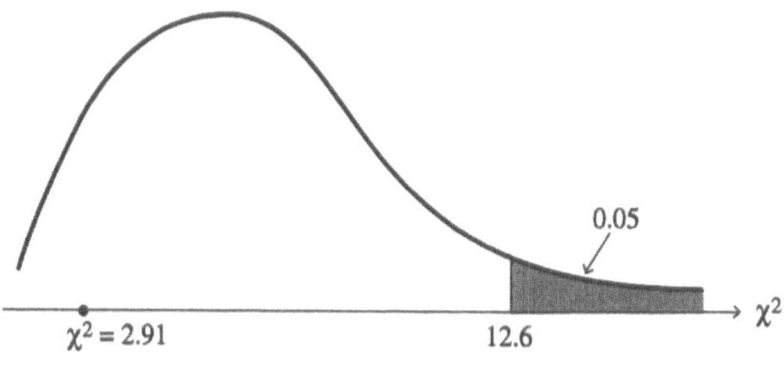

Figure 7

Since $\chi^2 = 2.91 < 12.6$, we accept the null hypothesis that the data are normally distributed with mean 4.6 and standard deviation 1.6, or reserve judgment.

Some Noteworthy Observations

1. The Kolmogorov-Smirnov and Lilliefors test discussed in Section 16.4 do not have a requirement that each expected value e be at least 5.

2. When applied to "small" samples or populations of data, the chi-square test usually fails to reject the hypothesis of normality when the data are symmetrically distributed and mound shaped, even though nonnormal.

   The K-S and Lilliefors tests are more powerful tests in cases where the number of data is "small." As a guideline take "small" here to be less than 100.

3. The K-S and Lilliefors tests require individual sample values, which are not available if you only have access to grouped data.

EXERCISES

1. In genetics, when two particular types of peas are bred the Mendelian Principle of Segregation states that the following four varieties of peas should occur in the ratio 9:3:3:1: round and yellow (RY), round and green (RG), wrinkled and yellow (WY), and wrinkled and green (WG). Dwight Frazier, horticulturist for the Plentiful Peas Company, carried out an experiment with randomly selected peas and found 1775, 570, 635 and 220 of the RY, RG, WY and WG varieties, respectively. At the 0.025 level of significance, do these frequencies follow Mendel's predicted ratio?

2. In an earlier adventure of Dapper Dan involving his friend Gullible Gus, Dan handed Gus a die governed by the behavior, he claimed, that the even numbered faces are favored to show over the odd numbered ones by 2 to 1. Gus appreciated that he would be wise to look into this further.

   (a) Set up a probability model for Dan's die based on his claim (see Section 6.4).

   (b) Gus tossed Dan's die 225 times and observed the outcomes summarized in Table 34.

Table 34

No. of dots on face	1	2	3	4	5	6
Observed frequency	22	48	27	54	23	51

At the 0.05 significance level, do these data support Dan's claim about the nature of the die he handed to Gus? Explain.

3. The Colossal Corporation employs thousands of people on an hourly basis. Joan Rivera, head of Colossal's payroll department, believes that hourly wages at this corporation are normally distributed. To check this hypothesis, she looked up the hourly wage of 2000 randomly selected workers and obtained the frequency distribution shown in Table 35.

Table 35

Hourly wage ($)	Frequency
4-6	19
7-9	135
10-12	512
13-15	757
16-18	418
19-21	143
22-24	16
	2000

(a) Find the sample mean and sample standard deviation of hourly wages.

(b) Test Joan's hypothesis about the underlying distribution of hourly wages at the 0.10 level of significance.

4. Tables of random digits are supposed to be constructed in such a way that each digit is a value of a random variable which takes the values of 0, 1, 2, 3, 4, 5, 6, 7, 8, and 9 with equal probabilities of 0.10. Determine, at the 0.05 level, whether the 150 digits given in Table 36 were randomly generated.

Table 36

5	2	3	6	0	4	6	6	5	8	6	6	5	1	1
0	4	1	7	2	7	3	0	8	5	1	1	7	9	5
5	5	5	9	4	0	4	1	5	7	5	0	0	7	9
6	1	3	4	3	6	4	3	1	5	7	0	8	3	6
8	2	8	5	7	3	5	3	3	5	8	6	0	0	0
6	0	0	7	0	6	6	2	4	1	3	2	8	3	6
2	7	5	7	3	1	1	4	7	9	9	4	1	1	4
4	1	2	6	8	8	0	1	8	7	2	0	3	5	1
0	9	6	3	6	8	4	6	6	8	4	2	4	8	6
7	1	3	0	3	6	1	4	7	1	4	5	3	2	2

5. The Delicious Ice Cream Company enlisted the aid of the Stats-R-Us marketing research firm to determine whether there is a preference in Newtown between the Company's ice cream and that of their top three competitors—Heavenly, Passion Supreme, and Sinful Treat. A sample of each ice cream brand was given to 600 randomly selected Newtown residents who were asked to identify the brand they like best. The data collected by Stats-R-Us is shown in Table 37.

Table 37

Ice Cream	Number of Customers
Delicious	161
Heavenly	142
Passion Supreme	174
Sinful Treat	123

At the 0.025 level of significance, test the null hypothesis of no preference between the four brands.

6. Alex Wisdom, an observer of the statistics scene, commented as follows: "For the data of Example 3, which make up a population, to be normally distributed the χ^2 statistic 2.91 should be zero; there should be no difference between the o and e values. Therefore, it is not tenable to accept the null hypothesis that the population is normally distributed with $\mu = 4.6$ and $\sigma = 1.6$." Do you agree with Alex? Explain. In this connection it might be useful to review Section 8.4.

7. The final exam grades of 165 students who took statistics at the Vilnius Technical Institute in a recent semester are given in Exercise 6 of Section 16.4 (page 589).

   (a Set up a frequency distribution for the exam grades based on the grade intervals 30-39, 40-49, 50-59, 60-69, 70-79, 80-89, 90-99 (see Exercise 4 of Section 3.1, page 93)

   (b) Based on this frequency distribution determine, at the 0.05 significance level, whether the population is normally distributed.

Tables

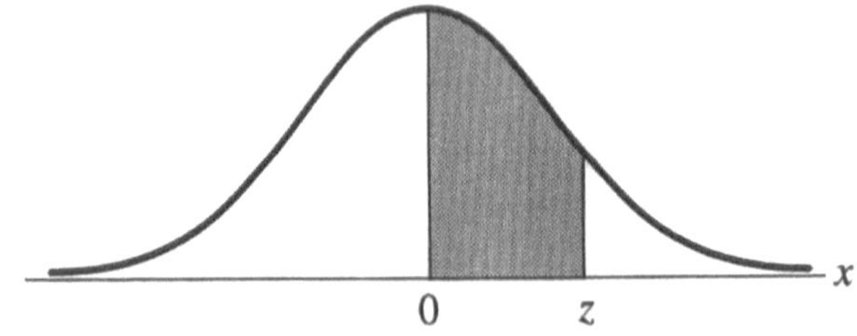

Table A Standard Normal Curve

z	.00	.01	.02	.03	.04	.05	.06	.07	.08	.09
.0	.0000	.0040	.0080	.0120	.0160	.0199	0.239	.0279	.0319	.0359
.1	.0398	.0438	.0478	.0517	.0557	.0596	.0636	.0675	.0714	.0753
.2	.0793	.0832	.0871	.0910	.0948	.0987	.1026	.1064	.1103	.1141
.3	.1179	.1217	.1255	.1293	.1331	.1368	.1406	.1443	.1480	.1517
.4	.1554	.1591	.1628	.1664	.1700	.1736	.1772	.1808	.1844	.1879
.5	.1915	.1950	.1985	.2019	.2054	.2088	.2123	.2157	.2190	.2224
.6	.2257	.2291	.2324	.2357	.2389	.2422	.2454	.2486	.2518	.2549
.7	.2580	.2612	.2642	.2673	.2704	.2734	.2764	.2794	.2823	.2852
.8	.2881	.2910	.2939	.2967	.2995	.3023	.3051	.3078	.3106	.3133
.9	.3159	.3186	.3212	.3238	.3264	.3289	.3315	.3340	.3365	.3389
1.0	.3413	.3438	.3461	.3485	.3508	.3531	.3554	.3577	.3599	.3621
1.1	.3643	.3665	.3686	.3708	.3729	.3749	.3770	.3790	.3810	.3830
1.2	.3849	.3869	.3888	.3907	.3925	.3944	.3962	.3980	.3997	.4015
1.3	.4032	.4049	.4066	.4082	.4099	.4115	.4131	.4147	.4162	.4177
1.4	.4192	.4207	.4222	.4236	.4251	.4265	.4279	.4292	.4306	.4319
1.5	.4332	.4346	.4357	.4370	.4382	.4394	.4406	.4418	.4429	.4441
1.6	.4452	.4463	.4474	.4484	.4495	.4505	.4515	.4525	.4535	.4545
1.7	.4554	.4564	.4573	.4582	.4591	.4599	.4608	.4616	.4625	.4633
1.8	.4641	.4649	.4656	.4664	.4671	.4678	.4686	.4693	.4699	.4706
1.9	.4713	.4719	.4726	.4732	.4738	.4744	.4750	.4756	.4761	.4767
2.0	.4772	.4778	.4783	.4788	.4793	.4798	.4803	.4808	.4812	.4817
2.1	.4821	.4826	.4830	.4834	.4838	.4842	.4846	.4850	.4854	.4857
2.2	.4861	.4864	.4868	.4871	.4875	.4878	.4881	.4884	.4887	.4890
2.3	.4893	.4896	.4898	.4901	.4904	.4906	.4909	.4911	.4913	.4916
2.4	.4918	.4920	.4922	.4925	.4927	.4929	.4931	.4932	.4934	.4936
2.5	.4938	.4940	.4941	.4943	.4945	.4946	.4948	.4949	.4951	.4952
2.6	.4953	.4955	.4956	.4957	.4959	.4960	.4961	.4962	.4963	.4964
2.7	.4965	.4966	.4967	.4968	.4969	.4970	.4971	.4972	.4973	.4974
2.8	.4974	.4975	.4976	.4977	.4977	.4978	.4979	.4979	.4980	.4981
2.9	.4981	.4982	.4982	.4983	.4981	.4981	.4985	.4985	.4986	.4986
3.0	.4987	.4987	.4987	.4988	.4988	.4989	.4989	.4989	.4990	.4990

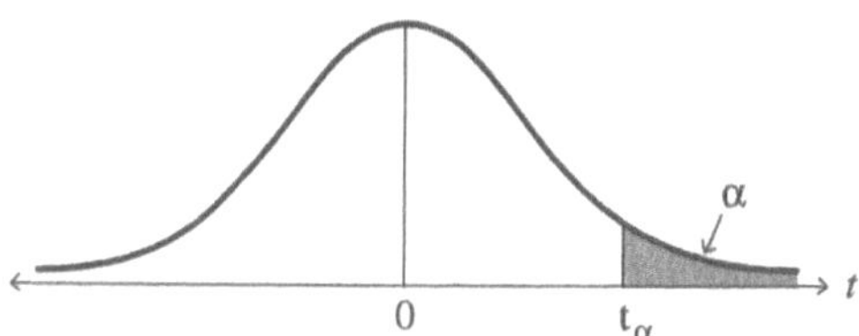

Table B t Distribution

Degrees of Freedom	$t_{.10}$	$t_{.05}$	$t_{.025}$	$t_{.01}$	$t_{.005}$
1	3.078	6.314	12.706	31.821	63.657
2	1.886	2.920	4.303	6.965	9.925
3	1.638	2.353	3.182	4.541	5.841
4	1.533	2.132	2.776	3.747	4.604
5	1.476	2.015	2.571	3.365	4.032
6	1.440	1.943	2.447	3.143	3.707
7	1.415	1.895	2.365	2.998	3.499
8	1.397	1.860	2.306	2.896	3.355
9	1.383	1.833	2.262	2.821	3.250
10	1.372	1.812	2.228	2.764	3.169
11	1.363	1.796	2.201	2.718	3.106
12	1.356	1.782	2.179	2.681	3.055
13	1.350	1.771	2.160	2.650	3.012
14	1.345	1.761	2.145	2.624	2.977
15	1.341	1.753	2.131	2.602	2.947
16	1.337	1.746	2.120	2.583	2.921
17	1.333	1.740	2.110	2.567	2.898
18	1.330	1.734	2.101	2.552	2.878
19	1.328	1.729	2.093	2.539	2.861
20	1.325	1.725	2.086	2.528	2.845
21	1.323	1.721	2.080	2.518	2.831
22	1.321	1.717	2.074	2.508	2.819
23	1.319	1.714	2.069	2.500	2.807
24	1.318	1.711	2.064	2.492	2.797
25	1.316	1.708	2.060	2.485	2.787
26	1.315	1.706	2.056	2.479	2.779
27	1.314	1.703	2.052	2.473	2.771
28	1.313	1.701	2.048	2.467	2.763
29	1.311	1.699	2.045	2.462	2.756
30	1.310	1.697	2.042	2.457	2.750
40	1.303	1.684	2.021	2.423	2.704
60	1.296	1.671	2.000	2.390	2.660
120	1.289	1.658	1.980	2.358	2.617

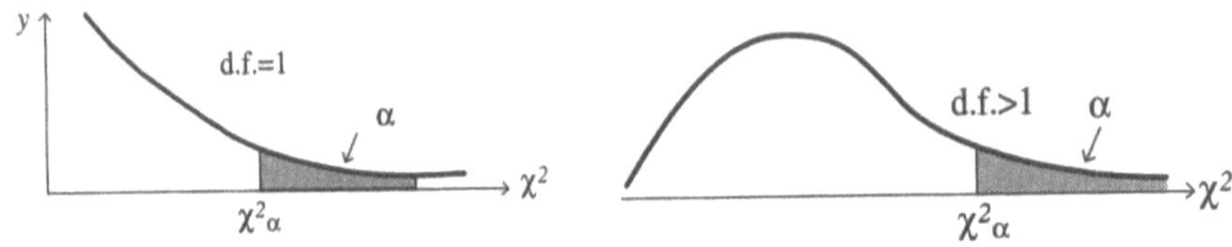

Table C Chi-Square Curve

d.f.	$\chi^2_{.995}$	$\chi^2_{.99}$	$\chi^2_{.975}$	$\chi^2_{.95}$	$\chi^2_{.90}$	$\chi^2_{.10}$	$\chi^2_{.05}$	$\chi^2_{.025}$	$\chi^2_{.01}$	$\chi^2_{.005}$
1	.0000393	.000157	.000982	.00393	.01579	2.706	3.841	5.024	6.635	7.879
2	.0100251	.0201007	.0506356	.102587	.210720	4.60517	5.99147	7.37776	9.21034	10.5966
3	.0717212	.114832	.215795	.351846	.584378	6.25139	7.81473	9.34840	11.3449	12.8381
4	.206990	.297110	.484419	.710721	1.063623	7.77944	9.48773	11.1433	13.2767	14.8602
5	.411740	.554300	.831211	1.145476	1.61031	9.23635	11.0705	12.8325	15.0863	16.7494
6	.675727	.872085	1.237347	1.63539	2.20413	10.6446	12.5916	14.4494	16.8119	18.5476
7	.989265	1.239043	1.68987	2.16735	2.83311	12.0170	14.0671	16.0128	18.4753	20.2777
8	1.344419	1.646482	2.17973	2.73264	3.48954	13.3616	15.5073	17.5346	20.0902	21.9550
9	1.734926	2.087912	2.70039	3.32511	4.16816	14.6837	16.9190	19.0228	21.6660	23.5893
10	2.15585	2.55821	3.24697	3.94030	4.86518	15.9871	18.3070	20.4831	23.2093	25.1882
11	2.60321	3.05347	3.81575	4.57481	5.57779	17.2750	19.6751	21.9200	24.7250	26.7569
12	3.07382	3.57056	4.40379	5.22603	6.30380	18.5494	21.0261	23.3367	26.2170	28.2995
13	3.56506	4.10691	5.00874	5.89186	7.04150	19.8119	22.3621	24.7356	27.6883	29.8194
14	4.07468	4.66043	5.62872	6.57063	7.78953	21.0642	23.6848	26.1190	29.1413	31.3193
15	4.60094	5.22935	6.26214	7.26094	8.54675	22.3072	24.9958	27.4884	30.5779	32.8013
16	5.14224	5.81221	6.90766	7.96164	9.31223	23.5418	26.2962	28.8454	31.9999	34.2672
17	5.69724	6.40776	7.56418	8.67176	10.0852	24.7690	27.5871	30.1910	33.4087	35.7185
18	6.26481	7.01491	8.23075	9.39046	10.8649	25.9894	28.8693	31.5264	34.8053	37.1564
19	6.84398	7.63273	8.90655	10.1170	11.6509	27.2036	30.1435	32.8523	36.1908	38.5822
20	7.43386	8.26040	9.59083	10.8508	12.4426	28.4120	31.4104	34.1696	37.5662	39.9968
21	8.03366	8.89720	10.28293	11.5913	13.2393	29.6151	32.6705	35.4789	38.9321	41.4010
22	8.64272	9.54249	10.9823	12.3380	14.0415	30.8133	33.9244	36.7807	40.2894	42.7958
23	9.26042	10.19567	11.6885	13.0905	14.8479	32.0069	35.1725	38.0757	41.6384	44.1813
24	9.88623	10.8564	12.4011	13.8484	15.6587	33.1963	36.4151	39.3641	42.9798	45.5585
25	10.5197	11.5240	13.1197	14.6114	16.4734	34.3816	37.6525	40.6465	44.3141	46.9278
26	11.1603	12.1981	13.8439	15.3791	17.2919	35.5631	38.8852	41.9232	45.6417	48.2899
27	11.8076	12.8786	14.5733	16.1513	18.1138	36.7412	40.1133	43.1944	46.9630	49.6449
28	12.4616	13.5648	15.3079	16.9279	18.9392	37.9159	41.3372	44.4607	48.2782	50.9933
29	13.1211	14.2565	16.0471	17.7083	19.7677	39.0875	42.5569	45.7222	49.5879	52.3356
30	13.7867	14.9535	16.7908	18.4926	20.5992	40.2560	43.7729	46.8792	50.8922	53.6720
40	20.7065	22.1643	24.4331	26.5093	29.0505	51.08080	55.7585	59.3417	63.6907	66.7659
50	27.9907	29.7067	32.3574	34.7642	37.6886	63.1671	67.5048	71.4202	76.1539	79.4900
60	35.5346	37.4848	40.4817	43.1879	46.4589	74.3970	79.0819	83.2976	88.3794	91.9517
70	43.2752	45.4418	48.7576	51.7393	55.3290	85.5271	90.5312	95.0231	100.425	104.215
80	51.1720	53.5400	57.1532	60.3915	64.2778	96.5782	101.879	106.629	112.329	116.321
90	59.1963	61.7541	65.6466	69.1260	73.2912	107.565	113.145	118.136	124.116	128.299
100	67.3276	70.0648	74.2219	77.9295	82.3581	118.498	124.342	129.561	135.807	140.169

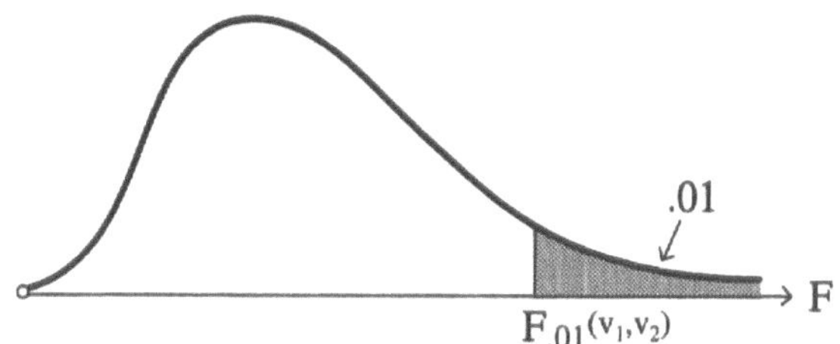

Table D $F_{.01}(v_1, v_2)$

v_1	DEGREES OF FREEDOM FOR NUMERATOR																
v_2	1	2	3	4	5	6	7	8	9	10	12	15	20	24	30	40	60
1	4,052	5,000	5,403	5,625	5,764	5,859	5,928	5,982	6,023	6,056	6,106	6,157	6,209	6.235	6,261	6,287	6,313
2	98.5	99.0	99.2	99.2	99.3	99.3	99.4	99.4	99.4	99.4	99.4	99.4	99.4	99.5	99.5	99.5	99.5
3	34.1	30.8	29.5	28.7	28.2	27.9	27.7	27.5	27.3	27.2	27.1	26.9	26.7	26.6	26.5	26.4	26.3
4	21.2	18.0	16.7	16.0	15.5	15.2	15.0	14.8	14.7	14.5	14.4	14.2	14.0	13.9	13.8	13.7	13.7
5	16.3	13.3	12.1	11.4	11.0	10.7	10.5	10.3	10.2	10.1	9.89	9.72	9.55	9.47	9.38	9.29	9.20
6	13.7	10.9	9.78	9.15	8.75	8.47	8.26	8.10	7.98	7.87	7.72	7.56	7.40	7.31	7.23	7.14	7.06
7	12.2	9.55	8.45	7.85	7.46	7.19	6.99	6.84	6.72	6.62	6.47	6.31	6.16	6.07	5.99	5.91	5.82
8	11.3	8.65	7.59	7.01	6.63	6.37	6.18	6.03	5.91	5.91	5.97	5.52	5.36	5.28	5.20	5.12	5.03
9	10.6	8.02	6.99	6.42	6.06	5.80	5.61	4.47	5.35	5.26	5.11	4.96	4.81	4.73	4.65	4.57	4.48
10	10.0	7.56	6.55	5.99	5.64	5.39	5.20	5.06	4.94	4.85	4.71	4.56	4.41	4.33	4.25	4.17	4.08
11	9.65	7.21	6.22	5.67	5.32	5.07	4.89	4.74	4.63	4.54	4.40	4.25	4.10	4.02	3.94	3.86	3.78
12	9.33	6.93	5.95	5.41	5.06	4.82	4.64	4.50	4.39	4.30	4.16	4.01	3.86	3.78	3.70	3.62	3.54
13	9.07	6.70	5.74	5.21	4.86	4.62	4.44	4.30	4.19	4.10	3.96	3.82	3.66	3.59	3.51	3.43	3.34
14	8.86	6.51	5.56	5.04	4.70	4.46	4.28	4.14	4.03	3.94	3.80	3.66	3.51	3.43	3.35	3.27	3.18
15	8.68	6.36	5.42	4.89	4.56	4.32	4.14	4.00	3.89	3.80	3.67	3.52	3.37	3.29	3.21	3.13	3.05
16	8.53	6.23	5.29	4.77	4.44	4.20	4.03	3.89	3.78	3.69	3.55	3.41	3.26	3.18	3.10	3.02	2.93
17	8.40	6.11	5.19	4.67	4.34	4.10	3.93	3.79	3.68	3.59	3.46	3.31	3.16	3.08	3.00	2.92	2.83
18	8.29	6.01	5.09	4.58	4.25	4.01	3.84	3.71	3.60	3.51	3.37	3.23	3.08	3.00	2.92	2.84	2.75
19	8.19	5.93	5.01	4.50	4.17	3.94	3.77	3.63	3.52	3.43	3.30	3.15	3.00	2.92	2.84	2.76	2.67
20	8.10	5.85	4.94	4.43	4.10	3.87	3.70	3.56	3.46	3.37	3.23	30.9	2.94	2.86	2.78	2.69	2.61
21	8.02	5.78	4.87	4.37	4.04	3.81	3.64	3.51	3.40	3.31	3.17	3.03	2.88	2.80	2.72	2.64	2.55
22	7.95	5.72	4.82	4.31	3.99	3.76	3.59	3.45	3.35	3.26	3.12	2.98	2.83	2.75	2.67	2.58	2.50
23	7.88	5.66	4.76	4.26	3.94	3.71	3.54	3.41	3.30	3.21	3.07	2.93	2.78	2.70	2.62	2.54	2.45
24	7.82	5.61	4.72	4.22	3.90	3.67	3.50	3.36	3.26	3.17	3.03	2.89	2.74	2.66	2.58	2.49	2.40
25	7.777	5.57	4.68	4.18	3.86	3.63	3.46	3.32	3.22	3.13	2.99	2.85	2.70	2.62	2.53	2.45	2.36

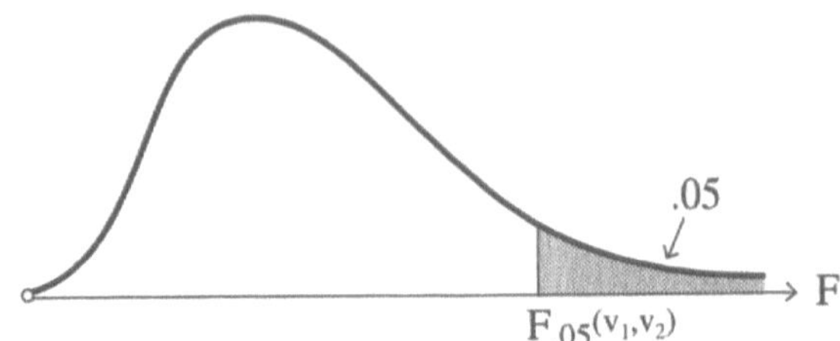

Table D $F_{.05}(\nu_1, \nu_2)$

ν_2 \ ν_1	DEGREES OF FREEDOM FOR NUMERATOR															
	1	2	3	4	5	6	7	8	9	10	12	15	20	24	30	40	60
1	161	200	216	225	230	234	237	239	241	242	244	246	248	249	250	251	252
2	18.5	19.0	19.2	19.2	19.3	19.3	19.4	19.4	19.4	19.4	19.4	19.4	19.4	19.5	19.5	19.5	19.5
3	10.1	9.55	9.28	9.12	9.01	8.94	8.89	8.85	8.81	8.79	8.74	8.70	8.66	8.64	8.62	8.59	8.57
4	7.71	6.94	6.59	6.39	6.26	6.16	6.09	6.04	6.00	5.96	5.91	5.86	5.80	5.77	5.75	5.72	5.69
5	6.61	5.79	5.41	5.19	5.05	4.95	4.88	4.82	4.77	4.74	4.68	4.62	4.56	4.53	4.50	4.46	4.43
6	5.99	5.14	4.76	4.53	4.39	4.28	4.21	4.15	4.10	4.06	4.00	3.94	3.87	3.84	3.81	3.77	3.74
7	5.59	4.74	4.35	4.12	3.97	3.87	3.78	3.73	3.68	3.64	3.57	3.51	3.44	3.41	3.38	3.34	3.30
8	5.32	4.46	4.07	3.84	3.69	3.58	3.50	3.44	3.39	3.35	3.28	3.22	3.15	3.12	3.08	3.04	3.01
9	5.12	4.26	3.86	3.63	3.48	3.37	3.29	3.23	3.18	3.14	3.07	3.01	2.94	2.90	2.86	2.86	2.79
10	4.96	4.10	3.71	3.48	3.33	3.22	3.14	3.07	3.02	2.98	2.91	2.85	2.77	2.74	2.70	2.66	2.62
11	4.84	3.98	3.59	3.36	3.20	3.09	3.01	2.95	2.90	2.85	2.79	2.72	2.65	2.61	2.57	2.53	2.49
12	4.75	3.89	3.49	3.26	3.11	6.00	2.91	2.85	2.80	2.75	2.69	2.62	2.54	2.51	2.47	2.43	2.38
13	4.67	3.81	3.41	3.18	3.03	2.92	2.83	2.77	2.71	2.67	2.60	2.53	2.46	2.42	2.38	2.34	2.30
14	4.60	3.74	3.34	3.11	2.96	2.85	2.76	2.70	2.65	2.60	2.53	2.46	2.39	2.35	2.31	2.27	2.22
15	4.54	3.68	3.29	3.06	2.90	2.79	2.71	2.64	2.59	2.54	2.48	2.40	2.33	2.29	2.25	2.20	2.16
16	4.49	3.63	3.24	3.01	2.85	2.74	2.66	2.59	2.54	2.49	2.42	2.35	.228	2.24	2.19	2.15	2.11
17	4.45	3.59	3.20	2.96	2.81	2.70	2.61	2.55	2.49	2.45	2.38	2.31	2.23	2.19	2.15	2.10	2.06
18	4.41	3.55	3.16	2.93	2.77	2.66	2.58	2.51	2.46	2.41	2.34	2.27	2.19	2.15	2.11	2.06	2.02
19	4.38	3.52	3.13	2.90	2.74	2.63	2.54	2.48	2.42	2.38	2.31	2.23	2.16	2.11	2.07	20.3	1.98
20	4.35	3.49	3.10	2.87	2.71	2.60	2.51	2.45	2.39	2.35	2.28	2.20	2.12	2.08	2.04	1.99	1.95
21	4.32	3.47	3.07	2.84	2.68	2.57	2.49	2.42	2.37	2.32	2.25	2.18	2.10	2.05	2.01	1.96	1.92
22	4.30	3.44	3.05	2.82	2.66	2.55	2.46	2.40	2.34	2.30	2.23	2.15	2.07	2.03	1.98	1.94	1.89
23	4.28	3.42	3.03	2.80	2.64	2.53	2.44	2.37	2.32	2.27	2.20	2.13	2.05	2.01	1.96	1.91	1.86
24	4.26	3.40	3.01	2.78	2.62	2.51	2.42	2.36	2.30	2.25	2.18	2.11	2.03	1.98	1.84	1.89	1.84
25	4.24	3.39	2.99	2.76	2.60	2.49	2.40	2.34	2.28	2.24	2.16	2.09	2.01	1.96	1.92	1.87	1.82

Table E Kolmogorov-Smirnov Test Bounds

Sample	Significance Level α				
Size n	.20	.10	.05	.02	.01
1	.900	.950	.975	.990	.995
2	.684	.776	.842	.900	.929
3	.565	.636	.708	.785	.829
4	.493	.565	.624	.689	.734
5	.447	.509	.563	.627	.669
6	.410	.468	.519	.577	.617
7	.381	.436	.483	.538	.576
8	.358	.410	.454	.507	.542
9	.339	.387	.430	.480	,513
10	.323	.369	.409	.457	,489
11	.308	.352	.391	.437	.468
12	.296	.338	.375	.419	.449
13	.285	.325	.361	.404	.432
14	.275	.314	.349	.390	.418
15	.266	.304	.338	.377	.404
16	.258	.295	.327	.366	.392
17	.250	.286	.318	.355	.381
18	.244	.279	.309	.346	.371
19	.237	.271	.301	.337	.361
20	.232	.265	.294	.329	.352
21	.226	.259	.287	.321	.344
22	.221	.253	.281	.314	.337
23	.216	.247	.275	.307	.330
24	,212	.242	.269	.301	.323
25	.208	.238	.264	.295	.317
26	.204	.233	.259	.290	.311
27	.200	.229	.254	.284	.305
28	.197	.225	.250	.279	.300
29	.193	.221	.246	.275	.295
30	.190	.218	.242	.270	.290
31	.187	.214	.238	.266	.285
32	.184	.211	.234	.262	.281
33	.182	.208	.231	.258	.277
34	.179	.205	.227	.254	.273
35	.177	.202	.224	.251	.269
36	.174	.199	.221	.247	.265
37	.172	.196	.218	.244	.262
38	.170	.194	.215	.241	.258
39	.168	.191	.213	.238	.255
40	.165	.189	.210	.235	.252
Over 40	$\frac{1.07}{\sqrt{n}}$	$\frac{1.22}{\sqrt{n}}$	$\frac{1.36}{\sqrt{n}}$	$\frac{1.52}{\sqrt{n}}$	$\frac{1.63}{\sqrt{n}}$

Table F Lilliefors Test Bounds

Sample Size n	Significance Level α				
	.20	.10	.05	.02	.01
4	.300	.319	.352	.3814	.417
5	.285	.299	.315	.337	.405
6	.265	.277	.294	.319	.364
7	.247	.258	.276	.300	.348
8	.233	.244	.261	.285	.331
9	.223	.233	.249	.271	.311
10	.215	.224	.239	.258	.294
11	.206	.217	.230	.249	.284
12	.199	.212	.223	.242	.275
13	.190	.202	.214	.234	.268
14	.183	.194	.207	.227	.261
15	.177	.187	.201	.220	.257
16	.173	.182	.195	.213	.250
17	.169	.177	.189	.206	.245
18	.166	.173	.184	.200	.239
19	.163	.169	.179	.195	.235
20	.160	.166	.174	.190	.231
25	.142	.147	.157	.173	.200
30	.131	.136	.144	.161	.187
Over 30	$\frac{.736}{\sqrt{n}}$	$\frac{.768}{\sqrt{n}}$	$\frac{.805}{\sqrt{n}}$	$\frac{.886}{\sqrt{n}}$	$\frac{.1031}{\sqrt{n}}$

Table G Random Numbers

05 90 35 89 95	01 61 1696 94	50 78 13 69 36	37 68 53 37 31	71 25 35 03 71	04 31 17 21 56	33 73 99 19 87
44 43 80 69 98	46 68 05 14 85	90 78 50 05 62	77 79 13 57 44	59 60 10 39 66	61 06 98 03 91	87 14 77 43 96
61 81 31 96 82	00 57 25 60 59	46 72 60 18 77	55 66 12 62 11	08 99 55 64 57	86 93 85 56 88	72 87 08 62 40
42 88 07 10 05	24 98 65 53 21	47 21 61 88 32	27 80 30 21 60	10 92 35 36 12	21 74 32 47 45	73 96 07 94 52
77 94 30 05 39	28 10 99 00 27	12 73 73 99 12	49 99 57 94 82	96 88 57 17 91	15 69 53 82 80	79 96 23 53 10
78 83 19 76 16	94 11 68 84 26	23 54 20 86 85	23 86 66 99 07	36 37 34 92 09	02 89 08 04 49	20 21 14 68 86
87 76 59 61 81	43 63 64 61 61	65 76 36 95 90	18 48 27 45 68	27 23 65 30 72	87 18 15 89 79	85 43 01 72 73
91 43 05 96 47	55 78 99 95 24	37 55 85 78 78	01 48 41 19 10	35 19 54 07 73	98 83 71 94 22	59 97 50 99 52
84 97 77 72 73	09 62 06 65 72	87 12 49 03 60	41 15 20 76 27	50 47 02 29 16	10 08 58 21 66	72 68 49 29 31
87 41 60 76 83	44 88 96 07 80	83 05 83 38 96	73 70 66 81 90	30 56 10 48 59	47 90 56 10 08	88 02 84 27 83
22 17 68 65 84	68 95 23 92 35	87 02 22 57 51	61 09 43 95 06	59 24 82 03 47	22 85 61 68 90	49 64 92 85 44
19 36 27 59 46	13 79 93 37 55	39 77 32 77 09	85 52 05 30 62	47 83 51 62 74	67 80 43 79 33	12 83 11 41 16
16 77 23 02 77	09 61 87 25 21	28 06 24 25 93	16 71 13 59 78	23 05 47 47 25	27 62 50 96 72	79 44 61 40 15
78 43 76 71 61	20 44 90 32 64	97 67 63 99 61	46 38 03 93 22	69 81 21 99 21	33 78 80 87 15	38 30 06 38 21
03 28 28 26 08	73 37 32 04 05	69 30 16 09 05	88 69 58 28 99	35 07 44 75 47	13 13 92 66 99	47 24 49 57 74
93 22 53 64 39	07 10 63 76 35	87 03 04 79 88	08 13 13 85 51	55 34 57 72 69	10 27 53 96 23	71 50 54 36 23
78 76 58 54 74	92 38 70 96 92	52 06 79 79 45	82 63 18 27 44	69 66 92 19 09	28 41 50 61 88	64 85 27 20 18
23 68 35 26 00	99 53 9361 28	52 70 05 48 34	56 65 05 61 86	90 92 10 70 80	34 21 42 57 02	59 19 18 97 48
15 39 25 70 99	93 86 52 77 65	15 33 59 05 28	22 87 26 07 47	86 96 98 29 06	61 81 77 23 23	82 82 11 54 08
58 71 96 30 24	18 45 23 34 27	85 13 99 24 44	49 18 09 79 49	74 16 32 23 02	61 15 18 13 54	16 86 20 56 88
57 35 27 33 72	24 53 63 94 09	41 10 76 47 91	44 04 95 49 66	39 60 04 59 81	91 76 21 64 64	44 91 13 32 97
48 50 86 54 48	02 06 34 72 52	82 21 15 65 20	33 29 94 71 11	15 91 29 12 03	00 97 79 08 06	37 30 28 59 85
61 96 48 95 03	07 16 39 33 99	98 56 10 56 79	77 21 30 27 12	90 49 22 23 62	36 46 18 34 94	75 20 80 27 77
36 93 89 41 26	29 70 83 63 51	99 74 20 52 36	87 09 41 15 09	98 60 16 03 03	88 98 99 60 50	65 95 79 42 94
18 87 00 42 31	57 90 12 02 07	23 47 37 17 31	54 08 01 88 63	39 41 88 92 10	04 37 5987 21	05 02 03 24 17
88 56 53 27 59	33 35 72 67 47	77 34 55 45 70	08 18 27 38 90	16 95 86 70 75	63 62 06 34 41	94 21 78 55 09
09 72 95 84 29	49 41 31 06 70	42 38 06 45 18	54 84 73 31 65	52 53 37 97 15	78 47 23 53 90	34 41 92 45 71
12 96 88 17 31	65 19 69 02 83	60 75 86 90 68	24 64 19 35 51	56 61 87 39 12	87 68 62 15 43	53 14 36 59 25
85 94 57 24 16	92 09 84 38 76	22 00 27 69 85	29 81 94 78 70	21 94 47 90 12	47 60 92 10 77	88 59 53 11 52
38 64 43 59 98	98 77 87 68 07	91 51 67 62 44	40 98 05 93 78	23 32 65 41 18	56 88 87 59 41	65 28 04 67 53
53 44 09 42 72	00 41 86 79 79	68 47 22 00 20	35 55 31 51 51	00 83 63 22 55	02 57 45 86 67	73 43 07 34 48
40 76 66 26 84	57 99 99 90 37	36 63 32 08 58	37 40 13 68 97	87 64 81 07 83	31 54 14 13 17	48 62 11 90 60
02 17 79 18 05	12 59 52 57 02	22 07 90 47 03	28 14 11 30 79	20 69 22 40 98	28 50 16 43 36	28 97 85 58 99
95 17 82 06 53	31 51 10 96 46	92 06 88 07 77	56 11 50 81 69	40 23 72 51 39	63 29 62 66 50	02 63 45 52 38
35 76 22 42 92	96 1183 44 81	34 68 35 48 77	33 42 40 90 60	73 96 53 97 86	45 65 58 26 51	76 96 59 38 72
26 29 13 56 41	85 47 04 66 08	34 72 57 59 13	82 43 80 46 15	38 26 61 70 04	39 65 36 63 70	77 45 85 50 51
77 80 20 75 82	72 82 32 99 90	63 95 73 76 63	89 73 44 99 05	48 67 26 43 18	73 741 98 16 04	29 18 94 51 23
46 40 66 44 52	91 36 74 43 53	30 82 13 54 00	78 45 63 98 35	55 03 36 67 68	72 20 56 20 11	72 65 71 08 86
37 56 08 18 09	77 53 84 46 47	31 91 18 95 58	24 16 74 11 53	44 10 13 85 57	75 17 26 99 76	89 37 20 70 01
61 65 61 68 66	37 27 47 39 19	84 83 80 07 48	53 21 40 96 71	95 06 79 88 54	37 48 60 82 29	91 30 15 39 14
93 43 69 64 07	34 18 04 52 35	56 27 09 24 86	61 85 53 83 45	19 90 70 99 00	68 08 02 80 72	83 71 46 30 49
93 43 69 64 07	34 18 04 53 35	56 27 09 24 83	61 85 53 83 45	19 90 70 99 00	68 08 02 80 72	83 71 46 30 49
21 96 60 12 99	11 20 99 45 18	48 13 93 55 34	18 37 79 49 90	65 97 38 20 45	14 23 98 61 67	70 52 85 01 50
95 20 47 97 97	27 37 83 28 71	00 06 41 41 74	45 89 09 39 84	51 67 11 52 49	49 08 96 21 44	25 27 99 41 28
97 86 21 78 73	10 65 81 92 59	58 76 17 14 97	04 76 62 16 17	17 95 70 45 80	78 37 06 08 43	63 61 62 42 29
69 92 06 34 13	59 71 74 17 32	27 55 10 24 19	23 71 82 13 74	63 52 52 01 41	37 21 34 17 68	68 96 83 23 56

Answers to Selected Exercises

Section 1.5 (page 24)

1. Reliable data can be obtained by keeping careful records on how many partients a doctor operated on and how many survived for a specified period of time. Such data are not well-chosen because they do not reflect how ill the patients were going into surgery.

Section 2.1 (page 37)

2. Differences have meaning. However, there is no meaningful zero. The lowest score is 200 and even that score does not imply to knowledge of the subject.

3. (a) nominal, (b) nominal, (c) ordinal, (d) interval, (e) ratio, (f) ratio, (g) ordinal, (h) ordinal, (i) interval

4. No; SAT scores are interval data and have no meaningful zero. Thus, ratios are not meaningful.

Section 2.2 (page 46)

1. (a) In probability sampling one can objectively estimate the precision of the sampling results.

   (b) No. It depends on circumstances. In a preliminary marketing study, for example, one might want to test market a product on a very specific, chosen group.

6. Random Sampling:

   advantage—completely non biased.

   disadvantage—a component subgroup (such as Administration) might not be included in the sample.

   Stratified Sampling:

   advantage—Guarantees all sectors are represented.

disadvantage—it is biased in the sense that it is designed to include representation of all subgroups.

Quota Sampling:

advantage—Guarantees all sectors represented with proper weights.

Disadvantage—non probability method.

Section 2.3 (page 53)

1. No. Make-up of busiest street not necessarily same as make-up throughout city.

3. Mix up an A, 2, 3, . . . , 10 from a deck of playing cards and let 10 count as zero. Deal off the top card (say, a 7) and go across top row of random numbers (see Table 2) to that column. Then go down column to the first 3 digit numbers between 1 and 500. This gives 326, 088, 020, 267, and 391.

6. Assign a number to each name going down the first column and then the next, etc. Repeat the card selection process of Exercise 3 (say a 3 is picked). Go to column 3 and pick the first 5 appropriate 2-digit numbers between 1 and 24. This gives Eller, Levy, Morgan, O'Brien and Santora.

Section 2.4 (page 56)

1. No. Random sample should include all five schools. This sample applies only to the Business School.

4. Depends on how thoroughly the tickets were mixed. If the first tickets tended to remain on the bottom then Reynolds has a strong point.

Section 2.5 (page 57)

2. No. It does not guarantee that the pitfalls will not occur. Refer to *Literary Digest* situation.

6. Mistakes did not surface in 1932 poll but did in 1936. *Literary Digest* got lucky in the 1932 poll in the sense that all segments of society favored Roosevelt as representing badly needed change.

Section 3.1 (page 91)

1. (a), (c) Table 1

Table 1

Response Times	Frequency	Percentages
4-6	8	22.2
7-9	10	27.8
10-12	10	27.8
13-15	2	5.6
16-18	3	8.3
19-21	3	8.3
	36	100.0

(b) Table 2

Table 2

Class	LB	LCL	UCL	UB	CM	CW
4-6	3.5	4	6	6.5	5	3
7-9	6. 5	7	9	9.5	8	3
10-12	9.5	10	12	12.5	11	3
13-15	12.5	13	15	15.5	14	3
16-18	15.5	16	18	18.5	17	3
19-21	18.5	19	21	21.5	20	3

Class width = 3

(d) Table 3

Table 3

Response Times	Cumulative Frequency	Percentages
< 4	0	0
< 7	8	22.2
< 10	18	50.0
< 13	28	77.8
< 16	30	83.4
< 19	33	91.7
< 22	36	100.0

(e) 0| 9, 6, 8, 8, 5, 5, 8, 8, 7, 8, 9, 5, 7, 6, 9, 4, 6, 4
1| 2, 3, 7, 0, 5, 0, 0, 1, 9, 2, 6, 1, 0, 0, 8, 0
2| 0, 1

(f) Most data between 4 and 12, average about 10, not symmetrical.

(g) No. Overlapping intervals. (h) No. Does not allow for 4.

(i) Yes. No overlapping, accounts for all data. Advantage—only two classes. Disadvantage—loss of too much information.

(j) Yes. No overlapping, accounts for all data. Advantage—fewer classes. Disadvantage—unequal class widths.

6. (a) Yes

(b) No. Do not know how many data equal 50.

(c) No. Do not know how many data equal 50. (d) Yes

Section 3.2 (page 99)

1. (a)

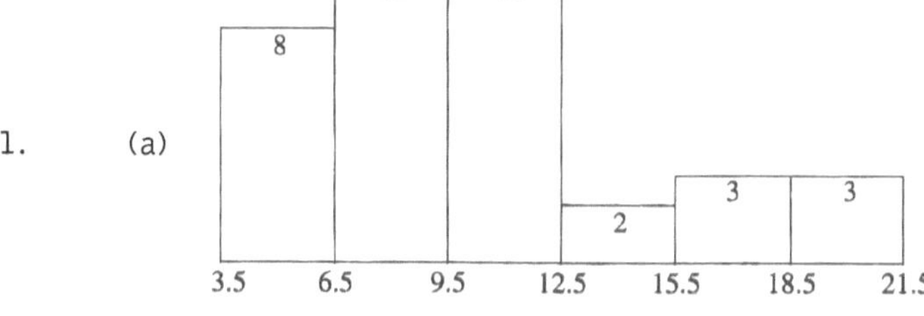

Figure 1

(b)

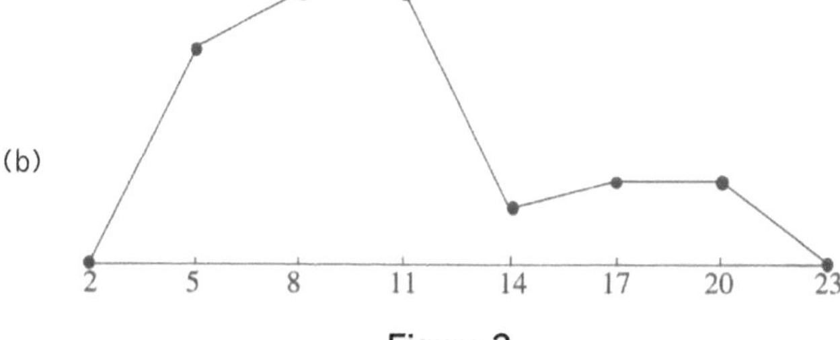

Figure 2

(c) 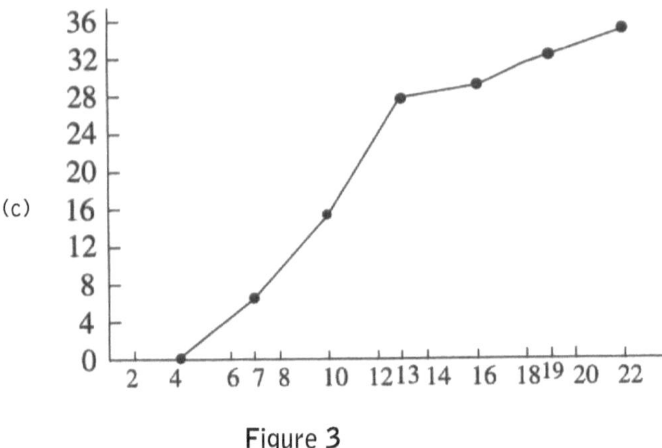

Figure 3

Section 3.3 (page 102)

1. Accuracy of scale can be misleading, i.e., the rectangle for class 1.0-1.49 is not four times higher than for class 4.0-4.49.

Section 4.3 (page 114)

1. (a) 53.3 (b) 51.5 (c) 51

4. (a) (i) 80.4 (ii) 85 (iii) 85, 93

   (b) To Prof. Martin it is the population of all the midterm grades in her accounting class. To the department it is a sample of midterm grades in accounting.

7. The statement must be true by the definition of median.

9. $43.75 10. (a) 5.826 (b) 6.12%

10. (a) No, because the grades are concentrated in two clusters, very low and very high.

    (b) The median grade, 55, tells us that half the grades are less than or equal to 55 and half are greater than or equal to 55, which gives us a better overall picture of the grade distribution.

12. (a) 5.83 (b) 6.12%

Section 4.4 (page 122)

1. 3.43

4. (a) 6 (b) 53

7. (a) $s = 6.7$ (b) $s = 6.7$

Section 5.2 (page 127)

1. (a) $\bar{x} = 10$ (b) $\bar{x} = 10$

(b) Data within each class are, to a close approximation, symmetrically spread about the class mark.

(c) $\bar{x}$ from data equals 10 (e) 10

4. (a) $\bar{x} = 37.77$ (b) Data within each class are, to a close approximation, symmetrically spread about the class mark.

(page 136)

9. (a) $Q_1 = 6.8$ (ii) $D_7 = 11.7$ (iii) $M = 9.5$

(b) (i) 25% of the data are less than or equal to 6.8 and 75% are greater than or equal to 6.8.

(ii) 70% of the data are ≤11.7 and 30% are ≥11.7.

(iii) 50% of the data are ≤9.5 and 50% are ≥9.5.

(c) Data, to a close approximation, are uniformly spread out within the class of Q_1 and D_7.

8. (a) $Q_1 = 55.8$ (ii) $M = 67.4$ (iii) $D_8 = 81.6$

(b) (i) 25% of the data are less than or equal to 55.8 and 75% are greater than or equal to 55.8.

(ii) 50% of the data are less than or equal to 67.4 and 50% are greater than or equal to 67.4.

(iii) 80% of the data are less than or equal to 81.6 and 20% are greater than or equal to .81.6.

(c) Data are, to a close approximation, uniformly spread out within the classes of Q_1, M, and D_8.

(d) His grade was higher than 95% of the grades and 5% were higher than his.

(e) Since $\bar{x} = 67$ is close to $M = 67.4$, the distribution is approximately symmetrical.

Section 5.3 (page 140)

1. (a) $s^2 = 20.76$ (b) $s = 4.6$

(c) Data are approximately symmetrically distributed about the class mark within each class.

(d) The assumption in (c) is met. (e) Same: $s(\text{data}) = 4.6$

(f) $s^2 = 20.24$; $s = 4.5$

Section 6.1 (page 155)

3. $S_1 = \{c_1, c_2, \ldots, c_{52})$
$S_2 = \{\text{club, diamond, heart, spade}\}$
$S_3 = \{\text{red, black}\}$

Section 6.3 (page 162)

1. (a) 3/4 (b) In the long run, even appears approximately 75% of the time.

   (c) No. 3/4 was obtained by performing correct calculations. It is a valid conclusion for this reason.

   (d) No. Repeated trials should have shown about 750 even numbers.

   (e) No. Does not reflect relative frequency results.

Section 6.4 (page 165)

1. No. It's not that the probability of A will be 0.96 over the long run. The probability of A is 0.96, period. The relative frequency interpretation is that if the process is repeated over and over, event A will show with an approximate relative frequency of 0.96, or put another way, event A will show approximately 96% of the time.

Section 6.5 (page 167)

1. $5 \cdot 4 = 20$ 5. $5 \cdot 4 \cdot 3 \cdot 2 \cdot 1 = 120$ 6. $2^8 = 256$

(page 172)

10. $P(60,4)$ 14. $C(20,4)$

(page 175)

18. (a) $\dfrac{C(2,1)C(98,2)}{C(100,3)}$ (b) $\dfrac{C(2,2)C(98,1)}{C(100,3)}$

   (c) $\dfrac{C(2,1)C(98,2)+C(2,2)C(98,1)}{C(100,3)}$

Section 6.6 (page 180)

2. (a) Yes; 1/6 = (1/2)(1/3) (b) Yes; 1/18 = (1/3)(1/6)

Section 6.7 (page 187)

1. Assumptions: independent trials and fixed probability of head. Realistic. For a fair coin Bernoulli trial model: $n = 5$, E is the appearance of a head, $p = 0.5$, $q = 0.5$.

    (a) $P(2 \text{ heads}) = C(5,2)(0.5)^2(0.5)^3$

    (b) $P(3 \text{ or more heads}) = C(5,3)(0.5)^3(0.5)^2 + C(5,4)(0.5)^4(0.5)^1 + C(5,5)(0.5)^5(0.5)^0$

4. Assumptions: independent trials and fixed probability of correct answer. Realistic.

    Bernoulli trial model: $n = 10$, E is getting a correct answer, $p = 0.25$, $q = 0.75$

    $$P(\text{pass}) = C(10,6)(0.25)^6(0.75)^4 + C(10,7)(0.25)^7(0.75)^3 + \cdots + C(10,10)(0.25)^{10}(0.75)^0$$

Section 6.8 (page 191)

4. Subjective. The atomic bomb had never been exploded once, which rules out the relative frequency interpretation. 0.90 is a measure of Bethe's degree of belief that the atomic bomb would explode, which makes the interpretation one of subjective probability.

Section 6.9 (page 192)

1. 9:1 3. (a) 41:9 (b) 49:1 (c) 99:1 (d) 1:9

6. 20/21, 30/31

Section 6.10 (page 192)

1. (a) 4/5 (b) 1/4 (c) 1/5 (d) 2/3

4. (a) 1/3 (b) 1/3 (c) 2/3

6. (a) $S = \{b_1, b_2, \ldots, b_{1000}\}$ $P(b_1) = \ldots = P(b_{1000}) = 1/1000$ (b) 4/7

Section 7.1 (page 198)

1. (a) discrete (b) discrete (c) continuous (d) continuous

4. (a) 0.7 (b) 0.7 (c) 0.5 (d) 0.4

(e) 0.5 (f) 0.6

8. (a) -$15 (b) In the long run the company loses about $15 per man. (c) $30

10. No. Sum of the values is not equal to one.

Section 7.2 (page 203)

1. $\sigma^2 = 179.6$, $\sigma = 13.4$ 3. 3/4

Section 7.3 (page 204)

2. $p(x) = C(5,x)(0.5)^x(0.5)^{5-x}$ $\mu = 2.5,\ \sigma = 1.12$

5. $p(x) = C(10,x)(0.25)^x(0.75)^{10-x}$ $\mu = 2.5,\ \sigma = 1.37$

Section 8.1 (page 213)

1. 0.4913 5. 0.8849 11. 0.3083 13. 0.2426 18. 0.2050

21. 0.3632 22. 0.49 27. 2.58

Section 8.2 (page 220)

1. (a) 0.01 (b) 0.0015 (c) 0.9838 (d) 0.9738 (e) 0.5398

3. Normal approximation for Bernoulli trial model with $n = 10000$, $p = 0.05$, $q = 0.95$, E is an individual getting an allergic reaction.

(a) 0.0102 (b) 0.3300 (c) 0.4833

8. Normal approximation for Bernoulli trial model with $n = 300$, $p = 0.22$, $q = 0.78$, *E* is a viewer watching the program.

(a) 0.5832 (b) 0.9783 (c) 0.0934

9. Bernoulli trial model defined by $n = 10{,}000$, E is the event that a teenager has the measles, $p = 0.002$, $q = 0.99$, Assumption: whether or not one teenager contracts measles or not is independent of whether any other teenager contracts measles or not.

(a) Let X equal the number of successes (number of teenagers who contract measles). $p(x \le 25) = 0.890$ (normal curve estimate)

(b) If a large number of groups of 10,000 teens are selected, in approximately 89% of the cases at most 25 teens will have the measles.

(c) No; since measles is a highly contagious disease, the independence condition is seriously compromised as are the numbers that follow as valid consequences of this assumption.

Section 8.3 (page 224)

1. 0.6170; for many parts produced, approximately 61.7% of the parts will have to be rejected.

(a) 0.1498 (b) 0.3085 (c) 0.1592

Section 8.3 (page 230)

7. (a) 0.3256 (b) 0.5934 (c) 0.0950

(d) In the long run, approximately

32.56% of the errors will be between -0.05 and 0.05

59.34% of the errors will be between -0.42 and 0.42

9.50% of the errors will be greater than or equal to 0.2.

Section 8.4 (page 236)

2. 6.68% − A, 24.17% − B, 38.30% − C, 24.17% − D, 6.68% − F

5. (a) 143 (b) 84

Section 9.2 (page 244)

2. (a) $S = \{(15,6),(5,7),\ldots,(9,10)\}$, $P(\text{each sample}) = \frac{1}{C(6,2)} = \frac{1}{15}$

(b) $\bar{x}$ assigns to each sample of two from S the mean of the sample.

(c) see Table 4

Table 4

$\bar{x}$	5.5	6.0	6.5	7.0	7.5	8.0	8.5	9.0	9.5
$p(\bar{x})$	1/15	1/15	2/15	2/15	3/15	2/15	2/15	1/15	1/15

(d) 7.5 (e) 1.08 (f) (i) 0.4 (ii) 0.867

5. (c) 10 (d) 1.2

Section 9.3 (page 249)

2. (a) 0.2514 (b) 0.4706 (c) 0.8790 (d) 0.2789

4. $x_1 = 4.1$, $x_2 = 15.9$ (b) $x_1 = 3$, $x_2 = 17$

Section 10.2 (page 266)

2. $36.4 < \mu < 39.7$ with probability 0.98.

9. (a) $32,258 < \mu < 37,142$ with probability 0.98

(b) No. 98% chance that C.I. (confidence interval) in (a) will contain μ.

(c) 2% chance that C.I. in (a) will not contain μ.

(d) 1633

(e) Yes. 5% chance that error will be greater than 1633.

(f) 136

14. (a) $24 < \mu < 30$ with probability 0.85.

(b) 85% is a measure of our degree of belief that C.I. in (a) contains μ.

(c) 3.6

(d) Yes. 8% probability that it will. (e) 166

Section 10.3 (page 278)

1. (a) 1.341 (b) 2.896 (c) 2.045 (d) 4.032

5. (a) $12.6 < \mu < 15.6$ with probability 0.95.

(b) 95% is a measure of our degree of belief that C.I. in (a) contains μ.

(c) 1.8

(d) 98% is a measure of our degree of belief that error will not exceed 1.8.

7. (a) $90 < \mu < 101$ with probability 0.98.

(b) 0.98 is a measure of our degree of belief that C.I. in (a) contains μ.

(c) 6.1

(d) 0.99 is a measure of our degree of belief that the error does not exceed 6.1 if μ is estimated as 93.3 pounds.

(e) The normality, or approximate normality, of breaking strength values.

(f) The reliability of the result is seriously open to question because the major string attached has been broken.

Section 10.4 (page 286)

2. The budget of $150,000 allows for a test of at most 13 cars, which is a small sample. For a small sample test to be feasible Mr. Jackson should be concerned about the nature of the population of mileage values before transmission overhauls are required. If this population is approximately normal, then a small sample test is feasible; if it is highly skewed or bimodal, then a major string attached required for a small sample test is not in place.

   As to the trade-offs between sample size, reliability, and precision, Mr. Jackson should reflect on the discussion of this matter in Section 10.4.

Section 10.5 (page 295)

1. (a) Sufficiently large sample $(n \geq 200)$ and np and nq both ≥ 5. Yes.

   (b) $0.0013 < \pi < 0.0187$ with probability 0.95.

   (c) 95% is a measure of our degree of belief that C.I. in (b) contains π.

   (d) 0.01

   (e) 98% is a measure of our degree of belief that error will be at most 0.01 if π is estimated as 0.01.

6. Disagree. There are more than two medicines. The results of this section only deal with two-sided alternatives.

Section 10.7 (page 303)

1. (a) 20.4831 (b) 30.5779 (c) 3.57056 (d) 0.989265

Section 10.8 (page 305)

1. (b) (i) (0.29711, 13.2767 (ii) (3.05347, 24.7250)

(iii) (4.66043, 29.1413) (iv) (8.89720, 38.9321)

4. $119.9 < \sigma < 470.9$

7. (a) $31.6 < \sigma < 470.9$ with probability 0.90.

(b) 90% is a numerical measure of our degree of belief that C.I. in (a) contains σ.

Section 11.1 (page 313)

1. $H_a :> 0.6$ from consumer's point of view: $H_a : \mu \pm 6$ from Happy Light's point of view.

6. $H_o : \mu = 1400$; $H_a : \mu < 1400$; H_a is false only if $\bar{x}$ is significantly less than 1400.

10. $H_o : \mu = 1{,}000{,}000$; $H_a : \mu \neq 1{,}000{,}000$; H_o is false if $\bar{x}$ is significantly less than or greater than 1,000,000.

Section 11.2 (page 321)

1. $z = \dfrac{\bar{x} - 300}{\sigma / \sqrt{n}}$

$\alpha = 0.05$

see Fig. 3

Accept H_o/r.j. if $-1.96 \leq z \leq 1.96$
Reject H_o if $z < 1.96$ or $z > 1.96$

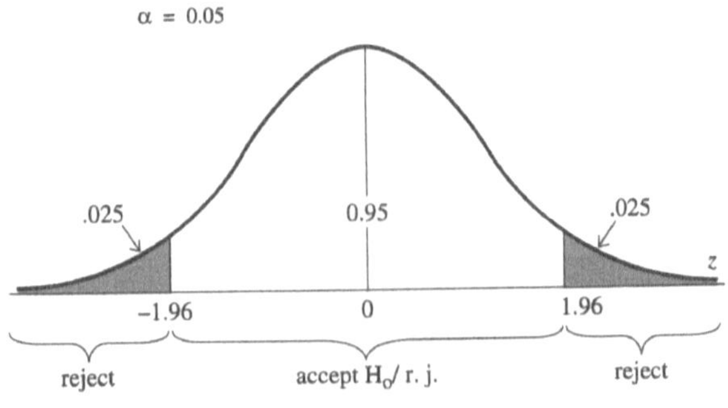

Figure 4

2. $z = \frac{\overline{x} - 100}{\sigma / \sqrt{64}}$

$\alpha = 0.05.$

see Fig. 5

Accept H_o/r.j. if $z \le 1.65$
Reject H_o if $z > 1.96$

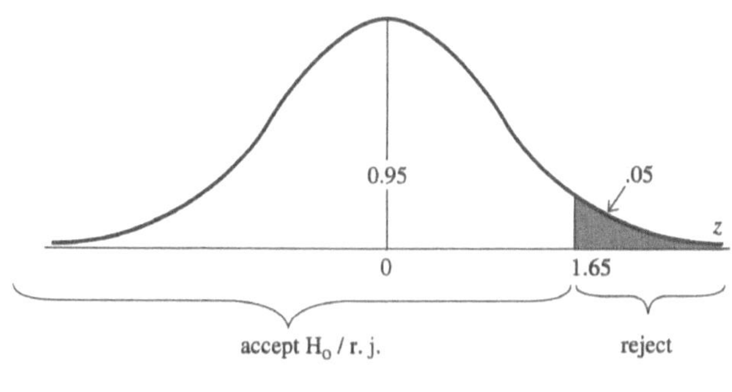

Figure 5

Section 11.2 (page 325)

5. $z = \frac{\overline{x} - 500}{s / \sqrt{60}}$

Reject: $z > 1.96$ or < -1.96
Accept: otherwise

9. $t = \frac{\overline{x} - 30}{s / \sqrt{8}}$

Reject: $t > 1.895$
Accept: otherwise

13. $H_o : \mu = 25$
$H_a : \mu < 25$

$t = \frac{\overline{x} - 25}{s / \sqrt{15}}$

Reject: $t < -2.624$
Accept: otherwise

17. $H_o : \mu = 600$
$H_a : \mu > 600$

$$t = \frac{\overline{x} - 600}{s/\sqrt{9}}$$

Reject: $t > 2.896$
Accept: otherwise

21. (a) $H_o : \mu = 1{,}000{,}000$; $H_a : \mu \neq 1{,}000{,}000$

(b) H_o is false if μ is significantly greater than or less than 1,000,000.

(c) $t = -1.45 > -3.182$; accept/reserve judgment.

(d) Normality or approximate normality of population of age values.

(e) Assuming normality, or approximate normality of the population of age values, the results are reliable.

Section 11.3 (page 334)

3. (a) 0.6170 (b) 0.0214

Section 11.4 (page 337)

2. (a)

μ	β	$1-\beta$
300 ± 0	0.0000	1.0000
$300 \pm .1$	0.8576	0.1424
$300 \pm .2$	0.5832	0.4168
$300 \pm .3$	0.2546	0.7454
$300 \pm .4$	0.0618	0.9382
$300 \pm .5$	0.0080	0.9920

(b) See Figure 6

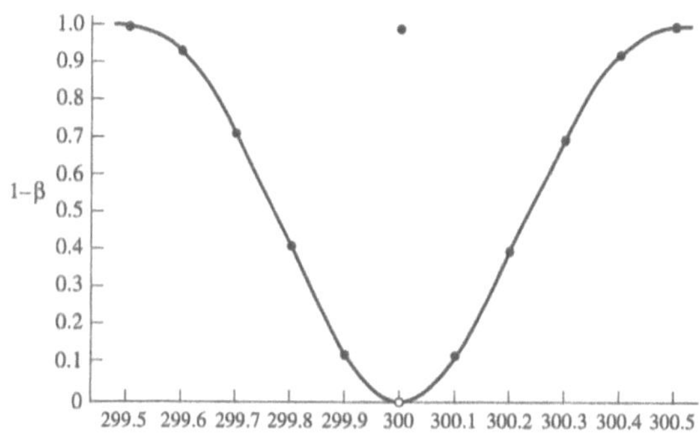

(c) See Figure 7

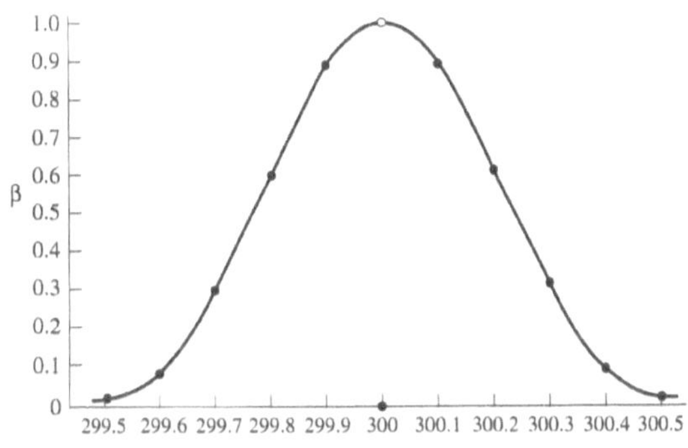

Figure 7

Section 11.5 (page 339)

1. (a) $H_o : \pi = 0.75$; $H_a : \pi < 0.75$

   (b) H_a reflects incorrect claim.

   (c) Reject: $z < -2.05$

   (d) $z = 0.69 > -2.05$; accept/reserve judgment.

   (e) Old medication abandoned.

   (f) Old medication better than new, yet it would be abandoned.

5. (a) $H_o : \pi = 0.75$; $H_a : \pi \neq 0.75$ (b) H_a contradicts H_o.

(c) Reject: if $t < 2.492$; Accept or reserve judgment otherwise.

(d) $t = -1.72 > -2.492$; accept/reserve judgment.

(e) No. It just supports the theory.

Section 11.6 (page 344)

1. (a) $H_o : \sigma = 0.85$; $H_a : \sigma > 0.85$ (b) H_a contradicts H_o.

(c) Reject: $\chi^2 > 26.119$; Accept otherwise.

(d) Population be normally, or approximately normally, distributed.

(e) $\chi^2 = 23.45 < 26.119$; cannot reject claim.

(f) Their claim has not been disproven.

Section 11.8 (page 359)

2. (a) Reject $z > 2.05$; Accept otherwise

(b) $z = 2.75 > 2.05$; reject H_o.

(c) Wells City's mean of tire lifetimes is greater than Jason City's.

5. (a) $H_o : \mu_1 - \mu_2 = 0$; $H_a : \mu_1 - \mu_2 < 0$

(b) H_a disproves the claim.

(c) Reject: $t < -2.492$; Accept otherwise

(d) $t = -1.72 > -2.492$; accept/reserve judgment.

(e) Their claim is not disproven.

(f) Yes. Small samples require normality or approximate normality of the populations; variances must be equal.

9. $H_o: \mu_1 - \mu_2 = 0$; $H_a: \mu_1 - \mu_2 > 0$

(b) Normality or approximate normality of the populations and equality of variances.

(c) $t = 0.2 < 1.812$; accept/reserve judgment.

(d) Neverdie's claim is not proven

Section 11.9 (page 372)

3. (a) $H_o: \mu_d = 0$; $H_a: \mu_d > 0$

(b) Normality or approximate normality of underlying populations.

(c) $t = 36.13 > 2.015$; reject H_o.

(d) Neverdie claim is proven.

(e) Paired differences sensitive to slight differences in data.

(f) Paired differences. Takes into account that Neverdie always did better than Longlife.

7. (a) Yes. Normality or approximate normality of sales figures.

(b) $H_o: \mu_d = 20$; $H_a: \mu_d > 20$.

(c) $t = -9.9 > -2.998$; reject H_o.

Section 11.10 (page 379)

1. (a) 9.15 (b) 15.2 (c) 3.37 (d) 4.41

(e) 0.0658 (f) 0.1093 (g) 0.2268

Section 11.11 (page 382)

2. (a) $F = 0.239 < 0.372$; reject H_o

(b) Normality or approximately normality of populations of battery lifetimes.

6. (a) $0.249 < F = 1.97 < 3.67$; accept/reserve judgment

(b) Yes. Violates assumption of population normality.

(c) For small samples we must have population normality.

10. (a) $H_o : \sigma_k^2 = \sigma_k^2;\ H_a : \sigma_k^2 > \sigma_r^2$.

(b) $F = 3.57 > 2.27$; reject H_o.

(c) Improve quality at Riverside plant.

Section 11.12 (page 392)

1. (a) 0.0793 (b) 0.1586 (c) 0.0294 (d) 0.0668 (e) 0.1336

(f) (i) 0.0152 (ii) reject H_o (g) (i) 0.0068 (ii) reject H_o

Section 11.13 (page 395)

1. (a) Accept/reserve judgment: $\bar{x} - 1.65(\sigma/\sqrt{n}) \le \mu_o \le \bar{x} + 1.65(\sigma/\sqrt{n})$. Reject otherwise.

(b) Accept/reserve judgment: $\bar{x} - 1.96(\sigma/\sqrt{n}) \le \mu_o \le \bar{x} + 1.96(\sigma/\sqrt{n})$. Reject otherwise.

(c) μ falling in the C.I. is equivalent to *z* falling in the acceptance/reserve judgment region.

Section 11.14 (page 398)

1. (a) A robust hypothesis test is one that remains applicable when its strings attached conditions are not fully attached.

(b) Determines whether test can be applied when assumptions are only approximately met.

Section 12.2 (page 415)

1. Strong direct linear relationship 4. Mild direct linear relationship

8. Strong direct linear relationship

Section 12.3 (page 423)

1. (a) $\hat{y} = -1.33 + 0.039x$ (b) 5.7

4. (a) $\hat{y} = 114.5 + 0.537x$ (b) 142

Section 12.4 (page 427)

1. (a) Yes—90 is within the range of observed x's.

   (b) No, 475 is outside this range.

Section 12.5 (page 434)

3. (a) $-3.27 < \alpha < 0.62$ $0.03 < \beta < 0.05$

   (b) 0.95 is a measure of our degree of certainty that α and β are in the C.I.'s constructed for them.

6. (a) $100.7 < \alpha < 128.3$ $0.23 < \beta < 0.85$

   (b) No. 99% is a measure of our degree of belief that they will not.

Section 13.1 (page 439)

1. (a) $\hat{y} = -1.14 + 2.235x$ (c) 45.3   (d) 42.5

   (e) 2.9   (f) 0.94   (g) large extent

5. (a) A linear fit is appropriate.   (b) $\hat{y} = -1.33 + 0.039x$ (c) 5%

12. (a) 472   (b) 354   (c) 118   (d) 25%

Section 13.2 (page 450)

1. (a) 0.97   (b) Strong direct relationship.

5. (a) -0.97   (b) Strong inverse relationship.

10. (a) $r = 0$ (b) No. Not a linear relationship   (c) $y = -x^2 + 1$

Section 13.4 (page 456)

1. (a) $t = 12.24 > 4.604$. Reject H_o.

   (b) Normality or approximate normality of x and y populations.

   (c) Yes. $r = 0.99$ means there is a strong linear relationship between x and y.

Section 14.1 (page 479)

3. New Orleans: 114 Omaha: 93

5. Billingstown: 114 New City: 90

Section 14.2 (page 486)

1. (a) 121 (b) 122 (c) (i) 150 (ii) 122.

   On average, the prices of the items are up 49.8% in 2002 over 1992 and up 22.2% in 2002 over 1997.

6. (a) (i) 106 (ii) 114 (b) 108

Section 14.3 (page 496)

1. (a) (i) 178 (ii) 116 (b) 190 (c) 201

   (d) 183 (e) 183

Section 14.7 (page 506)

1. (a) 106 (b) $80.19 (c) Yes; by $0.19.

Section 15.3 (page 525)

1. (a) 6.43, 6.57, 6.63, 6.57, 6.43, 6.4, 6.5, 6.67, 6.8, 6.83

   (b) 6.52, 6.54, 6.52, 6.50, 6.50, 6.54, 6.64, 6.74

(c) See Figure 7.

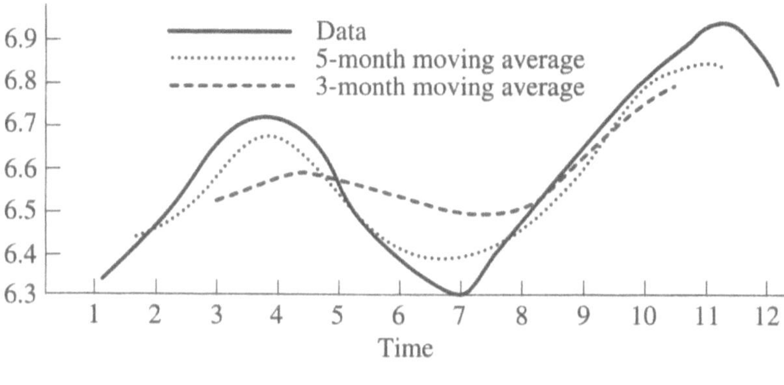

Figure 7

The five-month moving average has the lowest peaks, the highest troughs and is, in general, a smoother graph. In general, the greater the number of periods used to form the sequence of moving averages, the smoother the graph.

(d) (i) 0.02 (ii) 0.10

(e) The three-month moving average gives results closer to those of the data since less smoothing is used as compared to the five-month moving average, which is designed to focus more on the trend and less on fluctuations in the data due to seasonal, cyclical and random variation.

(f) Three-month—6.83, Five-month—6.74

6. (a) 68, 69.7, 72.3, 74.7, 75.7, 73.3, 68.7, 68

(b) 69.8, 72.6, 73.8, 73, 72, 71.2

(c) Three-year: 68,000, Five-year: 71,200

Section 15.4 (page 534)

2. (a) $S_1 = 2.06$, $S_2 = 2.02$, $S_3 = 1.92$, $S_4 = 1.97$, $S_5 = 2.01$, $S_6 = 2.07$
$S_7 = 2.00$, $S_8 = 2.00$, $S_9 = 2.04$, $S_{10} = 2.10$.

(b) $S_1 = 2.06$, $S_2 = 1.99$, $S_3 = 1.85$, $S_4 = 1.98$, $S_5 = 2.03$, $S_6 = 2.10$
$S_7 = 1.96$, $S_8 = 1.99$, $S_9 = 2.05$, $S_{10} = 2.13$.

(c) See Figure 8.

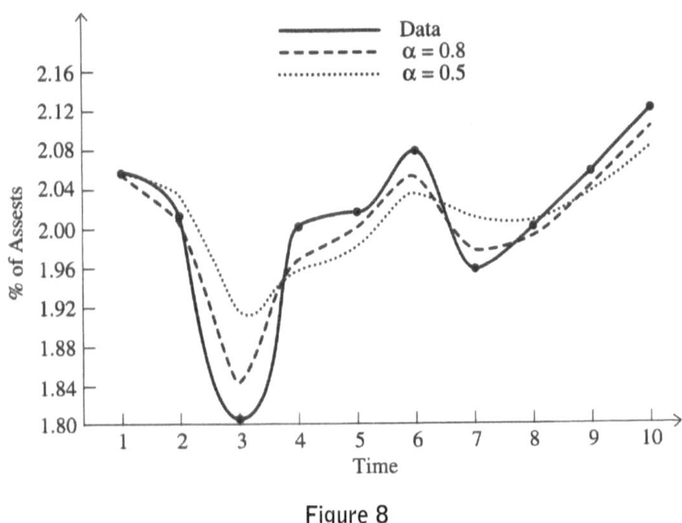

Figure 8

The exponentially smoothed data with $\alpha = 0.5$ is smoothest. In general, as α decreases, the resulting modified data becomes smoother.

(d) (i) \$2.13 (ii) \$2.15

Section 15.5 (page 540)

4. (a) $T_1 = 733.5 + 13.20t$

(b) (i) $T_{21} = 1010.7$, $T_{22} = 1023.9$,

$T_{23} = 1037.1$, $T_{24} = 1050.3$.

(ii) $T_{33} = 1169.10$, $T_{34} = 1182.3$,

$T_{35} = 1195.50$, $T_{36} = 1208.7$.

6. (a) $T_t = 40 + 0.56t$

(b) Upward. Yes. The trend line has positive slope.

(c) (i) $T_{37} = 60.7$, $T_{38} = 61.3$, $T_{39} = 61.8$,

$T_{40} = 62.4$, $T_{41} = 63$, $T_{42} = 63.5$,

(ii) $T_{43} = 64.1$, $T_{44} = 64.6$, $T_{45} = 65.2$,

$T_{46} = 65.8$, $T_{47} = 66.3$, $T_{48} = 66.9$,

Section 15.6 (page 549)

1. (a) $S_W = 1.080$, $S_{S_r} = 0.967$, $S_{S_u} = 0.921$, $S_F = 1.033$

(b) We eliminate the seasonal component in constructing the centered moving average.

3. (a) A season is each of the six consecutive two month periods of any year. A seasonal cycle is one year.

(b) $S_1 = 0.943$, $S_2 = 0.944$, $S_3 = 1.039$, $S_4 = 0.891$, $S_5 = 1.037$, $S_6 = 1.145$.

(c) We eliminated the seasonal component.

Section 15.7 (page 554)

1. (a) $\hat{y}_9 = 848$, $\hat{y}_{10} = 872$, $\hat{y}_{11} = 891$, $\hat{y}_{12} = 926$

(b) Not quite. The largest deseasonalized data value does occur in the fall but the smallest occurs in the winter rather than the summer. We would not expect the largest or smallest deseasonalized values to necessarily occur in the same time period as do the corresponding data values. This is because seasonally adjusted data take into account the expected fluctuations that are caused by the seasonal effects.

3. (a) $\hat{y}_7 = 44.54$, $\hat{y}_8 = 46.61$, $\hat{y}_9 = 45.24$, $\hat{y}_{10} = 47.14$, $\hat{y}_{11} = 46.29$, $\hat{y}_{12} = 46.29$.

(b) Not really. The largest data value occurs in November/December while the smallest data value occurs in both the January/February and July/August periods. In contrast, the largest and smallest deseasonalized values occur in July/August and January/February, respectively. We do not expect the peaks of the two data sets to occur at the same time, nor do we anticipate this for the troughs. Seasonal effects are accounted for in the deseasonalized data, which may alter the high and low points of such modified data.

Section 15.8 (page 559)

1. $C_1 = 1.010$, $C_2 = 0.978$, $C_3 = 1.007$, $C_4 = 1.007$

3. $C_1 = 1.008$, $C_2 = 1.014$, $C_3 = 0.977$

Section 15.9 (page 563)

1. (a) $\hat{y}_{17} = 1045$, $\hat{y}_{18} = 948$, $\hat{y}_{19} = 916$, $\hat{y}_{20} = 1041$,

   (b) $\hat{y}_{37} = 1291$, $\hat{y}_{38} = 1168$, $\hat{y}_{39} = 1124$, $\hat{y}_{40} = 1274$,

   (c) Assumptions include: (i) linearity of trend, (ii) use of a single cyclical index per year rather than per quarter, (iii) negligible error component, (iv) unchanged conditions.

3. (a) $\hat{y}_{31} = 52.8$, $\hat{y}_{32} = 53.3$, $\hat{y}_{33} = 59.3$,

   $\hat{y}_{34} = 51.3$, $\hat{y}_{35} = 60.3$, $\hat{y}_{36} = 67.2$

   (b) $\hat{y}_{43} = 61.2$, $\hat{y}_{44} = 61.7$, $\hat{y}_{45} = 68.6$,

   $\hat{y}_{46} = 59.3$, $\hat{y}_{47} = 69.6$, $\hat{y}_{48} = 77.5$

   (c) See 1. (c)

Section 16.2 (page 573)

2. (a) $z = -1.70 < -1.65$ Reject H_o (b) mean = median

4. $z=-1.60>-1.96$ Accept/reserve judgment

Section 16.3 (page 579)

1. $z=1.60<1.96$ Accept/reserve judgment

5. $z=3.50>2.33$ Reject H_o

Section 16.4 (page 587)

1. (a) $D=0.1271<0.337$ Accept/reserve judgment

   (b) Since each population of corn yields is "shown" to be normal, a hypothesis test comparing mean yields can be made.

3. (a) Accept/reserve judgment: $D=0.1736<0.285$

   (b) Since the underlying population is "shown" to be normal, we can construct 90% confidence limits for chlorine level.

4. (a) $D=0.2105<0.319$ Accept/reserve judgment

   (b) Since the assumption of normality is satisfied, we can test the H_o concerning the population standard deviation.

Section 16.5 (page 600)

1. (a) +1 for case I, -1 for case II since the rankings agree in case I and are diametrically opposite in case II.

   (b) $r_s=1$ in case I, $r_s=-1$ in case II.

3. (a) $z=-1.78<-1.65$ Reject H_o

   (b) No. The Spearman correlation coefficient does not measure the extent of linearity of the relationship.

6. $z=2.18>1.65$ Reject H_o

10. $z = 2.18 > 1.65$ Reject H_o

Section 16.6 (page 608)

1. $z = -2.07 < -1.96$ Reject H_o

5. $z = -0.93 > -2.33$ Accept/reserve judgment

7. (a) $z = -1.80 < -1.65$ Reject H_o

   (b) On average, men would finish a bit faster than women

Section 16.7 (page 617)

2. $z = -2.48 < -1.65$ Reject H_o

5. $z = -3.10 < -1.65$ Reject H_o

7. $z = -1.88 > -1.96$ Accept/reserve judgment

Section 17.1 (page 629)

1. No $\chi^2 = 7.84 > 7.37776$ Reject H_o

3. That there is a difference. $\chi^2 = 8.7325 > 6.25139$ Reject H_o

Section 17.2 (page 636)

1. Independent. $\chi^2 = 6.54 < 9.48773$ Accept/reserve judgment

6. Yes. $\chi^2 = 1.45 < 16.8119$ Accept/reserve judgment

Section 17.3 (page 648)

1. Yes. $\chi^2 = 5.89 < 9.3484$ Accept/reserve judgment

5. $\chi^2 = 9.93 > 9.3484$ Reject H_o

Answers to Selected Self-Test Questions

■ Self-Tests for Part 1

Self-Tests for Part I (page 141)

1. (a) $\overline{x} = \frac{\sum x}{n} = \frac{650}{25} = 26 \cdot$

(b) $s^2 = \frac{n\sum x^2 - (\sum x)^2}{n(n-1)} = \frac{25(20,770) - (650)^2}{25(24)} = 161,$

$s = \sqrt{161} = 13.$

(c) The assumption is made that the 25 salespersons were a sample of all the salespersons employed by Norden. If it were assumed that the 25 salespersons made up the entire staff, then they could have been considered a population and the variance and standard deviation would have been calculated by dividing by n^2 instead of $n(n-1)$.

(d) See Table 1.

Table 1

0	7, 3, 4
1	0, 5, 8, 4
2	5, 4, 1, 2, 7, 0, 8, 9
3	6, 2, 8, 9, 0, 2
4	7, 2, 2, 5

(e) $(n+1)/2 = (25+1)/2 = 13$. Referring to the stem-and-leaf plot, the 13th number is 27. Thus, the median is 27.

(f) Since the median is greater than the mean, the distribution is slightly skewed to the right.

(g) The mode is the most frequently occurring number, which are 32 and 42. Thus, the distribution is bi-modal with modes 32 and 42.

(h) The range is 47 - 3 = 44.

2. (a), (c). See Tables 2 and 3.

(a

Table 2

Sales	Frequency
0-9	3
10-19	4
20-29	8
30-39	6
40-49	4
	25

(b) Class Limits: 0 and 9; 10 and 19; 20 and 29; 30 and 39; 40 and 49.
Class Marks: 4.5, 14.5, 24.5, 34.5, and 44.5.
Class Widths: 10, 10, 10, 10, 10.

(c) See Table 3.

Table 3

Sales	Frequency	X	$x \cdot f$	$x^2 \cdot f$
0-9	3	4.5	13.5	60.75
10-19	4	14.5	58.0	841.00
20-29	8	24.5	196.0	4,802.00
30-39	6	34.5	207.0	7,141.50
40-49	4	44.5	178.0	7,921.00
			652.5	20,766.25

$$\bar{x} = \frac{652.5}{25} = 26.$$

(d) The distribution of the data within each class is approximately symmetrical about the class mark of the class.

(e) $s^2 = \frac{25(20,776.25)-(652.5)^2}{25(24)} = 155.7$, $s = \sqrt{155.7} = 12.5$.

(f) The distribution of the data within each class is approximately symmetrical about the class mark of the class.

(g) 0.5(25)=12.5. Since 3 + 4 = 7 and 7 + 8 = 15, the median is in the 20-29 class. Thus, $L = 19.5$, $f = 8$, $c = 10$, and $j = 12.5-(3+4) = 5.5$. Therefore,

$$M = 19.5 + 5.5\left(\frac{10}{8}\right) = 26.4 \cdot$$

(h) 0.60(25)=5. Since 3 + 4 + 8 = 15, P_{60} is in the 30—39 class. Thus, $L = 29.5$, $f = 6$, $c = 10$, and $j = 15-15 = 0$. Therefore,

$$P_{60} = 29.5 + 0\left(\frac{10}{6}\right) = 29.5 \cdot$$

(i) The data are uniformly distributed within the M and P_{60} classes.

(j) 60% of the data values are less than 29.5 and 40% are greater.

(k), (m). See Figures 1 and 2.

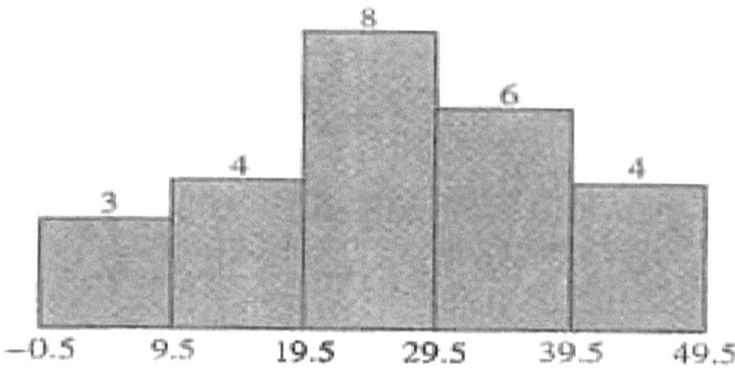

Figure 1

(l) See Table 4.

Table 4

Sales	Cumulative Frequency	Cumulative Percentages
< 0	0	0
< 10	3	12
< 20	7	28
< 30	15	60
< 40	21	84
< 50	25	100

(m) See Figure 2.

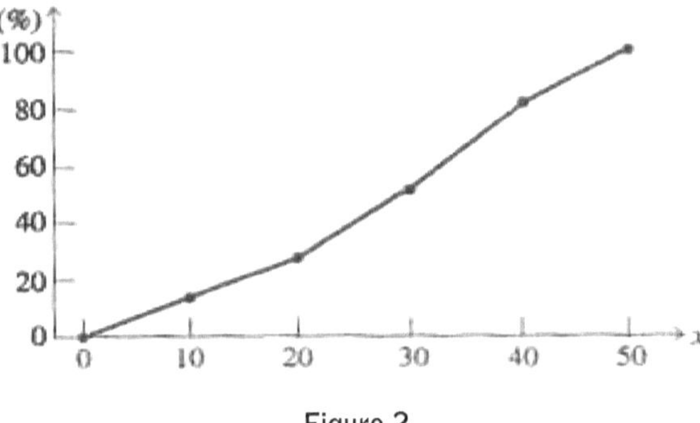

Figure 2

(n) 42%

Self-Test 2 (page 143)

1. (a) No. The data, simply interpreted, might give us a measure of the "efficiency" or "productivity" of each instructor for each class, but this does not translate to a measure of the over-all "efficiency" or "productivity" of the department.

   (b) No. This game of academic musical chairs will not change the department's over-all tuition revenue or costs, on which financial efficiency depends.

2. The statistician must have stepped into a deep spot. An average depth of one foot does not imply that the lake is shallow throughout; the statistician would have been wise to look at some measure of depth variation.

3. $\overline{x}_W = \dfrac{(20)(0.10)+(30)(0.30)}{20+30} = 0.22$ or 22%.

4. (a) Nominal (b) Ratio (c) Nominal (d) Ordinal

(e) Interval (f) Ordinal (g) Ratio (h) Interval

5. (a) In response to 5.(a) we refer the reader to Section 2.2.

(b) In posing questions Andy might want to strike a balance between manageability (which favors short answer questions) and insightfulness (which favors explanatory type questions). Andy might consider questions of the following type.

1. How would you rate your overall educational experience at Huxley? (1) Excellent, (2) Good, (3) Satisfactory, (4) Poor, (5) Very poor.

2. In terms of your major area of study, how would rate your educational experience at Huxley? (1) Excellent, (2) Good, (3) Satisfactory, (4) Poor, (5) Very poor.

3. If you could go back to the time that you were choosing a school, would you still choose Huxley knowing what you now know? (a) Yes, (b) Probably yes, (c) No, (d) Probably not, (e) Not sure.

4. What do you view as Huxley's strong points?

5. What do you view as Huxley's weak points?

6. What has been your most positive educational experience at Huxley thus far?

7. What has been your most negative educational experience at Huxley thus far?

8. How might Huxley improve its educational quality?

9. Would you recommend Huxley to a close friend or relative?
(a) Yes, (b) No, (c) Not sure.

10. Please briefly explain the basis for you answer to the preceding question.

Self-Test 3 (page 145)

1. (a) $\bar{x} = \frac{368}{16} = 23$.

(b) $s^2 = \frac{16(9888) - (368)^2}{16(15)} = 94.93$, $s = 9.7$.

(c) The number of homes sold is a sample of the number sold by a larger sales force of the Harmon Company.

(d)

0	9, 4
1	5, 0, 8
2	5, 4, 1, 2, 8, 8, 8
3	6, 2, 8, 9, 0

(e) $\frac{n+1}{2} = 8.5$

$$M = \frac{24+25}{2} = 24.5$$

2. See Table 5

Table 5

Sales	f_i
0-9	2
10-19	3
20-29	7
30-39	4
	16

(a) (i) 0 and 9, (ii) -0.5 and 9.5, (iii) 4.5, (iv) 10

(b) See Table 6; $\bar{x} = \dfrac{362}{16} = 22.6$

Table 6

Sales	f_i	x_i	$x_i f_i$	$x_i^2 f$
0-9	2	4.5	9.0	40.50
10-19	3	14.5	43.5	630.75
20-29	7	24.5	171.5	4201.75
30-39	4	34.5	138.0	4761.00
	16		362	9634

(c) In each class the data are approximately symmetrically distributed about the class mark of the class.

(d) From Table 6, $s^2 = \dfrac{16(9634) - (362)^2}{16(15)} = 96.25$; $s = 9.8$.

(e) Same as (c).

(f) $M = 19.5 + 10\left(\dfrac{3}{7}\right) = 23.8$

3. *The Literary Digest's* experience with its 1936 pre-presidential poll, which involved a sample size of nearly 10 million people, makes clear that the answer is no. It's not the size of the sample by itself that counts, but how the sample is chosen.

4. No; the average salary value is greatly influenced by the very high salaries (running into millions of dollars) of a small minority of players. The median salary figure gives us a better picture of what player salaries are like.

5. (a) No; the means of the ranking values, 74 and 72, may at best be viewed as consensus values which put the achievements of both nominees between Very Slippery (70) and Superb (80). Since the rankings are ordinal data, differences are not meaningful, and a conclusion that a mean of 74 expresses a higher degree of slipperiness than a mean of 72 is not tenable.

(b) The median ranking; Vera S. has a median ranking of 80, which means that two panelists ranked her achievements as superb or better and two ranked her achievements as superb or less. Joseph S. has a median ranking of 70, which means that two panelists ranked his achievements as very slippery or better and two ranked his achievements as very slippery or less.

Since 80 (Superb) is preferable to 70 (Very Slippery), the SSA Award should go to Vera S.

6. No; number of years driving accident free says nothing about the number of miles driven, which is the more telling figure.

7. The low response rate of 10% to the poll makes the conclusion unreliable. What were the other 90% thinking?

8. What went wrong? In 1932 voters of all economic *strata* tended to vote against President Hoover and the Republicans, holding them responsible for the Great Depression and its aftermath. Economic class differences were obscured, and *The Digest* got lucky in its poll. In 1936 the upper economic *strata*, disturbed by the direction of President Roosevelt's *New Deal*, were much more willing to return to the Republicans, as reflected by the *Digest's* 1936 poll. Spectacular success without an understanding of its origins bred spectacular arrogance followed by spectacular failure.

■ Self-Tests for Part 2

Self-Test 1 (page 250)

1. (a) $S = \{(f_1, f_2, f_3, f_4, f_5, f_6, f_7, f_8), \ldots, (f_{73}, f_{74}, f_{75}, f_{76}, f_{77}, f_{78}, f_{79}, f_{80})\}$
P assigns $1/C(80, 8)$ to each sample point.

(b) The committee is chosen at random from the 80 faculty. There is no bias or favoritism which favors the selection of certain faculty over others.

(c) $\dfrac{C(30,3)C(50,5)}{C(80,8)}$ (d) $\dfrac{C(78,6)}{C(80,8)}$

(e) $\dfrac{C(30,3)C(25,2)C(25,3)}{C(80,8)}$

(f) $\dfrac{C(30,3)C(50,5)}{C(80,8)}+\dfrac{C(25,2)C(55,6)}{C(80,8)}-\dfrac{C(30,3)C(25,2)C(25,3)}{C(80,8)}$

2. (a) Bernoulli trial model defined by $n=10,100$. E is the event a defective item is produced on a trial, $p=0.01$.

(b) The production of a defective or good item on any trial has no effect on the outcome of any other trial.

(c) $P(X \le 105) = P\left(z \le \dfrac{105.5-101}{10}\right) = P(z \le 0.45) = 0.6736$.

(3) (a) $P(A) = 8/13 = 0.615$.

(b) If the die in question is tossed a large number of times, an odd number will show approximately 61.5% of the time.

(c) No. This data is irrelevant to the conclusion's validity. The conclusion's validity follows from $P(1) = P(3) + P(5) = 8/13$.

(d) Yes; 795/1,500 = 0.612 which is "close" to the long run projection of 61.5% from (b).

(e) A: an odd number shows; B: a number greater than 3 shows. No; $P(A \cap B) = 0.23 \ne P(A)\cdot P(B) = (8/13)(7/13) = 0.33$.

4. 0.10 is a numerical measure of an individual's (or group's) degree of belief that event A will occur.

5. $\mu = 625(0.2) = 125$, $\sigma^2 = (625)(0.2)(0.8) = 100$, $\sigma = 10$, $P(124 \le X \le 130) = P(-0.15 \le z \le 0.55) = 0.2684$.

Self-Test 2 (page 251)

1. (a) $S = \{(6,8),(6,10),(6,12),(8,10),(8,12),(10,12)\}$;
P assigns 1/6 to each sample point.

(b) The sample is chosen at random from Q; that is, there is no bias which favors certain samples being chosen over others.

(c) $f(x) = P(\overline{x} = x)$; $f(7) = f(8) = f(10) = f(11) = 1/6$, $f(9) = 2/6$.

(d) $\mu_{\overline{x}} = 7f(7) + \cdots + 11f(11) = 54/6 = 9$.
Alternatively, $\mu_{\overline{x}} = \mu = (6+8+10+12)/4 = 9$.

(e) $\sigma_{\overline{x}}^2 = (7-9)^2(1/6) + \cdots + (11-9)^2(1/6) = 5/3; \sigma_{\overline{x}} = 1.29$.

Alternatively, $\sigma_{\overline{x}} = \dfrac{\sigma}{\sqrt{n}}\sqrt{\dfrac{N-n}{N-1}} = \dfrac{\sqrt{5}}{\sqrt{2}}\sqrt{\dfrac{4-2}{4-1}} = \sqrt{\dfrac{5}{3}} = 1.29$.

(f) $P(8 \le \overline{x} \le 10) = 1/6 + 2/6 + 1/6 = 2/3$.

2. (a) $Z(1) = Z(2) = Z(3) = -3$; $Z(4) = Z(5) = Z(6) = 2$

(b) $\mu_z = (-3)P(Z=-3) + (2)P(Z=2) = -21/12 + 10/12 = -0.917$

(c) If many tosses of Dan's die are made, Gus will lose approximately 92 cents per game.

(d) No; for the game to be fair we must have $\mu_z = 0$.

(e) Require Dan to pay to Gus 91.7 cents per play of the game.

(f) $\sum x_i^2 p(x_i) = 83/12$; thus, $\sigma_Z^2 = 83/12 - (-11/12)^2 = 6.08$; therefore $\sigma_Z = 2.5$.

3. Refer to Section 8.3

4. (a) $P(\text{rejection}) = P(z < -0.5) + P(z > 0.5) = 0.6170 \approx 0.62$.

(b) If D equals desired diameter, then $P\left(z < \dfrac{D-0.42}{0.02}\right) = 0.20$. Also,

$P(z < -0.84) = 0.20$. Therefore $\dfrac{D-0.42}{0.02} = -0.84$, and $D = 0.4032$.

5. $20 \cdot 19 \cdot 18 \cdot 17 = 116{,}280$ 6. $C(44, 6)$

Self-Test 3 (page 253)

1. (a) $P(\text{number greater than 3 shows}) = 1/16 + 2/16 + 3/16 = 3/8$.

(b) If Dan's die is tossed a large number of times, a number greater ??? show approximately 37.5% of the time.

(c) A: A number greater than 2 shows. B: An even number shows.
$P(A \cap B) = 4/16 \neq P(A) \cdot P(B) = (10/16) \cdot (7/16)$; not independent.

(d) $Z(1) = Z(2) = Z(3) = -1$; $Z(4) = Z(5) = Z(6) = 1$.

(e) $\mu_z = (-1)(10/16) + (1)(6/16) = -0.25$.

(f) If the game is played many times, Gus will lose, on average, approximately 25 cents per each play of the game.

(g) No; $\mu_z \neq 0$.

(h) $\Sigma x_i^2 p(x_i) = (-1)^2 \cdot (10/16) + (1)^2 \cdot (6/16) = 1$.
$\sigma_Z^2 = 1 - (-0.25)^2 = 0.9375$; $\sigma_Z = \sqrt{0.9375} = 0.97$.

(i) (1) No; the probability was calculated correctly based on the model. The behavior of the die is not relevant to the validity issue.

(2) Mathematical reasoning is precise in the sense that it yields valid conclusions with respect to the assumptions. What went wrong is that Dan's model was not a realistic description of the behavior of his die. This led to a valid conclusion, but one which was not realistic for Dan's die.

2. (a) Bernoulli trial model defined by $n = 60{,}000$, E is the event that the vaccine is ineffective on the person given it, $p = 0.001$.

(b) The effectiveness or non-effectiveness of the vaccine on any person does not influence the outcome on any other person given the vaccine.

(c) Yes. The effectiveness of the vaccine for any one person has to do with his/her internal biochemistry, which is not relevant to how any others will react to the vaccine.

(d) $P(X \le 70) = C(60,000,0)(0.001)^0(0.999)^{60,000} + \ldots +$
$C(60,000,70)(0.001)^{70}(0.999)^{59,930}$.

(e) Yes; since $np = 60 \ge 5$ and $n(1-p) = 59.940 \ge 5$.

(f) $\mu = 60$; $\sigma = \sqrt{(60,000)(0.001)(0.999)} = 7.74$,

$$P(X \le 70) = P\left(z \le \frac{70.5-60}{7.74}\right) = P(z \le 1.36) = 0.9131$$

3. (a) $S = \{(2,6),(2,10),(2,14),(2,16),(6,10),(6,14),(6,16),(10,14),(10,16),(14,16)\}$
P assigns to each sample point the value $1/C(5,2) = 0.1$.

(b) The sample is chosen without bias, at random, from the population.

(c) $\bar{x}(2,6) = 4$, $\bar{x}(2,10) = 6$, $\bar{x}(2,14) = 8$, $\bar{x}(2,16) = 9$, $\bar{x}(6,10) = 8$, $\bar{x}(6,14) = 10$, $\bar{x}(6,16) = 11$, $\bar{x}(10,14) = 12$, $\bar{x}(10,16) = 13$, $\bar{x}(14,16) = 15$. This yields $p(x) = P(\bar{x} = x)$ defined as follows: $p(4) = p(6) = p(9) = p(10) = p(11) = p(12) = p(13) = p(15) = 0.1$, $p(8) = 0.2$.

(d) $\mu_{\bar{x}} = (0.1)(4) + (0.1)(6) + \cdots + (0.1)(15) = 9.6$. Alternatively, $\mu_{\bar{x}} = \mu = 48/5 = 9.6$.

(e) $\sigma_{\bar{x}}^2 = (0.1)(4-9.6)^2 + (0.1)(6-9.6)^2 + \cdots + (0.1)(15-9.6)^2 = 9.84$. Thus, $\sigma_{\bar{x}} = \sqrt{9.84} = 3.14$. Alternatively, since $\sigma = 5.123$, $\sigma_{\bar{x}} = \frac{5.123}{\sqrt{2}}\sqrt{\frac{3}{4}} = 3.14$.

(f) (i) $P(8 \le \bar{x} \le 11) = 0.5$.

■ Self-Tests for Part 3

Self-Test 1 (page 400)

1. (a) A type I error is committed when you reject a true null hypothesis.

(b) A type II error is committed when you accept a false null hypothesis.

2. $E = 1.96\sqrt{(0.6)(0.4)/300} = 0.06$.

3. $z = \dfrac{\overline{x} - 800}{\sigma / \sqrt{36}}$, $\alpha = 0.05$. See Figure 1 for the accept/reserve judgment and reject regions.

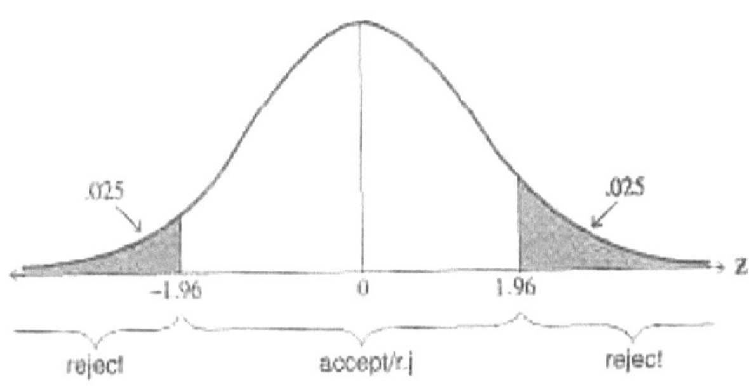

Figure 1

Type I error: H_o is true, $z > 1.96$ or $z < -1.96$, and H_o is rejected. Type II error: H_o is false, $-1.96 \leq z \leq 1.96$, and is accepted. Since the sample size is large $(n \geq 30)$, no conditions need be imposed on the population.

We should keep in mind, however, the caution noted in Section 9.3: If the population is highly non-normal (sharply skewed to the left or right, or bimodal), then the minimal sample size of $n = 30$ may not be large enough to yield a "good" normal curve approximation for the sampling distribution of $\overline{x}$.

4. $t = \dfrac{\overline{x} - 60}{s / \sqrt{10}}$, $\alpha = 0.05$, $\text{d.f} = 9$, see Figure 2.

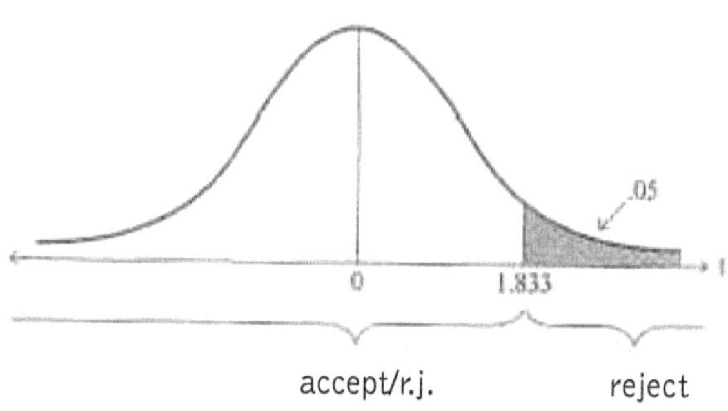

Figure 2

Type I error: H_o is true, $t > 1.833$ and H_o is rejected. Type II error: H_o is false, $t \leq 1.833$, and H_o is accepted. The population must be normally distributed or approximately so.

5. $z = \dfrac{\dfrac{x}{300} - 0.40}{0.028}$, $\alpha = 0.01$, see Figure 3.

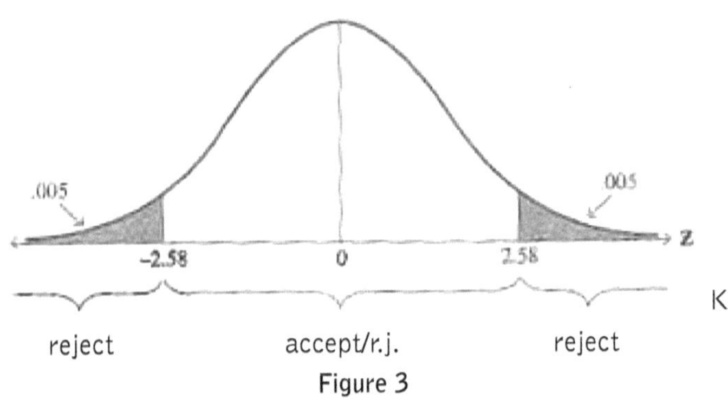

Figure 3

Type I error: H_o is true, $z > 2.58$ or $z < -2.58$, and H_o is rejected. Type II error: H_o is false, $-2.58 \leq z \leq 2.58$ and H_o is accepted.

6. $\chi^2 = s^2$, $\alpha = 0.01$, d.f. $= 9$, see Figure 4.

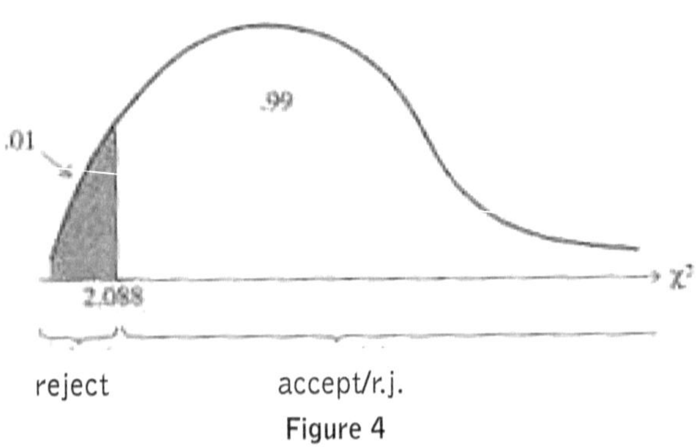

Figure 4

Type I error: H_o is true, $\chi^2 = s^2 < 2.088$ and H_o is rejected. Type II error: H_o is false, $\chi^2 = s^2 > 2.088$, and H_o is accepted. The population must be normally distributed or approximately so.

7. $z = \dfrac{\overline{x}_1 - \overline{x}_2}{\sqrt{\dfrac{\sigma_1^2}{32} + \dfrac{\sigma_2^2}{35}}}$, $\alpha = 0.02$, see Figure 5.

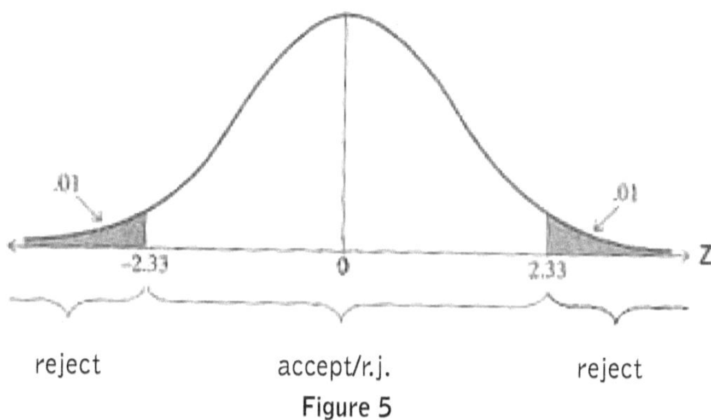

Figure 5

Type I error: H_o is true, $z > 2.33$ or $z < -2.33$, and H_o is rejected. Type II error: H_o is false, $-2.33 \le z \le 2.33$, and H_o is accepted.

8. $t = \dfrac{\overline{x}_1 - \overline{x}_2}{\sqrt{\dfrac{7s_1^2 + 10s_2^2}{15}\left(\dfrac{1}{8} + \dfrac{1}{11}\right)}}$, $\alpha = 0.05$, d.f.$=17$, see Figure 6.

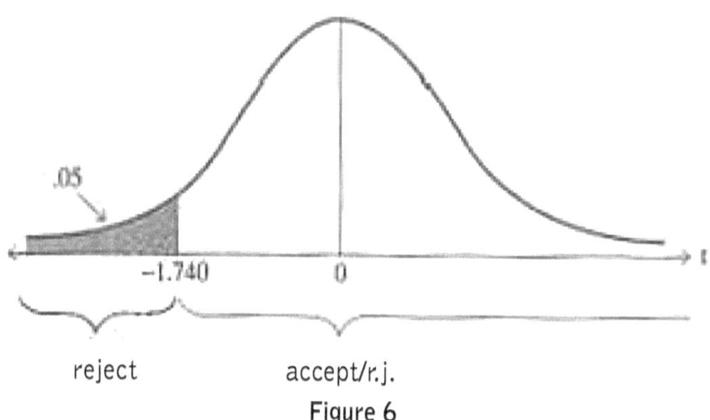

Figure 6

Type I error: H_o is true, $t < -1.740$, and H_o is rejected. Type II error: H_o is false, $t \ge -1.740$, and H_o is accepted. The population must be normally distributed or approximately so, and have the same standard deviation.

9. $F = s_1^2 / s_2^2$, $\alpha = 0.01$, see Figure 7.

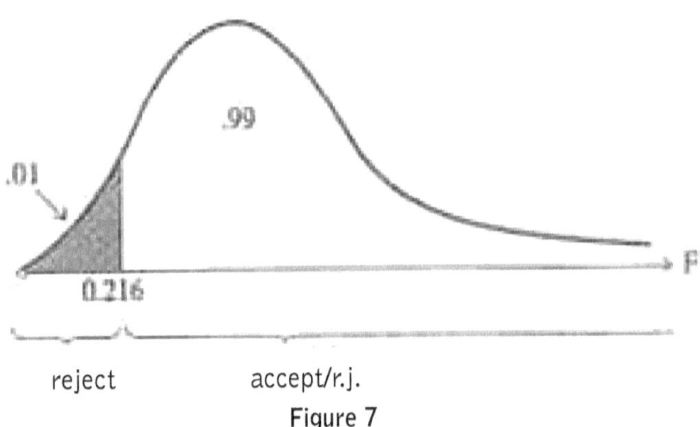

Figure 7

Type I error: H_o is true, $F < 0.216$, and H_o is rejected. Type II error: H_o is false, $F \geq 0.216$, and H_o is accepted. The population must be normally distributed, or approximately so.

10. $F = s_1^2 / s_2^2$, $\alpha = 0.01$, see Figure 8.

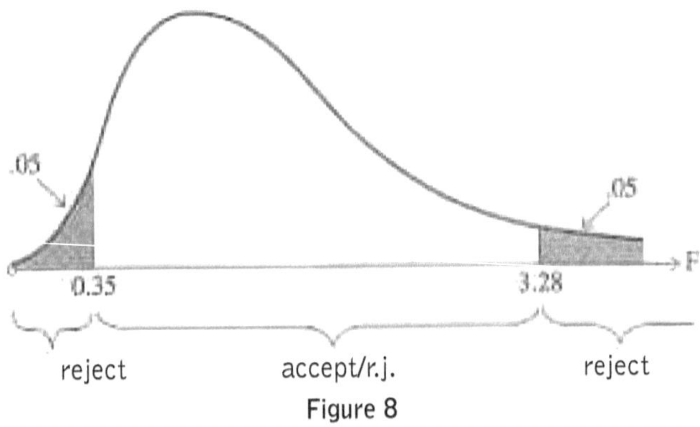

Figure 8

Type I error: H_o is true, $F < 0.35$ or $F > 3.28$, and H_o is rejected. Type II error: H_o is false, $0.35 \leq F \leq 3.28$, and H_o is accepted. The population must be normally distributed, or approximately so.

Self-Test 2 (page 401)

1. (a) $\bar{x}-1.65(\sigma/\sqrt{n})<\mu<\bar{x}+1.65(\sigma/\sqrt{n})$, with probability 0.90.

   (b) Considering the many, many population mean estimation problems that arise, in approximately 90% of these cases the population mean will be contained in the 90% confidence interval constructed for it.

   (c) We cannot say with certainty either way. In this, as in any such specific situation, 0.90 serves as a measure of our "collective" degree of belief, taken over from confidence interval analysis, that μ is between 25 and 30.

   (d) $\bar{x}-1.833(s/\sqrt{n})<\mu<\bar{x}+1.833(s/\sqrt{n})$, with probability 0.90.

2. $\Sigma x=1510$, $\bar{x}=302$, $s=19.2$

   (a) *t*-distribution, d.f. = 4, since the sample size is small (less than 30) and the underlying population of breakdown times is assumed to be approximately normal.

   (b) $302\pm3.747\dfrac{(19.2)}{\sqrt{5}}=270$ and 334, with probability 0.98.

   (c) 0.98 is a measure of degree of belief that μ is between 270 and 334 hours.

3. $H_o:\mu_1-\mu_2=0$. $H_a:\mu_1-\mu_2\neq0$. $z=\dfrac{18.50-19.34}{\sqrt{\dfrac{(4.6)^2}{200}+\dfrac{(5.2)^2}{100}}}=-1.37\geq-1.96\cdot$

   Accept/r. j. on H_o. No statistical evidence of a difference.

4. (a) The normality or approximate normality of the underlying population.

   (b) Reject H_o if $\chi^2=14s^2/100>26.119$ or $\chi^2<5.629$.
   Otherwise accept H_o or reserve judgment.

5. $E=2.33(5/\sqrt{50})=1.7$, with probability 0.98.

6. (a) $\frac{11(3)}{21.92} < \sigma^2 < \frac{11(3)}{3.816}$; $1.5 < \sigma^2 < 8.7$, with probability 0.95.

(b) $1.7 < \sigma^2 < 7.2$, with probability 0.90.

(c) Normality, or approximate normality, of the underlying population.

7. (a) $H_a : \mu < 500$ A consumer protection organization would be concerned with a manufacturer overstating the claim. A mean lifetime significantly less than 500 hours would be of concern.

(b) For a small sample test to be feasible the underlying population of battery lifetimes must be normally distributed or approximately so. If this condition is satisfied, then a small sample test, which is generally less expensive to carry out than a large sample one, is a possibility. Otherwise a large sample test should be employed.

(c) $H_o : \mu = 500$. $H_a : \mu < 500$
$z = -6.43 < -1.65$. Reject H_o. The claim is contradicted.

(d) $t = -2.86 < -1.753$. Reject H_o. The claim is contradicted.

Self-Test 3 (page 403)

1. (a) In terms of the responses received (408) the confidence limits are: $0.368 \mp 1.96\sqrt{\frac{0.368(0.632)}{408}}$; $0.321 < \pi < 0.415$, with prob. 0.95.

(b) 0.95 is a measure of our degree of belief that π is between 0.321 and 0.415.

(c) The response rate to a survey has an important effect on the reliability the results obtained (see Section 2.5). The high response rate of 85% to the Valenti Company's survey gives a high level of credibility to the confidence interval obtained.

2. (a) $H_o : \mu = 2{,}000$. $H_a : \mu < 2{,}000$.

(b) The consumers' concern is with whether $\mu < 2000$.

(c) 0.05

(d) $z = -5.81 < -1.65$. Reject H_o. Claim not supported.

(e) $t = -2.37 < -1.833$. Reject H_o. Claim not supported.

(f) Normality, or approximate normality, of the underlying population of Super-100 bulb lifetimes.

3. $n = \left[\frac{1.96(30)}{10}\right]^2 = 34.6$; therefore $n = 35$.

4. (a) $\sigma_{\bar{x}} = 2.8/\sqrt{100} = 0.28$

$$\beta = P(3.35 \le \bar{x} \le 4.65) = P\left(\frac{3.35 - 4.50}{0.28} \le z \le \frac{4.65 - 4.50}{0.28}\right)$$
$$= P(-4.11 \le z \le 0.54) = 0.7054.$$

(b) $\sigma_{\bar{x}} = 2.8/\sqrt{100} = 0.28$

$$\beta = P(3.35 \le \bar{x} \le 4.65) = P(1.25 \le z \le 5.89) = 0.1056.$$

5. (a) The variances of the two groups' test scores must be equal.

(b) $F = 70.3/90.1 = 0.78$, $F_{0.05}(15,12) = 2.62$, $F_{0.95}(15,12) = 1/2.48 = 0.40$. Since $0.40 \le 0.78 \le 2.62$ we accept/r. j. on H_o and conclude that H_o is supported.

(c) Yes

(d) $t = -1.81 \ge -2.052$. Accept / r.j. on H_o.

(e) The means are equal.

6. (a) 0.043 (see Fig. 9(a)) (b) 0.046 (see Fig. 9(b))

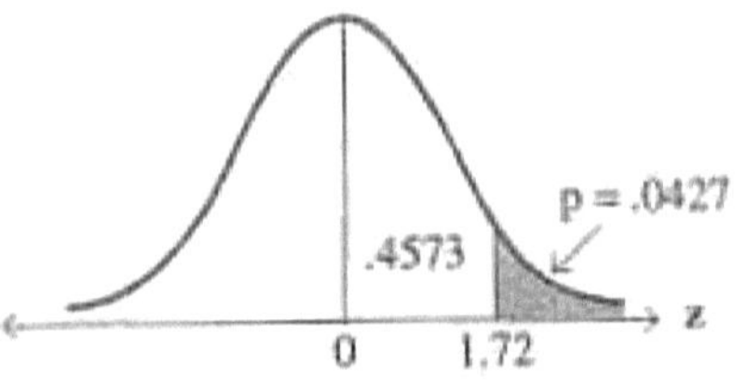

Figure 9(a)

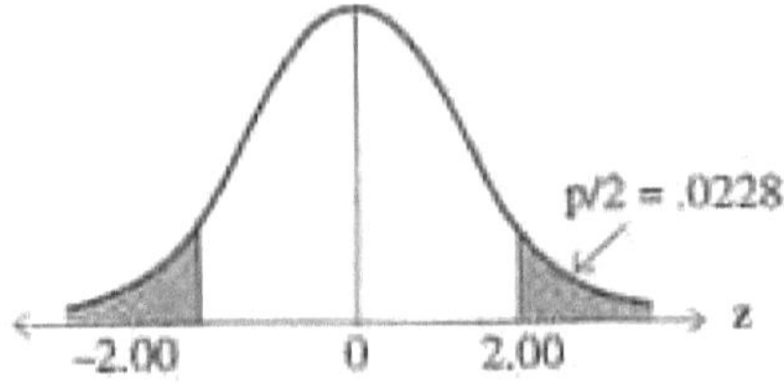

Figure 9(b)

Self-Test 4 (page 405)

1. (a) We define μ_1 as the mean of the population of WH—100 (new) yields. We define μ_2 as the mean of the population of WH—50 (old) yields. $H_o : \mu_1 - \mu_2 = 0$. $H_a : \mu_1 - \mu_2 > 0$.

(b) If we reject H_o, then we must accept H_a which proves the claim.

(c) Equality of variances of the two populations.

(d) $\Sigma x_1 = 662$, $\Sigma x_1^2 = 49,950$, $\bar{x}_1 = 73.6$, $s_1^2 = 157.0$. $\Sigma x_2 = 644$, $\Sigma x_2^2 = 47,300$, $\bar{x}_2 = 71.6$, $s_2^2 = 152.3$. $F = 152.3/157.0 = 0.97$. $F_{0.95}(8,8) = 0.29 \le 0.97 \le 3.44 = F_{0.05}(8,8)$. Accept/r.j. on H_o.

(e) Yes

(f) $t = 0.34 \le 1.746$. Accept/r. j. on H_o.

(g) Both hybrids have the same yield.

(h) Yes. Extraneous factors such as differences in soil conditions in the test plots would be eliminated.

(i) $\Sigma x_d = 18$, $\Sigma x_d^2 = 40$, $\bar{x}_d = 2$, $s_d = 0.707$.
$t = 8.48 > 1.86$. Reject H_o.

(j) Results of the two approaches are different.

(k) By pairing the data the effect of extraneous factors is eliminated. The pairing approach focused on the fact that for each pairing, the new out-performed the old.

2. (a) A robust hypothesis test is one that may still be employed when the prerequisite conditions for its application are not "quite" satisfied.

(b) Robustness is an important issue because it allows us to apply a hypothesis test and get reliable results when the assumptions are not fully met.

Self-Tests for Parts 1-4

Self-Test 1 (page 461)

1. $H_o: \mu_1 - \mu_2 = 0$; $H_a: \mu_1 - \mu_2 \neq 0$; $n_1 = 50$, $\overline{x}_1 = 5.2$, $s_1 = 2.0$, $n_2 = 50$, $\overline{x}_2 = 4.7$, $s_2 = 2.3$.

$z = \dfrac{\overline{x}_1 - \overline{x}_2}{\sqrt{\dfrac{s_1^2}{n_1} + \dfrac{s_2^2}{n_2}}} = \dfrac{0.5}{0.186} = 1.16 < 1.96$. Accept/r.j. on H_o. No (statistical) difference in effectiveness.

2. (a) See Table 1.

Table 1

x_i	y_i	$x_i y_i$	x_i^2	y_i^2
0.0	8.2	0.00	0.00	67.24
0.0	8.0	0.00	0.00	64.00
0.5	7.9	3.95	0.25	62.41
1.1	7.6	8.36	1.21	57.76
1.5	7.4	11.10	2.25	54.76
2.3	7.1	16.33	5.29	50.41
2.5	7.3	18.25	6.25	53.29
5.0	6.9	34.50	25.00	47.61
12.9	60.4	92.49	40.25	457.48

$$r = \frac{8(92.49) - (12.9)(60.4)}{\sqrt{8(40.25) - (12.9)^2}\sqrt{8(457.48) - (60.4)^2}} = -0.92$$

An r value so "close" to -1 indicates that there is a strong inverse linear relationship between x and y.

(b) $b = \dfrac{8(92.49) - (12.9)(60.4)}{8(40.25) - (12.9)^2} = -0.25$

$a = \dfrac{60.4 - (-0.25)(12.9)}{8} = 7.95$. This yields: $\hat{y} = 8.0 - 0.25x$.

(c) For $x = 3$, $\hat{y} = 7.3$.

The mean weight of babies born to mothers who smoke three packs of cigarettes a day during pregnancy is approximately 7.3 pounds.

(d) $r^2 = 0.85$. The number of packs of cigarettes smoked per day during pregnancy explains 85% of the total variation in the weights of newborn babies.

(e) No; there is a strong linear correlation between number of packs smoked per day and baby birth weight, but we cannot conclude that the first determines the second in the sense of cause-and-effect.

3. $E = 2.33\sqrt{(0.10)(0.90)/200} = 0.5$. The (maximum) error inherent in the estimate is 5% with probability 0.98.

4. (a) $H_a: \mu < 40{,}000$. From the point of view of consumer protection a mean lifetime significantly less than 40,000 miles is of concern.

(b) Small sample tests are generally less expensive to carry out than their large sample counterparts, which is in their favor. They are not as accurate as their large sample counterparts, which is not in their favor. They require that the underlying population be normally distributed or approximately so, which must be given primary consideration.

(c) $H_o: \mu = 40{,}000$; $H_a: \mu < 40{,}000$.

$z = \dfrac{39{,}850 - 40{,}000}{500/\sqrt{100}} = -3.00 < -2.33$. Reject H_o.

(d) $t = \dfrac{39{,}850 - 40{,}000}{500/\sqrt{15}} = -1.16 > -2.624 = -t_{0.01}$. Accept/r.j. on H_o.

5. The cases discussed in Ch. 1 make clear that a fundamental string must be attached to this view. The figures must be reliable and relevant to the issue under study.

6. First of all, the 80 percent figure is just that; it does not, by itself, determine an interpretation. As to Bob Levy's interpretation, it's quite a stretch to go from the 80 percent figure to the conclusion that Plutonians have an innate ability for math. Another possible interpretation is that the Plutonians in the class, at least those who got A's, are usually well-prepared and hard working.

Self-Test 2 (page 463)

1. (a) A type I error is committed when a true H_o is rejected.

   (b) A type II error is committed when a false H_o is accepted.

2. $\bar{x} - 1.711(s/\sqrt{n}) < \mu < \bar{x} + 1.711(s/\sqrt{n})$, $t_{0.05} = 1.711$; $366 < \mu < 434$.

3. See Table 2.

Table 2

Amount Earned	f_i	x_i
10.00-19.99	16	14.995
20.00-29.99	9	24.995
30.00-39.99	10	34.995
40.00-49.99	7	44.995
50.00-59.99	8	54.995
	50	

(a) $0.75n = 37.5$; therefore, $Q_3 = 39.995 + \dfrac{(10)(2.5)}{7} = 43.5$

(b) $\bar{x} = 31.4$

(c) $s^2 = 215.35$, $s = 14.7$

(d) First, we assume that the 50 workers in the accounting department are a sample of the larger population of accounting department personnel who worked overtime. For the computations part we assume that the data in each class are approximately symmetrically distributed in the class.

4. (a) Bernoulli trial model defined by $n=50$, E is the event that O.S. lands on target on a jump, $p=0.40$.

(b) The outcome of any one jump does not affect the outcome of any other jump.

(c) $P(X=22)=C(50,22)(0.40)^{22}(0.60)^{28}$

(d) Yes, since both np and nq exceed 5.

(e) $\mu=np=20$, $\sigma=\sqrt{np(1-p)}=3.464$.
$P(X=22)=P(0.43\le z\le 0.72)=0.0978$.

5. (a) The standard deviations of the underlying populations of bulb lifetimes must be equal.

(b) $H_o:\sigma_1^2=\sigma_2^2$; $H_a:\sigma_1^2\ne\sigma_2^2$.
σ_1^2 and σ_2^2 are the variances for Plant A and Plant B, respectively.
$F=\frac{14^2}{9^2}=2.42$, $F_{0.05}=3.18$, $F_{0.95}=0.31$.

Since $0.31<2.42<3.18$, accept/r.j. on H_o.

(c) Yes, since the test carried out in (b) supports the assumption of equal standard deviations.

(d) $H_o:\mu_1-\mu_2=0$; $H_a:\mu_1-\mu_2\ne 0$.

$$t=\frac{427-400}{\sqrt{\left(\frac{9(14)^2+9(9)^2}{9+9}\right)\left(\frac{1}{10}+\frac{1}{10}\right)}}=5.13>2.101=t_{0.025}$$. Reject H_o.

(e) The mean lifetimes of the bulbs made at the two plants are different.

5. See Sec. 2.5 of ch. 2.

6. No; it's the wrong target audience. Those who voted in last year's municipal elections do not necessarily know anything about the problems of the poor.

Self-Test 3 (page 465)

1. (a) $H_o: \pi = 0.35$; $H_a: \pi < 0.35$.

(b) MERT's claim is that 35% or more of its calls are of a life threatening nature. We would take issue with this if the sample proportion were significantly less than 0.35.

(c) $z = \dfrac{(64/200) - 0.35}{\sqrt{(0.35)(0.65)/200}} = -0.89 > -1.65$. Accept/r.j. on H_o.

2. (a) $S = \{(3,7),(3,9),(3,11),(3,13),(7,9),(7,11),(7,13),(9,11),(9,13),(11,13)\}$. P assigns 1/10 to each sample point in S.

(b) We assume that the sample is drawn at random, as indicated; that is, there is no bias in the sampling which favors certain samples being drawn over others.

(c) $\bar{x}(3,7) = 5$, $\bar{x}(3,9) = 6$, $\bar{x}(3,11) = 7$, $\bar{x}(3,13) = 8$, $\bar{x}(7,9) = 8$, $\bar{x}(7,11) = 9$, $\bar{x}(7,13) = 10$, $\bar{x}(9,11) = 10$, $\bar{x}(9,13) = 11$, $\bar{x}(11,13) = 12$.

This yields $f(x) = P(\bar{x} = x)$ defined as follows; $f(5) = f(6) = f(7) = f(9) = f(11) = f(13) = 0.1$, $f(8) = f(10) = 0.2$.

(d) $\mu_{\bar{x}} = 5(0.1) + 6(0.1) + \cdots + 12(0.1) = 8.6$; alternatively, $\mu_{\bar{x}} = \mu = 43/5 = 8.6$.

(e) $\sigma_{\bar{x}}^2 = (0.1)(5-8.6)^2 + (0.1)(6-8.6)^2 + \cdots + (0.1)(12-8.6)^2 = 4.44$.
$\sigma_{\bar{x}} = \sqrt{4.44} = 2.1$; alternatively, $\sigma_{\bar{x}}^2 = \dfrac{3.44}{\sqrt{2}}\sqrt{\dfrac{5-2}{5-1}} = 2.1$.

(f) (i) $P(6 \le x \le 8) = 0.4$.

3. (a) (i) $\bar{x} = 304/16 = 19$. (ii) $\text{Median} = (16+20)/2 = 18$.

(iii) $\text{Mode} = 13$. (iv) $\text{Range} = 30 - 11 = 19$.

(v) $\sum x = 304$, $\sum x^2 = 6{,}316$,

$$s^2 = \frac{16(6{,}316) - (304)^2}{(16)(15)} = 36$$

(vi) $s = \sqrt{36} = 6$.

(b) See Table 3.

Table 3

Waiting Times	Frequency
10-14	5
15-19	3
20-24	5
25-29	2
30-34	1
	16

(c) (i) $LCL = 15$, $UCL = 19$

(ii) Class Mark = 17

(iii) Class Width = 5

(iv) Class boundaries, 14.5 and 19.5

(d) See Figure 1.

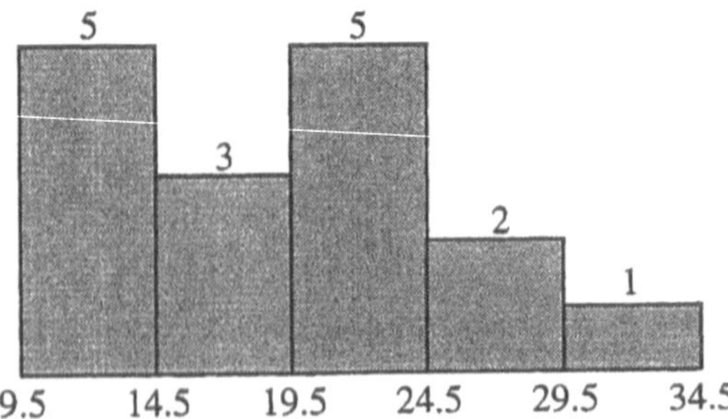

Figure 1

(e) See Table 4.

Table 4

Waiting Times	Cumulative Frequency
< 10	0
< 15	5
< 20	8
< 25	13
< 30	15
< 35	16

(f) See Figure 2.

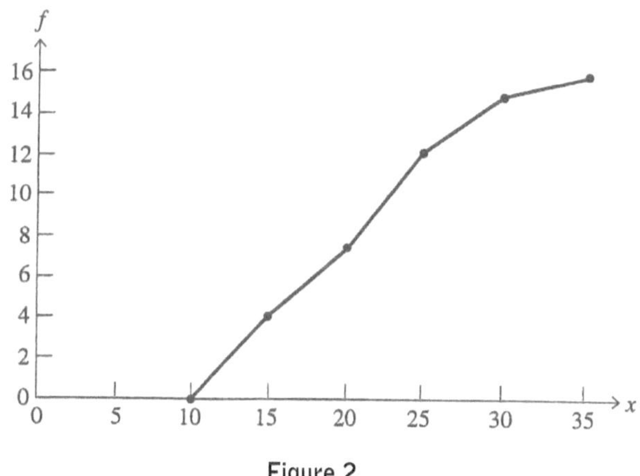

Figure 2

4. (a) With $n=16$, $s^2=36$, we have:

$$\frac{15(36)}{27.488}<\sigma^2<\frac{15(36)}{6.2621},\ 19.6<\sigma^2<86.2.$$

(b) $\bar{x}-2.131(s\sqrt{n})<\mu<\bar{x}+2.131(s\sqrt{n});\ 16<\mu<22.$

(c) The population of waiting times is approximately normally distributed.

5. (a) The hypothesis of elementary errors views an error E which arises in a measurement process as the sum of a large number of elementary

errors $e_1, e_2, \ldots, e_n$ which arise independently from different sources, where each elementary error is negligible in comparison to the sum E.

(b) If a random variable behaves according to the aforenoted hypothesis of elementary errors then its probability behavior can be described by a normal curve. See Section 8.3 for further discussion.

6. "Male teen-agers engage in risky behavior associated with H.I.V. infection more than previously thought, and they are also much more willing to report such behavior if the survey is administered by computer rather than on paper, a new study suggests.

The research appears to call into question much of the data that has been gathered on sensitive subjects like drug use and sexual habits, and to confirm what researchers have long assumed to be under-reporting in such areas."

7. (a) Even if the figures were reliable (see discussion (b)) Dr. Shalala's interpretation of them can only be seen as one point of view. It is not at all "obvious" that most of our young people are getting the message she describes. Considering her position, we can appreciate the pressure she must have felt to say something that strikes a positive note. It would have been more realistic to say something along the lines that while any interpretation of the data would be premature, we can only hope that this means that most of our young people are . . .

(b) The fact that the study did not take its sample from the same population of young people in 1999 as it did in 1997 by itself is enough to call into question the reliability of the figures. And then there is the finding that computers elicit more honest responses about sex behavior than human beings and the question that this finding raises about a sensitive subject like drug use. Consider the afore comments concerning question 6.

8. Yes; the temperature reading cited is interval data, which renders meaningless the statement "It's four times as warm there as it is here"? To be meaningful the underlying data would have to be ratio data.

Self-Test 4 (page 467)

1. $t = \dfrac{\overline{x} - 250}{s/\sqrt{8}}$, $\alpha = 0.05$, $\text{d.f.} = 7$, see Figure 3.

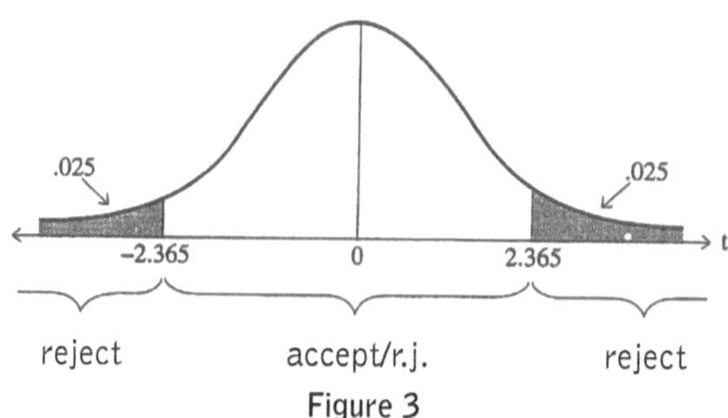

Figure 3

Type I error: H_o is true, $t > 2.365$ or $t < -2.365$, and H_o is rejected. Type II error: is false, $-2.365 \le t \le 2.365$, and H_o is accepted. The underlying population from which the sample is drawn is assumed to be normally distributed or approximately so.

2. $z = \dfrac{\dfrac{x}{500} - 0.30}{0.021}$, $\alpha = 0.01$, see Figure 4.

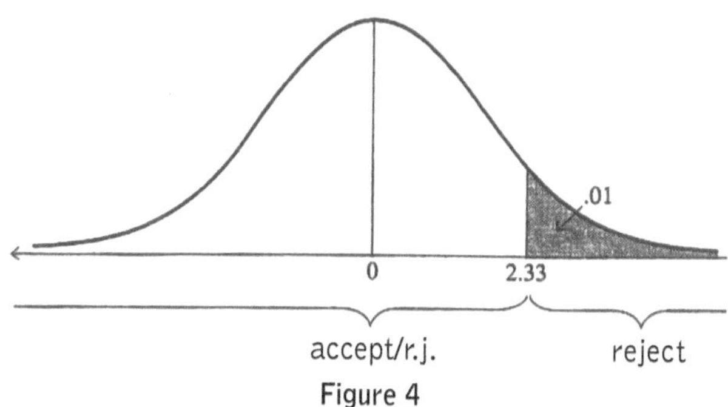

Figure 4

Type I error: H_o is true, $z > 2.33$ and H_o is rejected. Type II error: H_o is false, $z \le 2.33$, and H_o is accepted.

3. $\chi^2 = \frac{11s^2}{25}$, $\alpha = 0.01$, $\text{d.f.} = 11$, see Figure 5.

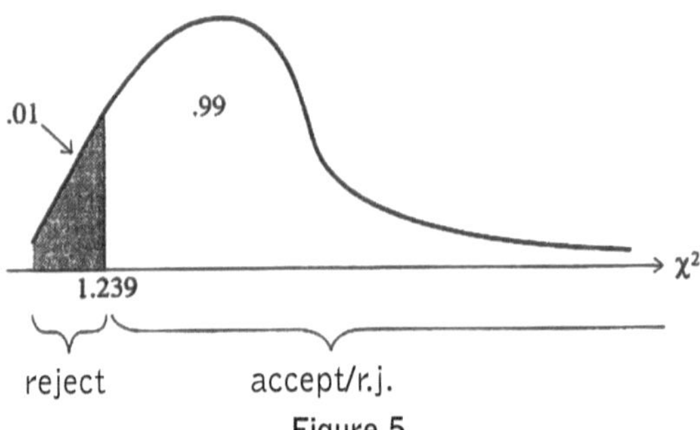

Figure 5

Type I error: H_o is true $\chi^2 < 1.239$ and H_o is rejected. Type II error: H_o is false, $\chi^2 \geq 1.239$. and H_o is accepted. The underlying population is assumed to be normally distributed or approximately so.

4. $z = \frac{\bar{x}_1 - \bar{x}_2}{\sqrt{\frac{\sigma_1^2}{32} + \frac{\sigma_2^2}{36}}}$, $\alpha = 0.05$, see Figure 6.

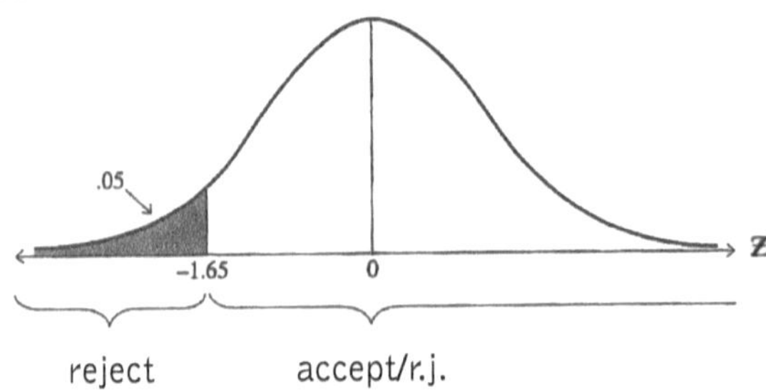

Figure 6

Type I error: H_o is true, $z < -1.65$, and H_o is rejected. Type II error: H_o is false, $z \geq -1.65$, and H_o is accepted.

5. $F = s_1^2 / s_2^2$, $\alpha = 0.05$, numerator d.f. = 10, denominator d.f. = 14, see Figure 7.

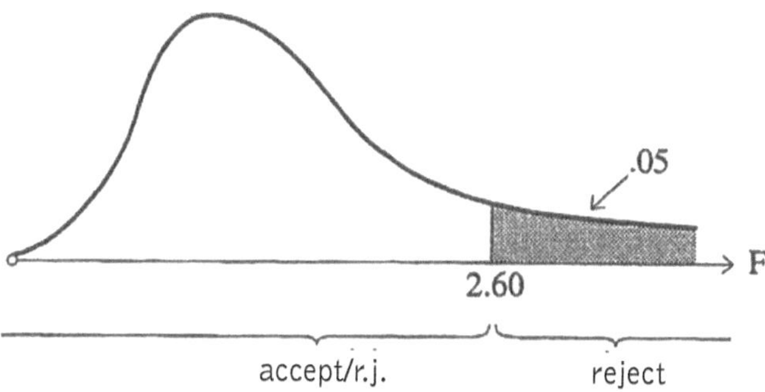

Figure 7

Type I error: H_o is true $F > 2.60$ and H_o is rejected. Type II error: H_o is false, $F \le 2.60$, and H_o is accepted. The underlying populations are assumed to be normally distributed, or approximately so, with equal variances.

6. Since the sample size of 30 is less than 5% of the population size of 850, we may proceed without introducing an adjustment factor (see Section 10.2). $25,000 \mp 1.96 \frac{(3200)}{\sqrt{30}} = 23,855$ and 26,145 with probability 0.95.

7. $\Sigma x = 355$, $\bar{x} = 55.8$

(a) $s = 13.2$. In this small sample case we have $\bar{x} \pm t_{0.05}(s/\sqrt{n})$, $\text{d.f.} = n-1 = 5$. This yields, $55.8 \pm 2.015(13.2/\sqrt{6}) = 55.8 \pm 10.9$; $44.9 < \mu < 66.7$.

(b) 0.90 is a numerical measure of degree of belief that μ is between 44.9 and 66.7.

(c) E=10.9 with probability 0.95.

(d) That the distribution of the client waiting times is normal, or approximately so.

(e) The results would be unreliable since they are based on the assumption of normality.

8. No, because it does not take into account the number of departures and the number of miles flown.

9. See Section 10.4

Self-Test 5 (page 469)

1. (a) $75 \mp 2.262\frac{(15)}{\sqrt{10}} = 64$ and 86 with prob. 0.95.

(b) 0.95 is a measure of our degree of belief that the mean price of the statistics books is between \$64 and \$86.

(c) Yes, that the population of book prices is normally distributed, or approximately so.

2. Let us first observe that the sample size of 36 is less than 5% of the population size of 800 so that we may proceed without introducing an adjustment factor (see Section 10.2).

(a) $E = 2.33\frac{(2.5)}{\sqrt{36}} = 1$ with prob. 0.98.

(b) 0.98 is a measure of degree of belief that in estimating the mean age of the taxi fleet as 5.5 years, the error of this estimate does not exceed 1 year.

(c) We take $s = 2.5$ as an estimate for σ since 2.5 was obtained through use of a large sample size. $n = \left(\frac{2.33(2.5)}{0.5}\right)^2 = 136$.

3. See Section 10.4.

4. The median value of the rankings, 10, which corresponds to Satisfactory. Four students ranked him Satisfactory or below and four ranked him Satisfactory or higher.

5. $\Sigma x = 9.4$, $\Sigma y = 959$, $\Sigma xy = 925$, $\Sigma x^2 = 9.3$, $\Sigma y^2 = 93{,}569$

(a) $\hat{y} = 48.2 + 50.7x$

(b) $\hat{y} = 98.9$; the mean sales volume for a large number of monthly periods in which $1000 was spent for advertising is approximately $98,900.

(c) $r = 0.86$; there is a strong, direct linear relationship between advertising expenditure and sales volume.

(d) $r^2 = 0.74$; advertising expenditures account for 74% of the total variation in sales volume.

(e) No; cause-and-effect is not established by a high value of *r*, just that there is a strong linear relationship between the variables.

(f) No, because 3 is outside of the spectrum of values for which the regression line was determined.

6. See Section 12.3, Exercise 6.

(a) $H_o: \rho = 0$, $H_a: \rho \neq 0$.
$n = 9, \Sigma x = 590, \Sigma x^2 = 55,204, \Sigma xy = 25,409, \Sigma y = 317, \Sigma y^2 = 12,829$,
$r = 0.88$, $t = 5 > 2.365$. Reject H_o.

(b) The conclusion in (a) would be unreliable since it was based on assumptions of normality that proved to be false.

(c) Yes. $\rho \neq 0$, and *r* has fairly high value which adds to our confidence.

7. This kind of data is not sufficient because it gives no indication about the quality of the research being done at Ecap U.

8. (a) $p/2 = 0.0409$, $p = 0.08$ (b) 0.01

(c) $p = 0.01 < 0.02$ means that $z = -2.23$ is in the reject region, which calls for rejection of H_o.

9. (a) A robust hypothesis test is one that may still be employed when the prerequisite conditions for its application are not "quite" satisfied.

(b) Robustness is an important issue because it allows us to apply a hypothesis test and get reliable results when the assumptions are not fully met.

10. A percentage growth of 25% occurs when the course registration goes from 4 to 5 students, which may or may not be sufficient to run the course. It would be useful to look at the actual number of registrants involved. Percentage growth figures are large when the underlying base is numerically small.

Self-Test 6 (page 473)

1. Please refer to Section 2.2 in the text.

2. Please refer to Section 2.1 in the text.

3. (a) $P(A) = 2/10 + 3/10 + 2/10 = 7/10 = 0.70$.

   (b) If Dan's die is tossed a large number of times, an even number will show approximately 70% of the time.

   (c) This evidence is irrelevant to the validity of the conclusion cited in (a). The validity of $P(A) = 0.70$ is established by adding $P(2)$, $P(4)$, and $P(6)$ in the model and obtaining 0.70.

   (d) The evidence cited in (c) tells us that the relative frequency with which an even number has shown for a large number of tosses of Dan's die is 0.40, which is sharply at variance with the relative frequency behavior of the die predicted in (b). This means that the conclusion cited in (a), interpreted in relative frequency terms, is false.

   (e) A: the event that a number greater than 2 shows; B: the event that an even number shows.
   $P(A \cap B) = 0.5 \neq P(A)P(B) = (0.7)(0.7) = 0.49$. Therefore, A and B are not independent events.

4. As a subjective probability. It is a numerical measure of A.W.'s degree of belief that he will receive an A in statistics this semester. Whether this belief is based on wishful thinking or exam grades or both, we cannot say.

5. (a) Considering the large number of population mean estimation problems that arise, in approximately 98% of the cases the population mean μ will lie within the 98% confidence interval constructed for it.

(b) (i), (ii) We cannot say for sure, one way or the other. The 0.90 probability value serves as a numerical measure of our degree of belief that μ is between 30 and 40 in this specific confidence interval situation.

6. No; focusing on poll results without an appropriate context is misleading. Information should be given about the date the poll was taken, sample size, survey design, percentage of respondents among those contacted, response options, and the wording of questions.

   Out-of-context poll results are best viewed with questioning skepticism.

7. No; standard online polls are a modern version of the Newspaper Poll (see Section 2.5). The responses received are representative of those who cared to respond.

INDEX

D

E

S

www.ingramcontent.com/pod-product-compliance
Lightning Source LLC
Chambersburg PA
CBHW031808170526
45157CB00001B/5